Nitrogen Metabolism and Excretion

Edited by
Patrick J. Walsh and Patricia Wright

CRC Press
Boca Raton New York London Tokyo

Library of Congress Cataloging-in-Publication Data

Nitrogen metabolism and excretion / editors, Patrick J. Walsh, Patricia Wright.
p. cm. -- (Animal physiology)
Includes bibliographical references and index.
ISBN 0-8493-8411-7
1. Nitrogen excretion. 2. Nitrogen--Metabolism. 3. Physiology, Comparative.
I. Walsh, Patrick J., Ph.D. II. Wright, Patricia (Patricia A.) III. Series.
[DNLM: 1. Hepatitis B virus. QW 710 G289h]
QP211.N58 1995
612.3′942—dc20
DNLM/DLC
for Library of Congress 94-46469
CIP

International Standard Book Number 0-8493-8411-7
Library of Congress Card Number 94-46469
Printed in the United States of America 2 3 4 5 6 7 8 9 0
Printed on acid-free paper

To
Our
Families

PREFACE

Although the basic evolutionary patterns of nitrogen metabolism and excretion have been outlined for decades, in the past 15 years these paradigms have been challenged and there has been a resurgence of research activity. For example, while it was widely accepted that bony fish excreted their nitrogenous waste primarily as ammonia, the past decade has seen rapid growth in the list of bony fish species which excrete urea as their prime waste nitrogen product. A second area where research in nitrogen metabolism has been stimulated is in the area of acid–base balance. An especially provocative hypothesis by Daniel Atkinson and his colleagues, namely that urea synthesis evolved primarily as a means to dispose of bicarbonate, has been extensively debated and tested. This debate continues among many of our authors; in the interest of objective scientific pursuit we have carefully limited our editorial prerogatives in this regard. Finally, the molecular revolution has touched the field as well. Recent studies on nitrogen metabolism and excretion now almost routinely use the tools of molecular biology, and of special interest are recent studies of evolutionary relationships between proteins of nitrogen metabolism.

This book features chapters on the most recent developments in nitrogen metabolism and excretion from a number of prominent researchers in the field. Although the individual chapters are detailed and specific, the scope of the book is rather broad and eclectic, covering taxonomic groups from protists to mammals, and subject areas from buoyancy in the oceans to molecular aspects of metabolism and excretion. Given this diversity, the chapters in this book could have been organized in a number of ways. We have chosen first to present the two chapters by Houlihan and colleagues (on protein turnover) and by Anderson (on carbamoyl phosphate synthesis) as information basic to all animals. The majority of chapters (3 through 14) are then arranged more or less in phylogenetic order. However, we have made some obvious exceptions in the hope the reader will glean that many adaptations of nitrogen biochemistry transcend phylogenetic stereotypes and boundaries. In the first chapter, nitrogen is viewed principally as a critical nutrient to conserve, but then as part of compounds conferring neutral buoyancy (Chapter 3 by Sanders and Childress), then as part of compounds used extensively in osmoregulation (Chapters 4 to 6 by Henry, Wright and Neufeld, and Goldstein and Perlman, respectively). In the phylogenetic treatment, a specific general chapter on fish was omitted as processes in this group have been extensively reviewed recently [e.g., see C.M. Wood (1993) Reference 2 in Chapter 15]; even though our authors chose more specialized topics, much general information on fish is contained in the initial sections of their chapters. Next follow chapters on phyla in which the transition to terrestriality occurred (Chapters 7 to 13 by O'Donnell and Wright, Harrison, Atkinson, Campbell, Dantzler, Meijer, and Knepper and Chou, respectively), and for which the selective pressures on nitrogen metabolism and excretion are (perhaps) clearer. The chapter by Campbell offers some particularly interesting evolutionary insights. With this phylogenetic background in mind, a group of three chapters on developmental (ontogenetic) aspects follows to recapitulate and reinforce aspects brought to light in the phylogenetic section (Chapters 14 to 16 by Dingemanse and Lamers, Kormanik, and Korsgaard, Mommsen, and Wright, respectively; Chapter 9 by Atkinson easily could have been placed here as well). Although all of our authors have presented forward looking chapters, we end with two chapters in particularly new research areas, namely nitrates and nitrites (Chapter 17 by Jensen) and nitric oxide (Chapter 18 by Fukuto). These chapters reinforce just how versatile is the nitrogen atom, and how many interwoven metabolic and excretory processes must be controlled simultaneously.

The stimulus for this book came as a request by Dr. Evans, coincident with planning for a symposium on "Nitrogen Metabolism and Excretion" at the 1993 meeting of the Society for Experimental Biology in Canterbury, England. Both editors (P.J. Walsh, symposium co-organizer; P.A. Wright, symposium keynote speaker) and a number of the authors participated in that symposium. Although the symposium organizers (P.J. Walsh and C.M. Wood) initially promised that no manuscripts would be required, clearly one of us went back on that promise. We are especially grateful to those symposium participants (Drs. Anderson, Henry, Houlihan, O'Donnell, Wright, Mommsen, Meijer, and Knepper) for their extra effort in contributing to this book, and to the Society for its sponsorship of the symposium. We wish to thank all of the authors for their hard work and adherence to our Draconian enforcement of page limits. You will notice in perusing the list of contributors that we must be in the (W)right field of study.

We also wish to extend thanks to Ms. Marva Clarke and Ms. Eva-Maria Kiene for their assistance with word processing and correspondence, to Dr. Dave Evans for inviting us to initiate this series, and to the many mentors we have had in our own research on nitrogen metabolism and excretion (Drs. Dave Evans, Paul Anderson, Mark Knepper, Tom Mommsen, Dave Randall, and Chris Wood). Lastly, we wish to thank our students and postdoctoral colleagues for keeping us on our toes (Lee Barber, Todd Hopkins, Bonnie Tucker, Pamela Wright, Erle Rahaman-Noronha, Andrew Felskie, and Michelle Land). We have learned a great deal by editing this book. We hope you enjoy reading it as much as we have.

THE EDITORS

Patrick J. Walsh, Ph.D., is Professor and Chairperson in the Division of Marine Biology and Fisheries at the Rosenstiel School of Marine and Atmospheric Science of the University of Miami.

Dr. Walsh received his B.S. (magna cum laude) in Biology and Chemistry from the University of Miami in 1975, and his Ph.D. in Marine Biology from the Scripps Institution of Oceanography of the University of California, San Diego in 1981. He held a postdoctoral position in the Department of Biology of the University of Ottawa, and later advanced to Assistant Professor, during 1981 to 1984. In 1984 he transferred to the University of Miami as an Assistant Professor, was promoted to Associate Professor in 1988, and to Professor in 1992. He has served as Division Chairperson since 1991.

Dr. Walsh is an Associate Director of the National Institute of Environmental Health Sciences, Marine and Freshwater Biomedical Sciences Center of the University of Miami, as well as Associate Director of the Ocean Pollution Research Center of the University of Miami. He currently serves on the Integrative Animal Biology Panel of the Division of Integrative Biology and Neurosciences of the National Science Foundation, and as an Associate Editor for the *Journal of Experimental Zoology*. He is a member of several scientific societies, including both the American and Canadian Societies of Zoologists, the Society for Experimental Biology, and the Society for the Protection Of Old Fishes.

Dr. Walsh has presented 10 invitational lectures at international meetings and has published over 90 papers and book chapters. He has been the recipient of research grants from the Natural Sciences and Engineering Research Council of Canada, the National Science Foundation, the National Institute of Environmental Health Sciences, the National Oceanographic and Atmospheric Administration, the National Geographic Society, and the U.S. Coast Guard. His current research interest is in the evolution of urea metabolism in fishes, the population genetics and metabolism of pelagic marine organisms, and the responses of marine organisms to xenobiotics.

Patricia A. Wright, Ph.D., is an Assistant Professor in the Department of Zoology, University of Guelph.

Dr. Wright received her B.Sc. (Honors) in Biology at McMaster University in 1982 and her Ph.D. in Animal Physiology from the Department of Zoology, University of British Columbia in 1987. She was awarded a Killam postdoctoral fellowship to study in the Department of Biology of the University of Ottawa in 1988. From 1989 to 1991 she held a Fogarty International Fellowship in the Laboratory of Kidney and Electrolyte Metabolism at the National Institutes of Health, Bethesda, Maryland. In 1991, she was appointed as an Assistant Professor in the Pathology Department at the University of Guelph, with a cross-appointment in Zoology. Her full appointment was transferred to the Zoology Department in 1994.

Dr. Wright is a member of the Canadian Society of Zoologists and served as chair of its Comparative Physiology and Biochemistry Section (1993 to 1994). She is also a member of the Society for Experimental Biology, Sigma Xi, and serves on the editorial board of the journal *Comparative Biochemistry and Physiology.*

Dr. Wright has been invited to speak at 6 international conferences, and has published over 30 papers and book chapters. Her research is funded by the Natural Sciences and Engineering Research Council of Canada, and she is currently studying nitrogen metabolism and excretion in fish and amphibia.

CONTRIBUTORS

Paul M. Anderson
Department of Biochemistry
and Molecular Biology
University of Minnesota
Duluth, Minnesota

Burr G. Atkinson
Molecular Genetics Unit
Western Science Centre
Department of Zoology
The University of Western Ontario
London, Ontario, Canada

James W. Campbell
Department of Biochemistry and Cell Biology
and
Institute of Biochemical and Genetic Engineering
Rice University
Houston, Texas

Chris G. Carter
Department of Aquaculture
University of Tasmania
Launceston, Tasmania, Australia

James Childress
Department of Biological Sciences
University of California
Santa Barbara, California

Chung-Lin Chou
Laboratory of Kidney and Electrolyte
Metabolism
National Heart, Lung, and Blood Institute
Bethesda, Maryland

William H. Dantzler
Department of Physiology
College of Medicine
University of Arizona
Tuscon, Arizona

Maria A. Dingemanse
Department of Anatomy and Embryology
Academic Medical Center
University of Amsterdam
Amsterdam,The Netherlands

Jon M. Fukuto
School of Medicine
Department of Molecular and Medical
Pharmacology
University of California
Los Angeles, California

Leon Goldstein
Department of Physiology
Brown University
Providence, Rhode Island

Jon F. Harrison
Department of Zoology
Arizona State University
Tempe, Arizona

Raymond P. Henry
Department of Zoology
Auburn University
Auburn, Alabama

Dominic F. Houlihan
Department of Zoology
University of Aberdeen
Aberdeen, Scotland

Frank B. Jensen
Institute of Biology
Odense University
Odense, Denmark

Bodil Korsgaard
Institute of Biology
Odense University
Odense, Denmark

Gregg A. Kormanik
Department of Biology
University of North Carolina at Asheville
Asheville, North Carolina
and
The Mount Desert
Island Biological Laboratory
Salsbury Cove, Maine

Mark A. Knepper
Laboratory of Kidney and Electrolyte
Metabolism
National Heart, Lung, and Blood Institute
Bethesda, Maryland

Wouter H. Lamers
Department of Anatomy and Embryology
Academic Medical Center
University of Amsterdam
Amsterdam, The Netherlands

Ian D. McCarthy
Department of Zoology
University of Aberdeen
Aberdeen, Scotland

Alfred J. Meijer
E.C. Slater Institute
Department of Biochemistry
Academic Medical Center
University of Amsterdam
Amsterdam, The Netherlands

Thomas P. Mommsen
Department of Biochemistry and
Microbiology
University of Victoria
Victoria, British Columbia, Canada

Douglas S. Neufeld
Department of Physiology
College of Medicine
University of Arizona
Tucson, Arizona

Michael J. O'Donnell
Department of Biology
McMaster University
Hamilton, Ontario, Canada

Deborah F. Perlman
Department of Physiology
Brown University
Providence, Rhode Island

Nancy Sanders
Division of Science
Northeast Missouri State University
Kirksville, Missouri

Jonathan C. Wright
Department of Mathematics
and Natural Sciences
Northern State University
Aberdeen, South Dakota

Patricia A. Wright
Department of Zoology
University of Guelph
Guelph, Ontario, Canada

Stephen H. Wright
Department of Physiology
College of Medicine
University of Arizona
Tucson, Arizona

TABLE OF CONTENTS

Nitrogen Metabolism and Excretion

Chapter 1

Protein Turnover In Animals

Dominic F. Houlihan, Chris G. Carter, and Ian D. McCarthy

CONTENTS

INTRODUCTION

Protein synthesis can be defined as the process of translation (i.e., polypeptide synthesis), and many factors are known to determine the rate of polypeptide elongation on the ribosomes.[1] On the other hand, when we look at a growing animal, we see the balance between protein synthesis and protein degradation. Of prime importance is the proportion of synthesized proteins which are retained as growth, and protein synthesis thus becomes paramount to our understanding of growth performance in farmed animals and the role of protein turnover in clinical nutrition.[2,3] The molecular and whole animal views of protein synthesis are linked by the energy cost of the process which can be estimated from our knowledge of translation mechanisms and which may have a significant effect on energy balances in animals.[4,5]

Making proteins consumes energy; the question is, how much? Polypeptide synthesis proceeds through three stages; (1) the formation of the initiation complex that contains two ribosomal subunits, (2) the process of peptide chain elongation, and (3) the process of termination. It is generally accepted that charging the tRNA molecule with an amino acid results in the formation of AMP and the utilization of 2 ATPs; the entry of the aminoacyl-tRNA into the ribosome results in the hydrolysis of GTP to GDP (Figure 1). The translocation of the newly formed peptidyl-tRNA also results in the hydrolysis of GTP to GDP and phosphate. Thus the energy requirement for the formation of one peptide bond includes the equivalent of the hydrolysis of 2 ATP molecules to ADP and 2 GTP molecules to GDP.

From this ATP utilization, it can be estimated that 40 mmol of ATP are needed for the synthesis of each gram of protein.[6] Thus in place of the conventional energy budget where the emphasis has been on the components such as energy used in maintenance, specific dynamic action, activity,

0-8493-8411-7/95/$0.00+$.50

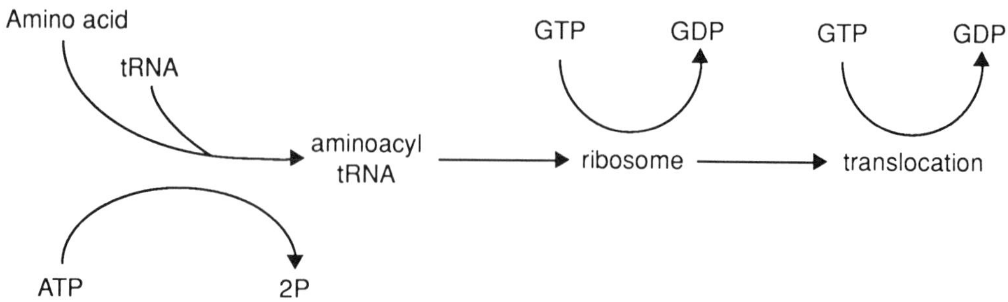

Figure 1 Diagram of the ribosomal cycle and ATP utilization. (Redrawn from Martin, N. B., in *Harper's Review of Biochemistry*, Lange Medical Publications, Norwalk, CT, 1985, 416.)

growth and reproduction, it is now possible to calculate at least one of the costs of living, namely the proportion of the ingested energy that is used for the synthesis of proteins. In inactive animals it turns out that protein synthesis represents between 20 and 40% of the total oxygen consumption. In ectotherms, where 55% of the absorbed energy is used for respiration,[7] protein synthesis may account for 15% of the total energy ingested (calculated from Reference 8).

Protein synthesis also presents us with an interesting contrast in time scales. Each ribosomal cycle of peptide-chain translation requires about only 1/20 of a second in a bacterium, and the synthesis of an average sized protein containing 400 amino acids is achieved in about 20 sec. In mammalian liver, the elongation rate is around 0.15 amino acid residues per second and polypeptide assembly time is around 1 min.[9] In fish at 30°C assembly time is around 2 min.[10] In contrast, protein accretion, the net effect of protein synthesis, is often only evident over days, and in the case of fish, over several weeks.

These differences in time scale have imposed a severe limitation on methods for measuring protein synthesis and relating synthesis to growth rates. The time scale used to measure rates of protein synthesis ranges from minutes[11-13] up to 24 h[14] depending on the species. However, these measurements must be compared with growth rates which may be measured in time periods over a hundred times longer. In fact the question of methodology still dominates our understanding of protein metabolism in animals; it is probably true that there is not one method for measuring protein synthesis which can be safely applied as a "gold standard." When it comes to considering the relative rates of protein synthesis and growth rates, the problems increase considerably.

There is a large amount of literature available on protein turnover in humans and farmed animals and critical discussions of theoretical and methodological problems.[3,15-18] The study of protein turnover in humans, especially as an aid to understanding clinical conditions, has been reviewed[3] and two papers have recently reviewed the different methods for measuring protein synthesis in humans.[19,20] The effects of diet on protein turnover in farmed animals has been examined[2] and an overview of protein and amino acid flux in farmed animals has been presented.[21] Both of these reviews made comparisons between species. Reviews of protein turnover/synthesis in nonmammalian species are fewer but are available.[4,5,22-26]

In framing this chapter we have included a section on methodology and placed it at the end of the chapter. This leaves us free to introduce protein/nitrogen flux in farmed animals and to compare flux between endotherms and ectotherms. We follow this with sections on comparative aspects of protein turnover, the effects of body size on protein metabolism, and the energy cost of protein synthesis. This chapter places emphasis on ectothermic animals, particularly fish. New studies of these animals are helping us understand the variety of tactics exhibited within the strategy of protein turnover.

PROTEIN-NITROGEN FLUX

ECTOTHERMS AND ENDOTHERMS

Simple models of protein-nitrogen flux provide a useful framework for comparing the relationships between protein intake, turnover, and retention between different animals (Figure 2). In order to clarify such flux models the nitrogen must be expressed in the same units. We have chosen to use nitrogen calculated as a protein equivalent value; thus although nitrogen is mainly excreted as ammonia in fish and as urea in mammals in Figure 2, we have taken the nitrogen values and converted them all to grams of protein. This has been done assuming that 16% of the mass of an average protein is elemental nitrogen.[21] The protein equivalent of amino acid nitrogen has been calculated assuming there are 9 mmol of amino acid per gram of protein and taking into account the loss of a water molecule due to peptide bond formation.[27] In Figure 2, each flux model is separated into protein-nitrogen consumed (PC) and protein nitrogen which is lost either via the feces (F), metabolic waste (U), or in products such as hair, feathers, or mucus (Pl), or retained as growth (PG) in new tissue. We have used a two compartment model where the absorbed dietary amino acid nitrogen enters a metabolic pool from which amino acid nitrogen is removed for protein synthesis into a tissue protein pool or excreted via amino acid nitrogen degradation products.

Although we will refer to protein and amino acid nitrogen, it should be remembered that animals do not use elemental nitrogen, but rather metabolize nitrogen-containing compounds. Therefore, it is incorrect to speak of "nitrogen requirements" in terms of an absolute amount of elemental nitrogen, but instead consideration must be given to the compounds absorbed.

The data in Figure 2 are an amalgam from different sources and with different degrees of uncertainty — not least because of problems with measuring whole animal protein synthesis and degradation. The nitrogen fluxes for some species are incomplete (this is the explanation for data missing from metabolic or tissue protein pools). Nevertheless, a comparative approach to protein metabolism has highlighted important differences and similarities between animals and may be particularly important at the practical level for improving efficiency of retention of ingested nitrogen.

First, although it is not evident from Figure 2, there are large differences in the ratio of the protein to energy requirement between ectothermic and endothermic animals. Monogastric farm animals are reported to require between 9.6 and 14.3 g digestible protein (DP) per kJ digestible energy (DE) to achieve maximum growth rates.[28] This amount of protein is considerably less than that required by fish which need between 19.4 and 28 g DP/DE; for example, the salmon in Figure 2 were fed a diet with a DP/DE ratio of 27.8.[29] A low digestible protein and energy ratio diet may also explain the low synthesis and growth rates in the goat and wallaroo.[30]

Despite differences in protein to energy requirements, there is no obvious difference in the protein growth efficiencies (protein growth*100/protein intake) between endotherms and ectotherms. For example, once differences in absorption have been taken into account, the salmon and pig have very similar protein growth efficiencies of 33 and 31%, respectively. The apparent nitrogen absorption efficiencies (nitrogen absorbed*100/nitrogen intake) were 70 and 81% for the pig[31] and salmon,[29] respectively, in Figure 2. Indeed, maximum protein growth efficiencies of 50% reported for trout[32] are similar to those of 58% for pigs,[33] 59% for chicks,[34] and 58% calculated for the lamb.[35] Thus *Octopus*, with a protein growth efficiency of 48%, is not particularly efficient.[36] These results contrast with the proposition that conversion efficiencies of ingested protein into protein growth are lower in endotherms than ectotherms.[37]

One potentially useful indicator of protein metabolism in animals is the rate of protein synthesis correlated with a given rate of protein ingestion (i.e., the anabolic stimulation ratio, protein synthesis*100/protein intake). It is notable in Figure 2 that the endotherms demonstrate anabolic stimulation ratios of between 1.7 and 3.5 whereas ectotherms exhibit much smaller ratios (0.48 to 0.74). Clearly the arrival of a meal in endotherms engenders much more protein synthesis, gram for gram, than in ectotherms. Based on experimental data for cod,[38] it has been suggested that in fish approximately 1 gram of protein was synthesized for every gram of protein consumed.[5] Subsequent studies have found roughly similar proportions in other fish species.[14]

In endotherms high anabolic stimulation ratios are part of the high turnover strategy, whereas in *Octopus* a low ratio can still result in relatively high growth rates. Thus, in terms of achieving high

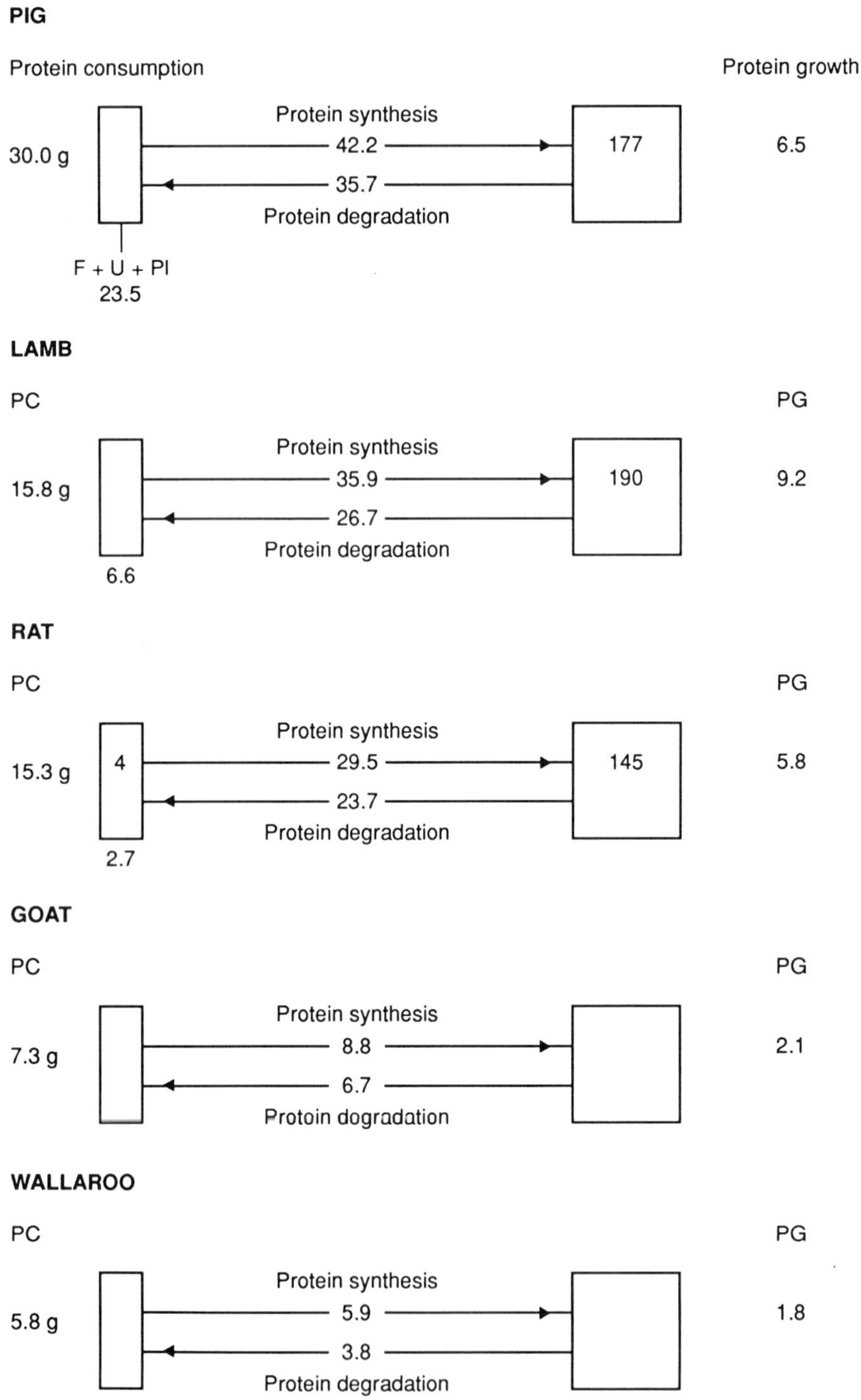

Figure 2 Protein-nitrogen flux in endotherms and ectotherms. The flux model divides protein nitrogen consumed (PC) into protein nitrogen retained as growth in new tissue (PG) or lost as fecal (F) or metabolic waste (U) or lost in products such as hair, feathers, or mucus (Pi). The data for this figure come from the following sources: 17- to 47-kg pigs fed an adequate lysine diet[31]; 12.2-kg male lambs fed reconstituted milk[35]; 211-g rats fed ad-lib[12] and the assumption has been made that the animals had the same growth patterns/nitrogen budgets as the rats investigated by Pullar and Webster[150]; the nitrogen flux measurements for goats and wallaroos were from animals fed a low protein diet (9.4% crude protein); the chickens were grown from 67 to 138 g over 10 d; the Atlantic salmon data are from animals growing from 96 to 180 g at 13°C and the data for *Octopus* are from animals growing at 22°C.

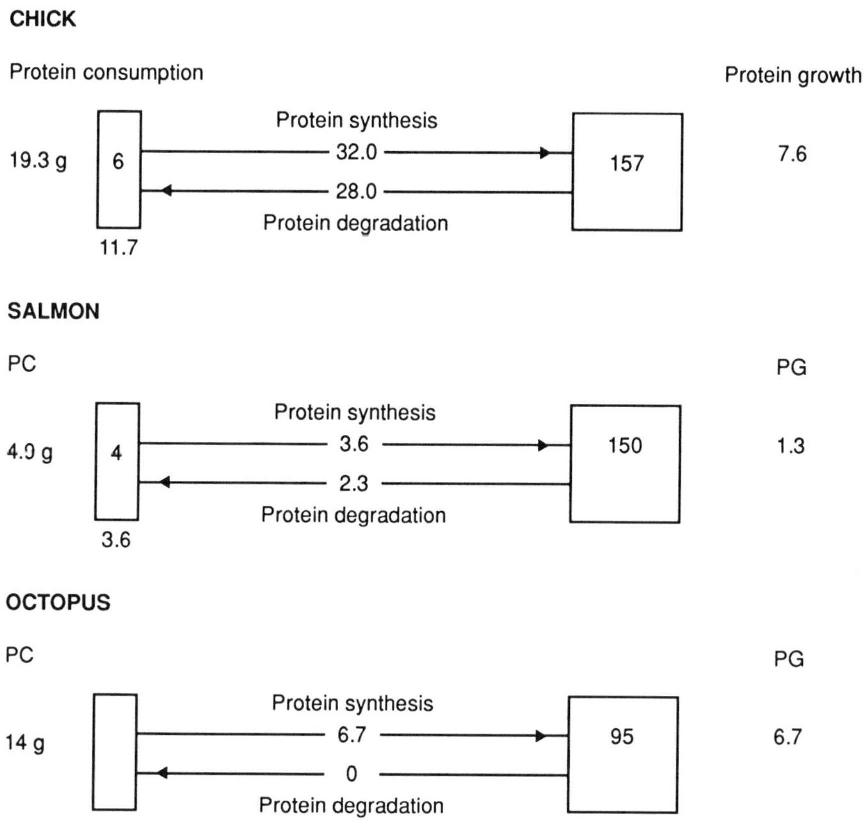

Figure 2 continued Data were selected for young growing animals (i.e., with a positive protein nitrogen balance) for which measurements of protein synthesis had been made. The units are grams of crude protein (or protein equivalent) per unit of metabolic weight per day (g protein/kg$^{0.75}$/d). Although a variety of weight exponents have been used for mammals, birds, and fish, for simplicity 0.75 was selected for all the examples. The whole-body protein (% wet weight) content of the pig, rat, chick, salmon, and octopus are from Simon,[21] Goldspink and Kelly,[12] Kita et al.,[149] Carter et al.,[29] and Houlihan et al.,[36] respectively.

rates of growth, the crucial feature of protein metabolism may not be anabolic stimulation but recycling rates from protein degradation and the difference between synthesis and degradation. In fact, we have recently suggested with data from fish that there is an inverse correlation between anabolic stimulation efficiency and the efficiency of retention of synthesized proteins.[26] However, it would be very interesting to test this hypothesis in a terrestrial ectotherm.

The efficiency of retention of synthesized proteins (protein growth*100/protein synthesis) is a useful measure of the excess of protein synthesis over protein degradation. In thc animals presented in Figure 2 this ratio varies from 15% in the pig to 100% in the *Octopus*. Generally, endotherms have lower retention efficiencies than ectotherms; in other words endotherms are turning over their proteins at higher rates than ectotherms. Thus the high growth rates of *Octopus* are achieved not by particularly high protein synthesis rates, but by very low or zero protein degradation rates.[36] Explanations for different growth rates in invertebrate animals will be discussed later.

Where the data are available, tissue free amino acid pools are simplified in Figure 2 and represented as a single metabolic pool which expresses the whole body pool in terms of a protein equivalent. This is clearly an oversimplification but allows comparisons between the relative sizes of the protein-nitrogen flux components and the metabolic and protein pool sizes. In most animals, the total free amino acids are around 2 to 4% of the amino acids in proteins (Figure 2).[21] Thus the free amino acid pools are not a reserve; if input of amino acids from protein degradation and other

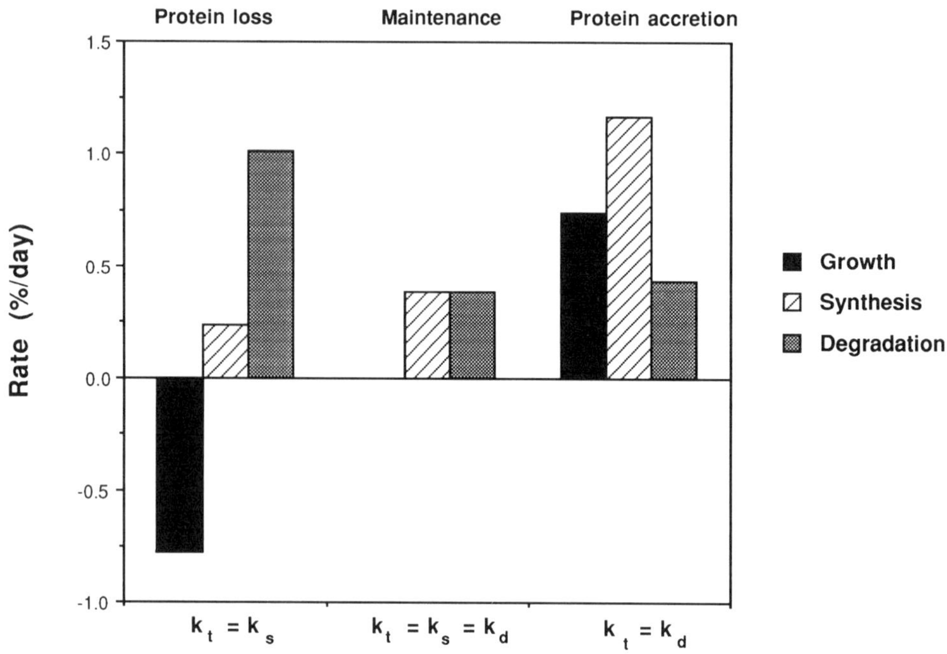

Figure 3 Whole-body fractional rates of protein growth, synthesis (k_s), and degradation (k_d) for juvenile sea bass (*Dicentrarchus labrax* L.) under different nutritional conditions. When the animals are starving (protein loss), the rate of turnover (k_t) equals the rate of synthesis. Under maintenance conditions where there is no net loss or gain of body protein k_t equals synthesis or degradation. In growing animals (protein accretion), the rate of turnover equals the rate of degradation. (From McCarthy, I. D., Langar, H., Houlihan, D. F., and Guillaume, J., unpublished data, 1992.)

metabolic pathways were to cease, the free amino acids would be rapidly metabolized.[39] In the case of endotherms the turnover time of the amino acids is extremely rapid; for some amino acids in the order to 20 to 30 min.[40] In juvenile fish it has recently been pointed out that the amino acid input from a moderately sized meal is approximately equal to, or as much as double, the size of the metabolic pool,[25] whereas in larval fish the free pools are relatively large compared with the meal size.[41] In both endotherms and ectotherms the free amino acid pools in the majority of tissues appear to remain constant after a meal.[42-45] It has been suggested that in mammals there are few species differences in protein synthesis rates.[46] Simon[71] analyzed the data from a number of studies (using a variety of techniques to measure whole body protein synthesis) and concluded that in growing mammals there is a linear relationship between weight and the absolute rate of protein synthesis, suggesting an average protein synthesis rate of 30 g protein/kg$^{0.75}$/d. In fish it was recently concluded that there are considerable differences in protein synthesis rates between different species, as might be expected in view of the wide range of environments to which the animals are adapted.[25]

Protein turnover, which has been used several times already in this chapter, was used by Waterlow et al.[40] to describe both protein synthesis and degradation, depending on conditions. The definition of protein turnover thus depends on nutritional status. Under maintenance conditions where there is no net loss or gain of body protein, fractional rates of protein synthesis (k_s) and degradation (k_d) will be equal and equivalent to the rate of turnover (k_t), i.e., $k_s = k_d = k_t$ (Figure 3). Under conditions of protein loss, the rate of degradation exceeds synthesis and the rate of turnover equals the synthesis rate, i.e., $k_s = k_t$. During weight gain, *de novo* synthesis and resynthesis of degraded proteins contribute to synthesis, so turnover will be equal to degradation, $k_d = k_t$.[25,47] Thus endotherms retaining protein as new tissue have high rates of protein turnover due to high rates of protein degradation.

DIURNAL CYCLING

Until now we have been using static descriptions of protein turnover or flux with a single level of protein intake. Variable food intake is the more normal condition, however, and there have been a number of descriptions of changes in protein turnover with variable food consumption. These changes have been reviewed recently.[26] Generally, increases in protein consumption result in increases in protein synthesis but there appears to be less consistency in the responses of protein degradation.[48]

As an example of an analysis of possible daily cycles of protein metabolism, we have used the model of Millward and Rivers[45] to describe the amino acid flux in a 100-g rainbow trout on 2 d following consumption of a meal weighing 1% of the body weight.[14] The food contained 7.4% nitrogen[14] and we assumed an apparent absorption efficiency of 85%.[49] Rates of nitrogen excretion (ammonia-nitrogen plus urea-nitrogen) were measured on day 1 and day 2. The obligatory losses were estimated from the nitrogen excretion of rainbow trout that had not been fed for two days (He, Z.-Y., unpublished data, 1991). Rates of protein synthesis and rates of nitrogen excretion on days 1 and 2 were measured at the same time and from the same fish.[14] Protein growth was calculated as the difference between absorption and excretion on day 1 and equal to excretion on day 2 to represent the diurnal cycling from the protein pool to satisfy maintenance requirements in the absence of absorbed dietary amino acids.

The 100-g trout used in the model were assumed to have a protein pool equivalent to 180 mmol amino acids, calculated from protein being 20% of the wet weight[43] and a metabolic pool of 3.24 mmol.[44] On day 1, the influx of absorbed amino acids doubled the metabolic pool of amino acids (Figure 4), emphasizing the point regarding the potential for dietary amino acids to alter tissue free concentrations as described above. However, amino acid pools of the majority of tissues were relatively stable and changed little following a meal.[42-44] Protein synthesis plays an important role in the regulation of the size of the tissue amino acid pools. On day 1, protein synthesis was 128% of the pool size and indicated recycling of amino acids derived from protein degradation. Amino acid nitrogen was also lost via nitrogenous excretion. Regulatory losses (i.e., postprandial excretion), were higher on day 1 than day 2. By day 2, postprandial increases in metabolism had nearly returned to maintenance levels. On day 2 the rate of protein degradation was greater than synthesis and about 60% of the amino acids retained as protein over day 1 were lost via nitrogen excretion. The net retention of absorbed dietary protein was about 15% over the two days but would have been higher if the fish had been fed on the second day.

This analysis suggests that at a high frequency of eating, protein degradation rates may be kept to a minimum and protein accretion maximized. However, it is by no means clear that the strategy of maximizing protein growth rates through the minimization of turnover is the only evolutionary stable strategy, and there may be disadvantages in reducing the repair of protein molecules if protein degradation is a selective process. Certainly in groups of conspecific fish which have so far been investigated, there are some highly efficient individuals and some that are much less efficient at retention of synthesized protein.[8,48] It remains to be seen whether it is the timing of the rise and fall in protein synthesis and possibly protein degradation that is the principal controlling process determining protein retention efficiency.

COMPARATIVE PROTEIN TURNOVER

STEADY-STATE CONDITIONS

A comparative account of protein turnover is hampered because there is only fragmentary information available on protein consumption, synthesis, growth, and degradation in some species. Often the data are for individual tissues and not the whole animal, especially in studies involving larger farm animals. Even for the examples in Figure 2, it is difficult to find single studies in which all of the required components to construct a complete protein-nitrogen flux model have been measured.

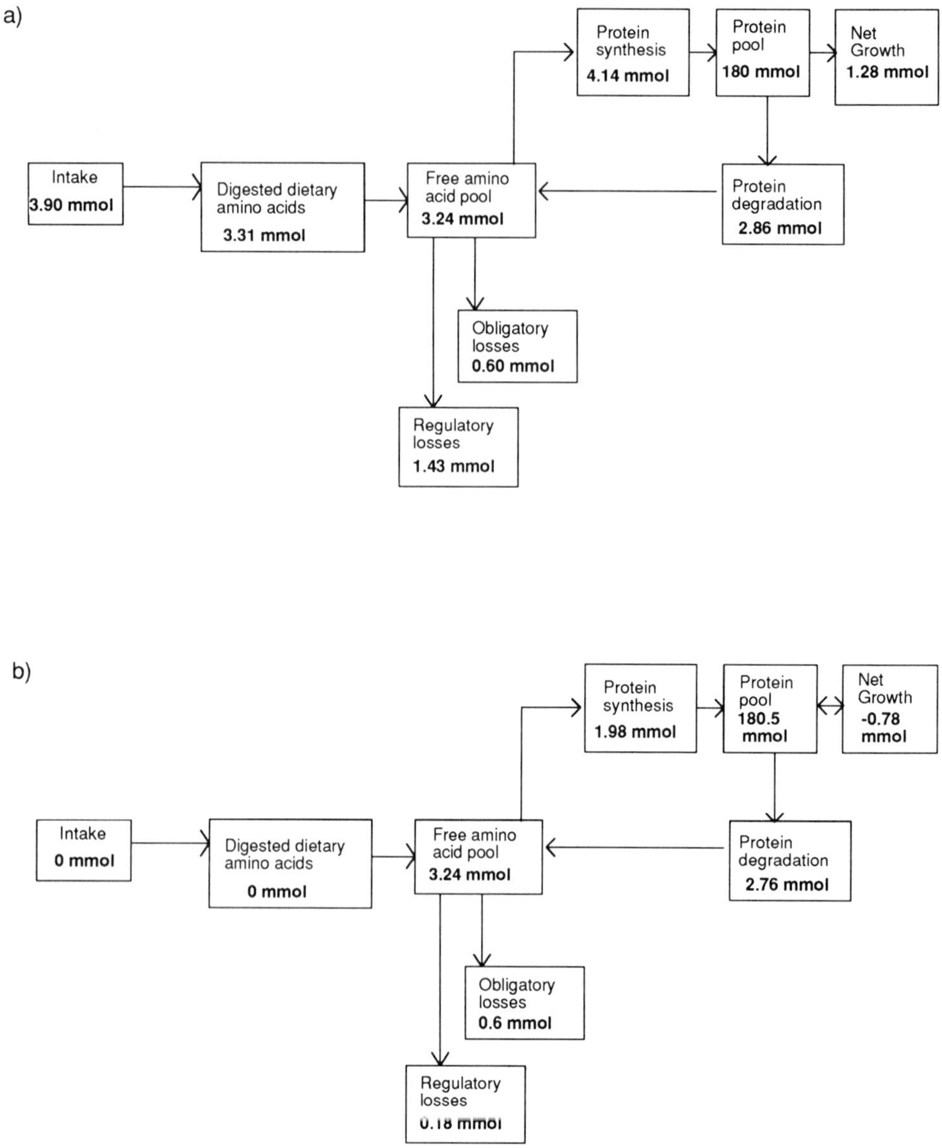

Figure 4 Amino acid flux for a 100-g rainbow trout, *Oncorhynchus mykiss*, at 10°C over 48 h following a single 1% body weight meal. The flux is based on the model of Millward and Rivers[45] and the various parameters used in the model are defined in the text. (a) Refers to the first 24 h period following a meal; (b) refers to the second 24 h period.

Furthermore, a single set of measurements at a single nutritional regime only provides part of the story.[26] A full analysis with variable protein consumption, for example in the scheme described by Houlihan et al.,[25] provides a much better description of protein consumption/synthesis relationships. Ideally, information would be available and the likely range of efficiencies under different environmental conditions would be known.

In the face of incomplete information on protein metabolism of different animals, we have concentrated on the efficiency of retention of synthesized protein (protein growth*100/protein synthesis) as a key indicator of strategies of protein metabolism (Table 1). High efficiencies of retention of synthesized proteins indicate reduced protein degradation rates and hence low turnover

Table 1 The Efficiency of Retention of Synthesized Protein (whole-body growth*100/whole-body synthesis, k_g/k_s) for Various Ectotherms

Species	k_g/k_s	g	Reference
Mussel (*Mytilus edulis*)	92	10	50
	70	10	139
Octopus (*Octopus vulgaris*)	95	150	36
Brown tiger prawn (*Penaeus esculentus*)	48–93	5	59
Tobacco hornworm (*Manduca sexta*)	88	5	Houlihan, D. F., unpublished data, 1994
Nase (*Chondrostoma nasus*)	50	0.045	70
Trout (*Oncorhynchus mykiss*)	35–69	0.2	71
Mossambique tilapia (*Oreochromis mossambicus*)	30–40	0.01	91
	30–40	0.1	
	30–40	1.0	
	30–40	10	
Sea bass (*Dicentrarchus labrax*)	30–60	2.3	128
	47	3.5	126
	77	8	43
Goldfish (*Carassius auratus*)	69	14	143
Grass carp (*Ctenopharyngodon idella*)	54	23	8
Turbot (*Scophthalmus maximus*)	33	50	143
Common carp (*Cyprinus carpio*)	25	30–60	103
Trout (*Oncorhynchus mykiss*)	45	51	122
	53	51	
	35	70	48
Plaice (*Pleuronectes platessa*)	51	60	143
Dab (*Limanda limanda*)	43	250	62
Salmon (*Salmo salar*)	32	180	29
Cod (*Gadus morhua*)	42	300	88

rates in growing animals. At first sight it looks as though invertebrate animals have higher efficiencies of retention of synthesized proteins as compared with most fish and with the mammalian examples cited in Figure 2. Further studies will reveal whether invertebrates have minimized protein turnover in order to maximize growth rates, presumably with the loss of the advantages that protein turnover brings.

The mussel *Mytilus edulis* grows relatively slowly. The extensive literature on its protein metabolism using ^{15}N-labeled food has demonstrated that increased growth rates are achieved through increased protein synthesis and decreased protein degradation. This results in protein retention efficiencies as high as 92% in the fastest growing animals.[50] Hawkins has also linked protein metabolism and energy expenditure and shown that lower rates of protein synthesis and protein degradation correlate with reduced energy expenditure in faster growing individual mussels (which are also more heterozygous).[51] Recently, a method has been described for the measurement of the rate of protein degradation using a decay curve of labelled amino acid (^{14}C-leucine) in the tissues of mussels exploiting the rapid uptake of soluble compounds from the water.[52] The use of this approach in conjunction with the ^{15}N technique may allow us to directly quantify protein synthesis and degradation, rather than calculating rates of protein degradation as the difference between synthesis and growth.

The common octopus, *Octopus vulgaris*, grows extremely rapidly, at rates equivalent to those of mammals of comparable size.[53] Rates of protein synthesis of *Octopus* tissues and whole bodies have been determined by a single injection of a high dose of labeled phenylalanine (see below). In fast growing animals protein synthesis rates were only slightly higher than growth rates (Table 1).[36] Part of this high efficiency of retention of synthesized protein may be explained by the high proportion of the animal's body being comprised of predominantly white muscle fiber types in the mantle and arms (77% of live weight). White muscle fibers are characterised by very low rates of protein turnover. The muscle fibers of *Octopus* are small (5 to 11 μm)[54] in comparison with muscle fibers of fish. It is possible that these obliquely striated muscles exhibit a much lower requirement for protein turnover compared with mammalian (or fish) muscle.

Kirkwood[55] asked whether semelparous organisms (e.g., Pacific salmon, *Octopus*, and other invertebrates) reduce repair when reproduction began. In such animals, damage to the body would be repaired during the growth phase, but it would be reduced during reproduction. If repair is taken to be protein degradation, then the answer to Kirkwood's question appears to be that even during the growth phase of a semelparous species, repair is reduced as compared with iteroparous species.

Although there have been many studies on the incorporation rate of amino acids into proteins in insects,[56] there seems to have been little interest in the relative rates of protein synthesis and degradation in rapidly growing insect larvae. The caterpillars of the tobacco hornworm, *Manduca sexta*, grow at between 26 and 75% day^{-1} in the last larval instar at 25°C.[57] By comparison, the fastest growth rates of altricial birds of around 10 g weight are about 10% day^{-1}. Mammals and precocial birds have growth rates of 5% day^{-1}.[58] Thus these animals are similar to *Octopus* in having high growth rates achieved by low turnover. If there are drawbacks to reduced protein degradation, such as the accumulation of "error proteins" and the increased possibility of mortality, then in holometabolous insects, the intervention of the pupal stage (when all the tissues are broken down and the amino acids reassembled into new adult structures), means that protein degradation can be "put off" until all the tissues are broken down and proteins synthesized *de novo*. Protein synthesis rates of whole *Manduca* bodies measured with the phenylalanine flooding dose technique have given values only slightly higher than protein growth rates resulting in efficiencies of retention of synthesized protein around 88% (Table 1).

In juvenile brown tiger prawns, *Penaeus esculentus*, protein synthesis rates determined with a single high dose phenylalanine injection revealed protein retention efficiencies as high as 93% (Table 1).[59] These preliminary results from invertebrates suggest that protein turnover rates (i.e., degradation) are much lower than in vertebrates. They may be just an artifact of the small data base or, alternatively, they may point to a general mechanism where high growth rates are achieved through low cost, efficient retention of synthesized proteins (see below).

Yolk-sac larvae of the African catfish, *Clarias gariepinus*, at 28°C may grow at 100% day^{-1} and weight-specific growth rates for many fish embryos and larvae range between 20 and 50% day^{-1}.[60] Protein synthesis rates accompanying such high growth rates have been estimated by bathing embryos in radiolabeled amino acid and indicate that they are around 1.5 times protein growth rates (Conceição, L. unpublished data, 1994). Thus, in this developing vertebrate, there is considerable protein turnover as compared with some invertebrates but not as compared with endothermic vertebrates. Similar results have been obtained from juvenile fish (see below).

Yolk-sac fish larvae are interesting because the yolk should present an ideal amino acid composition for the growth of the larvae. It is worthwhile therefore to quantify the flux of amino acids from the yolk into protein synthesis and oxidation. At a crude level we have found that the

Table 2 Efficiency of Amino Acid Utilization in Yolk-Sac Larvae of the African Catfish, *Clarias gariepinus*, at 28°C Calculated as ng Total Amino Acid at Time *t*/ng Amino Acid at Fertilization

	Time after hatching (h)				
Amino Acid	**+28 h**	**+48 h**	**+75 h**	**+92 h**	**+168 h**
GLY	0.95	1.07	1.20	1.20	0.80
HIS	0.96	0.91	0.81	0.69	0.35
PRO	1.02	0.75	0.78	0.82	0.48
ALA	0.87	0.78	0.55	0.45	0.28
VAL	0.89	0.83	0.65	0.55	0.32
ARG	0.93	0.86	0.74	0.65	0.40
THR	0.87	0.83	0.69	0.59	0.36
GLU	0.96	0.93	0.85	0.76	0.45
MET	0.90	0.86	0.75	0.59	0.34
PHE	0.94	0.94	0.81	0.70	0.42
ASP	0.89	0.90	0.83	0.71	0.42
CYS	0.99	0.96	0.96	0.85	0.51
SER	0.91	0.82	0.60	0.49	0.30
LEU	0.92	0.84	0.64	0.52	0.31
LYS	0.94	0.90	0.81	0.71	0.42
ILE	0.89	0.79	0.59	0.47	0.28
TYR	0.96	0.91	0.81	0.73	0.44

Note: Complete yolk absorption occurs at 92 h posthatching and at 168 h the larvae are starving.

From Polat, A. and Verreth, J., unpublished data, 1994.

efficiency of transfer of protein from the yolk to the larvae is approximately 75% in Atlantic salmon (McCarthy, I. D., Houlihan, D. F., and Hackney, L. D. unpublished data, 1994). Using the data on the rates of yolk absorption in the African catfish (Polat, A. and Verreth, J., unpublished data, 1994), it is suggested that fish larvae demonstrate very high efficiencies of amino acid utilization in the conversion of yolk sac amino acids into growing larvae (Conceição, L., personal communication, 1994). Calculation of the efficiency of transfer of amino acids from the yolk to the larva only reveal loss of amino acid when the yolk sac has been completely absorbed and amino acid oxidation is presumably occurring for energy production (Table 2). Presumably the fat stores in the yolk also provide the fuel for growth during the period of high amino acid efficiency.[152]

A similar situation of high efficiency transfer of amino acids from yolk to embryo has been found for chicks. Calculation of the efficiency of the conversion of individual amino acids from whole egg protein in 1-d-old chicks gives very high values (87%),[61] and for amino acids such as proline, histidine, and glycine, the calculated efficiency exceeds 100%. This observation suggests that there is recycling of nitrogen by *de novo* synthesis of nonessential amino acids.

In a recent review of protein metabolism in fish larvae, we concluded that, although weight-specific growth rates are high, rates of protein synthesis, amounts of free amino acids, and RNA concentrations are not exceptional, considering the small size of the organisms. In fact, fish larvae seem to be following scaling relationships established for larger fish.[41] There was little evidence that rapidly growing fish larvae or juvenile fish sacrifice protein turnover in order to maximize retention efficiencies of synthesized protein; fish larvae are not exhibiting the *Octopus* or insect larval strategy described earlier. However, it would be surprising if examples of reduced turnover growth were not found in some rapidly growing species.

In the remainder of the fish species in Table 1, retention efficiencies are mainly between 30 and 50%. Presently it is not possible to clearly relate these different efficiencies to different growth rates, environmental conditions, temperature, etc. However, the value of protein degradation may permit both rapid adaptation to environmental perturbations as well as the removal of damaged or abnormal proteins that may be harmful.[55] Proteins cannot be repaired but only replaced, and therefore, protein degradation rates may be correlated with stressful environments where proteins are damaged. Alternatively, restriction of a species to narrow environmental conditions may permit low levels of protein degradation. Retention of synthesized proteins as growth has the advantage of minimal cost of growth, and therefore, protein degradation may be seen as in direct conflict with least-cost growth. We do not know the costs and benefits of repair and future research should directly test the above hypotheses, particularly utilizing, for example, the wide range of environmental conditions experienced by some fish species.

PROTEIN TURNOVER AND ENVIRONMENTAL STRESS

Dab (*Limanda limanda*), a flatfish, exposed to environmental stress in the form of sewage sludge, exhibited no difference in protein synthesis rates as compared with control animals.[62] However, protein growth rates were reduced in the sewage sludge conditions and therefore the efficiencies of retention of synthesized proteins were lower in the stressful condition. This appears to be the first study which has demonstrated increased protein breakdown in an aquatic animal as a result of long-term stress, and indicates that there is a physiological cost, in the form of increased protein turnover, to the toxic compounds in the sewage sludge. The fish were feeding; in all such stress studies feeding rates of individuals should be determined so that protein metabolism may be related to protein intake. All of the tissues examined, white muscle, liver, spleen, and kidney had increased protein turnover indicating a generalized effect. It is interesting to note that there was some diminution in the immune response of the sewage-exposed fish[63,64] and therefore the possibility exists that increased protein turnover does not improve physiological performance. Alternatively, performance was maintained at the cost of increased turnover and hence energy allocation to protein turnover.

The rate of protein synthesis is known to be controlled by the level of energy supply, and, in conditions of hypoxia, mammalian tissues have been found to reduce rates of protein synthesis.[65] In isolated hepatocytes of rainbow trout, protein synthesis was found to be PO_2 independent between 20 and 10 kPa and started to fall below a PO_2 of 2 kPa.[66] During anoxia, protein synthesis rates of isolated liver cells of the turtle, *Chrysemys picta bellii*, were reduced by 92%.[67] Estimates of the costs of protein degradation vary from 1 to 14% of total ATP consumption. Under anoxia, 45% of total ATP turnover is believed to be directed to protein turnover.[68]

Recent experiments with crucian carp, *Carassius carassius*, living in complete anoxia have indicated that although liver protein synthesis rates may be reduced by 90% or more compared with normoxic animals, protein synthesis in the brain is maintained in anoxia at normoxic rates (Houlihan, D. F., Smith, R. W., and Nilsson, G. E., unpublished data, 1994). Thus the key to long-term survival of the whole animal, in the face of reduced energy supply from aerobic metabolism, may be the diversion of glucose to the brain in order to maintain protein synthesis, whereas other tissues can survive long-term reduction in protein synthesis rates.

Finally, it has been found that protein synthesis rates are temperature dependent with Q_{10} values in the range of 2 to 3 in various fish species.[23,69] Recently it has been demonstrated that the efficiencies of retention of synthesized proteins increase with temperature[70,71]; reduced protein degradation with increased temperature in growing animals is contrasted with increased degradation in starved animals.

SCALING

The rate of many physiological variables are proportional to body weight according to the allometric relationship $Y = aX^b$, where X is body weight, a is a constant, and b is the weight exponent.[72,73] Commonly these scaling relationships are examined following log transformation of the data ($\text{Log}Y = \text{Log}a + b{*}\text{log}X$). There are two approaches that can be used to obtain scaling relationships for protein synthesis and protein growth. First, data can be drawn from a number of different species in order to obtain a wide range of body weights. An example of such an approach for ectotherms (fish) and endotherms (birds and mammals) is presented in Tables 3 and 4 and shown in Figure 5.

As body size increases, fractional rates of protein synthesis of fish decline, with an overall exponent value of –0.30 (Figure 5). Fractional rates of protein growth decline with increasing body size with an exponent of –0.34 compared with values of –0.3 to –0.45 for various salmonids.[74] This scaling exponent for synthesis is much lower than that previously assumed for fish (–0.77).[75] However, this steep slope was a result of the high protein synthesis rates that were thought to be characteristic of larval and juvenile fish.[76-78] Fish show indeterminate growth throughout their life history (they do not reach a maximum size), and this type of growth rate is characterized by parallel lines for both protein growth and synthesis with increasing size (Figure 5). Although the lines are parallel, the ratios of growth rate to synthesis (k_g/k_s) decrease from 44% for a 10-mg fish to 31% for a 1-kg fish; efficiency decreasing with increasing size.

In contrast, the endotherm data reveal that the weight exponent for protein growth is larger than the exponent for protein synthesis (Figure 5); the lines diverge markedly as growth rates fall precipitously, while protein synthesis rates decline more gradually with increasing size (Figure 5). The retention efficiencies are 43% for a 10-g endotherm and 6% for a 200-kg endotherm (calculated from the regression equations in Figure 5). Thus, as in fish, protein retention efficiency falls with increasing size. We have recently obtained similar scaling exponents for the relationships between body weight and RNA:DNA and RNA:protein ratios, and between body weight and free pool amino acid concentrations in fish.[41] From this analysis it seems that the scaling relationships are robust enough to absorb variation due to species and differences in nutritional plane and, in the case of fish, the influence of water temperature. However, it is still preferable to construct allometric relationships for each species reared under identical conditions.

Recently in our laboratory we have collected data for rates of protein synthesis for immature rainbow trout ranging in size from 0.088 to 480 g and rates of protein growth for rainbow trout ranging in size from 0.202 to 480 g.[41] All fish were reared at 10°C in the same tank conditions and given similar feeding regimens. The scaling exponents for the fractional rates of protein synthesis and growth in rainbow trout are similar with values around –0.25 (Figure 6a). Similar scaling

Table 3 Whole-body Fractional Rates of Protein Synthesis (k_s) and Protein Growth (k_g) in Various Endotherms of Different Sizes (wet weight, g or kg)

	Weight	k_s	k_g	Reference
Rat	0.53 g	69.4	63.0	12
	1.62 g	53.1	46.9	
	50.0 g	30.6	7.3	
	211 g	18.0	1.88	
	584 g	12.0	0.23	
Chicken	80 g	34.0	12.2	144
	137 g	31.5	8.5	
	227 g	29.7	6.6	
	307 g	25.8	4.9	
	133 g	22.2	9.8	145
	295 g	20.4	10.2	
	135 g	31.6	7.2	146
	139 g	32.3	7.0	
	125 g	32.0	5.8	
	141 g	33.2	6.6	
Cow	204 kg	8.0	0.48	147
	263 kg	6.3	0.44	

Note: The data presented here are used to construct the scaling relationships shown in Figure 5b.

Table 4 Whole-body Fractional Rates of Protein Synthesis (k_s) and Protein Growth (k_g) in Fish of Different Sizes (wet weight, g) Reared Under Different Water Temperature (T°C) and Feeding Regimens

Species	g	T°C	k_s	k_g	Feeding	Reference
Nase (*Chondrostoma nasus*)	0.030	20	25.0	10.0	Ad-lib	70
	0.045	20	27.0	14.0		
	0.050	20	30.2	15.0		
Mossambique tilapia (*Oreochromis mossambicus*)	0.01	26	90.0	31.0	Ad-lib	91
	0.10	26	26.9	9.5		
	1.0	26	7.9	3.0		
	10.0	26	2.3	0.9		
	0.07	25	21.3	13.9	Ad-lib	143
	0.40	25	13.2	6.3		
	0.54	25	11.1	2.7		
	1.24	25	9.8	2.6		
	2.45	25	3.8	2.0		
	8.45	25	2.1	1.0		
Goldfish (*Carassius auratus*)	7.3	10	3.6	0.6	Ad-lib	143
	13.3	15	3.3	0.7	Ad-lib	
	14.3	25	3.2	1.3	Restricted	
	13.9	25	4.2	2.9	Ad-lib	
Grass carp (*Ctenopharyngodon idella*)	22.9	22	2.2	1.2	2 meals/d	8
Rainbow trout (*Oncorhynchus mykiss*)	51.2	11	4.4	1.8	1 meal/d	122
	69.0	10	2.4	0.8	1 meal/d	48
	64.5	10	2.8	0.4	1 meal/d	
	83.0	10	3.3	1.1	1 meal/d	43
Atlantic salmon (*Salmo salar*)	180	13	2.5	0.8	2.5% BW/d	29
Atlantic cod (*Gadus morhua*)	300	10	3.1	1.0	1 meal/d	88
	300	10	4.8	2.0		

Note: The data presented here are used to construct the scaling relationships shown in Figure 5a.

exponents were also found between body weight and whole-body RNA concentrations and whole-body RNA:protein ratio.[41] These exponents are similar to the scaling exponent for weight-specific oxygen consumption in trout (–0.2),[72] a correlation previously used as part of the argument suggesting that protein synthesis makes a significant contribution to oxygen consumption in ectotherms.[5] Our recent analysis shows that, although synthesis rates of larval and juvenile fish are higher than those reported for adults, they obey the conventional scaling relationships (Figure 6).[41] As expected, the synthesis retention efficiencies (k_g/k_s) calculated from the experimental data decreased with increasing body weight (50.5 ± 11.5% for a 0.160 g fry and 24.6 ± 2.8% in a 316-g fish). However, since the scaling exponents in Figure 5 for synthesis and growth are similar, the predicted synthesis retention efficiencies did not show the same pattern (20.3% for a 0.160-g fry and 22.1% for a 316-g fish). This may be explained by the absence of fractional protein growth data for the yolk-sac fry (0.088 to 0.131 g).

Goldspink and Kelly[12] studied rates of protein synthesis and protein growth in rats from the fetus to 105 weeks post partum. We have used their data on rats ranging in weight from 0.53 (16-d-old fetus) to 504 g (105 weeks post partum) to examine the relationship between body weight and protein synthesis and growth (Table 3 and Figure 6b). The scaling results indicate that fractional rates of protein synthesis decreased with a similar exponent value to that observed in the trout data (–0.24 compared to –0.26). However, the decline in fractional protein growth rates with increasing body size was much steeper (–0.74 compared to –0.25) and comparable to the results obtained with a range of endotherms (Figure 5). It seems as though individual species follow the same patterns as those obtained from a mixture of species (Figures 5 and 6).

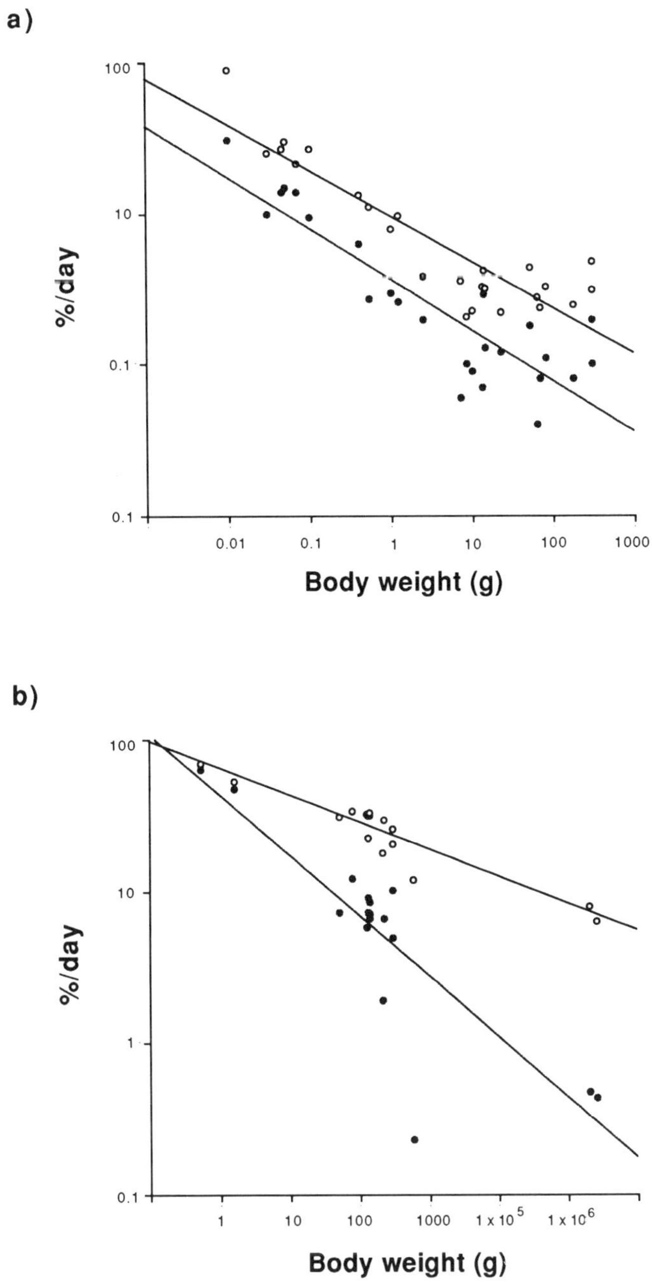

Figure 5 A summary of the relationship between body weight (g) and whole-body fractional rates of protein synthesis (k_s, %/d; open circles) and protein growth (k_g, %/d; closed circles) for (A) fish and (B) endotherms. The figure is calculated from the data presented in Tables 3 and 4. The relationships are described by the following linear regression equations: fish, $\log_{10}k_s = 0.983 - 0.304 \cdot \log_{10}\text{Weight}$ ($R^2 = 0.805$, $n = 25$, $p < 0.001$) and $\log_{10}k_g = 0.566 - 0.336 \cdot \log_{10}\text{Weight}$ ($R^2 = 0.784$, $n = 25$, $p < 0.001$); endotherms, $\log_{10}k_s = 1.811 - 0.180 \cdot \log_{10}\text{Weight}$ ($R^2 = 0.872$, $n = 17$, $p < 0.001$) and $\log_{10}k_g = 1.625 - 0.397 \cdot \log_{10}\text{Weight}$ ($R^2 = 0.698$, $n = 17$, $p < 0.001$).

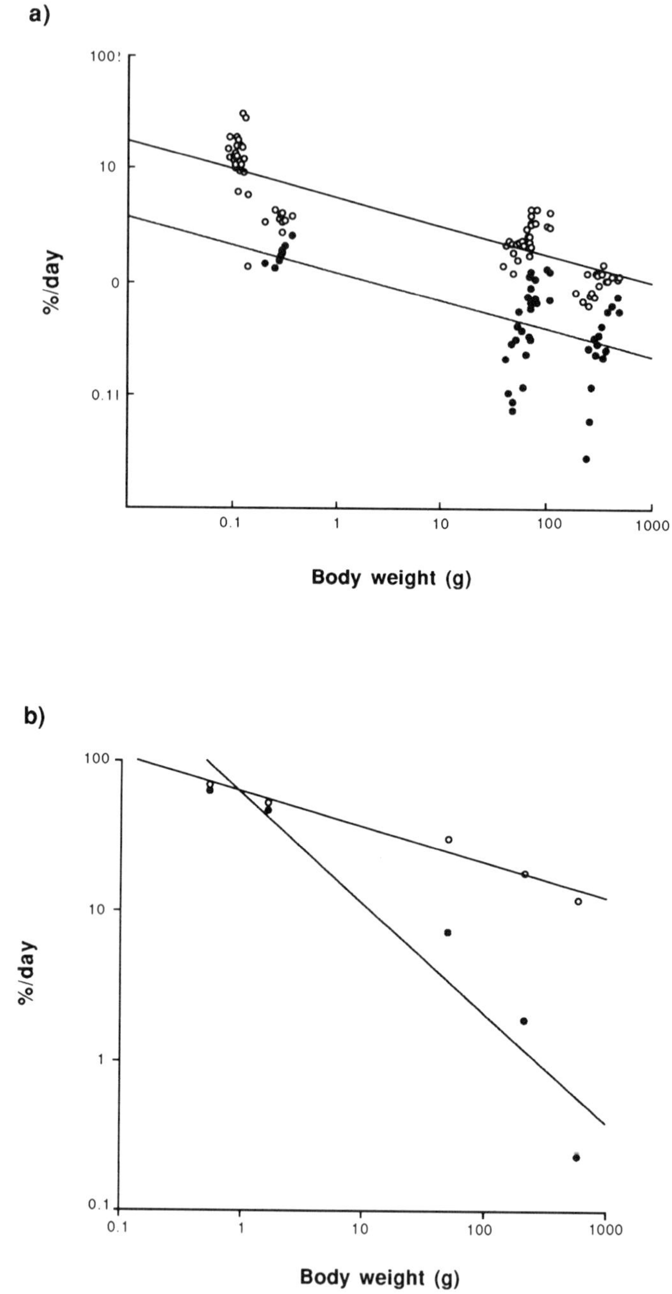

Figure 6 A summary of the relationship between body weight (g) and whole-body fractional rates of protein synthesis (k_s, %/d; open circles) and protein growth (k_g, %/d; closed circles) for (A) rainbow trout[41] and (B) rats (Table 3).[12] The relationships are described by the following linear regression equations: rainbow trout, $\log_{10}k_s = 0.765 - 0.259 \cdot \log_{10}\text{Weight}$ ($R^2 = 0.753$, $n = 79$, $p < 0.001$) and $\log_{10}k_g = 0.081 - 0.248 \cdot \log_{10}\text{Weight}$ ($R^2 = 0.364$, $n = 54$, $p < 0.001$); rat, $\log_{10}k_s = 1.794 - 0.235 \cdot \log_{10}\text{Weight}$ ($R^2 = 0.956$, $n = 5$, $p < 0.001$) and $\log_{10}k_g = 1.800 - 0.733 \cdot \log_{10}\text{Weight}$ ($R^2 = 0.891$, $n = 5$, $p < 0.001$).

ENERGY COST OF PROTEIN SYNTHESIS

Apportioning the costs of the different components of maintaining animal life is proving to be difficult. Attention has mainly focused on the cost of ion transport and protein synthesis. Estimates suggest that protein synthesis accounts for around 40% of a resting cell's oxygen consumption and ion transport accounts for around 10%.[5,6,79] Estimates of the cost of actively transporting mineral ions through membranes against gradients have varied widely, and different cell types may spend varying proportions of their energy production on ion transport.[66,80-84]

There have been a number of attempts at quantifying the energy cost of protein synthesis either directly, by measuring rates of protein synthesis and multiplying by a protein synthesis cost factor, or indirectly by comparing protein synthesis and oxygen consumption.[6,51,84-87] Minimum values of between 11 and 42% of oxygen consumption have been reported for fish.[8,88,89] In a variety of mammals, protein synthesis was calculated to account for 20 to 25% of the resting heat production.[90] Some of the data on energy costs of protein synthesis have been determined by measuring oxygen consumption and protein synthesis before and after application of a protein synthesis inhibitor such as cycloheximide. They suggest that from 80 to 87% of oxygen consumption is used for protein synthesis in trout hepatocytes and juvenile tilapia.[66,91]

Generally, aerobic costs of protein synthesis appear to vary widely and different tissues have different energetic costs for protein synthesis.[83,86,92,93] In fact, the energy cost of protein synthesis appears to be related to the rate of protein synthesis; the higher the rate of synthesis the lower the aerobic cost.[66,153]

Recent data from juvenile fish have also demonstrated that there is no further increase in oxygen consumption at growth rates above 8 to 10% d^{-1},[94,95] and it has been suggested that variable protein synthesis costs could contribute to this apparent cost-free growth.[25]

Other costs of maintaining cell function are difficult to quantify; constant activation of tRNA, as well as the production of rRNA which remains remarkably constant regardless of the rate of protein synthesis,[96] most probably form a part of this fixed energy expenditure. It has been suggested by Cooper and Gibson[97] that accumulation of rRNA in the cytoplasm occurs only at half the rate at which it is synthesized in the nucleus. This would mean that the cost of rRNA synthesis would be underestimated by at least 100%.

In summary, the aerobic costs of protein synthesis, even using minimum theoretical values, appear to make a large contribution to the oxygen consumption of inactive fish. Allocation of the costs to the individual tissues is complicated because it will be the product of the mass of protein in each tissues, the tissue-specific rate of protein synthesis, and the tissue-specific cost.[25] In terms of the whole animal, however, it is interesting to relate the cost of protein synthesis to the cost of growth. Thus, rainbow trout with similar rates of protein synthesis, and hence similar protein synthesis costs (unless there are large differences in body composition), will have widely different costs of growth when the retention efficiency of synthesized proteins vary.[48]

There is a great deal of interest in the cost of growth, and a review by Wieser[98] has emphasized that the cost of transforming food into growth (strictly metabolizing energy into production) in a wide range of organisms is about three times the theoretical cost of synthesizing the same amount of tissue.[98] Costs of protein synthesis which decline with increasing protein synthesis rates[66,153] may help to explain variability between organisms, but a new method of analyzing growth costs is needed. It is possible that quantification of protein synthesis rates, efficiencies of retention of synthesized protein, and rate-specific protein synthesis costs may provide this new method.

Costs of living are often divided into the energy needed for activity, costs of tissue synthesis, and costs of maintenance.[37] From the foregoing analysis of protein turnover, however, it does not seem logical to subdivide protein synthesis into maintenance and growth components since animals are synthesizing proteins at different rates and a variable proportion of these proteins are retained as growth. Thus an estimate of the cost of protein growth [M_{O2}(growth), mol oxygen/animal/d] can be obtained from: M_{O2} (growth) = rate of protein synthesis*cost of protein synthesis*(rate of protein synthesis/rate of protein growth). The rate of protein synthesis would be in grams of protein synthesized per day, most commonly calculated from the fractional rate of protein synthesis (percent of the protein mass synthesized per day, see below) and the protein content of the animal. Costs of protein synthesis can be estimated from rates of synthesis; generally at above several percent per day, costs are the theoretical minimum. The ratio of synthesis in excess of growth is the reciprocal of the ratio shown in Table 1. For animals with high efficiency of retention of synthesized proteins (approaching 100%), costs of growth may approach the cost of protein synthesis. However, for animals having high rates of protein turnover (low efficiency of retention, e.g., 50%), costs of growth may be more than double protein synthesis costs. In addition, the energetic costs of protein degradation could be added to the above formulation when estimates become available.

This calculation for the protein synthesis of whole animals may seem to be too much of an oversimplification in view of the different tissues and diversity of proteins. However, a relatively small number of proteins (mainly those in muscle) make up a large proportion of the total protein content of the animal's body.[6,26] Attempting to account for whole animal food consumption, growth rate and oxygen consumption from a detailed analysis of individual tissues is beyond our ability at the moment due to lack of information.

The question then arises regarding the optimum cost of growth keeping in mind that there may be a trade off between turnover rates and longevity. At least in the above formulation of the cost of growth, it is recognized that what is costing energy is not growth, but the synthesis of proteins. Both the synthesis costs and the retention efficiency are under physiological and selection pressures.

METHODOLOGY

Several methods are currently available for the *in vivo* measurement of rates of protein synthesis in whole bodies and tissues of endotherms and ectotherms. The majority of these measure the flux of an amino acid or nitrogen using a tracer substance (i.e., a specific amino acid or protein-bound amino acids labeled with an isotope), given in a single dose or by continuous infusion over a known period.[17,40] The techniques for measuring protein turnover can be divided into two broad categories: end point/product methods and plasma/precursor methods. The measurement of protein turnover in endotherms has been critically examined in several excellent reviews.[16-18,40,99-100] We will concentrate on the application of the available techniques to the study of protein synthesis in fish.

THE MEASUREMENT OF RATES OF PROTEIN SYNTHESIS IN FISH

In the majority of available literature, rates of protein synthesis in fish have been quantified *in vivo* by measuring the incorporation of a single radiolabeled essential amino acid into whole animal or tissue protein using either constant infusion[22,23,101] or a single flooding dose injection (see below) to administer the radiolabel. Since different proteins are synthesized at different rates, and may differ in amino acid composition, these measurements will represent the average synthesis rates for the tissue or whole animal.[102] However, recent studies have measured synthesis rates using protein uniformly labeled with ^{15}N,[14,103] enabling us to compare these results with those obtained using a single radiolabeled amino acid (see below).

Although *in vivo* incorporation is the most commonly used method to measure rates of protein synthesis, several studies have used isolated polyribosomes from fish epaxial muscle to make *in vitro* measurements of rates of protein synthesis using ^{14}C-phenylalanine[104-112] or ^{3}H-leucine.[113-115] This technique has been used to provide information on the incorporation rates of amino acids onto isolated polyribosomes under a number of nutritional conditions. However, it has not been widely used since it is not possible to compare incorporation rates, commonly expressed as pmol label incorporated per minute per gram wet weight or milligram RNA, with protein growth rates (k_g) to calculate degradation rates, and k_{RNA} and k_g/k_s values cannot be calculated.

Measurement of *In Vivo* Rates of Protein Synthesis in Fish by the 'Flooding Dose' Method

Protein synthesis rates of individual tissues or the whole body are really only of value when they are measured in animals whose food consumption, growth rate, and other physiological indices of condition have been measured. Typically the animals will be injected with a single flooding dose injection of ^{3}H-phenylalanine via the caudal blood vessels or as a single intraperitoneal injection. The measurements that are needed for the successful calculation of protein synthesis rates using the flooding dose technique using these two methods of introducing the radiolabel are presented in Figure 7. The data are taken from studies on the protein metabolism of 80-g rainbow trout at 10°C (caudal vein injection)[48] and 8-g seabass at 18°C (intraperitoneal injection) (McCarthy, I. D., Langar, H., Houlihan, D. F., and Guillaume, J., unpublished data, 1992). By measuring the normal background tissue or whole body free phenylalanine concentration, and comparing this value with the measured concentration following injection, it is possible to estimate the degree of flooding obtained.

The free phenylalanine concentration in the white muscle of rainbow trout is normally 74 nmol g^{-1} wet wt,[44] and following a flooding dose injection, a 12-fold increase above normal was achieved (Figure 7a), remaining stable over the range of incorporation times. In seabass, the free phenylalanine concentration in the white muscle is normally 77 nmol g^{-1} wet weight (McCarthy, I. D., Langar, H., Houlihan, D. F., and Guillaume, J., unpublished data, 1992) indicating an 11-fold increase above normal following intraperitoneal injection (Figure 7b), remaining stable over the range of incorporation times.

The mean free phenylalanine specific radioactivity in the rainbow trout white muscle free pool remained elevated and stable over the range of incubation times (Figure 7c), and reached a mean value of 87% of the specific radioactivity of the injection solution (1091 DPM/nmole compared to 1250 DPM/nmole).[48] In Figure 7d, the data for the seabass indicates that, following intraperitoneal injection, the mean free phenylalanine specific radioactivity in the white muscle free pool also remained elevated and stable, and reached a mean level of 72% of the specific radioactivity of the injection solution (1616 DPM/nmol compared to 2250 DPM/nmol) (McCarthy, I. D., Langar, H., Houlihan, D. F., and Guillaume, J., unpublished data, 1992).

The labeling of the body protein must be linear over the incorporation time (Figure 7c and d) and, from the intercept of the regression line, it is possible to estimate how soon after injection the radiolabel began to be incorporated into body protein. In the examples shown in Figures 7c and d, the intercept indicates that labeling began 15 to 20 min after introduction of ^{3}H-phenylalanine via either the caudal blood vessels or absorbed from the peritoneum. In these examples the intercept values are not significantly different from zero (rainbow trout $t = -0.15$, $p > 0.05$; seabass $t = -1.49$, $p > 0.05$). In some studies where a significant intercept has been recorded, this value has been subtracted from the incorporation time in order to calculate the fractional rate of protein synthesis.[41,70]

The equation commonly used to calculate fractional rates of protein synthesis is:

$$k_S\ (\%\ d^{-1}) = (S_b/S_a)*(1440/t)*100 \tag{1}$$

where S_b and S_a are the protein-bound and free phenylalanine specific radioactivities after an incorporation time of t min, with the requirement that the free phenylalanine specific radioactivity remains elevated and stable for the duration of the time course. In studies where, following injection, the free phenylalanine specific radioactivity is elevated and shows a slow linear decline with time, the following equation can be used to calculate fractional rates of protein synthesis:

$$k_S\ (\%\ d^{-1}) = (S_{b(t2)} - S_{b(t1)}/S_{a(t2-t1)})*(1440/t2 - t1)*100 \tag{2}$$

where $S_{b(t2)}$ is the protein-bound phenylalanine specific radioactivity after an incorporation time of $t2$ min, and $S_{b(t1)}$ is the average incorporation at an earlier time (usually 20 min). $S_{a(t2-t1)}$ is the average free phenylalanine specific radioactivity over the time period $t2 - t1$.

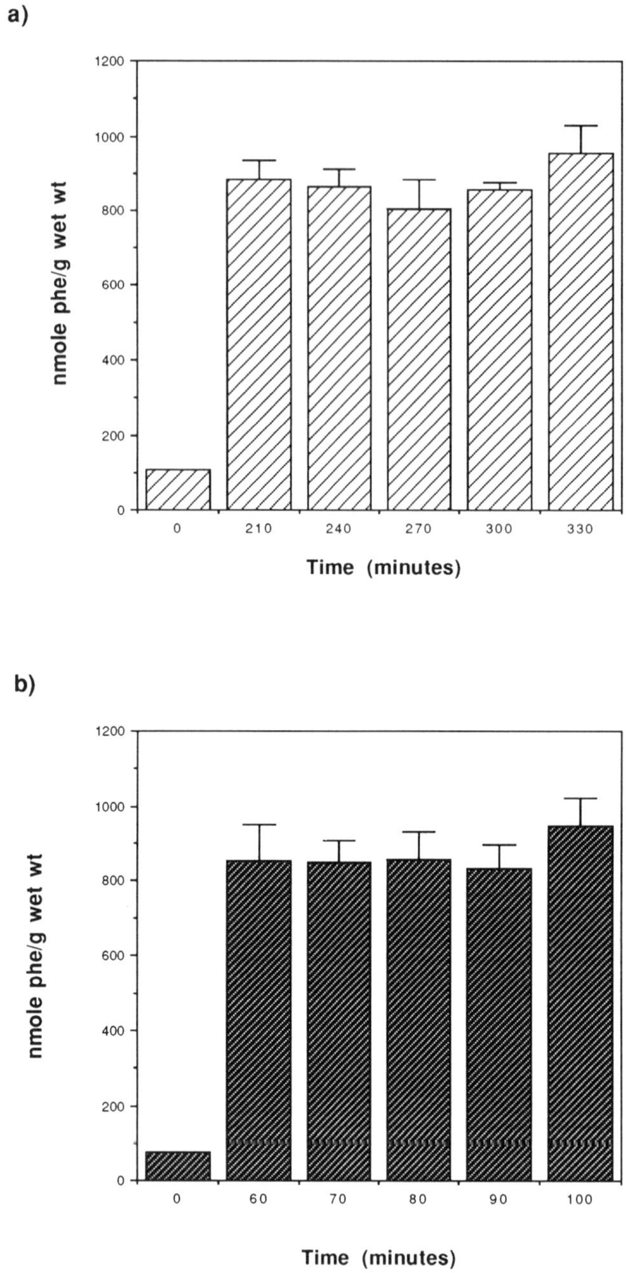

Figure 7 (a) The mean white muscle free pool phenylalanine concentration (nmol phe g wet wt^{-1}) in rainbow trout (*Oncorhynchus mykiss*) at various time intervals following a single caudal vein flooding dose injection of 135 m*M* phenylalanine and L-[2, 6-^{3}H] phenylalanine. The value at time 0 indicates the white muscle phenylalanine concentration prior to injection; (b) the mean white muscle free pool phenylalanine concentration (nmol phe g wet wt^{-1}) in sea bass (*Dicentrarchus labrax*) at various time intervals following a single intraperitoneal flooding dose injection of 135 m*M* phenylalanine and L-[2, 6-^{3}H] phenylalanine. The value at time 0 indicates the white muscle phenylalanine concentration prior to injection; (c) the mean white muscle free pool phenylalanine specific radioactivity (S_a, DPM nmol phe^{-1}) and the mean whole-body protein bound phenylalanine specific radioactivity (S_b, DPM nmol phe^{-1}) for rainbow trout at various time intervals following a single caudal vein flooding dose injection of 135 m*M* phenylalanine and L-[2, 6-^{3}H] phenylalanine[26,48]; (d) the mean white muscle free pool phenylalanine specific radioactivity (S_a, DPM nmol phe^{-1}) and the mean whole-body protein bound phenylalanine specific radioactivity (S_b, DPM nmol phe^{-1}) for sea bass at various time intervals following a single intraperitoneal flooding dose injection of 135 m*M* phenylalanine and L-[2, 6-^{3}H] phenylalanine. (Parts a and c redrawn from Refs. 26 and 48; parts b and d from McCarthy, I. D., Langar, H., Houlihan, D. F., and Guillaume, J., unpublished data, 1992).

c)

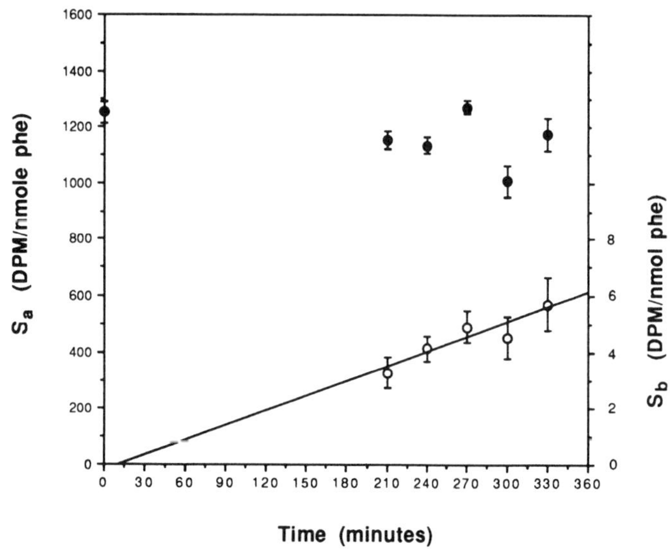

d)

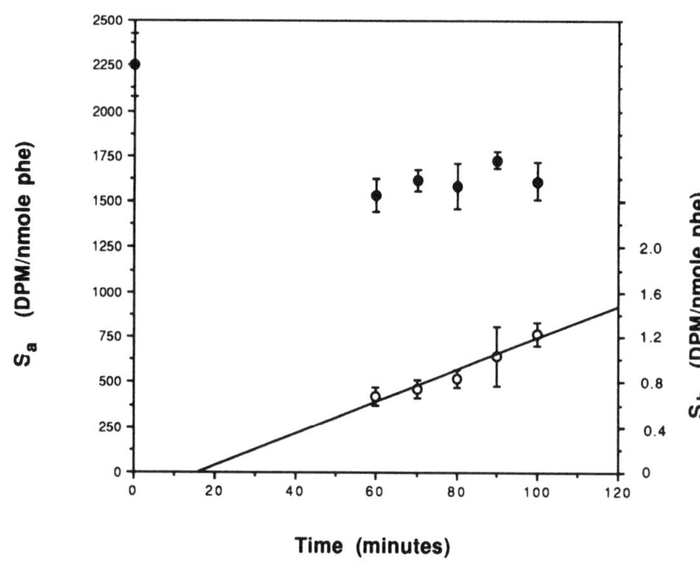

Figure 7 continued

The use of the flooding dose method to measure fractional rates of protein synthesis in fish has been validated for a range of water temperatures, using an incorporation time of between 40 min and 6 h.[8,48,73,88,116-118] This is in contrast to experimental procedures in endotherms where t needs to be shorter due to faster incorporation rates.[11-13] The technique has been used to measure rates of protein synthesis in fish from 1 to 500 g in weight, by either intraperitoneal or intravenous injection of ^{3}H-phenylalanine.[8,29,36,38,48,62,73,88,89,116-129] Recently it has been used to measure synthesis rates in fish of less than 1 g either by microinjection into the peritoneal cavity,[41,43,71] or by adapting the method of Fauconneau[76] and bathing juvenile fish in a flooding dose of ^{3}H-phenylalanine.[41,70,91,154] The ^{3}H-phenylalanine flooding dose method has also been applied to measure rates of protein synthesis in body organs/tissues[73,88,120-125,129] and cells[66] (Smith, R. W. and Houlihan, D. F., unpublished data, 1994).

Because of its ease of application and adaptability, coupled with a sensitive and selective method of extraction and measurement of phenylalanine (see below), the flooding dose technique has become a valuable tool in the study of protein synthesis in fish.

Several radiolabeled amino acids have been used in studies measuring the *in vivo* rates of protein synthesis in fish. The predominant ^{3}H-labeled amino acid used in flooding dose studies has been phenylalanine (see references above), but ^{3}H-leucine has also been used.[130] Several ^{14}C-labeled amino acids have been used in studies measuring the *in vivo* rates of protein synthesis in fish; ^{14}C-labeled amino acids have included leucine, arginine, phenylalanine, and tyrosine.[76-78,131,132] However, the choice of amino acid may be important since some amino acids (e.g., branched chain amino acids), are thought to stimulate protein synthesis. Phenylalanine and tyrosine are both suitable precursor molecules since both amino acids have a small tissue free pool in fish tissues[133-135] and since simple and reliable assays exist for selective extraction and measurement of these amino acids.[136,137] However, phenylalanine is the preferred radiolabel because its high solubility enables a larger dose to be given in the same injection volume.

Since this method was validated for measuring rates of protein synthesis in fish,[119] its ease of application has resulted in a marked increase in work on protein synthesis in fish (see above). The use of the flooding dose method is based on several assumptions: (1) that the presence of the high intracellular concentration of phenylalanine does not affect the rate of protein synthesis; (2) that the labelled amino acid equilibrates rapidly with the intracellular free pool; (3) that the specific radioactivity of the free amino acid in the intracellular free pool remains elevated over the course of the incorporation period; and (4) that the labelling of body protein with the radiolabel is linear over the incorporation time. With regard to the first criterion, a single flooding dose injection of phenylalanine has been shown not to affect the rate of protein synthesis in rats using ^{3}H-lysine or ^{14}C-threonine[11,138] or in fish using ^{14}C-leucine.[116] However, there has been some debate in studies of human protein metabolism about whether the constant fusion or the flooding dose method is the optimal approach to measure tissue synthesis rates.[19,20] Rennie et al.[20] recorded up to twofold higher synthesis rates obtained using the flooding dose method compared to constant infusion. However, this difference has been attributed by Garlick et al.[19] to changes in the isotopic enrichment of the precursor amino acid, which are not fully accounted for by the infusion method, but which are minimized by following a single flooding dose injection. With regard to the remaining three criteria, several studies have validated the use of the flooding dose method in fish at a range of water temperatures and using an incorporation time of between 40 min and 6 h.[8,48,73,88,116-118]

Figure 8 outlines our approach to measuring fractional rates of protein synthesis in fish following a single flooding dose injection of ^{3}H-phenylalanine. The methodology is set up for replicate 100 mg samples which are initially homogenized in 2 ml 0.2M PCA and centrifuged in order to separate the free pool (supernatant) from the protein pellet. The free pool and protein samples are then treated separately to selectively extract and measure the free phenylalanine and protein-bound specific radioactivity of the samples. The solubilization stage in the treatment of the protein pellet also allows the measurement of the protein and nucleic acid concentrations and allows the levels of radioactivity in the solubilized proteins to be counted (NaOH scintillation counts). The selective extraction procedure involves the decarboxylation of phenylalanine to β-phenylethylamine and the subsequent measurement of the concentration and specific radioactivity of β-phenylethylamine in the samples. This conversion step is 1:1 and therefore the results can be expressed in terms of the concentration and specific radioactivity of phenylalanine.

NONINVASIVE MEASUREMENTS OF PROTEIN TURNOVER IN FISH USING STABLE ISOTOPES

The measurement of rates of protein synthesis using end point stochastic analysis depends on the administration of an isotopically labeled precursor (usually ^{13}C or ^{15}N) and the collection and measurement of the label in a metabolic end product. Thus, either single amino acids or proteins

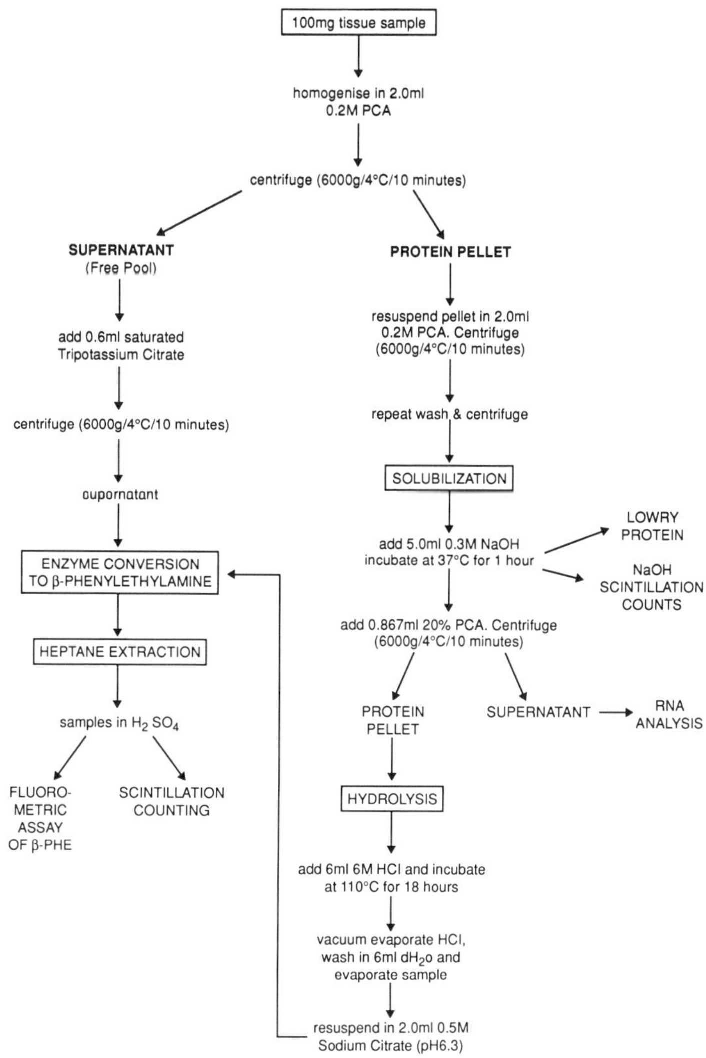

Figure 8 An outline of the steps and assays involved in measuring fractional rates of protein synthesis in tissue or whole body samples following a flooding dose injection of ^{3}H-phenylalanine.

in which all amino acids are labelled are introduced and a nitrogenous metabolic excretory product is collected over time. In previous work on mammals, the isotope has been applied using a single bolus or through constant infusion. The appearance of the ^{15}N isotope in a nitrogenous excretory product or products over a known period of time is then measured. The use of stable isotopes is the standard approach in studies of protein synthesis in humans and has also been widely applied to studies involving farm animals and other endotherms. However, there has been very little use of stable isotopes in the study of protein metabolism in ectotherms. The only available data are for mussels, *M. edulis*[4,50,51,139] and recent work on rainbow trout, *Oncorhynchus mykiss,*[14] and common carp, *Cyprinus carpio.*[103] Here we describe our adaptation of the stochastic endpoint model to measure rates of protein synthesis in fish following ingestion of ^{15}N protein, using the simple single pool model of Waterlow et al.[40] Using this experimental approach the following theoretical assumptions are made: (1) The end product is formed from the same precursor pool used for protein synthesis. If this assumption is met, any labeled end product can be used; (2) nitrogen flux through the nonprotein nitrogen pool is assumed to be negligible compared with flux through protein synthesis and oxidation/excretion. In fish, protein nitrogen accounts for as much as 98% of the total body nitrogen[140]; (3) the ^{15}N is uniformly distributed within the nitrogen pool. In endotherms it appears that compartmentation is likely and that a single pool model is simplistic.[99] However, since urea production in teleost fish is normally low and negligible, the use of a single pool model is justified for most teleost fish; (4) recycling of labeled nitrogen does not occur; however, recycling *does* occur and the longer the period of observation the greater the degree of recycling.[40] Clearly, the reintroduction of labeled amino nitrogen into the free amino acid pool will underestimate the synthesis rate.[16] Cumulative isotope excretion should reach a plateau; in the case of rainbow trout, we have found this to be about 24 h after administration at 14°C,[14] which indicates the negligible reintroduction of label into the free amino acid pool and the relative unimportance of recycling in fish. Waterlow et al.[40] also considered recycling to be quantitatively unimportant in humans because only 5% of the total excreted isotope was derived from labeled protein. Because fish have lower rates of protein turnover, recycling is probably less important for them than for mammals; (5) in studies involving the use of a single labelled amino acid tracer, the behavior of this tracer is assumed to be representative of the total amino acid nitrogen[50]; and (6) the subject is in steady state and there is no net loss or gain of nitrogen. In practice the change in the whole body protein pool will be relatively small over the period of measurement.[40] In fish, measured protein growth rates rarely exceed 5% of the body protein mass per day.[41]

Providing these assumptions can be met (or ignored), and if it is assumed that the majority of nitrogen flux can be accounted for by nitrogen intake (C), protein degradation (PD), protein synthesis (PS) and excreted nitrogen (E), then nitrogen flux (Q) can be described by:

$$Q = C + PD + PS + E \tag{3}$$

After its administration as a single dose, the isotope is excreted and the cumulative isotope excretion (e^*) reaches a plateau at which point all the isotope has been cleared from the metabolic and excretory pools. The cumulative excretion of isotope can then be calculated as the area under the curve. In practice it is easiest to collect the excretory products in even time periods over the time course and calculate an overall e^* from these data. The following equation can then be applied:

$$e^*/d^* = E_X/Q = E_X/(PS + E_{TN}) \tag{4}$$

so that

$$PS = Q - E_{TN} \tag{5}$$

where e^* (mmol ^{15}N) is the cumulative excretion of ^{15}N ammonia over t after a dose d^* (mmol ^{15}N) of ^{15}N-protein, E_x (mmol N) is the total ammonia excreted over time t, PS (mmol N) represents protein synthesis, Q (mmol N) the nitrogen flux, and E_{TN} (mmol N) the total nitrogen excretion over the total time period.[40] Detailed analysis of the assumptions and equations necessary to calculate rates of protein synthesis by end point modeling following constant infusion in endotherms has been presented by Reeds[17] and Wolfe.[18]

The approach that we have used in our laboratory to measure synthesis rates using uniformly labelled ^{15}N protein is outlined in Table 5. To date, this method has been used to measure rates of protein synthesis for individual rainbow trout,[14] flounder, *Platichthys flesus*, and eel, *Anguilla anguilla* (Carter, C. G., unpublished data, 1994; Owen, S. F., unpublished data, 1994). During initial studies, several water samples should be taken and the cumulative enrichment of ammonia due to ^{15}N excretion measured in order to determine the time interval during which ^{15}N ammonia excretion reaches a plateau and is constant. We have found that the time it takes to reach the plateau varies

Table 5 Methodology Associated With the *in vivo* Measurement of Fractional Rates of Protein Synthesis in Fish Following the Oral Administration of ^{15}N-labeled Protein

A. Sample Collection

1. Enriched food with 5 mg [^{15}N] yeast protein [98% atom percent excess (APE)] per gram food is fed to fish that are kept individually in 16 l tanks.
2. Duplicate 4 l initial water samples are taken when the experimental tanks are filled.
3. The total ammonia excretion over the first 24 h following feeding is measured. After 24 h duplicate 4 l water samples are taken.
4. After 24 h the fish is transferred to a known volume of fresh water. The total ammonia excretion over the second 24 h following feeding is measured. After this second 24 h period duplicate 4 l water samples are taken.

B. Isotope Collection

1. Each 4 l water sample is placed in a 5-l flask with 80 ml 0.1M EDTA in 10.0M NaOH and distilled into a flask containing 10 ml 1.0M Boric acid to give a final volume of about 150 ml.
2. This concentrated sample is frozen and freeze dried. The resulting white powder contains ^{15}N and ^{14}N ammonium borate.

C. Isotope Analysis

1. Replicate 20 mg borate samples are weighed into "tin capsules" for enrichment analysis which is carried out using an ANCA Roboprep-CN linked to a tracer mass isotope ratio mass spectrometer (Europa Scientific Ltd.). Food and fecal samples are treated in the same way.
2. The samples are combusted and the evolved nitrogen-containing compounds are reduced to nitrogen gas. The ^{15}N enrichment (APE) in the samples can be measured by running samples of known nitrogen enrichment alongside the experimental samples. Rates of protein synthesis are then calculated using the equations outlined in the text.

Note: Values are appropriate for routine sampling of 100-g rainbow trout, *O. mykiss*, at 14°C.

Adapted from Carter, C. G., Owen, S. F., He, Z.-Y., Watt, P. W., Scrimgeour, C., Houlihan, D. F., and Rennie, M. J., *J. Exp. Biol.*, 189, 279, 1994.)

with temperature and samples should be taken accordingly. In rainbow trout at 14°C, this plateau is reached after 24 h and therefore further water samples are taken at 24 and 48 h after feeding to measure the ^{15}N excretion (Table 5). Once the measurement period has been determined, water samples are collected at the start of the measurement period in order to measure the background ^{15}N levels in the water, total ammonia excretion is monitored with time, and samples are collected at the start and end of the measurement period to measure ^{15}N ammonia excretion. The collected water is used to concentrate the ammonia in preparation for analysis of the ^{15}N enrichment using the protocol outlined in Table 5. Urea excretion is also measured using the diacetyl monoxime method[141] to calculate the total nitrogen excretion.The absorption of ^{15}N can be calculated from the enrichment of feces and is used to obtain a more accurate value for the dose of the isotope.

COMPARING RATES OF PROTEIN SYNTHESIS OBTAINED USING ^{15}N-LABELED PROTEIN AND ^{3}H-PHENYLALANINE

Initial results from our laboratory on protein metabolism of rainbow trout indicate that protein synthesis rates obtained from feeding ^{15}N-labeled protein and collecting ammonia are similar to those obtained with radiolabeled amino acids (Table 6). The mean fractional rate of synthesis in 100-g rainbow trout measured over the first 24 h following a ^{15}N-labeled meal was $2.30 \pm 0.67\%\ d^{-1}$.[14] Following a flooding dose injection of L-[2,6-^{3}H]-phenylalanine to measure protein synthesis, the relationship between fractional rates of protein consumption (k_c, gram of protein consumed per gram of fish protein expressed as a percent per day) and synthesis (k_s) for feeding rainbow trout (40 to 100g:10°C) was described by $k_s = 1.69 + 0.45\ k_c$.[48] This relationship predicted a mean k_s of 2.60% day^{-1} for the trout used in the ^{15}N experiment, similar to that calculated from the stable isotope analysis. Similar rates of protein synthesis were also found in the white muscle of rainbow trout of similar size (Table 6). These synthesis rates were obtained following a flooding dose injection of ^{3}H-phenylalanine (incubation period of 3 to 6 h) and from a terminal measurement of the incorporation of ^{15}N-protein into white muscle protein 48 h after a meal. The similarity between whole

Table 6 A Comparison of Fractional Rates of Protein Synthesis in Rainbow Trout, *Oncorhynchus mykiss*, Measured by Different Methods

k_s (%/d)	Method	Wt	Temp	Meals (day^{-1})	Reference
2.3 ± 1.9[a] 5.1 ± 1.4[b]	Caudal vein infusion (L-U[^{14}C]-Leu)	80 g	10°C	2	131
2.6 ± 0.7	Flooding dose injection (L-[2,6 ^{3}H]-phe)	80 g	10°C	1	43
2.3 ± 0.7	Fed ^{15}N-protein	117 g	14°C	1	14
3.5 ± 0.6	Flooding dose injection (L-[2,6 ^{3}H]-phe)	105	10°C	1	McCarthy, I. D., unpublished data, 1991
3.8 ± 1.3	Fed ^{15}N-protein	117 g	14°C	1	Owen, S. F., unpublished data, 1993
1.5 ± 1.0*	Flooding dose injection (L-[2,6 ^{3}H]-phe)	100	10°C	1	McCarthy, I. D., unpublished data, 1991
2.5 ± 0.6*	Fed ^{15}N-protein	117 g	14°C	1	Owen, S. F., unpublished data, 1993

Note: All data presented are for whole-body synthesis rates except those indicated by an asterisk (*) which are white muscle synthesis rates.

[a] Synthesis rates calculated using the specific radioactivity of the plasma as an estimate of the free pool specific radioactivity.

[b] Synthesis rates calculated using the specific radioactivity of the tissues as an estimate of the free pool specific radioactivity.

animal and white muscle rates of protein synthesis measured over 2 d and over a few hours, using the flooding dose method, further supports the validity of calculating rates of protein degradation as the difference between synthesis and growth,[5,142] and therefore the estimates of protein turnover made for fish.[8,25,29,48]

Feeding ^{15}N-labeled protein is advantageous because it allows noninvasive and nondestructive measurements of protein synthesis which can be repeated on the same animal. Therefore, the use of stable isotopes should be used increasingly in the study of protein synthesis in fish. This would enable us to answer questions such as: How do changes in environmental conditions affect the protein turnover of the same fish? Furthermore, being able to "track" the same fish provides the opportunity to measure ontogenic changes in protein synthesis and construct synthesis/weight relationships for the same fish; the only assumption that we need to make is on the protein-nitrogen content of the fish.

ACKNOWLEDGMENTS

The research for this chapter arose from contracts from the Ministry of Agriculture Fisheries and Food and from grants from the Agriculture and Food Research Council, the British Council, the Natural Environment Research Council and the Science and Engineering Research Council. We are grateful to the following people for allowing us to quote their unpublished data: Dr Hassaan Heba, Richard Smith, Stewart Owen, Dr Karl Siegert, Dr Johan Verreth, Dr A. Polat, Luis Conceição, Dr Jean Guillaume, Dr Habib Langar, Zhong-Yang He, Dr Bendikte Pedersen, and Dr Göran Nilsson. The technical assistance of Carol Gray, Jim Brechin, Leyton Hackney, and Ian Davidson is gratefully acknowledged. We are grateful to Prof. Mike Rennie, Dr Peter Watt, and Dr Ayyub Patel for their assistance in developing the stable isotope technique.

REFERENCES

1. **Reeds, P. J. and Davis, T. A.,** Hormonal regulation of muscle protein synthesis and degradation, in *The Control of Fat and Lean Deposition*, Boorman, K. N., Buttery, P. J., and Lindsay, D. B., Eds., Butterworth-Heinemann, Oxford, 1992, chap 1.
2. **Lobley, G. E.,** Control of the metabolic fate of amino acids in ruminants: a review, *J. Anim. Sci.*, 70, 326, 1992.
3. **Rennie, M. J., Smith, K., Watt, P. W., Barua, J. M., and Bennet, W. M.,** Methods of assessment of muscle protein turnover and their application to problems in clinical nutrition, *Riv. Ital. Nutrizione Parenterale ed Enterale* 10, 63, 1992.
4. **Hawkins, A. J. S.,** Protein turnover: a functional appraisal, *Funct. Ecol.*, 5, 222, 1991.
5. **Houlihan, D. F.,** Protein turnover in ectotherms and its relationship to energetics, *Adv. Compar. Environ. Physiol.*, 7, 1, 1991.
6. **Reeds, P. J., Fuller, M. F., and Nicholson, B. A.,** Metabolic basis of energy expenditure with particular reference to protein, in *Substrate and Energy Metabolism*, Garrow, J. S. and Halliday, D., Eds., Libbey, London, 1985, 46.
7. **Jobling, M.,** *Fish Bioenergetics*, Chapman & Hall, London, 1994.
8. **Carter, C. G., Houlihan, D. F., Brechin, J., and McCarthy, I. D.,** The relationships between protein intake and protein accretion, synthesis and retention efficiency for individual grass carp, *Ctenopharyngodon idella* (Val.), *Can. J. Zool.*, 71, 392, 1993.
9. **Mathews, R. W. and Haschemeyer, A. E. V.,** Liver protein synthesis. Molecular weight distribution of pulse-labelled polypeptide chains in normal and thyroidectomized rats, *Biochim. Biophy. Acta*, 425, 220, 1976.
10. **Mathews, R. W. and Haschemeyer, A. E. V.,** Temperature dependency of protein synthesis in toadfish liver *in vivo*, *Comp. Biochem. Physiol.*, 61B, 479, 1978.
11. **Garlick, P. J., McNurlan, M. A., and Preedy, V. R.,** A rapid and convenient technique for measuring the rate of protein synthesis in tissues by the injection of ^{3}H phenylalanine, *Biochem. J.*, 217, 507, 1980.
12. **Goldspink, D. F. and Kelly, F. J.,** Protein turnover and growth in the whole body, liver and kidney of the rat from foetus to senility, *Biochem. J.*, 217, 507, 1984.
13. **Kita, K., Nakagawa, S., Pinontoan, R., Okumura, J., and Muramatsu, T.,** Protein synthesis measured in the whole body of the chick embryos cultured *in vitro*, *Br. Poult. Sci.*, 43, 1043, 1993.
14. **Carter, C. G., Owen, S. F., He, Z-Y., Watt, P. W., Scrimgeour, C., Houlihan, D. F., and Rennie, M. J.,** Determination of protein synthesis in rainbow trout, *Oncorhynchus mykiss* (Walbaum), using a stable isotope, *J. Exp. Biol.*, 189, 279, 1994.
15. **Garlick, P. J. and Fern, E. B.,** Whole body protein turnover: Theoretical considerations, in *Substrate and Energy Metabolism in Man*, Garrow, J. S. and Halliday, D., Eds., Libbey, London, 1985, 7.
16. **Assimon, S. A. and Stein, T. P.,** ^{15}N-glycine as a tracer to study protein metabolism *in vivo*, in *Modern Methods in Protein Nutrition and Metabolism*, Nissen, S., Ed., Academic Press, London, 1992, 275.
17. **Reeds, P. J.,** Isotopic estimation of whole-body protein content *in vivo*, in *Modern Methods in Protein Nutrition and Metabolism*, Nissen, S., Ed., Academic Press, London, 1992, 249.
18. **Wolfe, R. B.,** *Radioactive and Stable Isotope Tracers in Biomedicine: Principles and Practices of Kinetic Analysis*, Wiley-Liss, Chichester, 1992.
19. **Garlick, P. J, McNurlan, M. A., Essen, P., and Wernerman, J.,** Measurement of tissue protein synthesis rates *in vivo*: a critical analysis of contrasting methods, *Am. J. Physiol.*, 266, E287, 1994.
20. **Rennie, M. J., Smith, K., and Watt, P. W.,** Measurement of human tissue protein synthesis: an optimal approach, *Am. J. Physiol.*, 266, E298, 1994.
21. **Simon, O.,** Metabolism of protein and amino acids, in *Protein Metabolism in Farm Animals*, Bock, H.-D., Eggum, B. O., Low, A. G., Simon, O., and Zebrowska, T., Eds., Oxford University Press, Oxford, 1989, 273.
22. **Haschemeyer, A. E. V.,** The response of fish to a modified thermal environment, in *Responses of Fishes to Environmental Changes*, Chavin, W., Ed., Charles C Thomas, Springfield, IL, 1973, 3.
23. **Fauconneau, B.,** Protein synthesis and protein deposition in fish, in *Nutrition and Feeding in Fish*, Cowey, C. B., Mackie, A. M., and Bell, J. G., Eds., Academic Press, London, 1985, 17.
24. **Matty, A. J. and Lone, K. P.,** The hormonal control of metabolism and feeding, in *Fish Energetics: New Perspectives*, Tytler, P. and Calow, P., Eds., Croom Helm, London, 1985, 185.
25. **Houlihan, D. F., Mathers, E., and Foster, A.,** Biochemical correlates of growth rate in fish, in *Fish Ecophysiology*, Rankin, J. C. and Jensen, F. B., Eds., Chapman & Hall, London, 1993, 45.
26. **Houlihan, D. F., Carter, C. G., and McCarthy, I. D.,** Protein synthesis in fish. in *Fish Molecular Biology and Biochemistry*, Vol. 4, Hochachka, P. and Mommsen, T., Eds., Elsevier Press, Amsterdam, 1995, in press.
27. **Rønnestad, I., Fyhn, H., and Gravningen, K.,** The importance of free amino acids to the energy metabolism of eggs and larvae of turbot (*Scophthalmus maximus*), *Mar. Biol.*, 114, 517, 1992.
28. **National Research Council,** *Nutrient Requirements of Fish*, National Academy Press, Washington D.C., 1993.
29. **Carter, C. G., Houlihan, D. F., Buchanan, B., and Mitchell, A. I.,** Protein-nitrogen flux and protein growth efficiency of individual Atlantic salmon (*Salmo salar*), *Fish Physiol. Biochem.*, 12, 305, 1993.
30. **Freudenburger, D. and Nolan, J.,** Protein turnover in two kangaroo sub-species (*Macropus robustus* and *M.r. erubescens*) from divergent habitats and the sympatric feral goat (*Capra hircus*), *Comp. Biochem. Physiol.*, 104A, 443, 1993.

31. **Fuller, M. F., Reeds, P. J., Cadenhead, A., and Seve, B.,** Effects of the amount and quality of dietary protein on nitrogen metabolism and protein turnover of pigs, *Br. J. Nutr.*, 58, 287, 1987.
32. **Cho, C. Y. and Kaushik, S. J.,** Effects of protein intake on metabolism and net energy values of fish diets, in *Nutrition and Feeding in Fish*, Cowey, C. B., MacKie, A. M., and Bell, J. G., Eds., Academic Press, London, 1985, 95.
33. **Armstrong, D. G. and Mitchell, H. H.,** Protein nutrition and the utilization of dietary protein at different levels of intake by growing swine, *J. Anim. Sci.*, 14, 48, 1955.
34. **Solberg, G., Buttery, P. J., and Boorman, K. N.,** Effect of moderate methionine deficiency on food protein and energy utilization in the chick, *Br. Poult. Sci.*, 12, 297, 1971.
35. **Oddy, V. H., Lindsay, D. B., Barker, P. J., and Northrop, A. J.,** Effect of insulin on hindlimb and whole-body leucine and protein metabolism in fed and fasted lambs, *Br. J. Nutr.*, 58, 437, 1987.
36. **Houlihan, D. F., McMillan, D. N., Agnisola, C., Trara Genoino, I., and Foti, L.,** Protein synthesis and growth in *Octopus vulgaris*, *Mar. Biol.*, 106, 251, 1990.
37. **Sibly, R. M. and Calow, P.,** *Physiological Ecology of Animals*, Blackwell Scientific Publications, Oxford, 1986.
38. **Houlihan, D. F., Hall, S. J., and Gray, C.,** Effects of ration on protein turnover in cod, *Aquaculture*, 79, 103, 1989.
39. **Young, V. R.,** The role of skeletal and cardiac muscle in regulation of protein metabolism, in *Mammalian Protein Metabolism*, Vol. 4, Munro, H. N., Ed., Academic Press, New York, 1970, 586.
40. **Waterlow, J. C., Garlick, P. J., and Millward, D. J.,** *Protein Turnover in Mammalian Tissues and in the Whole Body*, Elsevier/North Holland Biomedical Press, Amsterdam, 1978.
41. **Houlihan, D. F., McCarthy, I. D., Carter, C. G., and Marttin, F.,** Protein turnover and amino acid flux in fish larvae, *ICES J. Mar. Sci.*, 1995, in press.
42. **Lyndon, A. R., Davidson, I., and Houlihan, D. F.,** Changes in tissue and plasma free amino acid concentrations after feeding in Atlantic cod, *Fish Physiol. Biochem.*, 10, 365, 1993.
43. **McCarthy, I. D.,** *Feeding Behaviour and Protein Turnover in Fish*, Ph.D. thesis, University of Aberdeen, 1993.
44. **Carter, C. G., He, Z.-Y., Houlihan, D. F., McCarthy, I. D., and Davidson, I.,** Effect of feeding on the tissue free amino acid concentrations in rainbow trout [*Oncorhynchus mykiss*(Walbaum)], *Fish Physiol. Biochem.*, 1995, in press.
45. **Millward, D. J. and Rivers, J.,** The nutritional role of indispensable amino acids and the metabolic basis for their requirements, *Eur. J. Clin. Nutr.*, 42, 367, 1988.
46. **Reeds, P. J. and Lobley, G. E.,** Protein synthesis: are there real species differences?, *Proc. Nutr. Soc.*, 39, 43, 1980.
47. **Weisner, R. J. and Zak, R.,** Quantitative approaches for studying gene expression, *Am. J. Physiol.*, 260, L179, 1991.
48. **McCarthy, I. D., Houlihan, D. F., and Carter, C. G.,** Individual variation in protein turnover and growth efficiency in rainbow trout, *Oncorhynchus mykiss* (Walbaum), *Proc. R. Soc. Lond.*, B257, 141, 1994.
49. **Kaushik, S. J. and Gomes, E. F.,** Effect of frequency of feeding on nitrogen and energy balance in rainbow trout under maintenance conditions, *Aquaculture*, 73, 207, 1988.
50. **Hawkins, A. J. S.,** Relationships between the synthesis and breakdown of protein, dietary absorption and turnovers of nitrogen and carbon in the blue mussel, *Mytilus edulis* L., *Oecologia* (Berlin), 66, 42, 1985.
51. **Hawkins, A. J. S., Bayne, B. L., Day, A. J., Rusin, J., and Worrall, C. M.,** Genotype-dependent interrelations between energy metabolism, protein metabolism and fitness, in *Reproduction, Genetics and Distribution of Marine Organisms*, Ryland, J. S. and Tyler, P. A., Eds., Olsen & Olsen, Fredensborg, 1989, 283.
52. **Viarengo, A., More, M. N., Pertica, M., Mancinelli, G., and Accomando, R.,** A simple procedure for evaluating the protein degradation rate in mussel (*Mytilus galloprovincialis* Lam.) tissues and its application in a study of phenanthrene effects on protein catabolism, *Comp. Biochem. Physiol.*, 103B, 27, 1992.
53. **Calow, P. and Townsend, C. R.,** Energetics, ecology and evolution, in *Physiological Ecology*, Townsend, C. R. and Calow, P., Eds., Blackwell, Oxford, 1981, 3.
54. **Kariya, Y., Ochiai, Y., and Hashimoto, K.,** Protein components and ultrastructure of the arm and mantle muscles of octopus, *Octopus vulgaris*, *Bull. Jap. Soc. Sci. Fish.*, 52, 131, 1986.
55. **Kirkwood, T. B. L.,** Repair and its evolution: survival versus reproduction, in *Physiological Ecology*, Townsend, C. R. and Calow, P., Eds., Blackwell, Oxford, 1981, 165.
56. **Makrides, S. C.,** Protein synthesis and degradation during aging and senescence, *Biol. Rev.*, 58, 342, 1983.
57. **Reynolds, S. E., Notingham, S. F., and Stephens, A. E.,** Food and water economy and its relation to growth in the fifth instar larvae of the tobacco hornworm, *Manduca sexta*, *J. Insect Physiol.*, 31, 119, 1985.
58. **Case, T. J.,** On the evolution and adaptive significance of postnatal growth rates in the terrestrial vertebrates, *Q. Rev. Biol.*, 53, 243, 1978.
59. **Hewitt, D. R.,** Response of protein turnover in the brown Tiger prawn *Penaeus esculentus* to variation in dietary protein content, *Comp. Biochem. Physiol.*, 103A, 183, 1992.
60. **Rombough, P. J.,** Energy partitioning during fish development: additive or compensatory allocation of energy to support growth?, *Funct. Ecol.*, 8, 178, 1994.
61. **Schreurs, V. V. A. M., ten Doeshate, R. A. H. M., Boekholt, H. A., and Koopmanschap, R. E.,** Physiological determinants to optimize metabolic utilization of dietary amino acids for animal production, in *Biological Basis of Sustainable Production, Proceedings of the Zodiac Symposium, Wageningen, Netherlands*, Huisman, E. A., Osse, J. W. M., van der Heide, D., Tamminia, S., Tolkamp, B. J., Scnoutew, W. P. E., Hollingwaede, C. E., and van der Winnkel, G. L., Eds., Wageningen Press, Wageningen, The Netherlands, 1994.

62. **Houlihan, D. F., Costello, M. J., Secombes, C. J., Stagg, R., and Brechin, J.,** Effects of sewage sludge exposure in growth, feeding and protein synthesis of dab, *Limanda limanda* L., *Mar. Environ. Res.*, 37, 331, 1994.
63. **Secombes, C. J., Fletcher, T. C., O'Flynn, J. A., Costello, M. J., Stagg, R., and Houlihan, D. F.,** Immunocompetence as a measure of the biological effects of sewage sludge pollution in fish, *Comp. Biochem. Physiol.*, 100C, 133, 1991.
64. **Secombes, C. J., Fletcher, T. C., White, A., Costello, M. J., Stagg, R., and Houlihan, D. F.,** Effects of sewage sludge on immune responses in the dab, *Limanda limanda* (L.), *Aquat. Toxicol.*, 23, 217, 1992.
65. **Milley, J. R.,** Protein synthesis during hypoxia in fetal lambs, *Am. J. Physiol.*, 252, E519, 1987.
66. **Pannevis, M. C. and Houlihan, D. F.,** The energetic cost of protein synthesis in isolated hepatocytes of rainbow trout (*Oncorhynchus mykiss*), *J. Comp. Physiol.*, B162, 393, 1992.
67. **Land, S. C., Buck, L. T., and Hochachka, P. W.,** Response of protein synthesis to anoxia and recovery in anoxia-tolerant hepatocytes, *Am. J. Physiol.*, 265, R41, 1993.
68. **Land, S. C. and Hochachka, P. W.,** Protein turnover during metabolic arrest in turtle hepatocytes:role and energy dependence of proteolysis, *Am. J. Physiol.*, 266, C1028, 1994.
69. **Haschemeyer, A. E. V., Persell, R., and Smith, M. A. K.,** Effect of temperature on protein synthesis in fish of the Galapogos and Perlas Islands, *Comp. Biochem. Physiol.*, 64B, 91, 1979.
70. **Houlihan, D. F., Wieser, W., and Foster, A. R.,** *In vivo* protein synthesis rates in larval nase, *Chondrostoma nasus* L., *Can. J. Zool.*, 70, 2436, 1992.
71. **Mathers, E. M., Houlihan, D. F., McCarthy, I. D., and Burren, L. J.,** Rates of growth and protein synthesis correlated with nucleic acid content in fry of rainbow trout, *Oncorhynchus mykiss* (Walbaum): effects of age and temperature, *J. Fish Biol.*, 43, 245, 1993.
72. **Rao, G. M. M.,** Oxygen consumption of rainbow trout (*Salmo gairdneri*) in relation to activity and salinity, *Can. J. Zool.*, 46, 781, 1968.
73. **Houlihan, D. F., McMillan, D. N., and Laurent, P.,** Growth rates, protein synthesis, and protein degradation rates in rainbow trout: Effects of body size, *Physiol. Zool.*, 59, 482, 1986.
74. **Jobling, M.,** Bioenergetics: feed intake and energy partitioning, in *Fish Ecophysiology*, Rankin, J. C. and Jensen, F. B., Eds., Chapman & Hall, London, 1993, chap 1.
75. **Dabrowski, K. R.,** Ontogenetical aspects of nutritional requirements of fish, *Comp. Biochem. Physiol.*, 85A, 639, 1986.
76. **Fauconneau, B.,** The measurement of whole body protein synthesis in larval and juvenile carp (*Cyprinus carpio*), *Comp. Biochem. Physiol.*, 78B, 845, 1984.
77. **Fauconneau, B., Aguirre, P., and Bergot, P.,** Protein synthesis in early life of coregonids: influence of temperature and feeding, *Arch. für Hydrobiol. Beih.*, 22, 171, 1986.
78. **Fauconneau, B., Aguirre, P., Dabrowski, K., and Kaushik, S. J.,** Rearing of sturgeon (*Acipenser baeri* Brandt) larvae. II. Protein metabolism: influence of fasting and diet quality, *Aquaculture*, 51, 117, 1986.
79. **Chinet, A. E.,** Energy cost of ionic homeostasis in mammalian skeletal muscle, in *Energy Transformations in Cells and Organisms*, Wieser, W. and Gnaiger, E., Eds., Thieme, Stuttgart, Germany, 1989, 58.
80. **Chinet, A., Clausen, T., and Girardier, L.,** Micro-calorimetric determination of energy expenditure due to active sodium-potassium transport in the soleus muscle and brown adipose tissue of the rat, *J. Physiol.* (*London*), 265, 43, 1977.
81. **Folke, M. and Sestoft, L.,** Thyroid calorigenesis in isolated perfused rat liver: minor role of active sodium-potassium transport, *J. Physiol.* (*London*), 269, 407, 1977.
82. **Gregg, V. A. and Milligan, L. P.,** Inhibition by ouabain of the O_2 consumption of mouse (*Mus musculus*) soleus and diaphragm muscles, *Gen. Pharmacol.*, 11,323, 1979.
83. **Gregg, V. A. and Milligan, L. P.,** *In vitro* energy costs of Na^+/K^+ ATPase activity and protein synthesis in muscle from calves differing in age and breed, *Br. J. Nutr.*, 48, 65, 1982.
84. **Krumschnabel, G., Schwarzbaum, P. J., and Wieser, W.,** Coupling of energy supply and energy demand in isolated goldfish hepatocytes, *Physiol. Zool.*, 67, 438, 1994.
85. **Jobling, M.,** Growth, in *Fish Energetics: New Perspectives*, Tytler, P. and Calow, P., Eds., Croom Helm, London, 1985, 213.
86. **Harris, P. M., Garlick, P. J., and Lobley, G. E.,** Interactions between energy and protein metabolism in the whole body and hind limb of sheep in response to intake, in *Energy Metabolism of Farm Animals*, van der Honing, Y. and Close, W. H., Eds., EAAP Publication, Wageningen, 1989, 167.
87. **Waterlow, J. C. and Millward, D. J.,** Energy cost of turnover of protein and other cellular constituents, in *Energy Transformations in Cells and Organisms*, Wieser, W. and Gnaiger, E., Eds., Thieme, Stuttgart, Germany, 1989, 277.
88. **Houlihan, D. F., Hall, S. J., Gray, C., and Noble, B. S.,** Growth rates and protein turnover in cod, *Gadus morhua*. *Can. J. Fish. Aquat. Sci.*, 45, 951, 1988.
89. **Lyndon, A. R., Houlihan, D. F., and Hall, S. J.,** The effect of short-term fasting and a single meal on protein synthesis and oxygen consumption in cod, *Gadus morhua*, *J. Comp. Physiol.*, B162, 209, 1992.
90. **Webster, A. J. F.,** Comparative aspects of the energy exchange, in *Comparative Nutrition*, Blaxter, K. and MacDonald, I., Eds., John Libbet, London, 1988, 37.
91. **Houlihan, D. F., Pannevis, M. C., and Heba, H. M. A.,** Protein synthesis in juvenile tilapia, *Oreochromis mossambicus*, *J. World Aquacult. Soc.*, 24, 145, 1993.

92. **Adeola, O., Young, L. G., McBride, B. W., and Ball, R. O.,** *In vitro* Na^+/K^+-ATPase (EC 3.6.1.3)-dependent respiration and protein synthesis in skeletal muscle in pigs fed at three dietary protein levels, *Br. J. Nutr.*, 61, 453, 1989.
93. **McBride, B. W. and Early, R. J.,** Energy expenditure associated with sodium/potassium transport and protein synthesis in skeletal muscle and isolated hepatocytes from hyperthyroid sheep, *Br. J. Nutr.*, 62, 673, 1989.
94. **Wieser, W. and Medgyesy, N.,** Aerobic maximum for growth of the larvae and juveniles of a cyprinid fish, *Rutilus rutilus* (L.): implications for energy budgeting in small poikilotherms, *Funct. Ecol.*, 4, 233, 1990.
95. **Wieser, W. and Medgyesy, N.,** Cost and efficiency of growth in the larvae of two species of fish with widely differing metabolic rates, *Proc. R. Soc. Lond.*, B242, 51, 1990.
96. **Oka, T., Kweon, S. H., Natori, Y., and Hasegawa, N.,** Effect of food intake on protein and RNA metabolism in neonatal chick liver, *Nutr. Res.*, 9, 785, 1989.
97. **Cooper, H. L. and Gibson, E. M.,** Control of synthesis and wastage of ribosomal ribonucleic acid in lymphocytes, *J. Biol. Chem.*, 246, 5059, 1971.
98. **Wieser, W.,** Cost of growth in cells and organisms: general rules and comparative aspects, *Biol. Rev.*, 68, 1, 1994.
99. **Young, V. R., Yu, Y-M., and Kremp, F. M.,** Protein and amino acid turnover using stable isotopes ^{15}N, ^{13}C and ^{2}H as probes, in *New Techniques in Nutritional Research*, Whitehead, R. G. and Prentice, A., Eds. Academic Press, London, 1991, 17.
100. **Calvert, C. C. and Baldwin, R. L.,** Mathematical models of protein metabolism, in *Modern Methods in Protein Nutrition and Metabolism*, Nissen, S., Ed., Academic Press, San Diego, 1992, 311.
101. **Smith, M. A. K.,** Estimation of growth potential by measurement of tissue protein synthetic rates in feeding and fasting rainbow trout, *J. Fish Biol.*, 19, 213, 1981.
102. **Sugden, P. H. and Fuller, S. J.,** Regulation of protein turnover in skeletal and cardiac muscle, *Biochem. J.*, 273, 21, 1991.
103. **Meyer-Burgdorff, K. -H. and Günther, K. D.,** N-Turnover of carp in relation to protein supply, in *Report of the EIFAC Workshop on Methodology for Determination of Nutrient Requirements of Fish* (EIFAC Occasional paper no. 29), Gropp, J. M. and Tacon, A. G. J., Eds., FAO, Rome, 1993.
104. **Lied, E., Lund, B., and Von der Decken, A.,** Protein synthesis *in vitro* by epaxial muscle polyribosomes from cod, *Gadus morhua*, *Comp. Biochem. Physiol.*, 72B, 187, 1982.
105. **Lied, E., Rosenlund, G., Lund, B., and von der Decken, A.,** Effect of starvation and refeeding on *in vitro* protein synthesis in white trunk muscle of atlantic cod (*Gadus morhua*), *Comp. Biochem. Physiol.*, 76B, 777, 1983.
106. **Lied, E., Lie, O., and Lambertsen, G.,** Nutritional evaluation in fish by measurement of *in vitro* protein synthesis in white trunk muscle tissue, in *Nutrition and Feeding in Fish*, Cowey, C. B., MacKie, A. M., and Bell, J. G., Eds., Academic Press, London, 1985, 169.
107. **Lied, E. and Rosenlund, G.,** The influence of the ratio of protein energy to total energy ratio in the feed on the activity of protein synthesis *in vitro*, the level of ribosomal RNA and the RNA-DNA ratio in the white trunk muscle of Atlantic cod (*Gadus morhua*), *Comp. Biochem. Physiol.*, 77A, 489, 1984.
108. **Rosenlund, G., Lund, B., Sandnes, K., Braekkan, O. R., and von der Decken, A.,** Muscle protein synthesis *in vitro* of saithe (*Pollachius virens*) correlated to growth and daily energy intake, *Comp. Biochem. Physiol.*, 77B, 7, 1984.
109. **Rosenlund, G. and Lied, E.,** Growth and muscle protein synthesis *in vitro* of saithe (*Pollachius virens*) and rainbow trout (*Salmo gairdneri*) in response to protein-energy intake, *Acta Agric. Scand.*, 36, 195, 1986.
110. **Hansen, P. E., Lied, E., and Borresen, T.,** Estimation of protein synthesis in fish larvae using an *in vitro* polyribosome assay, *Aquaculture*, 79, 85, 1989.
111. **Epse, M., Lied, E., and Torrissen, K. R.,** *In vitro* protein synthesis in muscle of Atlantic salmon (*Salmo salar*) as affected by the degree of proteolysis in feeds, *J. Anim. Physiol. Anim. Nutr.*, 69, 260, 1993.
112. **Epse, M. and Lied, E.,** Do Atlantic salmon (*Salmo salar*) utilize mixtures of free amino acids to the same extent as intact protein sources for muscle protein synthesis?, *Comp. Biochem. Physiol.*, 107A, 249, 1994.
113. **Junghahn, I. and Jürss, K.,** Effects of feeding and salinity on protein synthesis in the white epaxial muscle of the rainbow trout (*Salmo gairdneri*), *Comp. Biochem. Physiol.*, 89B, 229, 1988.
114. **Jürss, K., Junghahn, I., and Bastrop, R.,** Proteinsynthese und Fischwachstum, *Fischerei-Forschung Rostock*, 29, 30, 1991.
115. **Jürss, K., Junghahn, I., and Bastrop, R.,** The role of elongation factors in protein synthesis rate variation in white teleost muscle, *J. Comp. Physiol.*, B162, 345, 1992.
116. **Loughna, P. T. and Goldspink, G.,** Muscle protein synthesis rates during temperature acclimation in a eurythermal (*Cyprinus carpio*) and a stenothermal (*Salmo gairdneri*) species of teleost, *J. Exp. Biol.*, 118, 267, 1985.
117. **Watt, P. W., Marshall, P. A., Heap, S. P., Loughna, P. T., and Goldspink, G.,** Protein synthesis in tissues of fed and starved carp, acclimated to different temperatures, *Fish Physiol. Biochem.*, 4, 165, 1988.
118. **Foster, A. R., Houlihan, D. F., Hall, S. J., and Burren, L. J.,** The effects of temperature acclimation on protein synthesis rates and nucleic acid content of juvenile cod (*Gadus morhua* L.), *Can. J. Zool.*, 70, 2095, 1992.
119. **Pocrnjic, Z., Mathews, R. W., Rappaport, S., and Haschemeyer, A. E. V.,** Quantitative protein synthesis rates in various tissues of a temperate fish *in vivo* by the method of phenylalanine swamping, *Comp. Biochem. Physiol.*, 74B, 735, 1983.
120. **McMillan, D. N. and Houlihan, D. F.,** The effect of refeeding on tissue protein synthesis in rainbow trout, *Physiol. Zool.*, 61, 429, 1988.

121. **McMillan, D. N. and Houlihan, D. F.,** Short term responses of protein synthesis to refeeding in rainbow trout, *Aquaculture*, 79, 37, 1989.
122. **Foster, A. R., Houlihan, D. F., Gray, C., Medale, F., Fauconneau, B., Kaushik, S. J., and LeBail, P. Y.,** The effects of ovine growth hormone on protein turnover in rainbow trout, *Gen. Comp. Endocrinol.*, 81, 111, 1991.
123. **McMillan, D. N. and Houlihan, D. F.,** Protein synthesis in trout liver is stimulated by both feeding and fasting, *Fish Physiol. Biochem.*, 10, 23, 1992.
124. **Peragon, J., Ortega-Garcia, F., Barroso, J. B., de la Higuera, M., and Lupianez, J. A.,** Alterations in the fractional protein turnover rates in rainbow trout liver and white muscle caused by an amino acid based diet and changes in the feeding frequency, *Toxicol. Environ. Chem.*, 36, 217, 1992.
125. **Carter, C. G., Houlihan, D. F., Brechin, J., McCarthy, I. D., and Davidson, I.,** Protein synthesis in grass carp, *Ctenopharyngodon idella* (Val.) and its relation to diet quality, in *Fish Nutrition in Practice*, Kaushik, S. J. and Luquet, P., Eds., INRA, Paris, 1993, 673.
126. **Langar, H., Guillaume, J., Metailler, R., and Fauconneau, B.,** Augmentation of protein synthesis and degradation by poor dietary amino acid balance on European sea bass (*Dicentrarchus labrax*), *J. Nutr.*, 123, 1754, 1993.
127. **Martin, N. B., Houlihan, D. F., Talbot, C., and Palmer, R. M.,** Protein metabolism during sexual maturation in female Atlantic salmon (*Salmo salar* L.), *Fish Physiol. Biochem.*, 12, 131, 1993.
128. **Langar, H. and Guillaume, J.,** Effect of feeding pattern and dietary protein source on protein synthesis in European sea bass (*Dicentrarchus labrax*), *Comp. Biochem. Physiol.*, 108A, 461, 1994.
129. **Peragon, J., Barroso, J. B., Garcia-Salguero, L., de la Higuera, M., and Lupianez, J.A.,** Dietary protein effects on growth and fractional protein synthesis and degradation rates in liver and white muscle of rainbow trout, *Aquaculture*, 124, 35, 1994.
130. **Fauconneau, B., Breque, J., and Bielle, C.,** Influence of feeding on protein metabolism of Atlantic salmon (*Salmo salar*), *Aquaculture*, 79, 29, 1989.
131. **Fauconneau, B. and Arnal, M.,** *In vivo* protein synthesis in different tissues and the whole body of rainbow trout (*Salmo gairdneri* R.). Influence of environmental temperature, *Comp. Biochem. Physiol.*, 82A, 179, 1985.
132. **Fauconneau, B., Kaushik, S. J., and Blanc, J. M.,** Uptake and metabolization of dissolved compounds in rainbow trout (*Salmo gairdneri* R.) fry, *Comp. Biochem. Physiol.*, 93A, 839, 1989.
133. **Wilson, R. P. and Poe, W. E.,** Nitrogen metabolism in channel catfish, *Ictalurus punctatus*. III. Relative pool sizes of free amino acids and related compounds in various tissues of the catfish, *Comp. Biochem. Physiol.*, 48B, 545, 1974.
134. **Gras, J., Gudefin, Y., and Chagny, F.,** Free amino acids and ninhydrin-positive substances in fish. I. Muscle and skin of the rainbow trout (*Salmo gairdneri* Richardson), *Comp. Biochem. Physiol.*, 60B, 369, 1978.
135. **Van der Boon, J., Van den Thillart, G. E. E. J. M., and Addink, A. D. F.,** Free amino acid profiles of aerobic (red) and anaerobic (white) skeletal muscle of the cyprinid fish, *Carassius auratus* L. (Goldfish), *Comp. Biochem. Physiol.*, 94A, 809, 1989.
136. **Garlick, P. J. and Marshall, J.,** A technique for measuring brain protein synthesis, *J. Neurochem.*, 19, 577, 1972.
137. **Suzuki, O. and Yagi, K.,** A fluorometric assay for β-phenylethylamine in rat brain, *Anal. Biochem.*, 75, 192, 1976.
138. **McNurlan, M. A., Tomkins, A. M., and Garlick, P. J.,** The effect of starvation on the rate of protein synthesis in rat liver and small intestine, *Biochem. J.*, 178, 373, 1979.
139. **Hawkins, A. J. S., Bayne, B. L., and Day, A. J.,** Protein turnover, physiological energetics and heterozygosity in the blue mussel, *Mytilus edulis*: the basis of variable age-specific growth, *Proc. R. Soc. Lond.*, B229, 161, 1986.
140. **Niimi, A. J.,** Total nitrogen, nonprotein nitrogen, and protein content in large mouth seabass (*Micropterus salmoides*) with reference to quantitative protein estimates, *Can. J. Zool.*, 50, 1607, 1972.
141. **Price, N. M. and Harrison, P. J.,** Comparison of methods for the analysis of dissolved urea in seawater, *Mar. Biol.*, 94, 307, 1987.
142. **Millward, D. J., Garlick, P. J., Stewart, J. C., Nnanyelugo, D. O., and Waterlow, J. C.,** Skeletal-muscle growth and protein turnover, *Biochem. J.*, 150, 235, 1975
143. **Heba, H. M. A.,** *Protein Turnover in Temperate and Tropical Fish*, Ph.D. thesis, University of Aberdeen, 1992.
144. **Muramatsu, T. and Okamura, J.,** Whole-body protein turnover in chicks at early stages of growth, *J. Nutr.*, 115, 483, 1985.
145. **Muramatsu, T., Kita, K., Tasaki, I., and Okamura, J.,** Influence of dietary protein intake on whole-body protein turnover in chicks, *Br. Poult. Sci.*, 28, 471, 1987.
146. **Kita, K., Muramatsu, T., and Okamura, J.,** Effect of dietary protein and energy intakes on whole-body protein turnover and its contribution to heat production in chicks, *Br. J. Nutr.*, 69, 681, 1993.
147. **Lobley, G. E., Milne, V., Lovie, J. M., Reeds, P. J., and Pennie, K.,** Whole body and tissue protein synthesis in cattle, *Br. J. Nutr.*, 43, 491, 1980.
148. **Martin, N. B.,** Protein synthesis and the genetic code, in *Harper's Review of Biochemistry*, Martin, D. W., Mayes, P. A., Rodwell, V. W., and Granner, D. K., Eds., Lange Medical Publications, Norwalk, CT, 1985, 416.
149. **Kita, K., Muramatsu, T., Tasaki, I., and Okumura, J.,** Influence of dietary non-protein energy intake on whole-body protein turnover in chicks, *Br. J. Nutr.*, 61, 235, 1989.
150. **Pullar, J. D. and Webster, A. J. F.,** The energy cost of fat and protein deposition in the rat, *Br. J. Nutr.*, 37, 355, 1977.

151. **O'Dor, R. K. and Wells, M. J.,** Energy and nutrient flow, in *Cephalopod Life Cycles*, Vol. 2, Boyle, P. R., Ed., Academic Press, London, 1987, 109.
152. **Polat, A., Conceição, L., Sarihan, E., and Verreth, J.,** The protein, lipid and energy metabolism in eleuthero-embryos and starving larvae of the African catfish *Clarias gariepinus* (Burchell), *ICES J. Mar. Sci.*, 1995, in press.
153. **Smith, R. W. and Houlihan, D. F.,** Protein synthesis and oxygen consumption in fish cells, *J. Comp. Physiol. A*, 1995, in press.
154. **Houlihan, D. F., Pedersen, B. H., Steffensen, J. F., and Brechin, J.,** Protein synthesis growth and energetics in larval herring (*Clupea harengus*) at different feeding regimes, *Fish Physiol. Biochem.*, 13, 1995, in press.

Chapter 2

Molecular Aspects of Carbamoyl Phosphate Synthesis

Paul M. Anderson

CONTENTS

INTRODUCTION

Carbamoyl phosphate is the precursor metabolite for arginine and urea biosynthesis (via the urea cycle) and for pyrimidine nucleotide biosynthesis (Figure 1). In ureotelic species the complete urea cycle occurs only in liver and provides the major route for elimination of ammonia as the end product of nitrogen metabolism.[1,2] The presence of the first two enzymes of the urea cycle in intestinal tissue and the next two enzymes of the urea cycle in other extrahepatic organs provides for arginine biosynthesis. The urea cycle is also utilized for synthesis of urea in ureoosmotic elasmobranch fishes (sharks, skates, and rays); the urea formed is retained at concentrations as high as 0.4 M as an osmolyte against seawater.[3-5] Nearly all organisms and tissues are capable of *de novo* pyrimidine nucleotide biosynthesis by the basic pathway illustrated in Figure 1.[6] Thus, the enzymes catalyzing carbamoyl phosphate formation, the carbamoyl phosphate synthetases (CPSases), play an essential role in nitrogen metabolism in virtually all organisms. As might be expected given the diversity of organisms and metabolic circumstances (i.e., CPSase function may be specifically related to one pathway or the other or to both pathways) in which they function, CPSases exhibit a variety of regulatory, kinetic, and structural properties.[7,8] Nevertheless, studies during the past 10 years in particular have revealed mechanistic and structural themes common to all CPSases regardless of pathway specificity or organism in which the CPSase functions. The variations superimposed upon these basic themes have provided a basis for considering likely evolutionary relationships between the different types of CPSases relative to their biological function. The comparative structure and function relationships of the CPSases is the subject of this chapter.

In addition to these major pathways, carbamoyl phosphate is also utilized as a substrate in certain other unique circumstances, including glucose-6-phosphate synthesis catalyzed by a carbamoyl phosphate transferase activity of glucose-6-phosphatase,[9] phosphorylation of ADP to give ATP catalyzed by carbamate kinase in the arginine deiminase systems present in certain microorganisms and some plants,[10-12] formation of homocitrulline from lysine[13] and N-carbamyl putrescine from spermine,[14] and for a O-carbamyl transferase.[15] In addition to CPSases, carbamoyl phosphate formation occurs as the result of the phosphorolysis of citrulline,[16] oxamate,[17] and putrescine.[18]

0-8493-8411-7/95/$0.00+$.50

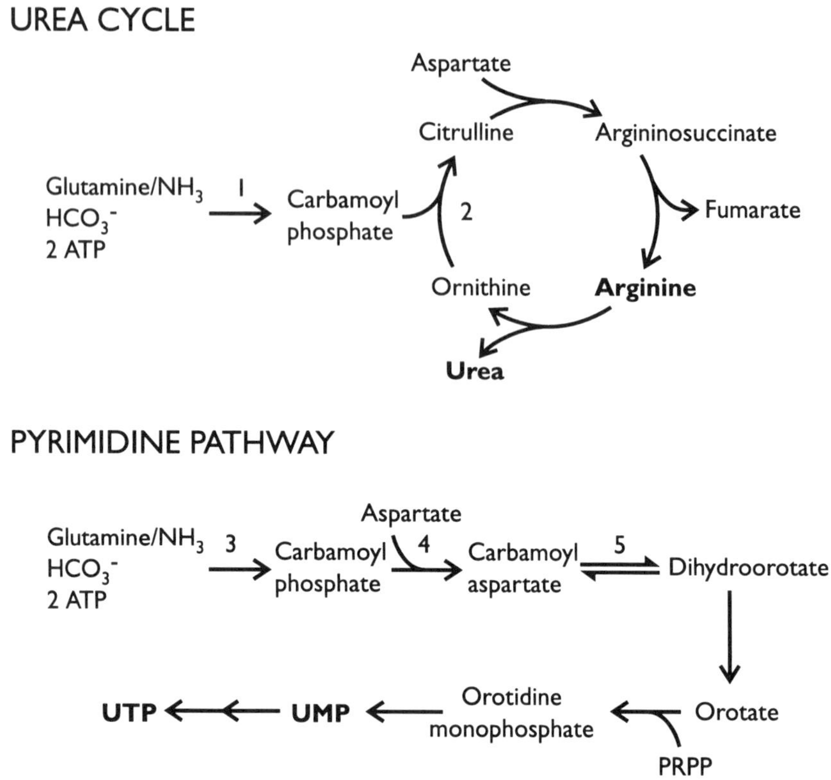

Figure 1 Illustration of urea cycle and pyrimidine nucleotide biosynthesis pathways in animals. End products are highlighted in bold. Enzymes discussed in this chapter are identified as follows: (1) CPSase I (NH_3) or CPSase III (glutamine); (2) ornithine transcarbamylase; (3) CPSase II; (4) ATCase; (5) DHOase.

Recent reviews relevant to the subject of this chapter include the structure and properties of the CPSase found in *Escherichia coli*,[8] the multifunctional CPSase related to pyrimidine biosynthesis found in higher eukaryotes,[7,19,20] the CPSase related to urea synthesis,[1] CPSase activity in fish,[4,5,21] and fungal CPSases.[22]

CARBAMOYL-PHOSPHATE SYNTHETASES

Three different types of CPSases have been recognized based on the nitrogen-donating substrate utilized and the requirement for the allosteric effector N-acetyl-L-glutamate (AcGlu) for activity. The salient properties and distribution of types I, II, and III CPSases are summarized in Table 1.

CPSase I utilizes only ammonia as the nitrogen-donating substrate and requires the presence of AcGlu for activity[1,23,24]:

$$NH_3 + 2\ ATP^{4-} + HCO_3^{-} \xrightarrow{\text{AcGlu, } Mg^{2+}} 2\ ADP^{3-} + Pi^{2-} + H^{+} + NH_2CO_2PO_3^{2-} \quad (1)$$

CPSase I is localized in the matrix of liver mitochondria of ureotelic terrestrial vertebrates and catalyzes incorporation of ammonia into carbamoyl phosphate as the first step of the urea cycle; CPSase I is also present in mitochondria of the small intestine. CPSase I is the only CPSase that cannot utilize glutamine as the nitrogen-donating substrate.

CPSase II utilizes glutamine instead of ammonia as the physiologically significant nitrogen-donating substrate and does not require AcGlu for activity[6,7,23]:

Table 1 Salient Features of Major Classes of CPSases

Class/Pathway	Nitrogen-donating substrate	Allosteric effectors	Glutaminase and synthetase domains/functions	Organisms
CPS I				
Urea, arginine	Ammonia	AcGlu (+)[a] (required)	Fused/monofunctional	Ureotelic terrestrial vertebrates
CPS II				
Arginine, pyrimidine	Glutamine	Ornithine, IMP (+), UMP (–)	Separate/monofunctional	Most prokaryotes
Arginine	Glutamine	None	Separate/monofunctional	Fungi, *Bacillus*
Pyrimidine	Glutamine	UTP (–)	Fused/bifunctional	Fungi
Pyrimidine	Glutamine	UTP (–), PRPP[b] (+)	Fused/multifunctional	Higher eukaryotes
CPS III				
Urea, arginine	Glutamine	AcGlu (+)	Fused/monofunctional	Invertebrates, fishes

[a] (+) Indicates activation; (–) indicates inhibition.
[b] PRPP, phosphoribosyl pyrophosphate.

$$\text{glutamine} + H_2O + 2\ ATP^{4-} + HCO_3^- \xrightarrow{Mg^{2+}} 2\ ADP^{3-} + Pi^{2-} + NH_2CO_2PO_3^{2-} + \text{glutamate} + 2\ H^+ \quad (2)$$

Bacteria generally have a single CPSase II which provides carbamoyl phosphate for both arginine and pyrimidine nucleotide biosynthesis.[8] An exception is *Bacillus subtilus* where there are two different types of CPSase, each specific to either the arginine or pyrimidine nucleotide pathways, respectively.[25] The single CPSase II present in *E. coli* is subject to allosteric feedback inhibition by UMP and activation by ornithine, reflecting its dual role. The enzyme is also activated by IMP and other purine nucleotides, which may reflect a means of regulating the ratio of purine and pyrimidine nucleotides. The bacterial enzymes normally consist of two nonidentical subunits, a synthetase (which can function with ammonia as substrate) and a glutaminase (required for glutamine-dependent activity) subunit.[8,26]

Fungi have two different types of CPSase II.[22] A pyrimidine-specific CPSase II is present as a bifunctional enzyme (single polypeptide chain) that includes the next activity of the pathway, aspartate transcarbamylase (ATCase), and is subject to feedback inhibition by UTP. This enzyme has been reported to be localized in the nucleus. An arginine-specific CPSase II is localized in the mitochondria (cytosol in some yeast) and, like the bacterial enzyme, consists of two different subunits.

In higher eukaryotes CPSase II is present in the cytosol of most cells as part of a multifunctional complex (single polypeptide chain) that includes the next two enzymes of the pyrimidine pathway, namely ATCase and dihydro-orotase (DHOase); it is subject to feedback inhibition by UTP and to allosteric activation by 5-phosphoribosyl-1-pyrophosphate (PRPP) and its function is related to pyrimidine nucleotide biosynthesis. This multifunctional complex is commonly referred to as CAD (from **C**PSase II, **A**TCase and **D**HOase).[6,7]

The third major group of CPSases, CPSase III, utilizes the amide group of glutamine as the nitrogen-donating substrate (like CPSase II) but requires AcGlu for activity (like CPSase I)[27,28]:

$$\text{glutamine} + H_2O + 2\ ATP^{4-} + HCO_3^- \xrightarrow{AcGlu,\ Mg^{2+}} 2\ ADP^{3-} + Pi^{2-} + NH_2CO_2PO_3^{2-} + \text{glutamate} + 2\ H^+ \quad (3)$$

CPSase III is found in the mitochondrial matrix of invertebrates and liver of some fishes.[21,27-30] The function of CPSase III, at least in ureoosmotic elasmobranch fishes, is clearly related to the urea cycle.[4]

GLUTAMINE-DEPENDENT CARBAMOYL PHOSPHATE SYNTHETASES

Glutamine amidotransferases are a family of enzymes that catalyze transfer of the amide group of glutamine to an acceptor unique to each enzyme in a variety of biosynthetic reactions.[26] A characteristic structural feature of all amidotransferases is that they are comprised of a "synthetase" domain (catalyzes formation of an acceptor, which reacts with ammonia) unique for each type of amidotransferase, and a "glutamine amide transfer" (GAT) domain (required for the capacity to utilize glutamine as the nitrogen-donating substrate). In some amidotransferases these two domains are represented as two different subunits, whereas in others the two domains exist on a single polypeptide chain. Virtually all amidotransferases can utilize ammonia in place of glutamine as nitrogen-donating substrate, but very high (nonphysiological) concentrations of NH_3/NH_4^+ are normally required for significant activity. Although it has not been conclusively demonstrated in all cases, it appears that the mechanism of glutamine amide transfer is essentially identical in all of the known 16 different amidotransferases, including the glutamine-dependent CPSases. Glutamine reacts with a specific cysteine SH group in the GAT domain to form a γ-glutamyl thioester intermediate, releasing ammonia in an unknown form (e.g., concerted reaction, NH_3, or NH_4^+) which reacts with an activated "acceptor" formed by the synthetase domain; the γ-glutamyl thioester intermediate is subsequently hydrolyzed, releasing glutamate and regenerating the free SH group. The specific cysteine residue involved in this process is conserved in all amidotransferases whose sequences have been determined, including the glutamine-dependent CPSases.

HOMOLOGY AND DOMAIN STRUCTURE

The amino acid sequences deduced from the corresponding cloned cDNA sequences have been determined for a number of CPSases in recent years, including rat,[31] human,[32] and tadpole (*Rana catesbiana*) CPSase I,[33] the pyrimidine-related trifunctional CPSase II from *Drosophila melanogaster*[34] and hamster,[35,36] the yeast CPSases related to arginine biosynthesis[37-39] and to pyrimidine biosynthesis,[40] the pyrimidine-related CPSases from *Dictyostelium discoideum*[41] and *B. subtilus*,[25] *E. coli* CPSase II,[42,43] and CPSase III from spiny dogfish shark (*Squalus acanthias*).[44] These and other studies have revealed a high degree of identity in the amino acid sequences and apparent structural domains in CPSases from a phylogenetically broad range of organisms. As illustrated in Figure 2, all CPSases are apparently comprised of two basic domains — a N-terminal glutaminase domain and a C-terminal synthetase domain. Where the CPSase is part of a multifunctional complex the additional polypeptide sequences that comprise the other enzyme activities are fused to the C-terminal end of the CPSase. In the case of CPSases targeted for mitochondrial import (i.e., CPSase I, CPSase III, and presumably the arginine pathway-specific CPSase II in fungi), the initially translated protein has a N-terminal sequence that is removed in the process of transfer into the mitochondria.

The glutaminase domain consists of distinct N- and C-terminal domains.[38] The C-terminal half is homologous to the GAT domain present in all known trp G-type amidotransferases (one of two distinct classes of amidotransferases) and is characterized by a sequence of 11 highly conserved amino acids that includes the cysteine residue essential for amide transfer in the glutamine-dependent CPSases.[26,45-47] However, as noted in Figure 2, this cysteine residue is replaced by a serine residue in human and rat CPSase I and several other residues are replaced as well. Although this has, until recently, been cited as an explanation for why glutamine cannot be utilized as the nitrogen-donating substrate by CPSase I, the recent report by Helbing and Atkinson[33] that this cysteine residue is conserved in the tadpole CPSase I (see Figure 2) indicates that other residues besides this cysteine are clearly required for glutamine-dependent activity. In addition, this observation suggests that these amino acids underwent replacement prior to loss of the cysteine residue in the proposed evolutionary transition from CPSase III to CPSase I. Thus, the tadpole CPSase I may represent an intermediate stage in this evolutionary

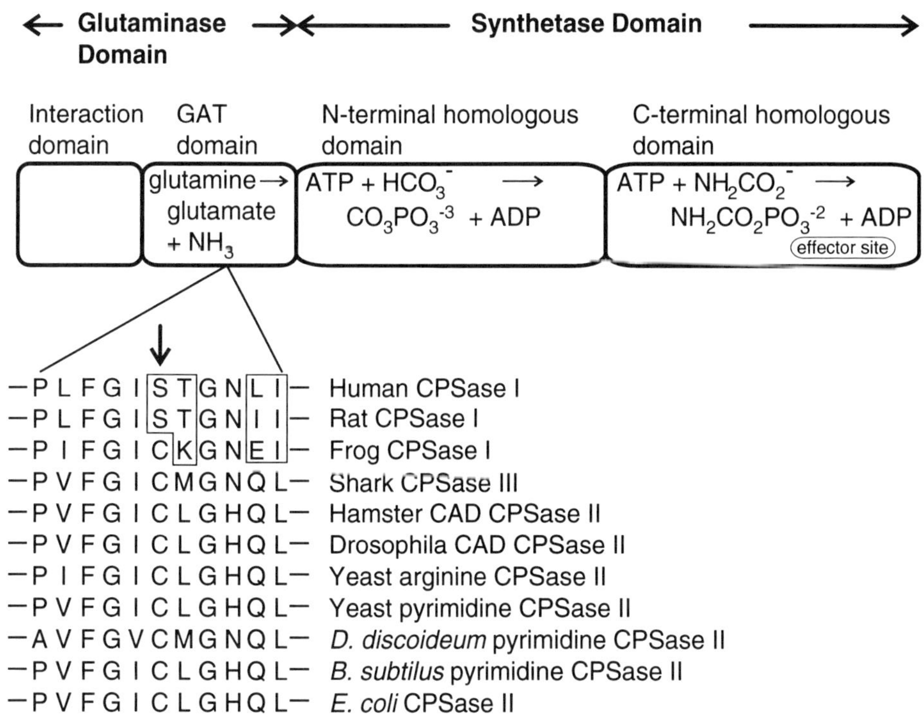

Figure 2 Illustration of the domain structure of CPSases and comparison of a highly conserved amino acid sequence in the GAT domain of various CPSases that includes the glutamine-binding site for CPSase II and CPSase III. Amino acid residues that appear to be unique to CPSase I are enclosed by boxes.

transition, as suggested by Helbing and Atkinson.[33] Identification of the residues other than the cysteine residue that are required for glutamine-dependent activity should prove to be particularly helpful in identifying the functional interactions between the synthetase and glutaminase domains and provide insight into the basis for the evolutionary origin of CPSase I.

The N-terminal subdomain of the glutaminase domain is present in all CPSases, but not in other amidotransferases; there is evidence that this domain participates in interactions between the glutaminase and synthetase domains.[19,48] In *E. coli* and certain other CPSases as cited above the glutaminase domain exists as a separate polypeptide chain coded by a separate gene.

The larger C-terminal domain (synthetase domain) is comprised of two halves with high sequence identity, suggesting that these parts of the gene arose from gene duplication and fusion, and are therefore homologous.[35,42] Although there is high sequence identity between the two halves, there is significantly less identity toward the C-terminal ends, which is consistent with accumulating evidence that the regulatory binding sites for the different CPSases (e.g., AcGlu for CPSase I and UMP for *E. coli* CPSase) are comprised of amino acids in the C-terminal end of the C-terminal subdomain of the synthetase domain.[19,49-53]

In addition to the highly conserved 11 amino acid residues in the GAT domain, a number of other highly conserved residues or regions have been identified that may have functional significance, including: (1) three conserved histidine residues in the glutaminase domain[54]; (2) three conserved histidine residues in the synthetase domain[55]; (3) two postulated ATP binding sites and two glycine-rich loop regions considered to be homologous to the proposed glycine-rich regions of other adenine nucleotide-binding proteins, one of each in each homologous half[37,56-59]; and (4) a glutamate residue that appears to play a role in the interaction between the two major domains.[60,61]

Geschwill and Lumper[50] identified two SH groups in the C-terminal subdomain of porcine CPSase I that can be alkylated at an enhanced rate if AcGlu is present resulting in loss of activity;

however, the presence of MgATP protects against alkylation. Potters and Powers-Lee[57] also showed that the ATP analog 5′-*p*-fluorosulfonylbenzoyladenosine reacts with one or both of these cysteine SH groups and that the rate of the reaction is affected by the presence of AcGlu. These two cysteines are conserved in CPSase III and in rat, human, and tadpole CPSase I, but are absent in all other CPSases that are not activated by AcGlu (with one known exception: one of the cysteines is present in the pyrimidine-specific CPSase II from *D. discoideum*). These observations suggest that these two SH groups may be a unique feature of AcGlu-dependent CPSases.[44]

CATALYTIC REACTION PATHWAY AND DOMAIN INTERACTIONS

The catalytic mechanism of the *E. coli* glutamine-dependent CPSase, recently reviewed in detail by Meister,[8] has been investigated more extensively than for any other CPSase. Although structural, regulatory, or functional characteristics unique to the different CPSases influence their kinetic properties, the basic reaction pathway as described for the *E. coli* enzyme appears to be essentially the same for all CPSases — at least for all glutamine-dependent CPSases.

Carbamoyl phosphate formation from ammonia proceeds via the following reaction pathway involving three identifiable steps:

$$ATP^{4-} + HCO_3^- \underset{}{\overset{Mg^{2+}}{\rightleftharpoons}} CO_3PO_3^{3-} + ADP^{3-} + H^+ \quad (4)$$

$$CO_3PO_3^{3-} + NH_3 \longrightarrow NH_2CO_2^- + Pi^{2-} \quad (5)$$

$$NH_2CO_2^- + ATP^{4-} \overset{Mg^{2+}}{\rightleftharpoons} NH_2CO_2PO_3^{2-} + ADP^{3-} \quad (6)$$

There appear to be two ATP binding sites on the synthetase subunit, one for each of the two ATPs utilized in the reaction, but a single site for the sequential formation of carboxyphosphate, carbamate, and carbamoyl phosphate.

In addition to the overall synthetase reaction from either ammonia or glutamine, the *E. coli* CPSase (and other CPSases) catalyzes several partial reactions:

$$ATP^{4-} + HCO_3^- + H_2O \xrightarrow{Mg^{2+}} ADP^{3-} + Pi^{2-} + HCO_3^- + H^+ \quad (7)$$

$$ADP^{3-} + NH_2CO_2PO_3^{2-} + H_2O \xrightarrow{Mg^{2+}} NH_3 + HCO_3^- + ATP^{4-} \quad (8)$$

$$glutamine + H_2O \longrightarrow NH_4^+ + glutamate \quad (9)$$

These partial reactions reflect different steps of the reaction pathway and proceed at rates that are normally much slower than the overall synthetase reaction. In the absence of glutamine or ammonia, carboxyphosphate formed in the first step of the reaction pathway apparently dissociates or is hydrolyzed, resulting in bicarbonate-dependent ATP hydrolysis (Reaction 7). The second step of the reaction pathway (Reaction 5) is irreversible and carbamate formed in the reverse of the third step (Reaction 6) dissociates along with ATP, resulting in formation of one ATP per carbamoyl phosphate and ADP utilized (Reaction 8).

In *E. coli* the synthetase and glutaminase domains are on different subunits. The synthetase subunit catalyzes all reactions that do not require glutamine, including carbamoyl phosphate formation from ammonia as illustrated by the reaction pathway (Reactions 4 to 6) and the two partial reactions discussed above, and contains the allosteric binding sites for ornithine, UMP, and IMP.[8,62] When associated with the glutaminase subunit, carbamoyl phosphate formation by the synthetase subunit proceeds with glutamine as the nitrogen-donating substrate. Utilization of the amide group of glutamine for biosynthesis of carbamoyl phosphate in the glutamine-dependent CPSases involves reaction with glutamine to form the γ-glutamyl thioester intermediate, releasing ammonia which reacts with the activated bicarbonate (carboxyphosphate); the thioester intermediate is subsequently hydrolyzed. This reaction pathway is reflected by the glutaminase activity (Reaction 9) observed in glutamine-dependent CPSases in the absence of ATP.

Significant interaction between the synthetase and glutaminase subunits of *E. coli* CPSase has been well documented and appears to reflect a functional linkage between formation and hydrolysis of the γ-glutamyl thioester intermediate on the glutaminase subunit and carboxyphosphate formation and/or other partial reactions on the synthetase subunit. For example, the presence of ATP and bicarbonate dramatically increases the rate of hydrolysis of γ-glutamyl hydroxamate (without transfer of the released hydroxylamine to the synthetase subunit or reaction with carboxyphosphate) and reaction of the SH group that normally forms the γ-glutamyl thioester intermediate with site-directed irreversible inhibitor analogs of glutamine or with cyanate stimulates the bicarbonate-dependent ATPase activity. Thus, wasteful hydrolysis of glutamine is prevented by linking rapid hydrolysis of the γ-glutamyl thioester intermediate on the glutaminase subunit to formation of carboxyphosphate on the synthetase subunit. Likewise, bicarbonate-dependent hydrolysis of ATP may be minimized due to the fact that a more reactive form of carboxyphosphate is linked to binding of glutamine. Similar interactions occur between the synthetase and glutaminase domains of other glutamine-dependent CPSases where the domains are on the same polypeptide chain.[46] It will be of interest to establish if the glutaminase domain in CPSase I, which cannot utilize glutamine as the nitrogen-donating substrate, plays a role in modulating the activity of the synthetase domain. The observations that the binding of AcGlu dramatically affects glutamine binding in CPSase III[63] and that the Km for ammonia is significantly lower in CPSase I than in CPSase II suggests that this interaction may be important for AcGlu binding and the binding of ammonia.

Earlier studies indicated that the synthetase domain (or subunit) has two physically distinct sites for binding one or the other of the two molecules of ATP utilized in the reaction pathway for carbamoyl phosphate formation. As noted in the previous section, sequence comparisons and inhibition studies with the ATP analogs 5′-*p*-fluorosulfonylbenzoyladenosine and 8-azido ATP (which can react with amino acid side chains presumed to be in the ATP binding site) identified likely ATP binding regions in both the N- and C-terminal homologous halves of the synthetase domain. Post et al.[56] showed that substitution of specific glycine residues with isoleucine within each of these two putative nucleotide binding regions by site-directed mutagenesis gave mutant enzymes in which either the bicarbonate-dependent ATPase or the ATP synthesis partial reactions were selectively inhibited, providing evidence that one ATP is associated with each of the subdomains and that the N- and C-terminal homologous halves are primarily involved in catalyzing the first and third steps of the reaction pathway, respectively (see Figure 2). A number of other site-directed mutagenesis studies (primarily with the *E. coli* enzyme) involving conserved residues identified as potentially important in binding or catalysis have been reported and have provided confirming evidence for the reaction pathway described above.[54,55,60,61,64,65] The identity of the specific cysteine residue required for formation of the γ-glutamyl thioester intermediate has also been confirmed by site-directed mutagenesis.[45] Although the mutant enzymes involving substitution of other amino acids for the cysteine residue cannot react with glutamine to form the thioester intermediate and are therefore inactive with glutamine as substrate, glutamine can still bind to the glutaminase domain and induce the conformational change that affects the synthetase subunit, indicating that binding of glutamine *per se*, and not thioester formation, is the driving force for the interaction. Using truncated and deletion mutants of the *E. coli* CPSase, Guillou et al.[48] provided evidence that the N-terminal domain of the glutaminase subunit in *E. coli* is essential for interaction with the synthetase subunit and that the synthetase subunit has two specific regions, one in each homologous half, required for interaction with the glutaminase subunit. A similar observation has been reported for the N-terminal "interaction" domain of CAD CPSase II.[19] These observations may be important when considering the role of the glutaminase domain in the function of CPSase I.

Guy and Evans[66] recently cloned and expressed the portion of the mammalian CAD CPSase II identified as the glutaminase domain and showed that (1) it complemented the synthetase subunit in an *E. coli* mutant in which the glutaminase subunit was absent and (2) it formed a complex with the isolated *E. coli* synthetase subunit that was fully active with glutamine as nitrogen-donating substrate. Thus, the CAD glutaminase domain apparently folds independently of the synthetase domain and the subunit interfaces must be nearly identical in the *E. coli* and mammalian CAD CPSase II. Evans et al.[19] have also reported that splicing a sequence of the mammalian CAD CPSase II thought to carry the binding sites for the allosteric effectors PRPP and UTP into the corresponding region of the *E. coli* synthetase subunit and expression of the hybrid enzyme in *E. coli* results in a hybrid *E. coli* CPSase that is activated by PRPP and inhibited by UTP; the hybrid enzyme was no longer affected by the normal *E. coli* negative allosteric effector UMP. These results provide direct evidence for the view illustrated in Figure 2 that although different CPSases are regulated by different effectors, the regulatory binding sites are located in a C-terminal region of the synthetase domain of all CPSases. Studies such as these offer considerable promise for further defining the evolutionary relationships between the different CPSases and for establishing the nature of functional interactions between domains. For example, the question of the requirement for glutamine-dependent activity, besides the essential cysteine residue, may be addressed by such hybridization experiments, perhaps combined with selective site-directed mutagenesis. Many useful combinations could be considered, e.g., hybridization of the glutaminase domain of the AcGlu-dependent frog CPSase I (inactive with glutamine, but retains the cysteine residue essential for glutamine-dependent activity) or shark CPSase III (active with glutamine) with the *E. coli* synthetase subunit, or hybridization of the synthetase domain of each of these two AcGlu-dependent CPSases with the glutaminase domain of *E. coli* or CAD CPSase II.

CPSASE I

The specific role of CPSase I in catalyzing the initial step of the urea cycle in liver, in which ammonia is initially assimilated as carbamoyl phosphate and subsequently converted to urea, is reflected in its localization in the mitochondrial matrix where free ammonia is generated and in the direct relationship between the rate of protein catabolism and the rate of carbamoyl phosphate synthesis. CPSase I and the enzyme which catalyzes the next step of the urea cycle, ornithine transcarbamylase (OTCase), are loosely associated with the inner mitochondrial membrane.[67] Although the two enzymes appear to associate with membrane independently, it is possible that these associations may facilitate functional association between the two enzymes. Ornithine transported into the mitochondria is utilized in preference to intramitochondrial ornithine for citrulline synthesis, implying that carbamoyl phosphate formed by CPSase I is channeled to OTCase for reaction with this compartmentalized pool of ornithine.[68,69] Fahien et al.[70] have also reported specific interactions of CPSase I with mitochondrial glutamate dehydrogenase and aspartate aminotransferase, enzymes which are essential for providing ammonia, the substrate for CPSase I. If these complex multiple associations are critical for physiological function, variations in the amino acid sequence may be constrained by requirements to conserve certain residues critical for these associations. Elucidation of the molecular nature of these interactions in detail will likely provide important insights into the structure–function relationships unique to CPSase I as well as provide a better understanding of urea cycle control.

Short-term regulation of CPSase I activity in response to an increased rate of protein catabolism is primarily mediated by increases in ammonia and AcGlu concentration in the mitochondria.[1,71] AcGlu-dependent CPSase I is not subject to feedback inhibition by metabolic end products characteristic of pyrimidine pathway-related CPSase II, which presumably reflects its function of efficient removal of ammonia as an end product of nitrogen metabolism. This property of CPSase I is manifested when the activity of urea cycle enzymes other than CPSase I are inhibited or ornithine concentration is limiting. Mitochondrial carbamoyl phosphate synthesis catalyzed by CPSase I continues unabated, resulting in leakage of accumulating carbamoyl phosphate into the cytosol where it serves as a substrate for ATCase in the pyrimidine nucleotide biosynthetic pathway. The pyrimidine pathway is not regulated under these circumstances (because CPSase II activity as the normal source of carbamoyl phosphate for the pyrimidine pathway has been bypassed) resulting in orotic aciduria.[72-74]

Although the mechanism of activation by AcGlu has not been clearly established, several studies have demonstrated strong interactions between the AcGlu binding site and the ATP binding site involved in activation of bicarbonate but not with the ATP binding site involved in carbamate phosphorylation, i.e., the effect of AcGlu binding on the C-terminal subdomain seems to be transmitted to the ATP binding site of the N-terminal subdomain of the synthetase domain.[57,75,76] An effect of AcGlu binding on the conformation of CPSase I is also manifested by its effect on the reactivity of two SH groups in the synthetase domain.[50,57,77] As noted above, the observation that the binding of AcGlu by CPSase III is significantly affected by the binding of glutamine suggests that the glutaminase domain might play a role in manifesting the effects of AcGlu in CPSase I. An approach for investigating this possibility might be to characterize the effects of AcGlu on the activity of a modified CPSase I in which the N-terminal glutaminase domain is missing; such a truncated enzyme would be analogous to the *E. coli* synthetase subunit, which is active with ammonia as the nitrogen-donating substrate.

CPSASE III

The distribution, properties, and function of CPSase III in fish has been recently reviewed by Anderson[21,78] and by Campbell and Anderson.[4] Except for the fact that glutamine is utilized as the nitrogen-donating substrate, the properties, amino acid sequence, and function of CPSase III are very similar to CPSase I. High levels of CPSase III are present in liver mitochondria of marine ureoosmotic elasmobranchs, which synthesize and retain urea in their tissues for the purpose of osmoregulation. CPSase III from spiny dogfish shark (*Squalus acanthias*), a representative elasmobranch, is subject to significant inhibition by physiological concentrations (0.2 to 0.4 M) of urea.[63] This may represent a regulatory feedback mechanism for maintaining normal physiological concentrations of urea. The observation that mitochondrial citrulline synthesis, but not respiration under the same conditions, is inhibited by urea is consistent with this view.[79] The additional observation that the CPSase III from largemouth bass (*Micropterus salmoides*), a freshwater teleost which does not synthesize and retain urea for the purpose of osmoregulation, is not significantly inhibited by urea, is also consistent with the view that feedback inhibition of the shark enzyme by urea is a unique property related to its function in osmoregulation.[80]

Most fish are ammonotelic and the level of CPSase III activity in most species is very low or undetectable. Nevertheless, CPSase III activity has been reported in several teleost species.[5,29,80,81] The presence of CPSase III in teleost fishes appears to be related to unique environmental circumstances or evolutionary adaptations that favor ureogenic metabolism (e.g., amphibious fishes, high alkalinity, or extended confinement).[21] Recent studies have suggested that CPSase III and other urea cycle enzymes may be expressed only during embryogenesis in some species.[82]

The discovery and characterization of CPSase III established that the AcGlu binding site of CPSase I probably did not evolve from the glutamine-binding site of CPSase II.[31,83] A rather dramatic synergistic relationship exists between these two sites in CPSase III; the apparent Km for each decreases significantly as the concentration of the other increases.[63,80] Thus the binding of AcGlu, presumably at a site on the synthetase domain, significantly influences the glutaminase domain. Although the glutaminase domain in CPSase I does not function to provide ammonia from glutamine, given the high degree of sequence and structural similarity between CPSase III and

CPSase I, it seems likely that AcGlu binding in CPSase I also influences the glutaminase domain. Perhaps the effects of AcGlu on activity in both CPSase I and CPSase III depend, in part, on interactions between these two domains. As indicated above, the availability of truncated forms of either of these enzymes in which the glutaminase domain has been deleted would be helpful in establishing if there is a functional relationship of the glutaminase domain to activity of the synthetase domain. Is the glutaminase domain essential for activity? Would AcGlu have similar effects on activity? What would be the effect on the Km for ammonia?

CAD CPSASE II

CAD is a trifunctional protein catalyzing the first three reactions of the pyrimidine pathway and is localized in the cytosol of most animal tissues.[7,84] The structural organization of CAD has been elucidated in considerable detail by limited proteolysis and sequencing studies.[7,19,20,35,85,86] The three enzymes are arranged in a simple domain structure in the sequence CPSase-DHOase-ATCase. The CPSase domain of CAD is comprised of a glutaminase and synthetase domain, analogous to other CPSases. Each of the enzymatic domains can be readily separated from each other by limited proteolysis of the protease-sensitive interdomain polypeptide linker sequences and each can function in the absence of the other. A number of studies involving expression in *E. coli* or mammalian cells of portions of the CAD gene representing one or the other of the three activities using appropriate expression vectors has demonstrated that each domain can fold into its catalytically active conformation independent of the other portions of the CAD protein.[86]

These observations are consistent with the view that CAD evolved by fusion of individual genes for each of the enzyme activities, which in prokaryotes are expressed as separate proteins. The order of the cistrons in the single operon for the separate glutaminase and synthetase subunits of the *E. coli* CPSase is identical to the corresponding domains in CAD CPSase II. Davidson and co-workers[87] have suggested that fusion of these two genes might occur by a single mutational event that eliminated the stop codon of the glutaminase cistron if this mutation left the downstream synthetase sequence in-frame. They tested this hypothesis by fusing the *E. coli* glutaminase and synthetase genes using the sequence encoding the hamster CPSase bridge region to replace the intercistronic region of the *E. coli* cistron. The fused product was expressed in an *E. coli* mutant lacking CPSase activity and was found to be fully active and stable. Similar mechanisms involving intron-intron rearrangement have been suggested by Davidson et al.[86] for the subsequent addition of DHOase and ATCase genes to give CAD.

The association of these three enzyme activities on a single polypeptide chain apparently does not result in channeling of substrates. Davidson et al.[86] have suggested that an alternative explanation for the evolution of CAD as a multifunctional entity may be the selective advantage of coordinate expression of enzymes catalyzing steps in a particular biosynthetic pathway, which would preserve the advantage of the analogous structure in bacteria — the operon. Evidence has been reported that the three enzyme activities represented in CAD are present as separate proteins in extracts prepared from frog[88] and dogfish shark[89] liver and from several protozoal parasites.[90-92] The above considerations do not rule out the possibility that specific posttranslational proteolysis of a CAD-like protein might occur in some species to give three separate enzymes. In the case of shark, recent studies have shown that the three enzymes are encoded by a single gene; a single large transcript has been demonstrated by sequencing and hybridization with probes specific for the cDNA of each enzyme.[93] Allosteric regulation of CAD is mediated solely through binding of PRPP and UTP at the extreme C-terminal end of the synthetase domain of the CPSase domain. A recent report suggests that polyamines also give significant inhibition at physiological concentrations and that this may be important in counteracting the effect of elevated PRPP concentrations in tumor tissues, when polyamine concentrations are known to increase.[94] CAD CPSase II is also subject to phosphorylation and dephosphorylation at a site within this "allosteric domain"; phosphorylation decreases the effect of UTP and is prevented by the presence of UTP.[20] In addition to allosteric regulation, the expression of CAD is subject to hormonal control, which is particularly evident in diabetes[95] and tumor tissues.[96]

Both the reaction pathway and the kinetic mechanism of the CPSase activity of CAD have been shown to be very similar to that of the *E. coli* CPSase.[46,97] Meek et al.[97] have pointed out that this may be in contrast to CPSase I in which the kinetic mechanism is apparently different and there is evidence that enzyme-bound bicarbonate may be activated by both molecules of ATP prior to reaction with the bound ammonia. They have suggested that this may reflect a difference between the requirement for ammonia transfer from glutamine in CPSase II and the direct reaction with ammonia in the AcGlu-dependent CPSase I. Elucidation of the basis for the mechanistic properties unique to CPSase I, if the mechanism is in fact significantly different, will be important for understanding the evolutionary relationship of CPSase I to various forms of CPSase II. Mechanistic studies with the glutamine- and AcGlu-dependent CPSase III similar to those done with CAD CPSase II, the *E. coli* CPSase II, and CPSase I would be helpful in addressing this question, as would construction of hybrids between the CPSase I synthetase domain and the glutaminase domain from CAD or *E. coli* CPSase II or CPSase III.

EVOLUTIONARY RELATIONSHIPS

The hypothetical evolutionary scheme shown in Figure 3 summarizes proposals and speculation by many investigators based on the general common structural and mechanistic themes discussed in previous sections of this chapter together with the obvious phylogenetic relationships of the different CPSases.[4,7,26,31,38,42,44,98] The prokaryotic CPSases are thought to have evolved from (1) an ancestral

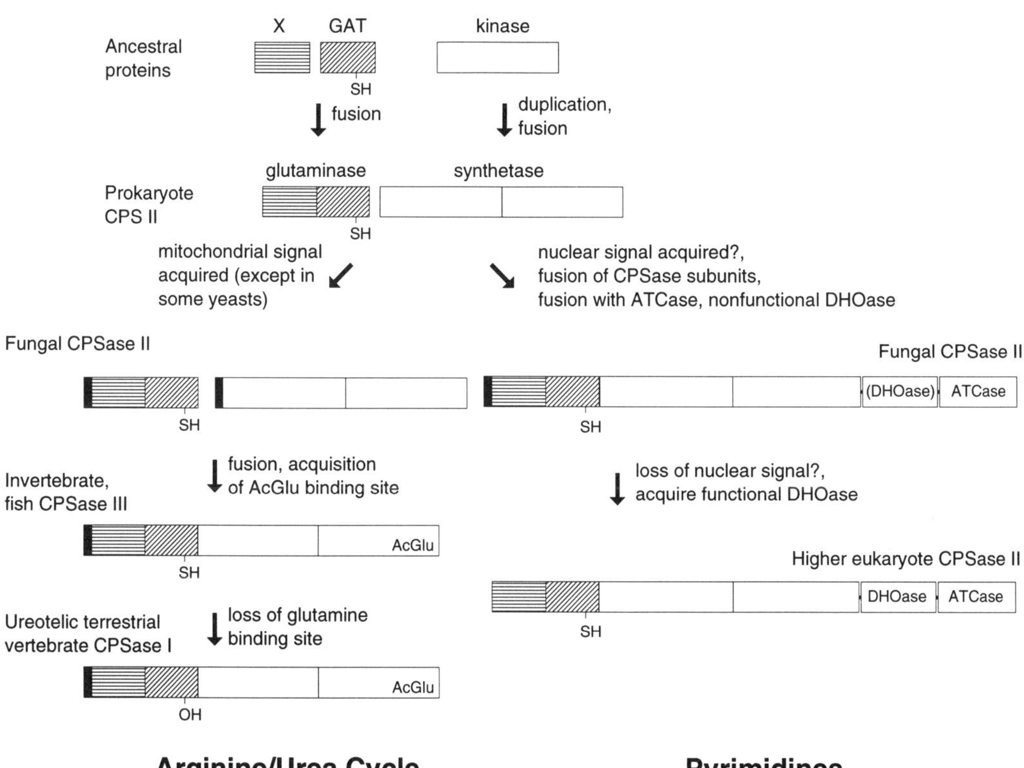

Figure 3 Illustration of a hypothetical scheme for steps in the evolution of the CPSases.

GAT gene common to all amidotransferases fused at the 5′ end with a small gene of unknown origin to give a gene for the CPSase glutaminase subunit, and (2) an ancestral kinase gene which was duplicated and then fused to give the synthetase gene. Association of the products (i.e., glutaminase and synthetase subunits, respectively) of these two genes represents the glutamine-dependent CPSase activity characteristic of prokaryotic CPSase II today.

It has been suggested that the ancestral kinase gene was a carbamate kinase, which is found in prokaryotes as part of the arginine deiminase system, providing energy from carbamoyl phosphate derived from arginine.[10,42] Carbamate kinase catalyzes the following reaction, which is essentially the same as the ATP synthesis partial reaction catalyzed by CPSases (Reaction 5, representing partial reversal of the third step of the reaction pathway)[99]:

$$ADP^{3-} + NH_2CO_2PO_3^{\ 2-} \xrightarrow{Mg^{2+}} NH_2CO_2^{\ -} + ATP^{4-} \qquad (10)$$

Thus, duplication and fusion of this gene would appear to provide the features characteristic of the synthetase domain, i.e., two homologous halves (each of which has an ATP binding site), one of which catalyzes the same reaction as carbamate kinase (ATP synthesis) and the other a very similar reaction (bicarbonate-dependent ATPase). Baur et al.[99] determined the amino acid sequence of carbamate kinase from *Pseudomonas aeruginosa* and reported that comparison with the C-terminal homologous half of the *E. coli* synthetase domain showed only 15.5% identity and 23.9% additional similarity of amino acid sequences, with even less similarity to the N-terminal domain (7.1% identity). They concluded that these results did not support the hypothesis. However, alignment comparison between this carbamate kinase amino acid sequence and the N-terminal homologous half of the synthetase domains of shark CPSase III, CPSase I, hamster CAD CPSase II, yeast arginine-specific CPSase II, and *E. coli* CPSase II using the BestFit program available in the UWGCG software, gave scores of ≈20% identity and an additional ≈28% similarity, respectively; somewhat lower scores were obtained for the C-terminal homologous half of these same CPSases (≈18 and 44%, respectively).[100] These observations suggest that more detailed comparative sequence and structural analyses may be warranted before carbamate kinase should be eliminated as a possible evolutionary precursor for the synthetase domain of CPSases. A possible factor to consider is that there may have been greater restraint on the mutation rates for CPSase than for carbamate kinase due to specific requirements for domain interactions and spatial arrangements between the ATP, glutamine, and bicarbonate binding sites in CPSases.

Evolution of the prokaryote CPSase II involved the acquisition of new functional sites, modification of regulatory binding sites, and divergence of function at an early date to give the two different CPSases characteristic of higher vertebrates, i.e., pyrimidine- and arginine/urea pathway-related CPSases, respectively. In the arginine pathway both subunits of prokaryote CPSase II must have acquired a mitochondrial targeting signal in the fungi.

CPSase III in fish and invertebrates is considered to be an evolutionary intermediate between the arginine-specific, glutamine-dependent CPSase II in fungi and the ammonia- and AcGlu-dependent CPSase I in ureotelic terrestrial vertebrates.[4,5] An AcGlu binding site was acquired and the separate glutaminase and synthetase subunits were fused in the transition from the fungal arginine-specific CPSase II to CPSase III of invertebrates and primitive fishes. Subsequent events led to loss of the amino acid residues required for utilization of glutamine as nitrogen-donating substrate in the transition from CPSase III to CPSase I.

In the pyrimidine pathway, the genes for DHOase and ATCase were apparently fused to the 3′ end of the synthetase subunit and the glutaminase and synthetase genes became fused at an early date to give a multifunctional enzyme. Although the DHOase protein is present in the yeast multifunctional CPSase, it is not catalytically functional and there is a second gene that codes for a separate DHOase activity.[40] Although pyrimidine pathway-related CPSase II in fungi has been reported to be localized in the nucleus, which would necessitate a nuclear targeting signal, recent observations suggest that this should be reevaluated.[84,101]

The scheme illustrated in Figure 3 is primarily based on the amino acid sequences deduced from the sequences of the corresponding cDNAs. Evans and co-workers[35] showed that dendograms based on nucleotide rather than amino acid alignments of the glutaminase and synthetase domains gave quite different results, suggesting that CPSase I may have arisen instead by defusion from the pyrimidine pathway-related multifunctional CPSase II. Schofield[102] has pointed out that CPSase fulfills the criteria as a model gene from which to study the structural evolution of large vertebrate genes, i.e. it is an ancient gene (present in each of the major lineages — Bacteria, Archaea, and Eukarya) with a tandem duplication; in addition, the gene is large, universally distributed, and is highly conserved over a great phylogenetic distance. Schofield observed that since the duplication characteristic of the synthetase domain is present in all three lineages, the event occurred in their common ancestor (i.e., the progenate). This provided the opportunity to test the hypothesis that if introns were present in the common progenotic ancestral gene before duplication, then at least some introns should be in concordant positions when comparing the two halves of intron-containing genes of different CPSases. Analysis of the rat CPSase I and the hamster CPSase II gene within the CAD multigene complex revealed a variable number of introns in each homologous half, including two in rat and one in hamster that were concordant with the homologous position in each half, and three introns in the hamster CPSase II that were concordant with rat CPSase I introns. These observations suggested that (1) the progenotic gene contained several introns before duplication and (2) since CPSase I and CAD CPSase II have different intron-exon structures, one is not simply a duplication of the nuclear-encoded gene of the other (i.e., the two genes were acquired independently).

The scheme in Figure 3 will likely be revised or refined as more sequence information becomes available and as the nucleotide sequences and the intron-exon structure of the corresponding genes for different CPSases are subjected to these kinds of analyses. Studies on the mechanism of gene regulation may also provide insights into the evolutionary relationships of the different CPSases.[103-105]

ACKNOWLEDGMENTS

I thank the the National Science Foundation for research grant support (DCB-91057997), C. Helbing and D. E. Atkinson for supplying a copy of their manuscript in press, and to Drs. David Evans and Wilmar Salo for reviewing the manuscript.

REFERENCES

1. **Meijer, A. J., Lamers, W. H., and Chamuleau, R. A. F. M.,** Nitrogen metabolism and ornithine cycle function, *Physiol. Rev.*, 70, 701, 1990.
2. **Campbell, J. W.,** Excretory nitrogen metabolism, in *Environmental and Metabolic Animal Physiology*, Prossker, C. L., Ed., Wiley-Liss, New York, 1991, chap. 7.
3. **Perlman, D. F. and Goldstein, L.,** Nitrogen metabolism, in *Physiology of Elasmobranch Fishes*, Shuttleworth, T. J., Ed., Springer-Verlag, Berlin and Heidelberg, 253.
4. **Campbell, J. W. and Anderson, P. M.,** Evolution of mitochondrial enzyme systems in fish: the mitochondrial synthesis of glutamine and citrulline, in *Biochemistry and Molecular Biology of Fishes*, Hochachka, P. W. and Mommsen, T. P., Eds., Elsevier, Amsterdam, 1991, chap. 3.
5. **Mommsen, T. P. and Walsh, P. J.,** Urea synthesis in fishes: evolutionary and biochemical perspectives, in *Biochemistry and Molecular Biology of Fishes*, Vol. 1, Hochachka, P. W. and Mommsen, T. P., Eds., Elsevier, Amsterdam, 1991, chap. 6.
6. **Jones, M. E.,** Pyrimidine nucleotide biosynthesis in animals: genes, enzymes, and regulation of UMP biosynthesis, *Annu. Rev. Biochem.*, 49, 253, 1980.
7. **Evans, D. E.,** CAD, a chimeric protein that initiates de novo pyrimidine biosynthesis in higher eukaryotes, in *Multidomain Proteins-Structure and Function*, Hardie, D. G., Ed., Elsevier, Amsterdam, 1986, chap. 9.
8. **Meister, A.,** Mechanism and regulation of the glutamine-dependent carbamyl phosphate synthetase of *Escherichia coli*, *Adv. Enzymol. Relat. Areas Mol. Biol.*, 62, 315, 1989.

9. **Nordlie, D. C., Bode, A. M., and Foster, J. D.**, Recent advances in hepatic glucose 6-phosphatase regulation and function, *Proc. Soc. Exp. Biol. Med.*, 203, 274, 1993.
10. **Abdelal, A. T.**, Arginine catabolism by microorganisms, *Annu. Rev. Microbiol.*, 33, 139, 1979.
11. **Baur, H., Luethi, E., Stalon, V., Mercenier, A., and Haas, D.**, Sequence analysis and expression of the arginine-deiminase and carbamate-kinase genes of *Pseudomonas aeruginosa*, *Eur. J. Biochem.*, 53, 179, 1989.
12. **Ludwig, R. A.**, *Arabidopis* chloroplasts dissimulate L-arginine and L-citrulline for use as N source, *Plant Physiol.*, 101, 429, 1993.
13. **Carter, A. L., Eller, A. G., Rufo, S., Metoki, K., and Hommes, F. A.**, Further evidence for a separate enzymic entity for the synthesis of homocitrulline, distinct from the regular ornithine transcarbamylase, *Enzyme*, 32, 26, 1984.
14. **Stalon, V.**, Putrescine carbamoyltransferase (*Streptococcus faecalis*), *Meth. Enzymol.*, 94, 339, 1983.
15. **Brewer, S. J., Taylor, P. M., and Turner, M. K.**, An adenosine triphosphate-dependent carbamoylphosphate-3-hydroxymethylcephem O-carbamoyltransferase from *Streptomyces clavuligerus*, *Biochem. J.*, 185, 555, 1980.
16. **Marcq, S., Diaz-Ruano, A., Charlier, P., Dideberg, O., Tricot, C., Piérard, A., and Stalon, V.**, Molecular size and symmetry of *Pseudomonas aeruginosa* catabolic ornithine carbamoyltransferase, *J. Mol. Biol.*, 220, 9, 1991.
17. **Wauven, C. V., Simon, J.-P., Slos, P., and Stalon, V.**, Control of enzyme synthesis in the oxalurate catabolic pathway of *Streptococcus faecalis* ATCC 11700: evidence for the existence of a third carbamate kinase, *Arch. Microbiol.*, 145, 386, 1986.
18. **Prasad, G. L. and Adiga, P. R.**, Purification and characterization of putrescine synthase from cucumber seedlings. A multifunctional enzyme involved in putrescine biosynthesis, *J. Biosci.*, 10, 373, 1986.
19. **Evans, D. R., Bein, K., Guy, H. I., Liu, X., Molina, L. A., and Zimmerman, B. H.**, CAD gene sequence and the domain structure of the mammalian multifunctional protein CAD, *Biochem. Soc. Trans.*, 21, 186, 1993.
20. **Carrey, E. A.**, Phosphorylation, allosteric effectors and inter-domain contacts in CAD; their role in regulation of early steps of pyrimidine biosynthesis, *Biochem. Soc. Trans.*, 21, 191, 1993.
21. **Anderson, P. M.**, Urea cycle in fish: molecular and mitochondrial studies, in *Fish Physiology*, Hoar, Randall, and Farrell, Series Eds., Vol. 14, *Ionoregulation: Cellular and Molecular Appproaches*, Wood, C. M. and Shuttleworth, T. J., Eds., Academic Press, New York, 1995, 57–83.
22. **Davis, R. H.**, Compartmental and regulatory mechanisms in the arginine pathways of *Neurospora crassa* and *Saccharomyces cerevisiae*, *Microbiol. Rev.*, 50, 280, 1989.
23. **Ratner, S.**, Enzymes of arginine and urea synthesis, *Adv. Enzymol.*, 39, 1, 1973.
24. **Marshall, M.**, Carbamyl phosphate synthetase I from frog liver, in *The Urea Cycle*, Grisolea, S., Báquena, R., and Mayor, F., Eds., John Wiley, New York, 1976, 133.
25. **Quinn, C. L., Stephenson, B. T., and Switzer, R. L.**, Functional organization and nucleotide sequence of the *Bacillus subtilus* pyrimidine biosynthetic operon, *J. Biol. Chem.*, 266, 9113, 1991.
26. **Zalkin, H.**, The amidotransferases, *Adv. Enzymol. Relat. Areas Mol. Biol.*, 66, 203, 1993.
27. **Trammel, P. R. and Campbell, J. W.**, Carbamyl phosphate synthesis in a land snail, *Strophocheilus oblongus*, *J. Biol. Chem.*, 245, 6634, 1970.
28. **Trammel, P. R. and Campbell, J. W.**, Carbamyl phosphate synthesis in invertebrates, *Comp. Biochem. Physiol.*, 40B, 395, 1971.
29. **Anderson, P. M.**, Glutamine- and N-acetylglutamate-dependent carbamoyl phosphate synthetase in elasmobranchs, *Science*, 208, 291, 1980.
30. **Mommsen, T. P. and Walsh, P. J.**, Evolution of urea synthesis in vertebrates: the piscine connection, *Science*, 243, 72, 1989.
31. **Nyunoya, H., Broglie, K. E., Widgren, E. E., and Lusty, C. J.**, Characterization and derivation of the gene coding for mitochondrial carbamyl phosphate synthetase I of rat, *J. Biol. Chem.*, 260, 9346, 1985.
32. **Haraguchi, Y., Uchino, T., Takiguchi, M., Endo, F., Mori, M., and Matsuda, I.**, Cloning and sequence of a cDNA encoding human carbamyl phosphate synthetase I: molecular analysis of hyperammonemia, *Gene*, 107, 335, 1991.
33. **Helbing, C. and Atkinson, B. G.**, 3,5,3′-Triiodothyronine-induced carbamyl-phosphate synthetase gene expression is stabilized in the liver of *Rana catesbeina* tadpoles during heat shock, *J. Biol. Chem.*, 269, 11743, 1994.
34. **Freund, J. N. and Jarry, B. P.**, The rudimentary gene of *Drosophila melanogaster* encodes four enzymic functions, *J. Mol. Biol.*, 193, 1, 1987.
35. **Simmer, J. P., Kelly, R. E., Rinker, A. G., Scully, J. L., and Evans, D. R.**, Mammalian carbamyl phosphate synthetases (CPS). cDNA sequences and evolution of the CPS domain of the syrian hamster multifunctional protein CAD, *J. Biol. Chem.*, 265, 10395, 1990.
36. **Bein, K., Simmer, J. P., and Evans, D. R.**, Molecular cloning of a cDNA encoding the amino end of the mammalian multifunctional protein CAD and analysis of the 5' flanking region of the CAD gene, *J. Biol. Chem.*, 266, 3791, 1991.
37. **Lusty, C. J., Widgren, E. E., Broglie, K. E., and Nyunoya, H.**, Yeast carbamyl phosphate synthetase. Structure of the yeast gene and homology to *Escherichia coli* carbamyl phosphate synthetase, *J. Biol. Chem.*, 258, 14466, 1983.
38. **Nyunoya, H. and Lusty, C. J.**, Sequence of the small subunit of yeast carbamyl phosphate synthetase and identification of its catalytic domain, *J. Biol. Chem.*, 259, 9790, 1984.
39. **Werner, M., Feller, A., and Piérard, A.**, Nucleotide sequence of yeast *CPA1* encoding the small subunit of arginine-pathway carbamoyl-phosphate synthetase. Homology of the deduced amino acid sequence to the other glutamine amidotransferases, *Eur. J. Biochem.*, 146, 371, 1985.

40. **Souciet, J. L., Nagy, M., LeGouar, M., Lacroute, F., and Potier, S.,** Organization of the yeast URA2 gene: identification of a defective dihydroorotase-like domain in the multifunctional carbamoyl phosphate synthetase-aspartate transcarbamylase complex, *Gene*, 79, 59, 1989.
41. **Faure, M., Carmonis, J. H., and Jacquet, M.,** Molecular characterization of a *Dictyostelium discoideum* gene encoding a multifunctional enzyme of the pyrimidine pathway, *Eur. J. Biochem.*, 179, 345, 1989.
42. **Nyunoya, H. and Lusty, C. J.,** The *carB* gene of *Escherichia coli*: a duplicated gene coding for the large subunit of carbamoyl-phosphate synthetase, *Proc. Natl. Acad. Sci. U.S.A.*, 80, 4629, 1983.
43. **Piette, J., Nyunoya, H., Lusty, C. J., Cunin, R., Weyens, G., Crabeel, M., Charlier, D., Glansdorff, N., and Piérard, A.,** DNA sequence of the *carA* gene and the control region of *carAB*: tandem promoters, respectively controlled by arginine and the pyrimidines, regulate the synthesis of carbamoyl-phosphate synthetase in *Escherichia coli* K-12, *Proc. Natl. Acad. Sci. U.S.A.*, 81, 4134, 1984.
44. **Hong, J., Salo, W. L., Lusty, C. A., and Anderson, P. M.,** Carbamyl phosphate synthetase III, an evolutionary intermediate in the transition between glutamine-dependent and ammonia-dependent carbamyl phosphate synthetases, *J. Mol. Biol.*, 243, 131, 1994.
45. **Rubino, S. D., Nyunoya, H., and Lusty, C. J.,** Catalytic domains of carbamyl phosphate synthetase. Glutamine-hydrolyzing site of *Escherichia coli* carbamyl phosphate synthetase, *J. Biol. Chem.*, 261, 11320, 1986.
46. **Chaparian, M. G. and Evans, D. R.,** The catalytic mechanism of the amidotransferase domain of the Syrian hamster multifunctional protein CAD. Evidence for a CAD-glutamyl covalent intermediate in the formation of carbamyl phosphate, *J. Biol. Chem.*, 266, 3387, 1991.
47. **Lusty, C. J.,** Detection of an enzyme bound γ-glutamyl acyl ester of carbamyl phosphate synthetase of *Escherichia coli*, *FEBS Lett.*, 314, 135, 1992.
48. **Guillou, F., Rubino, S. D., Markovitz, R. S., Kinney, D. M., and Lusty, C. J.,** *Escherichia coli* carbamoyl-phosphate synthetase: domains of glutaminase and synthetase subunit interaction, *J. Biol. Chem.*, 86, 8304, 1989.
49. **Carrey, E. A. and Hardie, D. G.,** Mapping of catalytic domains and phosphorylation sites in the multifunctional pyrimidine-biosynthetic protein CAD, *Eur. J. Biochem.*, 171, 583, 1988.
50. **Geschwill, K. and Lumper, L.,** Identification of cysteine residues in carbamoyl-phosphate synthase I with reactivity enhanced by N-acetyl-L-glutamate, *Biochem. J.*, 260, 573, 1989.
51. **Rodriguez-Aparacio, L. B., Guadalajara, A. M., and Rubio, V.,** Physical location of the site for N-acetyl-L-glutamate, the allosteric activator of carbamoyl phosphate synthetase, in the 20-kilodalton COOH-terminal domain, *Biochemistry*, 28, 3070, 1989.
52. **Rubio, V., Cervera, J., Lusty, C. J., Bendala, E., and Britton, H. G.,** Domain structure of the large subunit of *Escherichia coli* carbamoyl phosphate synthetase. Location of the binding site for the allosteric inhibitor UMP in the COOH-terminal domain, *Biochemistry*, 30, 1068, 1991.
53. **Cervera, J., Conejero-Lara, F., Ruiz-Sanz, J., Galisteo, M. L., Mateo, P. L., Lusty, C. J., and Rubio, V.,** The influence of effectors and subunit interactions on *Escherichia coli* carbamoyl-phosphate synthetase studied by differential scanning calorimetry, *J. Biol. Chem.*, 268, 12504, 1993.
54. **Miran, S. G., Chang, S. H., and Rauschel, F. M.,** Role of the conserved histidine residues in the amidotransferase domain of carbamoyl phosphate synthetase, *Biochemistry*, 30, 7901, 1991.
55. **Miles, B. W., Mareya, S. M., Post, L. E., Post, D. J., Chang, S. H., and Raushel, F. M.,** Differential roles for three conserved histidine residues within the large subunit of carbamoyl phosphate synthetase, *Biochemistry*, 32, 232, 1993.
56. **Post, L. E., Post, D. J., and Rauschel, F. M.,** Dissection of the functional domains of *Escherichia coli* carbamoyl phosphate synthetase by site-directed mutagenesis, *J. Biol. Chem.*, 265, 7742, 1990.
57. **Potter, M. D. and Powers-Lee, S. G.,** Location of the ATP γ-phosphate-binding sites on rat liver carbamoyl-phosphate synthetase I. Studies with the ATP analog 5′-*p*-fluorosulfonylbenzoyladenosine, *J. Biol. Chem.*, 267, 2023, 1992.
58. **Powers-Lee, S. G. and Corina, K.,** Photoaffinity labeling of rat liver carbamoyl phosphate synthetase I by 8-azido-ATP, *J. Biol. Chem.*, 262, 9052, 1987.
59. **Kim, H. S., Lee, L., and Evans, D. R.,** Identification of the ATP binding sites of the carbamyl phosphate synthetase domain of the Syrian hamster multifunctional protein CAD by affinity labeling with 5′-[*p*-(fluorosulfonyl) benzoyl]adenosine, *Biochemistry*, 30, 10322, 1991.
60. **Guillou, F., Liao, M., Garcia-Espana, A., and Lusty, C. J.,** Mutational analysis of carbamyl phosphate synthetase. Substitution of glu841 leads to loss of functional coupling between the two catalytic domains of the synthetase subunit, *Biochemistry*, 31, 1656, 1992.
61. **Lusty, C. J. and Liao, M.,** Substitution of glu841 by lysine in the carbamate domain of carbamyl phosphate synthetase alters the catalytic properties of the glutaminase subunit, *Biochemistry*, 32, 1278, 1993.
62. **Rubino, S. D., Nyunoya, H., and Lusty, C. J.,** *In vivo* synthesis of carbamyl phosphate from NH_3 by the large subunit of *Escherichia coli* carbamyl phosphate synthetase, *J. Biol Chem.*, 262, 4382, 1987.
63. **Anderson, P. M.,** Purification and properties of the glutamine- and N-acetyl-L-glutamate-dependent carbamoyl phosphate synthetase from liver of *Squalus acanthias*, *J. Biol. Chem.*, 256, 12228, 1981.
64. **Mullins, L. S., Lusty, C. J., and Raushel, F. M.,** Alterations in the energetics of the carbamoyl phosphate synthetase reaction by site-directed modification of the essential sulfhydryl group, *J. Biol. Chem.*, 266, 8236, 1991.
65. **Mareya, S. M. and Raushel, F. M.,** A molecular wedge for triggering the amidotransferase activity of carbamoyl phosphate synthetase, *Biochemistry*, 33, 2945, 1994.

66. **Guy, H. I. and Evans, D. R.,** Cloning, expression, and functional interactions of the amidotransferase domain of mammalian CAD carbamyl phosphate synthetase, *J. Biol. Chem.*, 269, 7702, 1994.
67. **Powers-Lee, S. G., Mastico, R. A., and Bendayan, M.,** The interaction of rat liver carbamoyl phosphate synthetase and ornithine transcarbamoylase with inner mitochondrial membranes, *J. Biol. Chem.*, 262, 15683, 1987.
68. **Cohen, N. S., Cheung, C.-W., and Raijman, L.,** Channeling of extramitochondrial ornithine to matrix ornithine transcarbamylase, *J. Biol. Chem.*, 262, 203, 1987.
69. **Cheung, C.-W., Cohen, N. S., and Raijman, L.,** Channeling of urea cycle intermediates *in situ* in permeabilized hepatocytes, *J. Biol. Chem.*, 264, 4038, 1989.
70. **Fahien, L. A., Kmiotek, E. H., Woldegiorgis, G., Evenson, M., Shrago, E., and Marshall, M.,** Regulation of aminotransferase-glutamate dehydrogenase interactions by carbamyl phosphate synthetase-I, Mg^{2+} plus leucine *versus* citrate and malate, *J. Biol. Chem.*, 260, 6069, 1985.
71. **Cohen, N. S., Kyan, F. S., Kyan, S. S., Cheung, C.-W., and Raijman, L.,** The apparent K_m of ammonia for carbamoyl phosphate synthetase (ammonia) *in situ*, *Biochem. J.*, 229, 205, 1985.
72. **Tremblay, G. C., Crandall, D. E., Knott, C. E., and Alfant, M.,** Orotic acid biosynthesis in rat liver: studies on the source of carbamoylphosphate, *Arch. Biochem. Biophys.*, 178, 264, 1977.
73. **Alonso, E. and Rubio, V.,** Orotic aciduria due to arginine deprivation: changes in the levels of carbamoyl phosphate and of other urea cycle intermediates in mouse liver, *J. Nutr.*, 119, 1188, 1989.
74. **Visek, W. J.,** Nitrogen-stimulated orotic acid synthesis and nucleotide imbalance, *Cancer Res.*, 52, 2082s, 1992.
75. **Rubio, V., Britton, H. G., and Grisolia, S.,** Mitochondrial carbamoyl phosphate synthetase activity in the absence of N-acetyl-L-glutamate. Mechanism of activation by this cofactor, *Eur. J. Biochem.*, 134, 337, 1983.
76. **Britton, H. G. and Rubio, V.,** Carbamoyl-phosphate synthetase I. Kinetics of binding and dissociation of acetylglutamate and activation and deactivation, *Eur. J. Biochem.*, 171, 615, 1988.
77. **Marshall, M. and Fahien, L. A.,** Proximate sulfhydryl groups in the acetylglutamate complex of rat carbamylphosphate synthetase I: their reaction with the affinity reagent 5′-*p*-fluorosulfonyl benzoyladenosine, *Arch. Biochem. Biophys.*, 241, 200, 1985.
78. **Anderson, P. M.,** Glutamine-dependent urea synthesis in elasmobranch fishes, *Biochem. Cell Biol.*, 69, 317, 1991.
79. **Anderson, P. M.,** Effects of urea, trimethylamine oxide, and osmolality on respiration and citrulline synthesis by isolated hepatic mitochondria from *Squalus acanthias*, *Comp. Biochem. Physiol.*, 85B, 783, 1986.
80. **Casey, C. A. and Anderson, P. M.,** Glutamine- and N-acetyl-L-glutamate-dependent carbamoyl phosphate synthetase from *Micropterus salmoides*. Purification, properties, and inhibition by glutamine analogs, *J. Biol. Chem.*, 258, 8723, 1983.
81. **Randall, D. J., Wood, C. M., Perry, S. F., Bergman, H., Maloiy, G. M. O., Mommsen, T. P., and Wright, P. A.,** Urea excretion as a strategy for survival in a fish living in a very alkaline environment, *Nature*, 337, 165, 1989.
82. **Wright, P., Felskie, A., and Anderson, P. M.,** Induction of ornithine-urea cycle enzymes and nitrogen metabolism and excretion in rainbow trout (*Oncorhynchus mykiss*) during early life stages, *J. Exp. Biol.*, 198, 127, 1995.
83. **Nyunoya, H., Broglie, K. E., and Lusty, C. J.,** The gene coding for carbamoyl-phosphate synthetase I was formed by fusion of an ancestral glutaminase gene and a synthetase gene, *Proc. Natl. Acad. Sci. U.S.A.*, 82, 2244, 1985.
84. **Chaparian, M. G. and Evans, D. R.,** Intracellular location of the multidomain protein CAD in mammalian cells, *FASEB J.*, 2, 2982, 1988.
85. **Kim, H., Kelly, R. E., and Evans, D. R.,** The structural organization of the hamster multifunctional protein CAD, *J. Biol. Chem.*, 267, 7177, 1992.
86. **Davidson, J. N., Chen, K. C., Jamison, R. S., Musmanno, L. A., and Kern, C. B.,** The evolutionary history of the first three enzymes in pyrimidine biosynthesis, *Bioessays*, 15, 1, 1993.
87. **Kern, C. B., Lusty, C. J., and Davidson, J. N.,** Evidence that mammalian glutamine-dependent carbamyl phosphate synthetase arose through gene fusion, *J. Mol. Evol.*, 35, 217, 1992.
88. **Kent, R. J., Lin, R.-L., Sallach, H. J., and Cohen, P. P.,** Reversible dissociation of a carbamoyl phosphate synthase-aspartate transcarbamoylase-dihydroorotase complex from ovarian eggs of *Rana catesbeiana*: Effect of uridine triphosphate and other modifiers, *Proc. Natl. Acad. Sci. U.S.A.*, 72, 1712, 1975.
89. **Anderson, P. M.,** Glutamine-dependent carbamoyl-phosphate synthetase and other enzyme activities related to the pyrimidine pathway in spleen of *Squalus acanthias* (spiny dogfish), *Biochem. J.*, 261, 523, 1989.
90. **Tampitag, S. and O'Sullivan, W. J.,** Enzymes of pyrimidine biosynthesis in *Crithidia luciliae*, *Mol. Biochem. Parasitol.*, 19, 125, 1986.
91. **Aoki, T. and Oya, H.,** Kinetic properties of carbamoyl-phosphate synthetase II (glutamine-hydrolyzing) in the parasitic protozoan *Crithidia fasciculata* and separation of the enzyme from aspartate carbamoyltransferase, *Comp. Biochem. Physiol.*, 87, 143, 1987.
92. **Mukherjee, T., Ray, M., and Bhaduri, A.,** Aspartate transcarbamylase from *Leishmania donovani*. A discrete, nonregulatory enzyme as a potential chemotherapeutic site, *J. Biol. Chem.*, 263, 708, 1988.
93. **Hong, J., Salo, W. L., and Anderson, P. M.,** Nucleotide sequence and tissue-specific expression of the multifunctional protein CAD mRNA in *Squalus acouthias*, submitted, 1995.
94. **Szondy, Z., Matyasi, G., and Elõdi, P.,** Effect of polyamines on the carbamoyl-phosphate synthase activity of CAD protein, *Acta Biochim. Biophys. Hung.*, 24, 107, 1989.

95. **Reardon, M. A. and Weber, G.**, Regulation by insulin of liver carbamoyl-phosphate synthase II (glutamine-hydrolyzing), *Biochem. J.*, 244, 345, 1987.
96. **Reardon, M. A. and Weber, G.**, Increased synthesis of carbamoyl-phosphate synthase II (EC 6.3.5.5.) in hepatoma 3924A, *Cancer Res.*, 46, 3673, 1986.
97. **Meek, T. D., Karsten, W. E., and DeBrosse, C. W.**, Carbamoyl-phosphate synthetase II of the mammalian CAD protein: kinetic mechanism and elucidation of reaction intermediates by positional isotope exchange, *Biochemistry*, 26, 2584, 1987.
98. **Souciet, J.-L., Potier, S., Hubert, J.-C., and Lacroute, F.**, Nucleotide sequence of the pyrimidine specific carbamoyl phosphate synthetase, a part of the yeast multifunctional protein encoded by the *URA2* gene, *Mol. Gen. Genet.*, 207, 314, 1987.
99. **Baur, H., Luethi, E., Stalon, V., Mercenier, A., and Haas, D.**, Sequence analysis and expression of the arginine-deiminase and carbamate-kinase genes of *Pseudomonas aeruginosa*, *Eur. J. Biochem.*, 179, 53, 1989.
100. **Hong, J. and Anderson, P. M.**, unpublished data, 1994.
101. **Hervé, G., Nagy, M., Gouar, M. L., Penverne, B., and Ladjimi, M.**, The carbamoyl phosphate synthetase-aspartate transcarbamoylase complex of *Saccharomyces cerevisiae*: molecular and cellular aspects, *Biochem. Soc. Trans.*, 21, 195, 1993.
102. **Schofield, J. P.**, Molecular studies on an ancient gene encoding for carbamoyl-phosphate synthetase, *Clin. Sci.*, 84, 119, 1993.
103. **Takiguchi, M., Matsubasa, T., Amaya, Y., and Mori, M.**, Evolutionary aspects of urea cycle enzyme genes, *Bioessays*, 10, 163, 1989.
104. **Farnham, P. J. and Kollmar, R.**, Characterization of the 5' end of the growth-regulated Syrian hamster CAD gene, *Cell Growth Differ.*, 1, 179, 1990.
105. **Goping, J. S., Lagace, M., and Shore, G. C.**, Factors interacting with the rat carbamyl phosphate synthetase promoter in expressing and nonexpressing tissues, *Gene*, 118, 283, 1992.

Chapter 3

Nitrogen and Buoyancy in Marine Organisms

Nancy Sanders and James Childress

CONTENTS

INTRODUCTION

More than 70% of the earth's surface is oceanic, with only a small fraction of this representing coastal habitat. The pelagic, open ocean realm, therefore, constitutes the single largest habitat on earth. The vast majority of the open ocean lies below depths of 1000 m, where high pressure (101.3 kPa for each 10 m depth), low temperatures (below 10°C throughout, except for the hydrothermal vent habitats), low oxygen partial pressures (approaching zero in areas of oxygen minimum layers), and low levels of light characterize the living space. It is in this setting where one can find a diversity of bizarre morphologies among the living organisms, often related to the animal's buoyancy strategy.

Many epipelagic and deeper living pelagic organisms have some type of strategy to reduce their density relative to seawater. The closer to neutral their buoyancy, the less energy expenditure will be needed to maintain their vertical position in the water column. Neutral density may also be an important predator–prey strategy because animals would be less detectable when they are not swimming. Numerous strategies have been evolved to achieve neutral buoyancy, and among these is the replacement of "heavier" ions with "lighter" ions, including nitrogenous ions. A number of reports on the use of nitrogenous ions as a mechanism for reducing density describe this strategy in pelagic, deep-sea organisms (e.g., cephalopods, chaetognaths and oplophorid crustaceans); in these three groups there is strong evidence for convergent evolution. After introductory remarks describing some of the buoyancy mechanisms known for pelagic organisms, this chapter will focus on fluid density reduction strategies that replace cations such as sodium with nitrogenous ions such as ammonium.

0-8493-8411-7/95/$0.00+$.50

BUOYANCY MECHANISMS OTHER THAN ION REPLACEMENT

Metabolism can be separated into basal or resting metabolism (maintenance condition), and active metabolism. In pelagic organisms that are not neutrally buoyant, constant activity (routine metabolism) is required simply to maintain the organism's position in the water column. The tissues of these organisms are more dense (1.03 to 1.11 specific gravity) than the seawater in the open ocean (1.02 to 1.028 specific gravity).[1] *Any* reduction in density, *via* a diversity of buoyancy mechanisms, reduces the energy output associated with this "swim or sink" phenomenon. Consequently, many different buoyancy strategies are found among marine organisms, and these are not restricted by taxonomic boundaries. A reduction in "heavy" components of the body, be it skeletal elements (e.g., carapace or bone), protein content, or "heavy" ions (e.g., SO_4^{2-}, Mg^{2+}, Ca^{2+}), will contribute to a reduction in density, as will any increase in "light" elements such as lipid or ions (e.g., NH^{4+}).

Buoyancy mechanisms in marine organisms are diverse and range from simple to elaborate. Organisms that are relatively small, such as foraminiferans and radiolarians, can reduce their sinking rate with projections of the skeleton that simply increase body surface area, and therefore drag.[1] Other representatives of these taxonomic groups use air bubbles or oil droplets to reduce density. Some species of copepods are able to hang motionless in the water column, partly due to the extension of their long antennae with plumose setae.[1] Some of the more elaborate mechanisms are found in animals living in the benthopelagic reaches of the deep sea, where neutral buoyancy is critical to facilitate the ability to swim and hover over the sea floor with minimal disturbance to the sediments below.[2]

WATER CONTENT

A high water content indicates a reduction of other materials and, all else being equal, the higher an animal's water content, the more closely it will approach neutral buoyancy. Animals with high water contents often need only utilize ion replacement or exclusion to achieve neutral buoyancy because they are already lightened markedly by the high water content. For example, the pelagic sea cucumber *Scotoanassa* sp. (common in the San Clemente Basin off Southern California,[4,5] and sighted from submersibles near the Galápagos hydrothermal vent sites; Childress and Sanders, personal observations) is more than 96% water.[4] Some crustaceans also have very high water content, including the amphipod *Phronima sedentaria* (92.6%) and the ostracod *Gigantocypris agassizii* (95.6%). One decapod species, the oplophorid *Notostomus* (Figure 1), also had a very high water content (91.3%).[7] Many gelatinous organisms have very high water contents (90% wet weight) in their tissues.[1,3-6] Gelatinous zooplankton (e.g., medusa, ctenophores, tunicates, pteropods, and heteropods) achieve near neutral buoyancy by maintaining relatively high water contents and low sulfate ion concentrations in their tissues.[3] The body fluids of these animals are isotonic to seawater, but have specific gravities less than that of seawater, mostly because of the exclusion of the sulfate ion.[3]

Figure 1 The oplophorid shrimp *Notostomus gibbosus*, showing the dorsally expanded carapace housing a low density fluid. (From Sanders, N. K. and Childress, J. J., *J. Exp. Biol.*, 138, 333, 1988. With permission.)

LIPID/PROTEIN CONTENT

Lipid contributes to buoyancy by virtue of its relatively low specific gravity (0.86). Protein, on the other hand, has a specific gravity (1.33) higher than that of seawater (1.26) and therefore it increases density. Some crustaceans, and some fish species, reduce buoyancy by increasing the lipid content, relative to the protein and skeletal components.[7] In one negatively buoyant species, the oplophorid *Acanthephyra curtirostris*, the lipid and protein contents are 7.5 and 35% of the wet weight, respectively. By comparison, two other oplophorid species are neutrally buoyant, with much higher lipid relative to protein contents: 19 and 30% (*Hymenodora frontalis*) and 12 and 35% (*Systellaspis cristata*) of the wet weight.

The relative buoyancies of nearly 30 species of crustaceans were also measured.[7] Relative buoyancy is estimated by weighing living specimens in seawater and in air, and is expressed in terms of the weight in seawater (mg) per weight in air (g). Relative buoyancy represents the weight that 1 g wet weight of animal has to support in order to maintain its position in the water column. A negative value indicates that the animal floats. The oplophorid *Notostomus gibbosus*, a species that has a low lipid content but a high water content, and uses nitrogenous ions to reduce density, has a negative relative buoyancy at sea level.[7,8] Two other species, the oplophorid *Hymenodora frontalis* (with high lipid levels) and the amphipod *Paracallisoma coecus* also have slightly negative values for relative buoyancy. *Systellaspis cristata*, and the ostracod *Gigantocypris agassizii* are nearly neutrally buoyant. All the other crustacean species measured had much higher positive values for relative buoyancy, indicating that they must swim or sink (Figure 2).[7]

Lipid content is also responsible for buoyancy in other taxa. For example, one family of oceanic squids, the Gonatidae, store large volumes of low density fats.[9] Larval fish may achieve some buoyancy from the presence of yolk due to its high lipid content.[10-13] Fish eggs vary in density as a result of the water content percentage and the presence or absence of oil globules,[14] or the chorion thickness.[15,16] Lipids may be important in density reduction in some midwater fish species.[17] The large, oily livers of five deep-sea sharks and one deep sea holocephalan were found to contribute to the near neutral buoyancy for these fish at depth.[2]

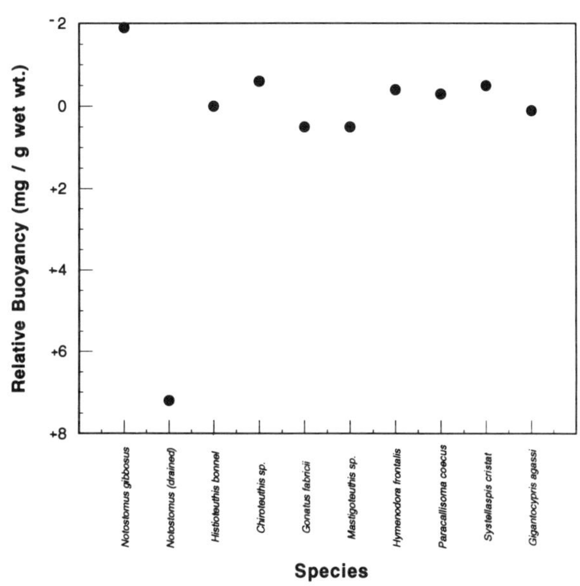

Figure 2 The relative buoyancies (weight under water in milligrams per wet weight in air in grams) of *Notostomus gibbosus*,[8] four species of deep sea squids,[9] and four species of midwater crustaceans.[7] The data points for *Notostomus gibbosus* and "*Notostomus* drained" are the mean of three individuals intact, and the same three after draining the low specific gravity carapace fluid.[8]

GAS SPACES

Many pelagic organisms utilize gas-filled spaces to achieve buoyancy; even a relatively small space/volume can provide the organism with tremendous lift. One major advantage conferred to animals using gas-filled spaces is that there does not need to be a reduction in the denser muscular (i.e., as would occur by increasing water content at the expense of protein) or skeletal system, so that a fish or a cephalopod with a well-developed gas-filled buoyancy chamber can still be an active swimmer. Two principal methods are used by animals to maintain gas-filled spaces. In fishes with swim bladders, (e.g., a number of pelagic and benthopelagic fishes), the gas-filled space does not have rigid walls and the gas pressure in the swim bladder is equal to the external hydrostatic pressure.[1] Elaborate physiological and morphological adaptations allow fish with swim bladders to concentrate specific gases high above external partial pressures.[49]

Pelagic cephalopods with reduced or absent shells use ion replacement strategies and gas spaces for buoyancy. Three extant shelled cephalopods use air spaces with rigid walls to achieve neutral buoyancy.[18] *Nautilus* has a robust, coiled shell in which the animal resides. *Spirula* has a fragile, coiled shell that is completely enclosed within the animal. The cuttlefish, *Sepia*, has an uncoiled shell enclosed within the animal. In all three, the shells are chambered and have a siphuncle that functions in the exchange of substances between the animal's tissues and the chambers of the shells. Buoyancy can be regulated by controlling the amount of liquid (*via* pumping mechanisms), the ionic composition of the liquid, and the gas present in the chambers. These types of buoyancy mechanism restrict these cephalopods to shallower waters, because at deeper depths their shells implode from the increased hydrostatic pressure.

Other organisms and accumulations of organisms also take advantage of air spaces for flotation. Large marine snow macroflocs in the North Sea were retained near the surface due to gas bubble formation.[19] Gas bubble formation correlated with periods of oxygen supersaturation in the surface waters; the macroflocs were dispersed by drastic increases in wind speed. Siphonophores (e.g., the Portuguese Man o' War) depend on buoyant members of the colony to provide buoyancy for the entire colony. In cyanobacteria, dilution of gas vesicles by growth, and ballasting with carbohydrates provide buoyancy in some species. Buoyancy regulation may contribute to diel vertical migration in some species.[20] Various strategies utilizing air spaces are also present in marine macroalgae.[21]

ION REPLACEMENT AS A BUOYANCY MECHANISM

A common trait pelagic organisms share is that their nonfluid components are more dense than seawater. Seawater has a specific gravity in the range of 1.020 to 1.028 g ml^{-1}, whereas protein has a specific gravity of about 1.33 g ml^{-1}. Skeletal materials vary in their specific gravity, and can be as high as 3 g ml^{-1}, but they are invariably more dense than seawater, and thus contribute to negative buoyancy.[1] Changes in the fluid content, or the ion concentrations of various body fluids, can bring about large changes in the overall buoyancy of the organism. Organisms that use ion replacement strategies generally have fluids which are isosmotic with seawater, but differ in the concentrations of specific ions. The two most significant ions for bringing about changes in fluid density are sulfate and ammonium. A reduction in the amount of sulfate, and/or a replacement of sodium by ammonium, results in a reduction in fluid density. Thus, an organism with a relatively high water content that also uses ion replacement strategies can markedly affect its buoyancy.

PARTIAL MOLAL VOLUME

The effect of various solutes on buoyancy depends on two factors: (1) the molecular weight of the solute which affects solution density as expected (e.g., heavier molecular weights make for more heavier solutions) and (2) the partial molal volume. Partial molal volume can be understood by visualizing a container of pure water so large that the addition of one mole of solute will not

Table 1 Partial Molal Volume and Molecular Weight Contributions to Density in Seawater and the Carapace Fluid of *Notostomus Gibbosus*

Ion	MW g mol⁻¹	V° cm³ mol⁻¹	Seawater mmol l⁻¹	Seawater g l⁻¹	Seawater ml l⁻¹	Carapace fluid mmol l⁻¹	Carapace fluid g l⁻¹	Carapace fluid ml l⁻¹
Na^+	23.0	−2.90	470.20	10.800	−1.370	62.1	1.428	−0.180
NH_4^+	18.0	+17.49	0.00	0.000	0.000	296.0	5.328	+5.176
K^+	39.1	+7.59	9.96	0.389	+0.076	10.8	0.422	+0.082
Me_3NH^+	60.0	+71.51	0.00	0.000	0.000	127.6	7.656	+9.284
Mg^{2+}	24.3	−21.66	53.57	1.302	−1.160	3.5	0.085	−0.076
Ca^{2+}	40.1	−19.26	10.23	0.410	−0.197	0.0	0.000	0.000
Cl^-	35.3	+16.65	548.30	19.465	+9.124	511.7	18.165	+8.520
SO_4^{2-}	96.1	+11.70	28.25	2.715	+0.331	1.2	0.115	+0.014
Totals				35.081	+6.804		33.199	+22.659

Note: Density was calculated as g ml^{-1} for sea water (1035.081 g l^{-1} / 1006.804 ml l^{-1} = 1.028 g ml^{-1}) and carapace fluid (1033.199 g l^{-1} / 1022.659 ml l^{-1} = 1.0103 g ml^{-1}), resulting in 17.7 mg ml^{-1} lift for carapace fluid in sea water. V°, partial molal volume; MW, molecular weight.

From Sanders, N. K. and Childress, J. J., *J. Exp. Biol.*, 138, 333, 1988. With permission.

significantly alter the concentration.[22] The change in the volume of water after adding that one mole of solute to the container of water is the partial molal volume (cm^3 mol^{-1}) of that substance (at a constant temperature, pressure, and moles of other solutes). A solute with a partial molal volume near zero will not affect the volume of the solution. A solute with a negative partial molal volume has a greater affinity for water molecules, resulting in a greater density of water molecules per volume of solution, and will therefore reduce the volume of water in the container. Conversely, a solute with a positive partial molal volume disrupts the structure of the water molecules in the solution and increases fluid volume. Therefore, a solution containing solutes with positive partial molal volumes, such as NH_4Cl (36.0 cm^3 mol^{-1}) and trimethylamine chloride (Me_3NHCl, 90 cm^3 mol^{-1}) will have fewer water molecules per unit volume than one containing relatively heavy solutes with small positive or negative partial molal volumes such as sodium (−3.51 cm^3 mol^{-1}), Na_2SO_4 (3.7 cm^3 mol^{-1}), $CaCl_2$ (13.05 cm^3 mol^{-1}), and $MgCl_2$ (12.7 cm^3 mol^{-1}), resulting in a less dense solution. Thus, if one measures the concentrations of the major ions in the body fluids of an organism, the density of that fluid can be calculated by taking into account both molecular weight and partial molal volume effects (Table 1). Although sulfate has a slightly positive partial molal volume, this is counteracted by its heavier molecular weight, making it an ion worth excluding to achieve neutral buoyancy. Thus, if an organism is able to reduce the concentration of sulfate, calcium, and magnesium in its body fluids relative to seawater concentrations, and/or replace sodium with ammonium, then substantial lift (mg ml^{-1}) can be gained.

NONNITROGENOUS IONS

Regulation of cellular density has long been a known or suspected mechanism of buoyancy in vacuolate marine plankton, particularly diatoms and dinoflagellates.[23] The large, mat forming diatom *Rhizosolenia*[24] appears to achieve positive buoyancy by ionically regulating its cellular density; a similar mechanism is proposed for the positively buoyant diatom *Ethmodiscus*.[25] Ionic control is also known for the diatom *Ditylum brightwellii* (West).[26] The nonmotile, positively buoyant marine dinoflagellate *Pyrocystis noctiluca* Murray ex Haeckel undergoes ontogenetic vertical migration that depends on ionic regulation of buoyancy.[27,28] The surf diatoms *Anaulus birostratus* achieve positive buoyancy during the day, enabling them to occupy the surface layers and foam found at the air/water interface.[29] In early morning, they adhere to air bubbles, but by late afternoon, they lose their attachment to air bubbles, sink and attach to sand particles that enable them to enter the sediment and be retained near shore during the night.

NITROGENOUS BUOYANCY MECHANISMS

Protists

One of the earliest published accounts of the use of the ammonium ion by a buoyant organism is the 1892 account by Goethard and Heinsius, cited by Krogh,[30] that reported on the osmotic properties of the dinoflagellate *Noctiluca miliaris*. This protist contains ammonia in its cell sap, and swells when placed in a solution hyposmotic to seawater. These early authors were able to deduce that the ionic composition of the cell sap was different than the surrounding seawater, and that the organism floated as a result of the presence of the ammonium ions. Gross and Zeuthen,[23] however, did not find appreciable ammonia in *Noctiluca*, and concluded that it achieved buoyancy by ion exclusion, primarily of sulfate. Denton[18] cites personal communication with Dr. K.J. Singarajah in confirming the presence of ammonium ions in *Noctiluca* in concentrations high enough to give the organism lift.

Positive buoyancy in the marine dinoflagellate *Pyrocystis noctiluca* is also achieved by ionic control, including the use of ammonium ions.[28] This species is an unusually large dinoflagellate with a large vacuole comprising up to about 95% of the cell volume. The cells achieve positive buoyancy in nature, and the highest population densities are found at the base of the euphotic zone at depths of 60 to 80 m. Although ammonium ions contribute to buoyancy in cells grown in nutrient rich media, the greatest source of buoyancy was from the exclusion of sulfate ions, followed by the reduction in concentrations of calcium and magnesium ions in cell sap, relative to seawater. The amount of ammonium in the vacuole increased when the cells had adequate supplies of nitrogen.[28]

Pelagic Sea Cucumber

The pelagic sea cucumber *Scotoanassa* sp. also appears to use ion replacement together with high water content to become more buoyant.[4] Measurements of ion concentrations in body fluids of 13 specimens showed elevated levels of ammonium ions (8.5 ± 2.3 mmol l^{-1}) and a slight reduction of sulfate ions (23.4 ± 9.4 mmol l^{-1}) relative to seawater, providing these animals with 1 mg ml^{-1} lift (N. Sanders, unpublished data).

Tunicate Eggs

Tunicate eggs are known to contain ammonium ions for flotation.[31] The ascidian *Corella willmeriana* produces floating eggs with large follicle cells that have vacuoles containing ammonium ions in their outer cellular layer that surrounds the egg. Egg density was calculated to be 1.018 g cm^3, based on the rate of egg rise in seawater with a density of 1.028 g cm^3. Eggs deprived of their flotation cells had a density of 1.080 g cm^3. The ammonia concentration within the follicle cells was 346 mmol l^{-1}. The source of the ammonium ions is unknown in *Corella*, but this species appears to secrete about half the ammonia of similar ascidians, while at the same time they also sequester uric acid. Since the concentration of ammonium ions in seawater is negligible, metabolic energy must be expended to maintain the high concentration of ammonium ions in follicle cells against a concentration gradient. The authors used metabolic inhibitors to conclude that glycolysis is necessary to provide the metabolic energy necessary to maintain the high internal concentration of ammonium ions in the follicle cells.

Cephalopods

Cephalopods with shells are restricted to shallower waters, while those that use ammonium ions live at much greater depths. Pelagic teuthids sort into two groups according to their density in seawater. One group is negatively buoyant and actively swims to avoid sinking. Species in the second group approach neutral buoyancy and, in most of these, ammonium ions are present sequestered in specific regions of the body. In certain body tissues and coelomic fluids the ammonium ion concentration has been measured at nearly 500 mmol l^{-1}.[9,34-36]

Squids in several families, including the Cranchiidae,[34,35] Histioteuthida, Chiroteuthidae, and Octopoteuthidae,[36] and Mastigoteuthidae and Enoploteuthidae[9] accumulate high concentrations of

Figure 3 A cranchiid squid, *Heliocranchia pfefferi* Massy (1907), demonstrating the coelomic chamber containing a low density fluid with a high ammonium ion concentration. (Photo courtesy of B. R. Robison.)

ammonium ions in acidic fluids that are isotonic with seawater. The ammonium ions are sequestered in specific chambers or regions of the body, and the ammonium ion concentration measured in the blood of some of these animals was found to be only a few mmol l^{-1}. Work on cranchiids and *Histioteuthis* sp. has shown that the larger nerves in these squids will not conduct impulses if exposed to high concentrations of ammonium ions found in some of their tissues,[9] suggesting that the animals effectively keep high levels of ammonia out of these sensitive tissues.

In the Cranchiidae, acidic fluid (pH 6 to 6.5) high in ammonium ions (470 mmol l^{-1}) is found in the expanding coelomic chamber (Figure 3). This fluid represents about two-thirds of the wet weight of these squids, with a specific gravity of 1.010 g ml^{-1}. The lift gained by the use of ammonium ions is about 16 mg buoyancy per ml of coelomic fluid. Ammonia accumulates as an end product of protein metabolism, and in Cranchiid squids it is accumulated and stored in the coelomic fluid. The acidity of the fluid may account for the retention of ammonium ions. Unionized ammonia in blood (pH 7 to 7.9) could pass through the cell membranes and once inside the low pH (5 to 6) coelomic chamber could be converted to ammonium ions and be effectively trapped due to the lower permeability of the ionized form. Using a gross growth efficiency value of 33% (a value found by Corner et al.[37] for *Calanus*), and assuming ammonia is the only end product of protein metabolism, Denton[18] predicted that a squid would have to retain 40% of its lifetime production of ammonia in order to maintain neutral buoyancy.

In the remaining buoyant families, various body parts contain ammonium ions in fluids. In some species, the removal of buoyant body parts, or fluids, results in a loss of buoyancy for the animal. For example, the two large ventral arms of *Mastigoteuthis* sp. were removed and were buoyant. After expressing about 1 ml of liquid from the arms, the arms sank. The density of the liquid was estimated to be 1.011 g ml^{-1}, comparable to that found in the coelomic chamber of cranchiid squids. In *Octopoteuthis danae* Joubin, about 75% of the lift is derived from the mantle, and the remaining 25% from the arms. The reverse is true for *Histioteuthis meleagroteuthis* (Chun). *Architeuthis* species are routinely found floating dead or dying at the surface. Analysis of fluids exuded from samples showed high levels of ammonium ions.

The ammonium ion concentration in fluids from the most buoyant parts of squids was always high, ranging from 276 to over 500 mmol l^{-1}. In contrast, the concentration of ammonia in the blood of these animals was always low, generally just a few mmol l^{-1}.[9] Sulfate concentrations were also greatly reduced, relative to seawater, in these squids. In all, 12 families of pelagic squids are known or predicted to use ammonium ions for buoyancy.[9]

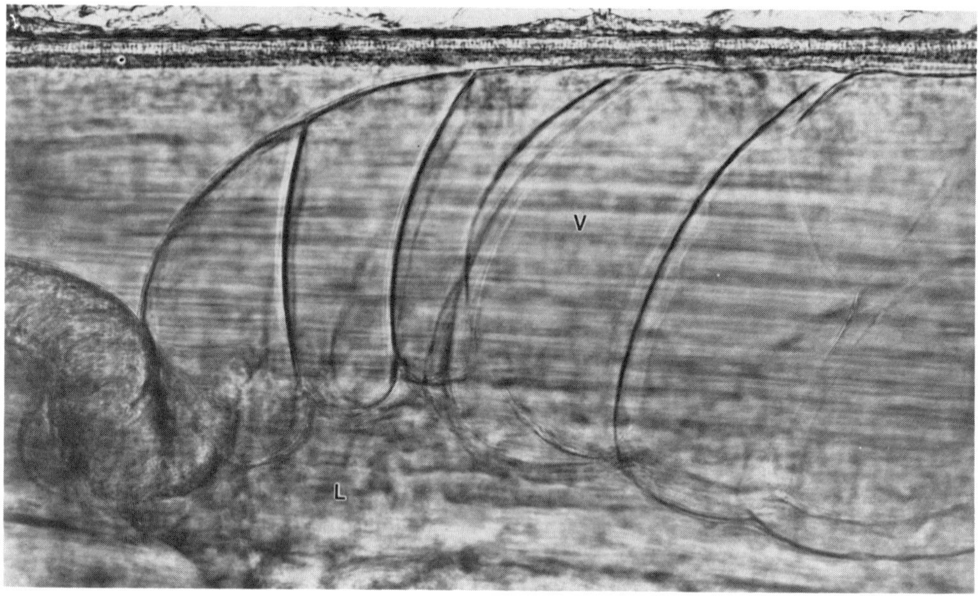

Figure 4 Vacuolated gut sacs in the chaetognath *Sagitta elegans*. V, vacuolated gut sac; L, lumen of gut. (Photo courtesy of G. Shinn.)

Chaetognaths

Chaetognaths are voracious predators, active swimmers, and possess more muscle and higher metabolic rates than many pelagic organisms.[32] Their flattened morphology, together with horizontally placed lateral fins, may provide some lift. One species, *Sagitta elegans*, has gut cells with vacuoles that contain NH_4^+ (Figure 4). This species is significantly less dense, and is less active, than a co-occurring species, *S. setosa*.[33] The gut cells in *S. elegans* are so large that they almost fill the coelomic cavity. In the living, transparent animal the conspicuous "septa" that divide up the space between the gut and the inner border of the muscle layer are actually the walls of these cells containing huge vacuoles. These "septa," and ammonium ions in cell vacuoles, are lacking in a closely related chaetognath species, *S. setosa*. The larger gut cells in *S. elegans* have vacuoles with acidic fluid (pH 6 to 6.5) and ammonium ion concentrations approaching 500 mmol l^{-1}. It is not known how the ammonium ions are secreted into the gut cells or how they are retained there. Unlike cranchiid squid, where the hemolymph pH is between 7 and 7.9, and the ammonium ion containing coelomic chamber has a pH of as low as 5, both the gut cells and gut sacs of *S. elegans* have a pH of 6 to 6.5. Therefore, the mechanism proposed for ammonium ion accumulation via an acidic trap in the coelom would not be as effective in *S. elegans*. Other species in the genus *Sagitta* are known to possess sacculated gut cells resembling those of *S. elegans*, but the possible buoyancy function of these sacs has not been investigated.

Oplophorid Shrimp

The caridean family Oplophoridae is exclusively pelagic. *Notostomus* is a common genus in the midwater worldwide, especially in the tropics. The oplophorid shrimp *Notostomus gibbosus* A. Milne-Edwards is nearly neutrally buoyant (Figure 1). It is found in depths from 700 to 1200 m and for our studies was collected off the Hawaiian island of Oahu. This shrimp has a strong odor of ammonia and trimethylamine when the carapace is disrupted, or when hemolymph is sampled. *N. gibbosus* possesses a dorsally expanded carapace filled with a low density fluid that comprises 43% of the animal's wet mass. This low pH fluid (6.6) contains high levels of two nitrogenous ions, ammonium (296 ± 51 mmol l^{-1}, $n = 9$) and trimethylamine (128 ± 20 mmol l^{-1}, $n = 9$), and provides the animal with 17.7 mg ml^{-1} lift. The relative buoyancies of live, intact *N. gibbosus* were –1 to –2 mg ml^{-1} wet wt, but were between 6 and 8 mg ml^{-1} in animals drained of their carapace fluid.[8] High levels of ammonium and trimethylamine ions were also found in oplophorid shrimps in the genus *Meningodora.*[8]

Ions in body fluids of *N. gibbosus* and other oplophorid species were measured using single column ion chromatography.[38] Ammonia electrodes and spectrophotometric techniques for determining ammonium ion concentrations do not differentiate among amines, so the initial determinations of "ammonia" in *N. gibbosus* (and perhaps earlier data for other animals, such as the cranchiid squids) did not account for the presence of the relatively large, heavy ion trimethylamine.[39] This ion, however, has a large, positive partial molal volume that contributes to the overall decrease in carapace fluid density relative to seawater (Table 1). A secondary problem in using ammonia electrodes is that trimethylamine appears to destroy the effectiveness of the electrode, making subsequent measurements of ammonia suspect.

Due to the presence of trimethylamine in the carapace fluid of *N. gibbosus*, the total solute mass calculated for this fluid (33.2‰) was little changed relative to seawater (34.1‰). The reduction in fluid density was primarily the result of the large, positive partial molal volumes of ammonium and trimethylamine ions, rather than a reduction in total mass of solutes.

High levels of ammonia (217 mmol l^{-1}, $n = 9$) and trimethylamine (113 mmol l^{-1}, $n = 9$) are also found in the hemolymph of *N. gibbosus*. The concentration of the respiratory protein hemocyanin was quite low, 8.2 mg ml^{-1}, and the hemolymph pH was relatively low (pH = 7.5 at 5°C). Both ammonia and trimethylamine directly affected the oxygen binding properties of the hemocyanin *in vitro*.[40] The presence of trimethylamine reduced oxygen affinity, decreased cooperativity, and suppressed the Bohr shift. This effect was strongly CO_2 dependent. The presence of physiological levels of ammonia, with or without CO_2, resulted in a decrease in oxygen affinity, no change in cooperativity, and the absence of a CO_2 or pH Bohr shift. The presence of ammonia masked any effect of trimethylamine or L-lactate, and suppressed any pH or CO_2 related shift in oxygen affinity. CO_2 alone caused a *decrease* in hemocyanin oxygen affinity, but this effect was masked in the presence of ammonia. This CO_2 effect on oxygen binding by hemocyanin is the reverse of what would be expected in most other crustaceans.[41,42] Hemolymph pH was unchanged by the presence of CO_2 — a stable, low pH was maintained by the high concentrations of ammonia and trimethylamine.

For *N. gibbosus*, one benefit of neutral buoyancy that results from the presence of nitrogenous compounds is a decreased oxygen consumption rate, reducing the reliance on an oxygen concentrating respiratory protein. Despite this, the presence of a functional hemocyanin in *N. gibbosus* does confer the ability to obtain and concentrate more oxygen from the low oxygen deep sea environment than would be possible without an oxygen binding protein. It is not known how *N. gibbosus* is able to tolerate hemolymph ammonia concentrations at least two orders of magnitude greater than what is toxic to most animals.

Excretion rates were measured for five *N. gibbosus* and for five specimens of the oplophorid *Acanthephyra curtirostris* Wood-Mason, a species with very low levels of ammonia[8] but sharing a common depth range with *N. gibbosus* (Figure 5). Ammonia production was measured in closed containers using ammonia electrodes for five consecutive d from capture. Recently captured specimens all showed signs of feeding in the capture nets, and would, therefore, be comparable to well-fed animals in diet-controlled experiments. The initial ammonia excretion rates were 5 to 6 (μmol/g wet wt/h), but by the sixth day without feeding the excretion rates had dropped to 0.1 to 0.5 for both *N. gibbosus* and *A. curtirostris* (Figure 5). These are comparable to excretion rates of the deep sea mysid *Gnathophausia ingens* Dohrn under starvation conditions.[43-45] It is not known how *N. gibbosus* obtains or retains the high concentrations of nitrogenous ions found in body fluids, but these excretion data indicate that it does not store all of its nitrogenous wastes.

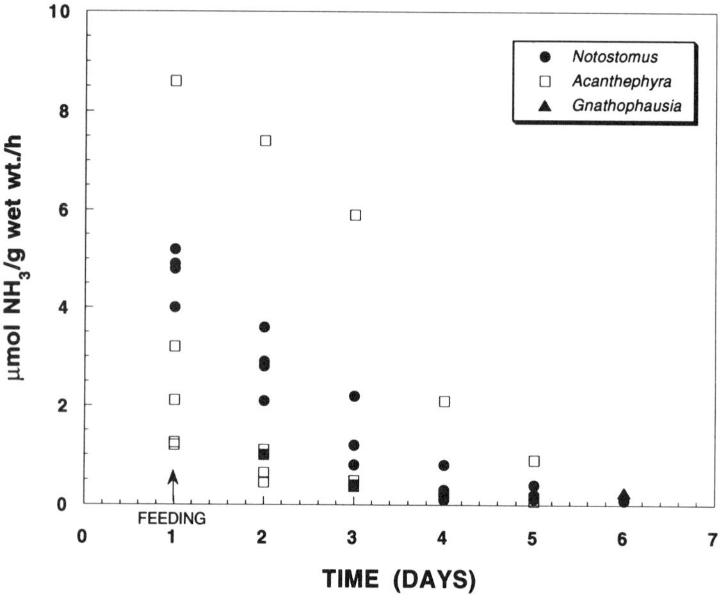

Figure 5 Ammonia excretion rates (in starved animals) for the midwater oplophorids *Notostomus gibbosus* ($n = 5$) and *Acanthephyra curtirostris* ($n = 5$), and the midwater mysid *Gnathophausia ingens*.[43]

Elasmobranchs

It has long been known that elasmobranchs store urea and trimethylamine oxide (TMAO) in high concentrations as an osmoregulatory mechanism.[46] Withers et al.[47,48] suggest that an additional function for urea and TMAO retention is buoyancy regulation. Plasma densities were calculated for three elasmobranch species, a whaler shark (*Carcharhinus obscurus*), a whiskey shark (*Furgaleus ventralis*), and two shovelnosed rays (*Rhinobatos banksii*). The specific gravity of plasma in *Carcharhinus* and *Rhinobatus* was 1.015, and 1.014 for *Furgaleus*, providing about 10 mg ml^{-1} lift from plasma. Urea contributes almost half of the lift, with a significant fraction due to TMAO. Although some lift is obtained in elasmobranchs from the presence of oily livers, in the three species studied the livers were small and body fluid buoyancy was estimated to be more significant.

CONCLUSION

Many fully pelagic organisms have some mechanism or combination of mechanisms to reduce their density, thereby reducing the energy expenditure necessary to remain in the same vertical position in the water column. The use of nitrogenous substances (i.e., ammonia and trimethylamine) to reduce fluid density is found in a diversity of organisms. Due to the relatively large, positive partial molal volumes of these substances, small changes in ion composition can result in substantial reductions in fluid density, providing the organism with the lift needed to overcome higher density body components (i.e., skeletal elements and proteins). Despite potential toxicity problems, the use of nitrogenous ions to reduce fluid density, and thereby approach neutral buoyancy, is an important buoyancy mechanism that has evolved independently in many groups.

REFERENCES

1. **Marshall, N. B.,** Against gravity, in *Developments in Deep Sea Biology*, Marshall, N. B., Ed., Blandford Press, Poole, Dorsett, 1979, chap. 9. .
2. **Corner, E. D. S., Denton, E. J., and Forster, G. B.,** On the buoyancy of some deep-sea sharks, *Proc. R. Soc. London B*, 171, 415, 1969.
3. **Denton, E. J. and Shaw, T. I.,** The buoyancy of gelatinous marine animals, *J. Physiol.*, 161, 14, 1961.

4. **Barnes, A. T., Quetin, L. B., Childress, J. J., and Pawson, D. L.,** Deep-sea macroplanktonic sea cucumbers: suspended sediment feeders captured from deep submergence vehicle, *Science*, 194, 1083, 1976.
5. **Childress, J. J., Gluck, D. L., Carney, R. S., and Gowing, M. M.,** Bethopelagic biomass distribution and oxygen consumption in a deep-sea benthic boundary layer dominated by gelatinous organisms, *Limnol. Oceanogr.*, 34, 913, 1989.
6. **Clarke, A. and Peck, L. S.,** The physiology of polar marine zooplankton, *Polar Res.*, 10, 355, 1991.
7. **Childress, J. J. and Nygaard, M.,** Chemical composition and buoyancy of midwater crustaceans as function of depth of occurrence off Southern California, *Mar. Biol.*, 27, 225, 1974.
8. **Sanders, N. K. and Childress, J. J.,** Ion replacement as a buoyancy mechanism in a pelagic deep-sea crustacean, *J. Exp. Biol.*, 138, 333, 1988.
9. **Clarke, M. R., Denton, E. J., and Gilpin-Brown, J. B.,** On the use of ammonium for buoyancy in squids, *J. Mar. Biol. Assoc. U.K.*, 59, 259, 1979.
10. **Coombs, S. H.,** A density-gradient column for determining the specific gravity of fish eggs, with praticular reference to eggs of the mackerel *Scomber scombrus*, *Mar. Biol.*, 63, 101, 1981.
11. **Coombs, S. H., Fosh, C. A. and Keen, M. A.,** The buoyancy and vertical distribution of eggs of sprat (*Sprattus sprattus*) and pilchard (*Sardin pilchardus*), *J. Mar. Biol. Assoc. U.K.*, 65, 461, 1985.
12. **Neilson, J. D., Perry, R. I., Valerio, P., and Waiwood, K. G.,** Condition of Atlantic cod *Gadus morhua* larvae after the transition to exogenous feeding: morphometrics, buoyancy and predator avoidance, *Mar. Ecol. Prog. Ser.*, 32, 229, 1986.
13. **Yin, M. C. and Blaxter, J. H. S.,** Temperature, salinity tolerance, and buoyancy during early development and starvation of Clyde and North Sea herring, cod and flounder larvae, *J. Exp. Mar. Biol. Ecol.*, 107, 279, 1987.
14. **Craik, J. C. A. and Harvey, S. M.,** The causes of buoyancy in eggs of marine teleosts, *J. Mar. Biol. Assoc. U.K.*, 67, 169, 1987.
15. **Solemdal, P. and Sundby, S.,** Vertical distribution of pelagic fish eggs in relation to species, spawning behaviour and wind conditions, *Int. Council Exploration of the Sea, C.M. 1981/G:77 (mimeo)*, 1981.
16. **Nissling, A. and Westin, L.,** Egg buoyancy of Baltic cod (*Gadus morhua*) and its implications for cod stock fluctuations in the Baltic, *Mar. Biol.*, 111, 33, 1991.
17. **Childress, J. J., Price, M. H., Favuzzi, J., and Cowles, D.,** Chemical composition of midwater fishes as a function of depth of occurrence off the Hawaiian Islands: food availability as a selective factor?, *Mar. Biol.*, 105, 235, 1990.
18. **Denton, E. J.,** Examples of the use of active transport of salts and water to give buoyancy in the sea, *Philos. Trans. R. Soc. London B*, 262, 277, 1971.
19. **Riebesell, U.,** The formation of large marine snow and its sustained residence in surface waters, *Limnol. Oceanogr.*, 37, 63, 1992.
20. **Villareal, T. A. and Carpenter, E. J.,** Diel buoyancy regulation in the marine diazotrophic cyanobacterium *Trichodesmium thiebautii*, *Limnol. Oceanogr.*, 35, 1832, 1990.
21. **Dromgoole, F. I.,** Gas-filled structures, buoyancy and support in marine macro-algae, in *Progress in Phycological Research*, Round and Chapman, Eds., Biopress Ltd., 1990, 170.
22. **Millero, F.,** The partial molal volumes of electrolytes in aqueous solutions, in *Water and Aqueous Solutions: Structure, Thermodynamics and Transport Processes*, Horne, R. A., Ed., Wiley-Interscience, New York, 1972, 519.
23. **Gross, F. and Zeuthen, E.,** The buoyancy of planktonic diatoms: a problem of cell physiology, *Proc. R. Soc. B*, 174, 382, 1948.
24. **Villareal, T. A.,** Positive buoyancy in the oceanic diatom *Rhizosolenia debayana* H. Peragallo, *Deep Sea Res.*, 35A, 1037, 1988.
25. **Villareal, T. A.,** Buoyancy properties of the giant diatom *Ethomdiscus*, *J. Plankton Res.*, 14, 459, 1992.
26. **Anderson, L. W. J. and Sweeney, B. M.,** Role of inorganic ions in controlling sedimentation rate of a marine centric diatom *Ditylum brightwelli*, *J. Phycol.*, 14, 204, 1978.
27. **Ballek, R. W. and Swift, E.,** Nutrient- and light-mediated buoyancy control of the oceanic non-motile dinoflagellate *Pyrocystis noctiluca* Murray ex Haeckel, *J. Exp. Mar. Biol. Ecol.*, 101, 175, 1986.
28. **Kahn, N. and Swift, E.,** Positive buoyancy through ionic control in the nonmotile marine dinoflagellate *Pyrocystis noctiluca* Murray ex Schuett, *Limnol. Oceanogr.*, 23, 649, 1978.
29. **Talbot, M. M. B. and Bate, G. C.,** The use of false buoyancies by the surf diatom *Anaulus birostratus* in the formation and decay of cell patches, *Estuarine Coastal Shelf Sci.*, 26, 155, 1988.
30. **Krogh, A.,** *Osmotic Regulation in Aquatic Animals*, Cambridge University Press, Cambridge, 1939, 242.
31. **Lambert, C. C. and Lambert, G.,** Tunicate eggs utilize ammonium ions for flotation, *Science*, 200, 64, 1978.
32. **Thuesen, E. V. and Childress, J. J.,** Enzymatic activities and metabolic rate of pelagic chaetognaths: lack of depth related decline, *Limnol. Oceanogr.*, 38, 935, 1993.
33. **Bone, Q., Brownlee, C., Bryan, G. W., Burt, G. R., Dando, P. R., Liddicoat, M. I., Pulsford, A. L., and Ryan, K. P.,** On the differences between the two 'indicator' species of chaetognath, *Sagitta setosa* and *S. elegans*, *J. Mar. Biol. Assoc. U.K.*, 67, 545, 1987.
34. **Denton, E. J., Gilpin-Brown, J. B., and Shaw, T. I.,** A buoyancy mechanism found in cranchid squid, *Proc. R. Soc. B*, 174, 271, 1969.

35. **Denton, E. J., Shaw, T. I., and Gilpin-Brown, J. B.,** Bathyscaphoid squid, *Nature*, 182, 1810, 1958.
36. **Clarke, M. R., Denton, E. J., and Gilpin-Brown, J. B.,** On the buoyancy of squid of the families Histioteuthidae, Octopoteuthidae and Chiroteuthidae, *J. Physiol.*, 203, 49, 1969.
37. **Corner, E. D. S., Cowey, C. B., and Marshall, S. M.,** On the nutrition and metabolism of zooplankton V. Feeding efficiency of *Calanus finmarchius*, *J. Mar. Biol. Assoc. U.K.*, 47, 259, 1967.
38. **Sanders, N. K. and Childress, J. J.,** The use of single column ion chromatography to measure ion concentrations in invertebrate body fluids, *J. Comp. Biochem. Physiol., A*, 98, 97, 1991.
39. **Sanders, N. K. and Childress, J. J.,** Ammonia as a buoyancy mechanism in the pelagic midwater crustacean *Notostomus gibbosus*, *Am. Zool.*, 26, 33A, 1986.
40. **Sanders, N. K., Morris, S., Childress, J. J., and McMahon, B. R.,** Effects of ammonia, trimethylamine, L-lactate and CO_2 on some decapod crustacean haemocyanins, *J. Comp. Biochem. Physiol.*, 101A, 511, 1992.
41. **Burnett, L. E. and Infantino, R. L.,** The CO_2 specific sensitivity of haemocyanin oxygen affinity in the decapod crustaceans, *J. Exp. Zool.*, 232, 59, 1984.
42. **Mangum, C. P. and Burnett, L. E.,** The CO_2 sensitivity of the haemocyanins and its relationship to Cl^- sensitivity, *Biol. Bull.*, 171, 248, 1986.
43. **Hiller-Adams, P. and Childress, J. J.,** Effects of feeding, feeding history, and food deprivation on respiration and excretion rates of the bathypelagic mysid *Gnathophausia ingens*, *Biol. Bull.*, 165, 182, 1983.
44. **Hiller-Adams, P. and Childress, J. J.,** Effects of season on the bathypelagic mysid Gnathophausia ingens: water content, respiration, and excretion, *Deep Sea Res.*, 30, 629, 1983.
45. **Hiller-Adams, P. and Childress, J. J.,** Effects of prolonged starvation on O_2 consumption, NH_4^+ excretion, and chemical composition of the bathypelagic mysid *Gnathophausia ingens*, *Mar. Biol.*, 77, 119, 1983.
46. **Holmes, W. N. and Donaldson, E. M.,** The body compartments and the distribution of electrolytes, in *Fish Physiology, Vol. 1, Excretion, Ionic Regulation, and Metabolism*, Hoar, W. S. and Randall, D. J., Eds., Academic Press, New York, 1969, 1.
47. **Withers, P. C., Morrison, G., Hefter, G. T., and Pang, T.-S.,** Role of urea and methylamines in buoyancy in elasmobranchs, *J. Exp. Biol.*, 188, 175-189, 1994.
48. **Withers, P. C., Morrison, G., and Guppy, M.,** Buoyancy role of urea and TMAO in an elasmobranch fish, the Port Jackson shark, *Heterodontus portusjacksoni*, *Physiol. Zool.*, in press, 1994.
49. **Pelster, B. and Scheid, P.,** Countercurrent concentration and gas secretion in the fish swim bladder, *Physiol. Zool.*, 65, 1, 1992.

Chapter 4

Nitrogen Metabolism and Excretion for Cell Volume Regulation in Invertebrates

Raymond P. Henry

CONTENTS

INTRODUCTION

The aquatic environment is neither uniform nor stable with respect to salinity. While open ocean waters are characterized by a relatively constant salinity of about 35 parts per thousand (ppt), the salinity in other bodies of water can range from 70 ppt in hypersaline lagoons and lakes to 0 ppt in rivers.[1] Furthermore, inshore and estuarine waters are characteristically unstable with respect to salinity, fluctuating with the tidal cycle, with seasonal variation in rainfall, and in response to singular events such as hurricanes.[2] These waters are, however, also very abundant in nutrient material and thus constitute a rich and exploitable environment for any organism that can adapt to the harsh physical conditions.[3] The physiological adaptations that allow marine and estuarine invertebrates to cope with fluctuating salinity are tied to patterns of nitrogen metabolism in these organisms.

PHYSIOLOGICAL ADAPTATIONS TO SALINITY

Marine organisms can be classified into two general categories with regard to their physiological adaptations to low and fluctuating ambient salinity: (1) osmoconformers, and (2) osmoregulators.[4,5] At full strength seawater (35 ppt) all marine invertebrates are osmoconformers. The total osmotic concentration of the extracellular fluid (ECF) of the organism passively reflects that of the ambient medium. In actuality, the ECF of marine invertebrates is slightly hyperosmotic to seawater by about 40 to 60 milliosmoles (mOsm), although that difference does not appear to be maintained by any active transport of substances on the part of the organism.[6] Furthermore, when the ambient salinity is decreased, the osmotic concentration of the ECF also decreases in a manner that parallels the isosmotic line.

Osmoregulators have the ability to maintain ECF osmotic concentrations significantly above those of the ambient medium, especially in dilute waters.[4,7] The physiological mechanisms of osmotic regulation do not operate at full strength seawater, however, but are activated at a critical salinity, usually around 25 ppt. Because of this, even these organisms behave as osmoconformers over the initial 10 ppt decrease in external salinity, with their ECF being diluted by a minimum of about 300 mOsm before regulation begins.

0-8493-8411-7/95/$0.00+$.50

Since osmoconformity is present to some extent in all marine invertebrates, all such organisms that experience any change in environmental salinity must therefore cope with the resultant physiological challenges. Dilution or concentration of the ECF (in response to low or high salinity, respectively) establishes an osmotic gradient across the cell membrane that will cause the tissue to gain or lose water, respectively.[8] For intact organisms, changes in cell volume have traditionally been documented by changes in whole-body wet weight; similar changes in suspensions of isolated cells have been measured directly. Cell swelling or shrinkage, as a result of intracellular volume changes, can be deleterious to normal cellular function. For most invertebrates, however, salinity-induced changes in intracellular volume are transient; the organisms possess physiological mechanisms for the readjustment of cell volume.[9] These mechanisms involve the regulation of the concentrations of the different substances that comprise the total intracellular osmotically active solute pool. A significant fraction of this pool is composed of small, nitrogen-containing compounds; thus, cell volume regulation is intimately tied to nitrogen metabolism and excretion in marine invertebrates.

NITROGEN COMPOUNDS AS INTRACELLULAR OSMOTIC EFFECTORS

The presence of nitrogenous compounds in the tissues of marine invertebrates has been known for almost 100 years, with taurine and glycine having been documented in significant concentrations in *Mytilus edulis* in 1904.[10] Early studies also showed that taurine was widely distributed among invertebrate species.[11-15] Further investigation revealed the presence of a large pool of nonprotein nitrogen that was composed of amino acids (including taurine), trimethylamine oxide (TMAO), and glycine betaine.[16-19] In 1955, Camien et al.[17] first suggested that these compounds have a role in the regulation of intracellular osmotic pressure, and the link between the size of the intracellular nitrogen pool and the ambient salinity was established soon after.[20]

Among the wide variety of marine and freshwater invertebrates examined, the tissue free amino acids (FAA) pool has been found to be dominated by one or a combination of a relatively few individual amino acids. These are alanine, glycine, taurine, and proline (Table 1). The distribution of the specific amino acids is consistant among different tissue types within a species,[21] but the pattern of the dominant amino acids does not appear to have either a taxonomic or an ecological basis. Closely related species of euryhaline bivalves utilize different amino acids. For example, in *Modiolus* (*Geukensia*) *demissus*, alanine and glycine make up over half of the FAA pool, but in *M. squamosus* taurine makes up 74% of the total pool.[22] On the other hand, *Mya arenaria* and *Spisula solidissima*, which are euryhaline and stenohaline, respectively, and which belong to different families, have very similar FAA profiles.[23,24]

While bivalves do not utilize proline, in taxonomic groups such as gastropods and polychaetes, proline is a significant component of the FAA pool (Table 1). The major amino acids in crustaceans are alanine, glycine, glutamate, arginine, and proline. Compared to other groups, the FAA pool among crustacean species appears more uniform, but there is a large departure in the pool composition in other aquatic arthropod species. *Limulus polyphemus*, a chelicerate, has a very small FAA pool that is composed mostly of taurine (Table 1). The bulk of the osmotically active pool of nitrogenous compounds is comprised of glycine betaine and homarine (Table 2). A similar situation exists in the gastropod, *Elysia chlorotica*, in which the FAA make up only 10% of the total intracellular free nitrogen pool, and proline betaine comprises virtually all of the remainder (Table 2).

While the presence of amino acids and quaternary ammonium compounds in invertebrate tissue and the changes in their concentrations in response to changes in ambient salinity have been well studied, there are relatively few reports that document the quantitative contribution of these compounds to the total intracellular osmotic concentration. This value is highly variable and depends on tissue type, species, and habitat of the organism. Among molluscs the FAA pool comprises between 4 and 34% of the total intracellular osmotic pressure. The low end is represented by a more primitive species, *Acanthochitona discrepans*, whose foot muscle has a very small intracellular FAA pool

Table 1 The Composition of the Free Amino Acid Pool in Various Tissues of Marine and Freshwater Invertebrate Species

Species	Salinity (ppt)	Tissue	Total FAA pool size	Percent composition of individual FAAs									Ref.
				ala	gly	tau	glu	arg	pro	ser	thr	asp	
Bivalves													
Mya arenaria	20	adductor muscle	248[a]	37	35	7	6	3	—	—	—	—	23
Rangia cuneata	20	adductor muscle	372[b]	69	7	—	8	—	6	—	—	—	21,25
			259[b]	74	12	—	9	—	—	—	—	—	
		gill	434[b]	47	17	2	3	—	3	—	—	—	21
		mantle	234[b]	61	14	—	3	—	3	—	—	—	21
Noetia ponderosa	28	red coelomocytes	138[c]	5	3	48	7	—	2	—	—	13	26
Modiolus demissus	36	ventricle	521[b]	15	38	31	5	—	—	4	—	7	29
Modiolus demissus granosissimus	41	mantle	77[d]	36	15	20	6	—	—	—	—	3	22
Modiolus squamosus	41	mantle	57[d]	5	10	74	—	—	—	—	—	6	22
Glycymeris glycymeris	39	adductor muscle	92[d]	8	—	51	6	7	—	—	—	10	28
Mytilus edulis	39	adductor muscle	257[d]	11	—	8	2	—	—	—	—	4	28
Venus gallina	31	whole animal	80[d]	15	22	40	7	—	—	—	—	8	30
Ligumia subrostrata	0	whole animal	23[b]	9	5	—	13	23	—	—	—	—	31
Gastropods													
Littorina littorea	36	foot	106[d]	17	9	38	4	6	18	—	—	—	32
Purpura lapillus	36	foot	74[d]	11	6	52	7	7	5	—	—	—	32
Thais haemostoma	35	foot	441[b]	26	14	5	8	—	9	7	—	7	33
Elysia chlorotica	31	whole animal	52[b]	<2	—	—	7	—	<2	—	—	—	34
Polychaetes													
Glycera dibranchiata	33	red coelomocytes	945[c]	19	—	11	—	—	20	5	—	—	27
	35	body wall	630[b]	24	—	—	—	—	22	13	—	37	35
Neanthes succinea	25	body wall	925[b]	6	44	10	2	—	—	7	—	—	36
Coelenterates													
Bunodosoma cavernata	36	whole animal	154[b]	5	7	55	7	—	—	—	—	—	37
Chelicerates													
Limulus polyphemus	31	heart	166[b]	1	—	10	3	3	3	—	—	—	38
	35	cardiac ganglion	89[b]	8	8	8	21	5	26	—	—	—	39
Crustaceans													
Eriochier sinensis	SW	muscle	282[d]	12	28	7	7	15	12	—	—	—	40
Carcinus maenas	SW	muscle	309[d]	7	33	—	12	12	32	—	—	—	40
Hemigrapsus edwardsi	SW	muscle	116[b]	4	22	29	2	28	11	—	—	—	41
Coenobita brevimanus	28	muscle	847[b]	27	43	—	—	10	10	8	—	—	42

[a]mmol/g Tissue water. [b]mmol/g Dry weight. [c]mmol/10^6 Cells. [d]mmol/g Wet weight.

Table 2 The Size and Composition of the Total Intracellular Pool of Osmotically Active Nitrogenous Compounds in Two Species of Marine Invertebrates

			Percent Composition					
Species	**Salinity (ppt)**	**Tissue**	**Total**	**proline-betaine**	**glycine betaine**	**homarine**	**FAA**	**Ref.**
Limulus polyphemus	31	heart	819*	—	63	17	20	38
Elysia chlorotica	31	whole animal	533*	88	—	—	10	34

Note: Superscript after total pool size indicates the units and is referred to at the end of the table.
* μmol/g dry weight.

even at high salinity (39 ppt).[28] The FAA pool in bivalve and gastropod muscle (adductor and foot, respectively) makes up between 10 and 34% of the total intracellular osmotic pressure, with more of the species being closer to the low end.[28,32]

The intracellular FAA pool makes up between 40 and 60% of the total intracellular osmotic concentration of nerve and muscle in marine crustaceans acclimated to 35 ppt sea water.[43-46,81] In general, nerve tissue has a higher intracellular concentration of FAA than does muscle tissue. In contrast, the gills of marine species (e.g., *Callinectes sapidus*) have an FAA pool size that is only about 15 to 20% as large as that in nerve and muscle tissues.[45] The tissues of terrestrial crustaceans have a reduced FAA pool size in comparison to their marine counterparts. The FAA pool in muscle of four terrestrial species makes up only between 11 and 16% of the total intracellular osmotic concentration, and in gills the contribution is even less (1 to 5%).[42] It is difficult, however, to draw any firm general conclusions about the quantitative contribution of the intracellular nitrogen pool to the total osmotic concentration. There are nitrogen compounds other than FAA that also function as intracellular osmotic effectors (e.g., homarine, glycine betaine, proline betaine),[34,38] and these and other potentially osmotically active solutes were not measured in the studies mentioned above. Therefore, it is likely that the reported values for the contribution of the total intracellular nitrogen pool are conservative estimates.

EVOLUTIONARY SIGNIFICANCE OF NITROGEN COMPOUNDS AS OSMOTIC SOLUTES

While the distribution of the major components of the FAA pool cannot be correlated with either taxonomy or ecology, there does appear to be a fundamental biochemical basis for the selection and conservation of specific nitrogen compounds as intracellular osmolytes. The specific types of compounds and their intracellular concentrations may be determined by their ability to conserve enzyme function in the presence of high and variable concentrations of salt. The most abundant salts in invertebrate body fluids and tissues (i.e., Na^+, K^+, Cl^-, Ca^{+2}, Mg^{+2}) are known to influence protein structure in a manner that is unfavorable for the maintenance of the protein's native structure.[47-49] Additionally, the concentrations of salts needed to alter protein structure (and therefore enzymatic function) appear to be within the physiological concentrations found in animals.[50,51] Amino groups, methylamines, and sulfates, on the other hand, appear to conserve native protein structure and function. Specifically, Clark and Zounes[52] showed that 500 m*M* concentrations of Na^+, K^+, and Cl^- reduced lactate dehydrogenase (LDH) activity from the polychaete, *Nereis succinea*, by as much as 75% compared to activity measured in buffer alone. Furthermore, physiological concentrations of KCl were found to triple the Km of pyruvate. In contrast, amino acids and methylamines (TMAO) had little or no effect on either LDH activity or Km.[52] This same trend was seen in other enzymes from other species, both vertebrate and invertebrate. Physiological concentrations of both NaCl and

KCl had inhibitory effects (decreased activity and increased Km) on pyruvate kinase (PK), whereas similar concentrations of the common amino acids in the intracellular FAA pool had no deleterious effect on activity or Km values.[53] In some cases, nitrogen osmolytes (e.g., TMAO and betaine) have been shown to counteract the inhibitory effects of high salt and urea on enzyme function.[54] These compounds have also been shown to protect enzymes from high temperature denaturation, cold inactivation, and freeze-thaw deactivation in a number of vertebrate examples. They also promote growth in bacteria, yeast, and mammalian cells in culture.[55]

The general consensus is that certain amino acids and methylamines were selected as intracellular solutes because they promote the stabilization of enzymes, nucleic acids, membrane-associated proteins, and intracellular organelles.[55-58] Energetically, these compounds are more costly to the organism; it takes more metabolic effort to synthesize these large and relatively complex molecules than it would to simply adjust intracellular osmotic concentration using available salts. Ultimately, however, this strategy may be more energetically efficient than selecting for proteins (enzymes) that are specifically adapted to function in high concentrations of salt. Some organisms, such as halophilic bacteria, have salt-adapted enzymes, but these live in stable, high-salt environments. They are not adapted to function in an environment in which salinity changes; when the salt concentration is experimentally lowered, the normal enzyme structure-function relationship is disrupted.[59] Marine invertebrates often face situations in which the environmental salt concentrations fluctuate quite widely, making proteins that were adapted to one particular salt regime disadvantageous in another. Therefore, it may be more energetically efficient to adjust the intracellular pool of organic osmolytes than it would to synthesize a new complement of functional proteins each time the organism faces a change in salinity. However, since the synthesis and catabolism of the FAA and methylamine pools do cost the organism something in terms of energy expenditure, it may be that the quantitative contribution to the total intracellular osmotic concentration is set by the minimum amounts of nonperturbing organic solutes needed to maintain protein integrity for a given salt concentration. That would allow for the stabilization of protein structure/function with mimimal energetic cost. This question has never been experimentally addressed.

CONTROL OF THE ORGANIC OSMOLYTE POOL

REDUCTION OF THE INTRACELLULAR FAA POOL IN LOW SALINITY

When marine invertebrates are exposed to low salinity the ECF becomes diluted. Since the intracellular fluid of tissues is in osmotic equilibrium with the ECF, a gradient is established which favors the movement of water into the cells. This results in an increase in cell volume and subsequent cell swelling.[8]

The initial response of organisms appears to be behavioral; they attempt to avoid osmotic stress. This is shown most clearly in bivalves. Valves clamp shut, valve movements decrease significantly, and oxygen uptake and ammonia excretion are both dramatically reduced during the initial few hours after a low salinity shock.[28,32,60-62] This "escape response" can delay the onset of ECF dilution in some species for as much as 24 h; in other species this behavior appears to slow the rate of ECF dilution and tissue swelling.

The increase in cell/tissue volume has usually been measured as an increase in wet weight. Intact organisms have been shown to gain from 8 to 80% of their original weight when transferred to low salinity; the magnitude of the change depends on the species and the magnitude of the salinity shock.[63-66] The weight gain, which results from a combination of extracellular and intracellular volume increases, usually occurs within 1 to 2 h after transfer, peaks for a variable time thereafter, and then gradually declines to near control values between 12 and 72 h.

Direct measurement of cell volume (on bivalve and polychaete red blood cells) using cell counters/sizers revealed that the increase in volume of isolated cells given a low salinity shock, while being within the same range of magnitude (130%) as that for the whole organism, occurs much more rapidly. The peak volume increase occurs within 2 to 5 min, and the decline takes place between 20 and 120 min.[67,68] When red cells are withdrawn from intact bivalves given a similar salinity shock, and then measured, the changes in cell volume occur over a period of hours-to-days.[69] It appears then that the behavioral responses of the organism buffer the time course of cell volume changes after an acute salinity shock. This could be important as rapid changes in cell volume were found to be more damaging, and more likely to result in fatality, than slow changes.[70]

Despite large changes in tissue hydration during acute exposure to low salinity, acclimated values of tissue water content change only slightly. For example, in a number of species acclimated to salinity ranges of as great as 12 to 40 ppt, tissue water content changes a maximum of 10%.[37,69,71,81] On the other hand, however, the intracellular FAA pool — the size of which is known to be reduced in low salinity — changes to a much greater extent than can be explained by dilution as a result of an increase in tissue hydration.[22,81] The current view of cell volume regulation holds that cell swelling is counteracted by a reduction of the intracellular osmolyte pool. FAA and ions (mostly K^+) are expelled from the cell, carrying with them osmotically obligated water and thus shrinking the cell.[8] While both inorganic and organic osmolytes are used in the readjustment of cell volume, it appears that ionic efflux takes place initially and rapidly, and that the efflux of FAA follows more slowly but has a more permanent effect.[72] In some cases, as the FAA pool is reduced, intracellular ion (K^+) concentrations are restored.

A number of the specific processes in the overall mechanism of FAA-mediated cell volume readjustment have been described. The FAA efflux appears to be caused by the decrease in total osmotic concentration, and not by a drop in the concentration of any single ion.[73] There is evidence that this activates a stretch-sensitive calcium channel, and that the influx of Ca^{+2} and subsequent increase in intracellular Ca^{+2} concentration is a critical step in the initiation of FAA efflux.[74-76] Calcium is believed to combine with calmodulin and phosphorylate proteins in the cell membrane, resulting in FAA (specifically taurine) efflux. The remaining steps in the mechanism, including the description of putative amino acid channels, are still to be elucidated.

While it has been well documented that individual tissues and cells excrete FAA intact in order to reduce intracellular volume, it is also well known that the majority of invertebrate organisms excrete primarily ammonia during the volume regulatory phase of low salinity adaptation.[77-79,82] The concentration of FAA in the ECF increases significantly by 4 h after low salinity transfer; this is followed by an increase in ECF ammonia concentrations and elevated ammonia excretion.[60,80] FAA excretion from the animal does not appear to be significant in most cases.

While the metabolic fate of the FAA is not known, there is circumstantial evidence supporting the role of the gill and/or mantle as the central organ of amino acid uptake, deamination, and ammonia excretion. Both gill and mantle tissue have been shown to oxidize a variety of amino acids that make up the FAA pool.[83,84] Furthermore, amino acid oxidation increases in dilute salinity. Gill tissue in particular shows increased cellular oxidative metabolism, as measured by O_2 consumption, in low salinity.[85-87] Crustacean gill tissue possesses high levels of amino acid transferase activity, and crustacean metabolism appears to be poised to catabolize amino acids preferentially over carbohydrates.[88,89] There are also high levels of activity of proline oxidase in crustacean gills that are about four- to fivefold higher in posterior than anterior gills.[90] Recently a basolateral amino acid uptake system was documented for the gill of *M. edulis*.[91] It is tempting to suggest that the intracellular FAA, which are released intact from tissues during low salinity shock, are taken up by the gill where they undergo a combination of transamination and oxidation in order to meet the energetic and transport demands of the gill. Metabolism of the amino acids could fuel ATP synthesis (which is needed to power ion transport pumps such as the Na/K ATPase) and supply NH_4^+ ions to serve as counterions for Na^+ uptake via Na^+/NH_4^+ exchange.[92] This idea has never been systematically investigated; existing work indicates that oxidative deamination of FAA in the gill could supply some, but not all, of the counterions for Na^+ uptake.[83,93] However, not all of the Na^+ uptake occurs via Na^+/NH_4^+ exchange; the majority is believed to be via Na^+/H^+ exchange.[94] Thus, it may be possible that intracellular deamination could supply the full complement of counterions for the fraction of Na^+ uptake that is transported in exchange for NH_4^+. These questions deserve more rigorous examination.

SYNTHESIS OF THE FAA POOL DURING HIGH SALINITY ADAPTATION

Over the past two decades three major hypotheses have been put forth to explain the increase in the FAA pool during high salinity adaptation. These are: (1) amino acid uptake from the ambient medium, (2) *de novo* synthesis involving the fixation of ammonia onto a keto acid through reductive amination and subsequent transamination, and (3) protein catabolism and transamination to the major constituents of the FAA pool. The bulk of the experimental evidence, which has been presented in a recent review,[95] favors protein turnover as the source for amino nitrogen. According to this scheme, alanine accumulation in bivalve tissue appears to depend on amino acid transferase activity, as inhibitors of these enzymes reduce alanine accumulation with only a slight reduction in the overall FAA pool size. The carbon skeleton for alanine synthesis can originate from either glycolysis or tricarboxylic acid cycle intermediates that feed into pyruvate. High salinity also results in an inhibition of pyruvate metabolism, thus facilitating the accumulation of alanine concentrations in the cell. The control of pyruvate and alanine metabolism appears to be the pyruvate dehydrogenase enzyme complex (PDH), which is located in the mitochondria of bivalve tissues. While the bulk of the supporting data comes from a few species of bivalves, circumstantial evidence points to this pathway as being more general among invertebrates. Interestingly, in crustaceans, in which most of the muscle is composed of fast glycolytic (white) fibers that lack mitochondria, alanine is only a minor component of the FAA pool.

Proline, which is either absent from bivalves or comprises only a small fraction of the FAA pool, is believed to be derived from ornithine via transamination.[95] Ornithine, in turn, originates from arginine and phosphoarginine. Proline accumulation during high salinity adaptation is transient, occurring early; as the FAA pool comes to equilibrium, however, proline concentrations decrease. A mitochondrial proline oxidase is believed to be responsible for the catabolism of proline and the subsequent reduction in its concentration. In other species (e.g., crustaceans), proline is a significant and permanent constituent of the FAA pool (Table 1). Its contribution may be due to the reliance on phosphoarginine as an energy source in crustacean muscle coupled with the absence of mitochondria (and associated pathways of proline catabolism) in white muscle fibers. Regardless, the metabolic pathways involved in proline synthesis and metabolism, outside of the bivalves, need further elucidation. Very little is known about the pathways involved in the accumulation of taurine, another important component of the FAA pool in bivalves and other groups of marine organisms.

Glycine betaine, an important osmotically active quaternary ammonium compound in the tissues of *Limulus* (Table 2), is synthesized from choline.[96] Choline is taken up into tissues where it is oxidized first to betaine aldehyde and then to glycine betaine. The latter step is accomplished by a betaine aldehyde dehydrogenase which is localized in the mitochondrial matrix. Hyperosmotic stress stimulates the transport of choline across both the cell and mitochondrial membranes.

SUMMARY

While there has been a lot of progress in the field of nitrogen metabolism and cell volume regulation over the last 30 years, there are still many unanswered questions, and the field remains a fertile area for new research. The metabolic fate of the intracellular FAA after they are released from the cell during hypoosmotic shock, and the potential relationship between cell volume adjustment, nitrogen metabolism and excretion, and epithelial salt transport is still not known. The mechanisms of activation and control of FAA transport and metabolism are not known. Likewise, the controlling factors that activate FAA synthesis are also not known. The organisms involved in nitrogen related cell volume regulation are excellent model systems for studying cellular and molecular processes of signal transduction, membrane transport, enzyme activation and induction, and metabolic control.

REFERENCES

1. **Sverdrup, H. U., Johnson, M. W., and Fleming, R. H.,** *The Oceans: Their Physics, Chemistry, and General Biology*, Prentice Hall, Inc., Englewood Cliffs, NJ, 1970, 1087.
2. **Remane, A. and Schlieper, C.,** *Biology of Brackish Water*, John Wiley and Sons, New York, 1971, 372.
3. **Kinne, O.,** Salinity: animals - invertebrates, in *Marine Ecology*, Vol. 1, Part 2, O. Kinne, Ed., Wiley Interscience, London, 1971, 821.
4. **Kirschner, L. B.,** Control mechanisms in crustaceans and fishes, in *Mechanisms of Osmoregulation in Animals*, R. Gilles, Ed., Wiley Interscience, London, 1979, 157.
5. **Schoffeniels, E.,** Adaptations with respect to salinity, *Biochem. Soc. Symp.*, 41, 179, 1976.
6. **Pierce, S. K.,** The water balance of *Modiolus* (Mollusca: Bivalvia: Mytukedae): osmotic concentrations in changing salinities, *Comp. Biochem. Physiol.*, 36, 521, 1970.
7. **Mantel, L. H. and Farmer, L. L.,** Osmotic and ionic regulation, in *The Biology of Crustacea*, Vol. 5, *Internal Anatomy and Physiological Regulation*, L. H. Mantel, Ed., Academic Press, New York, 1983, 54.
8. **Pierce, S. K. and Amende, L. M.,** Control mechanisms of amino acid-mediated cell volume regulation in salinity-stressed molluscs, *J. Exp. Zool.*, 215, 247, 1981.
9. **Pierce, S. K.,** Invertebrate cell volume control mechanisms: a coordinated use of intracellular amino acids and inorganic ions as osmotic solute, *Biol. Bull.*, 163, 405, 1982.
10. **Kelly, A.,** Beobachtungen uber das Vorkmenn von Atherschwefelsauren, von Taurin und Blycin bei niedern Tieren, *Beitr. Chem. Physiol. Pathol.*, 5, 377, 1904.
11. **Henze, M.,** Beitrage zur Muskelchemie der Octopoden, *Hoppe-Seyl. Zeitschr.*, 43, 477, 1905.
12. **Mendel, L. B.,** Uber das vorkmmen von Taurin in den Muskeln von Weichtieren, *Beitr. Zeitschr. Chem. Physiol. Pathol.*, 5, 582, 1904.
13. **Kossel, A. and Edlbacher, S.,** Beitrage zur chemischem Kenntnis der Echinodermen, *Hoppe Seyl. Zeitschr.*, 94, 264, 1951.
14. **Okuda, Y.,** Extractive matters of *Palinurus japonica* and *Loligo breekeri*, *Chem. Abstr.*, 14, 2827, 1920.
15. **Ackerman, D., Holtz, F., and Kutscher, K,** Extractives of *Eledone maxchata*, *Chem. Abstr.*, 18, 2387, 1924.
16. **Giordano, M. F., Harper, H. A., and Filice, F. P.,** The amino acids of a starfish and a sea urchin (Asteroidea and Echinoidea),*Wasmann*, *J. Biol.*, 8, 129, 1950.
17. **Camien, M. N., Sarlet, H., Duchateau, G., and Florkin, M.,** Non-protein amino acids in muscle and blood of marine and fresh water Crustacea, *J. Biol. Chem.*, 193, 881, 1951.
18. **Duchateau, G. and Florkin, M.,** Types de composition du pool des acides amines non proteiques des muscles, *Arch. Intern. Physiol.*, 62, 487, 1954.
19. **Kermack, W. O., Lees, H., and Wood, J. D.,** Some non-protein constituents of the tissues of the lobster, *Biochem. J.*, 60, 424, 1955.
20. **Simpson, J. W., Allen, K., and Awapara, J.,** Free amino acids in some aquatic invertebrates, *Biol. Bull.*, 117, 371, 1959.
21. **Henry, R. P., Mangum, C. P., and Webb, K. L.,** Salt and water balance in the oligohaline clam, *Rangia cuneata.* II. Accumulation of intracellular free amino acids during high salinity adaptation, *J. Exp. Zool.*, 211, 11, 1980.
22. **Pierce, S. K.,** A source of solute for volume regulation in marine mussels, *Comp. Biochem. Physiol.*, 38A, 619, 1971.
23. **Virkar, R. A. and Webb, K. L.,** Free amino acid composition of the soft shell clam *Mya arenaria* in relation to salinity of the medium, *Comp. Biochem. Physiol.*, 32, 775, 1970.
24. **DuPaul, W. D. and Webb, K. L.,** Salinity induced changes in the alanine and aspartic amino transferase activity in three marine bivalve molluscs, *Arch. Int. Physiol. Biochem.*, 82, 817, 1974.
25. **Allen, K.,** The effect of salinity on the amino acid concentration in *Rangia cuneata* (Pelecypoda), *Biol. Bull.*, 121, 419, 1961.
26. **Amende, L. M. and Pierce, S. K.,** Cellular volume regulation in salinity stressed molluscs: the response of *Noetia ponderosa* (Arcidae) red blood cells to osmotic variation, *J. Comp. Physiol.*, 138B, 283, 1980.
27. **Costa, C. J. and Pierce, S. K.,** Volume regulation in the red coelomocytes of *Glycera dibranchiata*: an interaction of amino acid and K^+ efflux, *J. Comp. Physiol.*, 151B, 133, 1983.
28. **Gilles, R.,** Osmoregulation in three molluscs: *Acanthochitona descrepans* (Brown), *Glycmeris glycmeris* (L) and *Mytilus edulis* (L), *Biol. Bull.*, 142, 25, 1972.
29. **Pierce, S. K. and Greenberg, M. J.,** The nature of cellular volume regulation in marine bivlaves, *J. Exp. Biol.*, 57, 681, 1972.
30. **Zurburg, W. and DeZwaan, A.,** The role of amino acids in anaerobiosis and osmoregulation in bivalves, *J. Exp. Zool.*, 215, 315, 1981.
31. **Hanson, J. A. and Dietz, T. H.,** The role of free amino acids in cellular osmoregulation in the freshwater bivalve *Ligumia subrostrata* (Say), *Can. J. Zool.*, 54, 1927, 1976.
32. **Hoyaux, J., Gilles, R., and Jeuniaux, C.,** Osmoregulation in molluscs of the intertidal zone, *Comp. Biochem. Physiol.*, 53A, 361, 1976.

33. **Kapper, M. A., Stickle, W. B., and Blakeney, E.,** Volume regulation and nitrogen metabolism in the muricid gastropod *Thais haemastoma*, *Biol. Bull.*, 169, 458, 1985.
34. **Pierce, S. K. and Rowland, L. M.,** Proline betaine and amono acid accumulation in sea slugs (*Elysia chlorotica*) exposed to extreme hyperosmotic conditions, *Physiol. Zool.*, 61, 205, 1988.
35. **Costa, C. J., Pierce, S. K., and Warren, M. K.,** The intracellular mechanism of salinity tolerance in polychaetes: volume regulation by isolated *Glycera dibranchiata* red coelomocytes, *Biol. Bull.*, 159, 626, 1980.
36. **Koenig, M., Powell, E. N., and Kasschau, M. R.,** The effects of salinity change on the free amino acid pools of two nereid polychaetes, *Neanthes succinea* and *Leonereis culveri.*, *Comp. Biochem. Physiol.*, 70A, 631, 1981.
37. **Kasschau, M. R., Ragland, J. B., Pinkerton, S. O., and Chen, E. C. M.,** Time related changes in the free amino acid pool of the sea anemone, *Bunodosoma cavernata*, during salinity stress, *Comp. Biochem. Physiol.*, 79A, 155, 1984.
38. **Warren, M. K., and Pierce, S. K.,** Two cell volume regulatory systems in the *Limulus* myocardium: an interaction of ions and quaternary ammonium compounds, *Biol. Bull.*, 163, 504, 1982.
39. **Prior, D. J., and Pierce, S. K.,** Adaptation and tolerance of invertebrate nervous systems to osmotic stress, *J. Exp. Zool.*, 215, 237, 1981.
40. **Schoffeniels, E.,** Biochemical approaches to osmoregulatory processes in crustacea, *Perspect. Exp. Biol.*, 1, 107, 1976.
41. **Bedford, J. J., Smith, R. A. J., Thomas, M., and Leader, J. P.,** ^{1}H-NMR and HPLC studies of the changes involved in volume regulation in the muscle fibres of the crab, *Hemigrapsus edwardsi*, *Comp. Biochem. Physiol.*, 100A, 145, 1991.
42. **Henry, R. P. and Cameron, J. N.,** A survey of blood and tissue nitrogen compounds in terrestrial decapods of Palau, *J. Exp. Zool.*, 218, 83, 1981.
43. **Lang, M. A. and Gainer, H.,** Isosmotic intracellular regulation as a mechanism of volume control in crab muscle fibers, *Comp. Biochem. Physiol.*, 30, 445, 1969.
44. **Evans, D. P.,** The free amino acid pool of the haemocytes of *Carcinus maenas* (L), *J. Exp. Biol.*, 56, 501, 1972.
45. **Gerard, J. F. and Gilles, R.,** The free amino acid pool in *Callinectes sapidus* (Rathbun) tissues and its role in the osmotic intracellular regulation, *J. Exp. Mar. Biol. Ecol.*, 10, 125, 1972.
46. **Bedford, J. J. and Leader, J. P.,** The composition of the haemolymph and muscle tissue of the shore crab *Hemigrapsus edwardsi* exposed to different salinities, *Comp. Biochem. Physiol.*, 57A, 341, 1977.
47. **vonHippel, P. H. and Schleich, T.,** The effects of neutral salts on the structure and conformational stability of macromolecules in solution, in *Structure and Stability of Biological Macromolecules*, Timasheff, S. N. and Fasman, G. D., Eds., Marcel Dekker, New York, 1969, 417.
48. **Warren, J. C. and Cheatum, S. G.,** Effects of neutral salts on enzyme activity nad structure, *Biochemistry*, 5, 1702, 1966.
49. **Warren, J. C., Stowring, L., and Morales, M.,** The effects of structure-disrupting ions on the activity of myosin and other enzymes, *J. Biol. Chem.*, 241, 309, 1966.
50. **Somero, G. N. and Low, P. S.,** Enzyme hydration may explain the catalytic efficiency differences among lactate dehydrogenase homologues, *Nature*, 266, 276, 1977.
51. **Somero, G. N., Neubauer, M., and Low, P. S.,** Neutral salt effects on the velocity and activation volume of the lactate dehydrogenase reaction: evidence for enzyme hydration changes during catalysis, *Arch. Biochem. Biophys.*, 181, 1977.
52. **Clark, M. E. and Zounes, M.,** The effects of selected cell osmolytes on the activity of lactate dehydrogenase from the euryhaline polychaete, *Nereis succinea*, *Biol. Bull.*, 153, 468, 1977.
53. **Bowlus, R. D. and Somero, G. N.,** Solute compatibility with enzyme function and structure: rationales for the selection of osmotic agents and end-products of anaerobic metabolism in marine invertebrates, *J. Exp. Zool.*, 208, 137, 1979.
54. **Yancey, P. H. and Somero, G. N.,** Methylamine osmoregulatory solutes of elasmobranch fishes counteract urea inhibition of enzymes, *J. Exp. Zool.*, 212, 205, 1980.
55. **Yancey, P. H.,** Compatible and counteracting solutes, in *Cellular and Molecular Physiology of Cell Volume Regulation*, Strange, K., Ed., CRC Press, Boca Raton, FL, 1994, 81.
56. **Yancey, P. H., Clark, M. E., Hand, S. C., Bowlus, R. D., and Somero, G. N.,** Living with water stress: evolution of osmolyte systems, *Science*, 217, 1214, 1982.
57. **Somero, G. N.,** Environmental adaptation of proteins: strategies for the conservation of critical functional and structural traits, *Comp. Biochem. Physiol.*, 76A, 621, 1983.
58. **Somero, G. N.,** Protons, osmolytes, and fitness of internal milieu for protein function, *Am. J. Physiol.*, 251, R197, 1986.
59. **Lanyi, J.,** Salt-dependent properties of proteins from extremely halophilic bacteria, *Bacteriol. Rev.*, 38, 272, 1974.
60. **Henry, R. P. and Mangum, C. P.,** Salt and water balance in the oligohaline clam, *Rangia cuneata*. III. Reduction of the free amino acid pool during low salinity adaptation, *J. Exp. Zool.*, 211, 25, 1980.
61. **Pierce, S. K.,** Volume regulation and valve movements by marine mussels, *Comp. Biochem. Physiol.*, 39A, 103, 1971.

62. **Shumway, S. E., Gabbot, P. A., and Youngson, A.,** The effect of fluctuating salinity on the concentrations of free amino acids and ninhydrin positive substances in the adductor muscles of eight species of bivalve molluscs, *J. Exp. Mar. Biol. Ecol.*, 29, 131, 1977.
63. **Dykens, J. A. and Mangum, C. P.,** The regulation of body fluid volume in the estuarine annelid *Nereis succinea*, *J. Comp. Physiol.*, 154B, 607, 1984.
64. **Gainey, L. F.,** The effect of osmotic hysteresis on volume regulation in *Mytilus edulis*, *Comp. Biochem. Physiol.*, 87A, 151, 1987.
65. **Henry, R. P., McBride, C. J., and Williams, A. H.,** Responses of the marsh periwinkle, *Littoraria* (*Littorina*) *irrorata* to temperature, salinity and desiccation, and the potential physiological relationship to climbing behavior, *Mar. Behav. Physiol.*, 24, 45, 1993.
66. **Reitze, M., and Schottler, U.,** The time dependence of adaption to reduced salinity in the lugworm *Arenicola marina* L. (Annelida: Polychaeta), *Comp. Biochem. Physiol.*, 93A, 549, 1989.
67. **Costa, C. J., Pierce, S. K., and Warren, M. K.,** The intracellular mechanism of salinity tolerance in polychaetes: volume regulation by isolated *Glycera dibranchiata* red coelomocytes, *Biol. Bull.*, 159, 626, 1980.
68. **Smith, L. H. and Pierce, S. K.,** Cell volume regulation by molluscan erythrocytes during hypoosmotic stress: Ca^{2+} effects on ionic and organic osmolyte effluxes, *Biol. Bull.*, 173, 407, 1987.
69. **Amende, L. M. and Pierce, S. K.,** Cellular volume regulation in salinity stressed molluscs: the respnse of *Noetia ponderosa* (Arcidae) red blood cells to osmotic variation, *J. Comp. Physiol.*, 138B, 283, 1980.
70. **Oglesby, L. C.,** Salt and water balance, in *Physiology of Annelids*, Mill, P. J., Ed., Academic Press, New York, 1978, 555.
71. **Pierce, S. K. and Greenberg, M. J.,** The nature of cellular volume regulation in marine bivalves, *J. Exp. Biol.*, 57, 681, 1972.
72. **Pierce, S. K.,** Invertebrate cell volume control mechanisms: a coordinated use of intracellular amino acids and inorganic ions as osmotic solute, *Biol. Bull.*, 163, 405, 1982.
73. **Pierce, S. K. and Greenberg, M. K.,** The initiation and control of free amino acid regulation of cell volume in salinity-stressed marine bivalves, *J. Exp. Biol.*, 59, 433,1973.
74. **Pierce, S. K. and Politis, A. D.,** Ca^{2+}-activated cell volume recovery mechanisms, *Ann. Rev. Physiol.*, 52, 27, 1990.
75. **Pierce, S. K.,** Osmolyte permeability in mulluscan red cells is regulated by Ca^{2+} and membrane protein phospholylation: the present perspective, *J. Exp. Zool.*, 268, 166, 1994.
76. **Gerard, J. F.,** Volume regulation and alanine transport. Response of isolated axons of *Callinectes sapidus* Rathbun to hypo-osmotic conditions, *Comp. Biochem. Physiol.*, 51A, 225, 1975.
77. **Allen, J. A. and Garrett, M. R.,** The excretion of ammonia and urea by *Mya arenaria* L. (Mollusca:Bivalvia), *Comp. Biochem. Physiol.*, 39A, 633, 1971.
78. **Emerson, D. N.,** Influence of salinity on ammonia ecretion rates and tissue constituents of euryhaline invertebrates, *Comp. Biochem. Physiol.*, 29, 1115, 1969.
79. **Harris, R. R. and Andrews, M. B.,** Total NPS pool and ammonia net efflux rate changes in *Carcinus maenas* during acclimation to low environmental salinity, *Comp. Biochem. Physiol.*, 82A, 301, 1985.
80. **Bartberger, C. A. and Pierce, S. K.,** Relationship between ammonia excretion rates and hemolymph nitrogenous compounds of a euryhaline bivalve during low salinity acclimation, *Biol. Bull.*, 150, 1, 1976.
81. **Shaw, J.,** Osmoregulation in the muscle fibers of *Carcinus maenas*, *J. Exp. Biol.*, 35, 920, 1958.
82. **Lang, M. A. and Gainer, K.,** Isosmotic intracellular regulation as a mechanism of volume control in crab muscle fibers, *Comp. Biochem. Physiol.*, 30, 445, 1969.
83. **Pressley, T. A. and Graves, J. S.,** Increased amino acid oxidation in the gills of blue crabs acclimated to dilute seawater, *J. Exp. Zool.*, 226, 45, 1983.
84. **Tkachuck, R. D., Rosenberg, G. D., and Hughes, W. W.,** Utilization of free amino acids by mantle tissue in the brachiopod, *Terebratalia transversa* and the bivalve mollusc, *Chlamys hastata*, *Comp. Biochem. Physiol.*, 92B, 747, 1989.
85. **Henry, R. P. and Mangum, C. P.,** Salt and water balance in the oligohaline clam, *Rangia cuneata*. I. Anisosmotic extracellular regulation, *J. Exp. Zool.*, 211, 1, 1980.
86. **Engel, D. W., Ferguson, R. L., and Eggert, L. D.,** Respiration rates and ATP concentrations in the excised gills of the blue crab as a function of salinity, *Comp. Biochem. Physiol.*, 52A, 669, 1975.
87. **King, E. N.,** The oxygen consumption of intact crabs and excised gills as a function of decreased salinity, *Comp. Biochem. Physiol.*, 15, 93, 1965.
88. **Walsh, P. J. and Henry, R. P.,** Activities of metabolic enzymes in the deep-water crabs *Chaceon fenneri* and *C. quinquedens* and the shallow-water crab *Callinectes sapidus*, *Mar. Biol.*, 106, 343, 1990.
89. **Henry, R. P., Booth, C. E., Lallier, F. H., and Walsh, P. J.,** Post-exercise lactate production and metabolism in three species of aquatic and terrestrial decapod crustaceans, *J. Exp. Biol.*, 186, 215, 1994.
90. **Schoffeniels, E.,** Biochemical approaches to osmoregulatory processes in crustacea, *Perspect. Exp. Biol.*, 1, 107, 1976.

91. **Neufeld, D. S. and Wright, S. H.,** Basolateral transport of tauring in epithelial cells of isolated, perfused *Mytilus* gills, *J. Exp. Biol.*, 198, 465, 1995.
92. **Kirschner, L. B.,** Control mechanisms in crustaceans and fishes, in *Mechanisms of Osmoregulation in Animals*, Gilles, R., Ed., Wiley-Interscience, New York, 1979, 157.
93. **Gilles, R. and Pequeux, A.,** Cell volume regulation in crustaceans: relationship between mechanisms for controlling the osmolality of extracellular and intracellular fluids, *J. Exp. Zool.*, 215, 351, 1981.
94. **Evans, D. H. and Cameron, J. N.,** Gill ammonia transport, *J. Exp. Zool.*, 239, 17, 1986.
95. **Bishop, S. H., Greenwalt, D. E., Kapperl M. A., Paynter, K. T., and Ellis, L. L.,** Metabolic regulation of proline, glycine and alanine accumulation as intracellular osmolytes in ribbed mussel gill tissue, *J. Exp. Zool.*, 268, 151, 1994.
96. **Dragolovich, J.,** Dealing with salt stress in animal cells: the role and regulation of glycine betaine concentrations, *J. Exp. Zool.*, 268, 139, 1994.

Chapter 5

Membrane Transport and Nitrogenous Solute Balance in Molluscs

Stephen H. Wright and Douglas S. Neufeld

CONTENTS

INTRODUCTION

Nitrogenous solutes play an important role in several aspects of the physiology of marine molluscs (and other marine invertebrates, as well). As necessary precursors to the synthesis of proteins, amino acids play a pivotal role in growth and development. Metabolic breakdown of amino acids can be a significant source of energy to support both growth and maintenance metabolism. In addition, the free amino acid and quaternary ammonium compounds found in molluscan tissues are quantitatively significant osmotic effectors for the maintenance of normal cell volume; regulated changes in the concentrations of these nitrogenous compounds are known to accompany the change in cell volume associated with the exposure of euryhaline molluscs to different salinities.

For the reasons outlined above, processes that influence the cytoplasmic concentration of nitrogenous solutes may play a critical role in the physiology of marine molluscs. Under steady-state conditions, the cytoplasmic concentration of nitrogenous solutes remains constant and processes that lead to the accumulation of amino acids and quaternary ammonium compounds are balanced by the loss of these solutes. Mechanisms by which nitrogenous solutes are accumulated include: *de novo* synthesis, protein degradation (for amino acids), and transport. Mechanisms by which these solutes are lost from cells include: catabolism, protein synthesis, and transport. The metabolic processes by which the concentration of nitrogenous solutes are changed have received considerable attention, particularly in the studies of Bishop,[1,2] Kohen,[3,4] and their colleagues. In particular, these studies have clearly shown that regulation of cell metabolism can play a central role in changes in the tissue concentration of amino acids.

The transport of nitrogenous solutes by marine molluscs has also been studied extensively, but the emphasis of these latter studies has typically been either on (1) the mechanism of the transport

0-8493-8411-7/95/$0.00+$.50

processes themselves,[5,6] or (2) the influence that they might have on the nutritional requirements of marine molluscs.[6-8] This chapter provides an overview of current ideas on these two issues. In addition, it will consider the potential role of membrane transport in the maintenance of cytoplasmic nitrogenous solute concentrations and suggest directions for future research on this subject.

HISTORY AND OVERVIEW OF INTEGUMENTAL TRANSPORT OF NITROGENOUS SOLUTES IN MARINE INVERTEBRATES

August Pütter[9] is generally credited with formulating at the turn of the century the hypothesis that aquatic animals accumulate dissolved organic material (DOM), e.g., free amino acids (FAA), directly from the surrounding medium into the general integument of the animal, although as noted recently by Manahan,[8] speculation on this subject dates back to the *Challenger* expedition. However, in 1931, August Krogh[10] reviewed the status of evidence on this issue and, drawing on his own studies of DOM accumulation in *freshwater* animals, concluded that they display little, if any, capacity to accumulate DOM from the medium. The subject received little attention until the late 1950s and early 1960s, which included the seminal contribution of Stephens and Schinske[11] presenting direct evidence showing that the capacity to accumulate amino acids from aqueous media is broadly distributed throughout marine, rather than freshwater, invertebrate phyla.

Grover Stephens et al. proceeded to document and characterize the nature of "integumental" DOM transport in marine invertebrates. These studies, as well as the earlier work by Pütter, Krogh, and their contemporaries, are outlined in Jorgensen's excellent review.[12] It is useful to summarize the general characteristics of integumental transport as defined by these studies:

1. All soft-bodied marine invertebrates are capable of accumulating FAA directly from sea water into the outwardly facing surface of the general integument. See References 13 to 15 for examples. Indeed, the Arthropoda appears to be the only marine invertebrate phylum which does not possess this capacity.[16,17]
2. In some cases, integumental accumulation may be dominated by specialized integumental organs (e.g., gills of polychaetes[18] and bivalves[19]).
3. As a process, integumental transport is carrier-mediated and adequately described by Michaelis-Menten saturation kinetics. See References 20 to 22 for examples.
4. Typically, separate transport pathways exist for the accumulation of glucose and several structurally distinct forms of amino acid. See References 23 and 24 for examples.
5. Integumental transport processes can accumulate substrates against chemical gradients that, in extreme cases, can exceed several million to one.[25] The energy for this concentrative transport appears to reside in the electrochemical gradient for Na^+,[26-28] and integumental transport appears fall into the general category of a "secondarily active, Na-cotransport process."[5,29]

CHARACTERISTICS OF NITROGENOUS SOLUTE TRANSPORT IN MOLLUSCS

Integumental transport in marine molluscs displays all the characteristics noted above. Owing to their ease in handling, most studies of transport in molluscs have used marine bivalves, particularly of the genus *Mytilus*. Following is a discussion of the current understanding of nitrogenous solute transport in *Mytilus* tissue, particularly in the gill which is the principal site of integumental transport.[30,31] There are at least four distinct pathways for nitrogenous solute transport in the apical (i.e., seawater-facing) aspect of *Mytilus* gill.[23] These include pathways that show specificity for: (1) taurine and other β-amino acids; (2) aspartate and other anionic amino acids; (3) alanine, other zwitterionic α-amino acids, and the cationic amino acid lysine; and (4) a distinct pathway for alanine, other zwitterionic α-amino acids, and the imino acid proline. This latter pathway (designated the "AP" pathway[28]) also transports the quaternary ammonium compound, betaine.[32] The gill also has at least one carrier-mediated pathway in the basolateral aspect of gill cells, i.e., a processes specific for taurine and other β-amino acids.[33] The mechanisms of all these processes appear to be similar.[5,6] Consequently, attention will be limited to a more detailed discussion of the pathways that handle the principal nitrogenous solutes of gill cells (i.e., taurine and betaine).

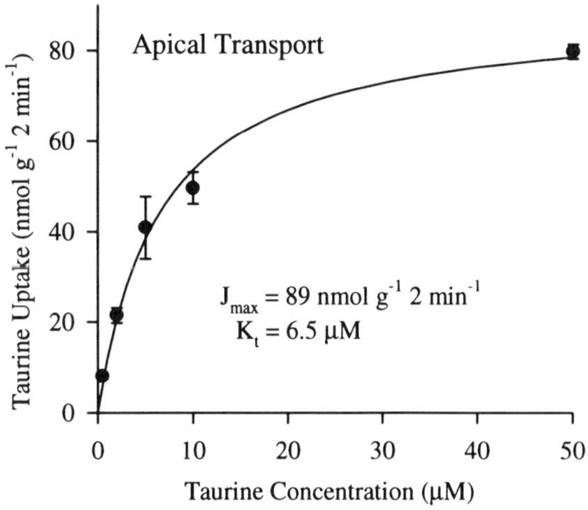

Figure 1 The effect of substrate concentration on the accumulation of taurine from seawater into isolated *M. californianus* gill tissue. Transport occurred across the apical (i.e., seawater-facing) aspect of the gill integumental epithelium. The curve describing this kinetic relationship was calculated from the Michaelis-Menten equation using the kinetic constant (J_{max} and K_t) presented in the figure. (Modified from Silva, A. L. and Wright, S. H., *J. Exp. Biol.*, 162, 265, 1992. With permission.)

NITROGENOUS SOLUTE TRANSPORT IN *MYTILUS* GILL

Apical Taurine Transport

Uptake of radiolabeled taurine into gills isolated from the California coastal mussel, *Mytilus californianus*, is a saturable process described by the Michaelis-Menten equation (Figure 1). The maximal rate (J_{max}) of apical taurine transport is between 1 and 4 μmol g^{-1} h^{-1}, and transport is half-saturated (K_t) at an ambient seawater concentration of 2 to 10 μM.[20,33,34] Transport of taurine is inhibited by β-amino acids (β-alanine and β-aminobutyric acid) and, to a lessor extent, by the γ-amino acid, γ-aminobutyric acid.[30] Alpha-amino acids and betaine have no inhibitory effect on apical taurine transport. In both its kinetics and apparent structural specificity, apical taurine transport appears to be quite similar to taurine transport processes in tissues in other animals.[35-39] There is no evidence for a nonsaturable "diffusional" component of apical taurine transport,[20] so it appears that taurine flux across the apical membrane of bivalve gill is effectively limited to a single carrier-mediated pathway.

The accumulation of radiolabeled taurine into integumental tissues represents a net uptake of taurine. When intact mussels are placed into seawater containing submicromolar levels of taurine, chromatographic analysis of the ambient medium reveals the net clearance of taurine at rates which parallel the clearance of the radioactively labeled substrate.[40] Indeed, *M. californianus* and *M. edulis* are capable of the net removal from seawater of taurine from external concentration of less than 10 nM.[40]

Integumental transport of taurine (and other nitrogenous solutes) is an example of "secondary active" transport because the energy for the accumulation of taurine against its electrochemical gradient comes from tapping the potential energy available in the electrochemical gradient of one (or more) other solute(s). The process as expressed in bivalve integument is qualitatively similar to that first described by Crane[41] for the coupled fluxes of glucose and Na^+ in the mammalian small intestine. In the gill, as in most animal cells, the cytoplasmic concentration of Na^+ is less than that in the surrounding solution (e.g., seawater). In addition, the electrical potential difference across the plasma membrane is inside-negative,[42] resulting in a substantial, inwardly directed electrochemical gradient for Na^+ that can drive the "uphill" movement of another solute into cells of the integument if the fluxes of Na^+ and the other solute are directly coupled.

The dependence of integumental taurine transport on Na^+ is evident from the effect on transport of removing Na^+ from seawater. As shown in Figure 2, when Na^+ in seawater is replaced with Li^+, taurine transport is reduced by ~95%. Increasing the concentration of Na^+ in seawater by readdition

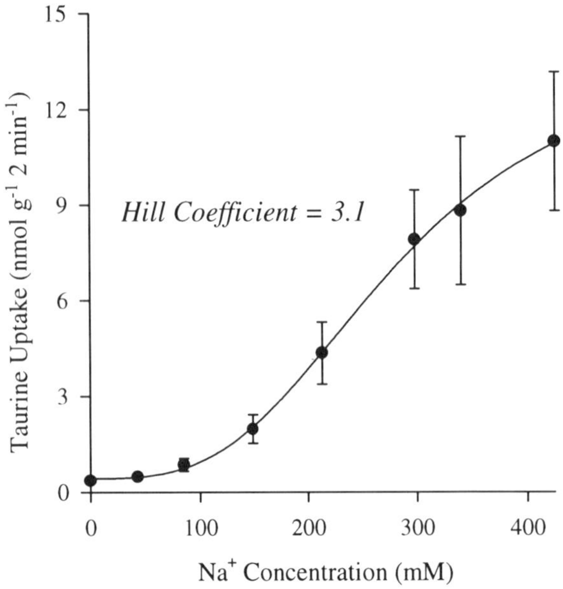

Figure 2 The influence of Na^+ concentration in seawater on the transport of taurine across the apical membrane of isolated *M. californianus* gill. The curve describing the activation of transport by taurine was calculated from the Hill equation (see text). The K_{50} was 370 μM. (Modified from Silva, A. L. and Wright, S. H., *J. Exp. Biol.*, 162, 265, 1992. With permission.)

produces a sigmoid activation of taurine transport. The Hill equation[43] provides a means to estimate the degree of interaction between a substrate and its "activating" ion(s). Integumental taurine transport is described by this relationship:

$$J^{[S]} = \frac{J^{[S]}_{max}[Na]^n}{(K_{0.5})^n + [Na]^n} \tag{1}$$

in which $J^{[S]}$ is the rate of substrate transport from a concentration of [S]; $J^{[S]}_{max}$ is the rate of transport from a concentration of [S] when the transporter is maximally activated; [Na] is the activator ion concentration; $(K_{0.5})^n$ is the concentration of activator that results in one-half $J^{[S]}_{max}$ and n is the Hill coefficient (i.e., a measure of the minimum number of Na ions involved in the activation of substrate transport). The Hill coefficient of 3.1[34] supports the hypothesis that (at least) three Na ions are coupled to uptake of each taurine molecule. In fact, direct evidence for a true "coupling" of Na^+ and taurine fluxes requires evidence beyond the analysis of the catalytic activation by Na^+ of taurine transport.[44] Nevertheless, direct evidence for the coupling of Na^+ to the integumental accumulation of nitrogenous solutes has been obtained for several transport pathways in *Mytilus* gill.[5] The similarities between integumental transport of taurine and the Na^+-taurine cotransporters in other tissues[38,45,46] support the contention that integumental taurine transport is coupled to Na^+ transport and, therefore, is driven by the Na^+ electrochemical gradient.

Taurine transport has in many tissues also been found to involve a cotransport with Cl^-. The role of Cl^- in transport of taurine was first noted for mammalian renal tissues,[47] but this characteristic of taurine transport now appears to be broadly distributed.[48] Apical transport of taurine in *Mytilus* gill also appears to require Cl^-; when Cl^- is replaced by a combination of isethionate and gluconate, taurine uptake is reduced by 90%.[33] In most systems, taurine transport has been found to involve a cotransport with a single Cl ion.[48] Taking into account the coupling of taurine transport to a cotransport with both Na^+ (three ions) and Cl^- (one ion), the equilibrium relationship between the developed electrochemical gradients of all three species is:

$$\frac{[\mathrm{Tau}]_i}{[\mathrm{Tau}]_o} \le \left(\frac{[\mathrm{Na}^+]_o}{[\mathrm{Na}^+]_i}\right)^{(n_{Na}/n_{Tau})} \left(\frac{[\mathrm{Cl}^-]_o}{[\mathrm{Cl}^-]_i}\right)^{(n_{Cl}/n_{Tau})} \exp\left(\frac{-\psi F}{RT}\right)\left(Z_{Na}\frac{n_{Na}}{n_{Tau}} + Z_{Cl}\frac{n_{Cl}}{n_{Tau}}\right) \tag{2}$$

where $[\mathrm{Tau}]_i$, $[\mathrm{Na}^+]_i$, and $[\mathrm{Cl}^-]_i$ are cytoplasmic activities of each solute, $[\mathrm{Tau}]_o$, $[\mathrm{Na}^+]_o$, and $[\mathrm{Cl}^-]_o$ are seawater activities of each solute; ψ is the electrical potential difference (in volts) between cytoplasm and seawater; Z is the valance of the indicated ion; n for each subscripted solute is the number of particles transported per turnover cycle of the transporter; and F/RT has its usual meaning (39.6 V^{-1} at 20°C). Direct estimates for all these parameters are not available for *Mytilus* gill (indeed, for most tissues). Nevertheless, reasonable approximations can be made for cytoplasmic Na^+ ([17,49]) and membrane potential (60 mV, inside negative[42]). For the purpose of this calculation, Cl^- can be assumed to be distributed at equilibrium. Inserting these values (and those for the activity of Na^+ and Cl^- in 100% seawater; i.e., ~300 m*M* for each[50]), into Equation 2 indicates that the maximum taurine gradient that can be supported by the apical taurine transporter is on the order of 7×10^6. Presuming an external taurine concentration (within the mantle cavity) of ~10 n*M**, the transporter can, in principal, support a net uptake of taurine into a maximum cytoplasmic concentration of 70 m*M*. Given the latitude inherent in the estimates of the critical parameters of this calculation, this value is quite close to measured values of ~200 m*M* for cytoplasmic taurine in gill cells acclimated to 100% ASW.

To place the transport "potential" of this pathway into perspective, working at J_{max} apical taurine transport could completely turn over the cytoplasmic taurine pool of gill cells approximately once each day. Of course, the transporter will not achieve such rates under physiological conditions. Indeed, taurine concentrations in near-shore waters are extremely low. Taurine has been measured in water samples taken from tidepools.[51] In samples of near-shore water, however, taurine concentrations are below the limit of resolution of current chromatographic techniques (<5 n*M*; J. R. Welborn and D. T. Manahan, personal communication). Thus, despite the large transport capacity of this process, actual rates of net taurine uptake into gill cells from seawater are likely to be low and the influence of this process on cell taurine levels remains unclear. This issue is addressed in more detail in an upcoming section.

Apical Betaine Transport

Betaine (trimethylglycine) is a quaternary ammonium compound found in large concentrations in the tissues of a number of marine molluscs.[52,53] In gill cells from *M. californianus*, the cytoplasmic betaine concentration is approximately 200 m*M*,[54] making it second only to taurine as an organic osmolyte. A related compound, proline betaine, has been identified in the marine gastropod, *Elysia chlorotica*,[55] and when exposed to hyperosmotic conditions, tissue proline betaine content can reach 650 µmol g^{-1} dry wt.

M. californianus gills can accumulate betaine directly from seawater by a process that shares many of the characteristics noted above for apical taurine transport. Betaine uptake is saturable with a half-saturation constant of 6 µ*M*.[32] The maximal rate of accumulation of betaine is 7 µmol/g wet wt^{-1} h^{-1}, slightly greater than the transport capacity of the taurine transporter. Betaine transport is not inhibited by taurine or other β-amino acids, or by anionic (aspartate) or cationic (lysine) amino acids. Betaine uptake is blocked by zwitterionic α-amino acids (e.g., alanine) and by proline, indicating that transport involves the "AP" amino acid transport pathway[28] of *Mytilus* gill. Uptake of betaine into intact *M. californianus* gill requires Na^+ in seawater[32] which is consistent with evidence showing that active accumulation of organic solutes via the AP pathway involves an obligatory cotransport with Na^+.[28] These characteristics parallel closely those for betaine transport in mammalian renal proximal tubule which also involves a Na-dependent transporter capable of handling proline.[56]

---

* When *Mytilus* is held in a confined volume, the taurine level frequently increases until a steady-state concentration of 3 to 20 n*M* is achieved,[40] leading to the choice of 10 n*M* as a representative concentration within the mantle cavity.

The stoichiometry of coupling between betaine and Na^+ transport in *Mytilus* gill is not clear; efforts to analyze the Na-activation kinetics of the AP pathway have been more equivocal than for other integumental transporters.[32] Consequently, it would be premature to estimate the maximum gradients for betaine that could be sustained by integumental transport. Nevertheless, it is likely that more than one Na^+ ion is coupled to the flux of each organic solute via the AP transporter,[28] suggesting that net flux of betaine (and the other substrates for this process) can occur against gradients upwards of 40,000.

Basolateral Taurine Transport

Efforts to characterize nitrogenous solute transport in invertebrate integument have focused entirely on examination of processes in the apical membrane of the cells of this tissue. The preoccupation with apical membrane events is largely a consequence of the interest of most investigators on the ability of these animals to accumulate materials from seawater, which entails a flux across the seawater-facing, or apical, aspect of the integument. The failure to consider basolateral transport processes stems not, however, from a lack of interest in these events. To the extent that material accumulated into the integument is made available to other tissues in the animal, substrates that enter across the apical membrane must leave the cell across the hemolymph-facing, or basolateral, aspect of the integument. Indeed, several studies have attempted to provide indirect evidence concerning the nature of the *trans*-integumental flux of exogenous solutes.[18,57,58] In addition, any effort to understand the maintenance of solute balance in integumental (or other) cells requires knowledge of all transport pathways in the cell, including basolateral processes. The concentration of nitrogenous material in hemolymph is large, as compared to that in seawater, so flux across the basolateral membrane could play a significant role in defining overall solute levels in integumental cells. The difficulty in studying basolateral transport has been technical; whereas it is comparatively simple to study apical transport processes because, until recently, there has been no tissue preparation available that has permitted the unambiguous examination of transport across the basolateral aspect of the integumental epithelium.

Gills isolated from *Mytilus* have recently proven to be a useful preparation for studying basolateral transport in integumental cells. The gill consists of thousands of individual filaments each of which is an elliptical, epithelial "tube" that is one cell layer thick; the apical aspect of the filament epithelium faces seawater while the basolateral aspect faces the branchial hemolymph. The filaments arise from and insert into afferent and efferent branchial vessels,[59] and isolation of the gills from the animal typically leaves the efferent branchial vessel (EBV) intact. Cannulation and retrograde perfusion of the EBV permits controlled exposure of the basolateral aspect of cells to test solutions.

Basolateral taurine transport has been characterized using this preparation.[33] As shown in Figure 3, basolateral taurine transport is saturable with a K_t and J_{max} of 35 μM and 0.35 μmol g wet wt^{-1} h^{-1}. The basolateral pathway is inhibited by β-alanine, but not by glutamate, α-alanine, or betaine. Taurine uptake is activated by increasing concentrations of Na^+ with a Hill coefficient of 2.3. Removal of Cl^- from the perfusate reduces transport by 70%. Comparison of these characteristics to those described earlier for the apical transport pathway reveals clear parallels: the affinity, capacity, specificity for substrate, and ionic dependence of the two poles of epithelium are similar. Although a firm conclusion about the exact identity of the two processes will require evidence of the kind provided by molecular cloning of the transport proteins, on a functional level at least the apical and basolateral taurine transporters appear to be virtually identical. Thus both aspects of the integument have carrier-mediated pathways for nitrogenous solute transport that can be expected to play a role in the maintenance of solute balance by providing pathways for solute uptake and/or loss.

IMPACT OF MEMBRANE TRANSPORT ON NITROGENOUS SOLUTE BALANCE

As avenues for the uptake (or loss) of solute, integumental transport processes have an obvious potential to influence several aspects of molluscan physiology. Most attention has focused on the

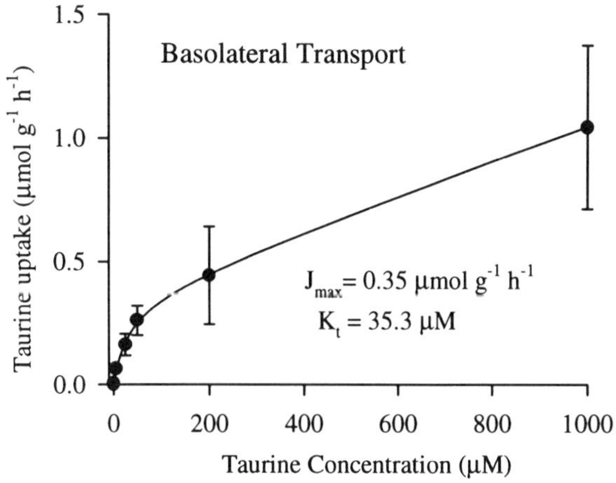

Figure 3 The effect of substrate concentration on the accumulation of taurine across the basolateral aspect of the integumental epithelium in gill tissue isolated from *M. californianus*. Uptake occurred from a seawater solution perfused through the branchial vasculature of the isolated gill. The curve describing this kinetic relationship was calculated from the Michaelis-Menten equation which also included a nonsaturable term that probably reflected the binding of taurine to the extracellular matrix of the branchial vasculature. (Modified from Neufeld, D. S. and Wright, S. H., *J. Exp. Biol.*, 1994. With permission.)

potential these processes can have in the nutrition of marine molluscs. This topic has been extensively reviewed[6-8] so the present report will provide only a synopsis of this issue and the current view of the nutritional impact of integumental transport in marine molluscs. A second topic that has received less attention will also be covered here, namely the possible influence of transport on the steady-state cytoplasmic concentration of nitrogenous solutes and, thereby, on the maintenance of normal cell volume.

NUTRITION

For integumental transport to play a role in whole animal nutrition the process must provide an input of substrate that can be funneled into appropriate catabolic or anabolic pathways at rates which represent some significant fraction of animal's requirements for such materials. The first of these two criteria is clearly met. Amino acids are metabolically versatile and many studies have shown that substrates accumulated directly from seawater rapidly find their way into a diverse set of metabolic pathways.[60,61]

The second of the criteria listed above, i.e., provision of substrate at rates capable of meeting a substantial fraction of the animal's requirements, is more problematic. The potential of integumental transport is, in fact, quite large. Working at maximal rates, integumental accumulation of amino acid by *Mytilus* can provide substrate sufficient to support in excess of 200% of the animal's oxidative metabolism.[62] However, working under conditions found in the environment, actual rates of integumental transport will be well below the maximal capacity. Concentrations of amino acid in near-shore waters are low. Although a few reports have found concentrations of total free amino acid exceeding 1 μ*M*,[63,64] recent studies suggest that, in temperate waters, concentrations of <0.1 μ*M* are more typical, (see Reference 65; J. R. Welborn and D. T. Manahan, personal communication). When exposed to such low levels of substrate, net accumulation of exogenous material could provide only a small fraction of the requirements for reduced carbon in most animals. Concentrations of free amino acid in pore water expressed from marine sediments are generally much higher, ranging from 1 to 100 μ*M*.[62] Therefore, molluscs that burrow in sediment (e.g., *Geukensia demissus*) or which forage on the benthos (e.g., *Hydrobia neglecta*; see Reference 66) may be exposed to higher levels of substrate. Firm conclusions about the quantitative role of integumental transport to the animal nutrition continue to require knowledge of the levels of substrate to which an animal is exposed in its normal habitat.

Recent work has shown that the concentration of dissolved nutrients, including amino acids, is much higher in some open water environments. During the annual spring thaw, levels of amino acid in Antarctic waters can reach concentrations of almost 0.5 m*M*.[67,68] Manahan and his colleagues have considered the potential role of amino acid uptake to the nutrition of larval echinoderms in these waters and have concluded that integumental transport can support as much as 33% of the oxidative requirements of some species when exposed to environmentally relevant concentrations of substrate.[69]

There is also speculation that the concentration of organic nutrients, including amino acids, may have an uneven spatial distribution, reaching very high levels in "micropatches" of small volume.[8,70] These micropatches may be effectively accessed by microorganisms and some planktonic invertebrate larvae.[8] Regardless of the nature of the source, larval gastropods reared in particulate-free media have been shown to gain dry organic mass when exposed to filtered natural seawater, whereas parallel control cultures exposed to artificial seawater lose mass at a rate commensurate with their metabolism.[71,72] The conclusion from these studies is that levels of DOM in near-shore waters, although low, are sufficient to contribute to the nutrition of (at least) some larval molluscs.

MAINTENANCE OF NORMAL CELL VOLUME

As with most soft-bodied marine invertebrates, the body fluids of molluscs are typically isoosmotic with the environment. In molluscs acclimated to 100% seawater, the intracellular concentration of nitrogenous solute is typically quite large, generally representing >50% of the osmotically active material in the cell. Consequently, the maintenance of the cytoplasmic pool of nitrogenous solute is a critical aspect of the maintenance of normal cell volume. Mediated transport pathways represent one mechanism by which cytoplasmic levels of solute can be modulated. Accumulation of solute from the hemolymph is quite likely. In *Mytilus*, for example, the concentration of nitrogenous solutes in the hemolymph is approximately 1 m*M* of which taurine is ~40%.[73] Within the branchial vasculature, the hemolymph taurine concentration may well be somewhat lower than this, owing to basolateral transport of taurine into gill cells and the low rate of hemolymph perfusion suspected to occur in *Mytilus* gill.[31] Nevertheless, the concentration of taurine in the branchial hemolymph is likely to be substantially greater than the 35 μ*M* apparent K_t of the basolateral taurine transporter.[33] If the basolateral transport worked at rates ranging from half to all of its transport capacity, the mediated input of taurine could provide an input each day equal to between 4 and 8% of the total cytoplasmic taurine pool.

Under steady-state conditions cytoplasmic solute concentrations are constant, implying a "loss" of taurine from gill cells, either through metabolism or transport, equal to the rate of total input. Metabolism of taurine (synthesis and degradation) in molluscan tissue is believed to be quite slow, although little data is available on this issue.[2] When exposed to taurine-free media, net loss of taurine from the apical face of the gill is low due to solute recycling (see below) and amounts to approximately 3% of the gill's total taurine pool per day.[40] The efflux of taurine across the basolateral membrane, measured by preloading gill tissue with radiolabeled taurine (D.S. Neufeld, unpublished observations), is approximately equal to that of the apical membrane. Thus, under steady-state conditions, a stable taurine concentration in gill cells could be maintained through the concerted activity of influx and efflux pathways.

Solute Recycling

The large cytoplasmic concentrations of molecules such as taurine and betaine favor the passive loss from cells to the extracellular space. The gradient supporting this loss is most extreme at the integumental interface between the animal and the surrounding seawater. Indeed, there has been speculation that passive losses of amino acids from marine invertebrates are substantial and routine.[74,75] However, studies on the net uptake of amino acid from seawater clearly show that net efflux of amino acids from soft-bodies marine invertebrates is very low, despite the outwardly directed gradients favoring such loss.[25,40]

Early studies on the transport of amino acids in bivalve molluscs led Bamford[76] to suggest that uptake could play a role in limiting the net loss of these compounds from the animal to the surrounding medium. In Gomme's studies on glucose transport in the integument of polychaetes [*Nereis diversicolor*;[77] and nematodes (*Ascaris*[78])], integumental transport was shown to be capable of "recycling" endogenous materials lost from the integument surface by passive processes (e.g., by diffusion across the apical membrane or through paracellular pathways), thereby reducing losses to the surrounding medium. Wright and Secomb[40] extended Gomme's work by considering the influence of integumental transport on the recycling of taurine in *Mytilus* gill in light of the structure and fluid flow characteristics of this tissue.[30] They developed a mathematical model relating the interaction of these parameters and found that apical taurine transport could recover up to 50% of the taurine that would otherwise be lost from the gill by passive processes. Independent estimates by Gomme[79] suggested that losses could be reduced by as much as 90%. Placed in different terms, ~5% of the gill's taurine pool would be lost each day if integumental taurine transport were not capable of recycling this material. The characteristics that permit taurine transporters to recover endogenous taurine are effectively duplicated by the several other transport processes in the gill, suggesting that loss of other endogenous solutes (e.g., betaine) is also limited through the activity of integumental transport.

Influence of Mediated Efflux on Cytoplasmic Solute Levels

As noted above, the balance between mediated influx and efflux of solute from gill cells could play an important role in maintaining steady-state osmolyte levels and, thereby, influence normal cell volume. Solute transport could also play a role in influencing the changes of cell solute levels that follow exposure to media of different salinities.

When euryhaline molluscs are shifted from high to low salinity, the ensuing osmotic challenge results in a rapid gain of tissue water.[80] This cell swelling is typically followed by a loss of solute and a concomitant loss of "osmotically-obligated" water, thereby achieving a return, albeit usually partial, toward normal cell volume. During long-term exposure to reduced salinity, nitrogenous solutes are frequently lost to a greater extent than are inorganic solutes such as K^+.[81] Yancey et al.[82] suggested that the selective advantage of this pattern of solute loss arises from the fact that the activity of cytoplasmic enzymes is less perturbed by changes in concentration of the so-called "organic osmolytes" (e.g., taurine and betaine) than by changes in inorganic ions (e.g., K^+ and Na^+). Subsequent studies in a number of experimental systems have shown that the regulation of organic osmolytes is a central and highly conserved feature of cell volume regulation in animals, plants, and bacteria.[83]

Florkin et al. first noted that a principal characteristic of the long-term response of marine molluscs to hypoosmotic stress is the specific reduction of tissue content of a few key nitrogenous solutes, i.e., taurine and betaine. For example, when *M. edulis* is shifted from 100% seawater (~980 mOsm) to 50% seawater, taurine concentration in the cytoplasm of adductor muscle decreases within 2 weeks from 100 to 67 m*M*, and tissue betaine decreases from 108 to 61 m*M*.[53] Similarly, when *M. californianus* is exposed for several weeks to 60% seawater, the cytoplasmic concentration of taurine and betaine decreases from approximately 220 and 190 m*M* to 154 and 109 m*M*, respectively.[54] Although the qualitative pattern of nitrogenous solute loss differs in other species (e.g., *G. demissus* displays a large loss of glycine in addition to taurine[84,85]), the general response of euryhaline molluscs to long-term hypoosmotic stress appears to be similar: a substantial loss of a few principal nitrogenous solutes as a means to effect a loss of osmotically obligated water.

The mechanism by which the cytoplasmic concentration of nitrogenous solutes is reduced in response to a reduction in external salinity has received little attention. Pierce and Greenberg[84] noted that acute exposure of ventricles isolated from *M. demissus* (*G. demissus*) to dilute seawater causes a rapid efflux of amino acid to the surrounding medium. Moreover, this loss appears to be influenced by the concentration of Ca^{2+} in the surrounding medium.[86] Subsequent studies of the volume regulatory responses of red cells from the clam, *Noetia ponderosa*, by Pierce and his colleagues[87-91] have further implicated a calcium sensitive set of processes in the mediated loss of taurine from these cells during acute exposure to dilute seawater.

While studies demonstrating an increased FAA efflux during an abrupt decrease in salinity clearly are consistent with the function of amino acids in osmotic balance, a quantitative analysis of the efflux rate casts doubt on its role in any rapid volume regulation. In *Mercenaria* ventricles, efflux was only ~5 μmol g^{-1} dry wt 2 h^{-1} in 100% seawater, but increased to 37 μmol g^{-1} 2h^{-1} after a switch to 50% seawater.[92] Given a total concentration of FAAs in ventricles at 100% seawater of 534 μmol g^{-1} dry weight, the total loss of FAAs in the 2 h after a switch to low salinity represents less than 10% of the total pool of FAAs. In red blood cells of *Noetia*, which clearly demonstrate a rapid RVD of approximately 50% in the 2-h period after a salinity change, measured efflux of amino acid during the 2-h period coincides with the drop in FAA content of cells.[93] However, using the reported volume of cells at 100% seawater (1.5×10^3 μm^3 per cell, equal to 1.5 μl/10^6 cells), cell volume increases to 2.1 μl/10^6 cells after exposure to 50% seawater followed by volume regulation back to 1.8 μl/10^6 cells. At 500 mOsm, the decrease in cell volume is equivalent to a 150 nmol loss of solute, which is much greater than the observed FAA efflux of 20 nmol/10^6 cells. Likewise, the FAA loss in *Geukensia* ventricles of ~125 μmol g^{-1} during the 2 h after an abrupt switch from 100 to 50% seawater[94] is a relatively small proportion of the total content of FAAs of 750 μmol g^{-1}.[95] The efflux rate in *Geukensia* gills of 35 μmol g^{-1} 2 h^{-1} is even lower, representing only 10% of the FAA content of gills that were acclimated for a short time in 100% seawater.[1]

These data show that FAA losses in the several hours after a salinity decrease, when any acute volume regulation would occur, represent a small fraction of the total pool of osmolytes. While the reported rates of FAA efflux appear too low to be responsible for any substantial short-term regulation of cell volume, it is likely that they are important in effecting long-term changes in FAA content. A lower rate of FAA loss is, in fact, consistent with the slow drop in FAA content normally caused by a salinity. FAA content takes many hours or days to fully change in red cells collected from hypotonically stressed *Noetia*,[93] *Geukensia* ventricle,[95,96] *Polymesoda* and *Corbicula* muscle,[97] *M. edulis* whole tissues,[98] and *Mytilus galloprovincialis* gill and muscle.[99]

Integumental transport (i.e., the mediated influx and efflux of solute), offers a potential mechanism by which euryhaline molluscs could effect a gradual loss of selected solutes following a shift to reduced salinity and then maintain a reduced concentration of these solutes during long-term exposure to the dilute medium. As noted earlier, the cytoplasmic osmolyte pool of most molluscs is typically dominated by very large concentrations of one or two nitrogenous solutes. In *Mytilus*, these are taurine and betaine. The transmembrane chemical gradients of these solutes are very large and, along integumental interfaces, are supported through the coupling to the energy in the electrochemical gradients for Na^+ (and, probably, Cl^-). In fact, within the limits of resolution available from currently available data, it appears that integumental transporters are "poised" at equilibrium, with the energy in the outwardly directed gradients of nitrogenous solute balanced by that available from coupling to the cotransported activator ions. This situation is presented in graphical form in Figure 4 which relates (1) the energy available in the chemical and electrical components of the forces supporting taurine transport to (2) the maximum gradient of taurine that can be sustained by activity of the integumental transporter.

It is evident from Figure 4 that any reduction in either the chemical or electrical driving forces that support taurine transport would reduce the maximum gradient of taurine that could be supported by activity of the transporter. Arrow 1 shows the maximum steady-state concentration of taurine (calculated using Equation 2) that could be sustained by integumental transport in tissue acclimated to 100% seawater. A reduction in either the chemical or electrical forces supporting this transport would result in a net efflux of taurine from the cell (via the transporter) until a new steady-state/equilibrium condition was achieved. For example, a decrease in external salinity from 100 to 60% ASW would result in a reduction in the external concentration (i.e., activity) of both Na^+ and Cl^-. The decrease in external concentration of these activator ions influences the rate of taurine transport.

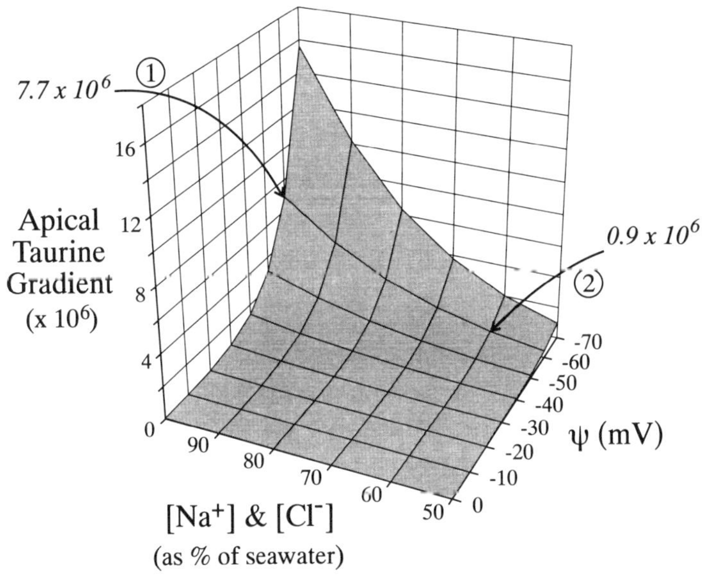

Figure 4 The influence of chemical (i.e., Na^+ and Cl^-) and electrical gradients on the maximum taurine gradient that can be sustained by transport across the apical membrane of *M. californianus* gill. The three-dimensional representation of the interplay between these parameters was calculated using Equation 2. The activities of Na^+ and Cl^- in seawater were assumed to be 0.70 and 0.62, respectively. For the purpose of these estimates, the intracellular Na^+ activity was taken to be 17 m*M*, and the intracellular Cl^- activity was taken to be 31 m*M*. Arrow 1 indicates the maximum taurine gradient that can be sustained by the apical membrane taurine transporter in gills of animals acclimated to 100% seawater. Arrow 2 indicates the maximum taurine gradient that can be sustained by the apical membrane taurine transporter if the ambient salinity was reduced to 60% (i.e., 40% decrease in external Na^+ and Cl^- activities). This estimate assumes that (1) neither intracellular Na^+ nor Cl^- changes as a consequence of the decrease in external salinity; and (2) the transmembrane PD does not change as a consequence of the change in external salinity (see text).

Indeed, when animals are acclimated to 60% seawater, the rate of taurine transport in gill tissue is reduced to the level predicted from the kinetics of Na-activation of the transporter.[34] The change in external ion concentration may well be expected to also influence the maximum gradient of taurine that can be supported by the transport. If, for example, intracellular concentrations of Na^+ and Cl^-, and the membrane PD, remained unchanged following exposure to 60% seawater, then the maximum sustainable taurine gradient would be reduced from 8×10^6 to 1×10^6 (Arrow 2 on Figure 4). If the effective external concentration of taurine remained unchanged, this would be equivalent to a reduction in the cytoplasmic concentration of taurine from ~80 to 10 m*M*. Moreover, this new concentration could be sustained through the activity of the transporter under this new set of energetic constraints.

UNRESOLVED ISSUES

Long-term exposure to reduced salinity does, in fact, result in decreases in the concentration of taurine (and/or other nitrogenous solutes) in integumental cells of some molluscs that approach those predicted above. Of course, the simplifying assumptions used to make these estimates are unlikely to hold when animals are exposed to changes in external salinity; both intracellular ion concentrations and the transmembrane PD may well change, each of which would influence the equilibrium distribution of a transported solute. There is no information currently available on the influence of salinity on these parameters in nonneural tissues.[100] Likewise, transporters in the basolateral membrane would be expected to have altered transport characteristics, but, again, there are no studies examining the effect of salinity on transport of solute from hemolymph into mollusc cells. Several recent studies have implicated volume-sensitive membrane channels in mediating the loss of organic solute, including taurine, in some vertebrate cells.[101] Such processes, if present in molluscs, could influence the rate of passive solute flux during long-term exposure to reduced salinity. Finally, rates of solute synthesis and/or degradation could change dramatically at reduced salinities and thereby influence the overall balance of solute input and output that defines the steady-state concentration of solutes in molluscan tissue (see Chapter 4). In fact, there are virtually no data on which to draw any conclusions concerning which process or combination of processes is most likely to regulate the concentration of nitrogenous solute in molluscan tissue during long-term exposure to different salinity. However, the point of the present discussion is to emphasize the potential role of mediated transport processes in influencing cytoplasmic concentration of osmolytes. Future studies must determine the relationship between rates of osmolyte transport and transmembrane chemical and electrical forces as a function of exposure to different salinities. Such studies may provide unique insights into the processes by which molluscan tissues manipulate nitrogenous solute content.

SUMMARY

Marine molluscs maintain large concentrations of nitrogenous solute in their cells. These solutes play an important role in molluscan metabolism and in the control of cell volume. Although the presence of transport processes capable of accumulating nitrogenous solute in molluscan tissue has been documented, the influence of these processes in the maintenance of nitrogenous solute balance in molluscs has received little attention. The transport potential of these processes is such that their activity could profoundly influence cytoplasmic levels of these compounds, including the regulation of their concentration during exposure to changing salinity. However, too little is known about the actual chemical and electrical forces that dictate actual rates of transport to draw firm conclusions about the role played by membrane transport in regulating nitrogenous solute balance. The study of these issues offers the potential of providing new insights into the mechanisms by which molluscs regulate the concentration of this important class of compounds.

ACKNOWLEDGMENTS

This work was supported in part by NSF grants DCB8819367 and IBN9407997 and by NIH Training Grant HL-07249.

REFERENCES

1. **Bishop, S. H., Greenwalt, D. E., Kapper, M. A., Paynter, K. T., and Ellis, L. L.,** Metabolic regulation of proline, glycine, and alanine accumulation as intracellular osmolytes in ribbed mussel gill tissue, *J. Exp. Zool.*, 268, 151, 1994.
2. **Bishop, S. H., Ellis, L. L., and Burcham, J. M.,** Amino acid metabolism in molluscs, in *The Mollusca*. I. *Metabolic Biochemistry and Molecular Biomechanics*, Hochachka, P. W., Ed., Academic Press, New York, 1983, 243.
3. **Deaton, L. E., Hilbish, T. J., and Koehn, R. K.,** Protein as a source of amino nitrogen during hyperosmotic volume regulation in the mussel *Mytilus edulis*, *Physiol. Zool.*, 57, 609, 1984.
4. **Deaton, L. E., Hilbish, T. J., and Koehn, R. K.,** Hyperosmotic volume regulation in the tissues of the mussel *Mytilus edulis*, *Comp. Biochem. Physiol.*, 80A, 571, 1985.
5. **Wright, S. H. and Pajor, A. M.,** Mechanisms of integumental amino acid transport in marine bivalves, *Am. J. Physiol.*, 257, R473, 1989.
6. **Wright, S. H. and Ahearn, G. A.,** Nutrient absorption in invertebrates, in *Handbook of Physiology. Comparative Physiology*, Dantzler, W. D., Ed., American Physiology Society, Bethesda, MD, 1995, in press.
7. **Wright, S. H. and Manahan, D. T.,** Integumental nutrient transport by aquatic organisms, *Annu. Rev. Physiol.*, 51, 585, 1989.
8. **Manahan, D. T.,** Adaptations by invertebrate larvae for nutrient acquisition from seawater, *Am. Zool.*, 30, 147, 1990.
9. **Pütter, A.,** *Die Ernährung der Wassertiere und der Stoffhaushalt der Gewässer*, Fischer, Jena, 1909,
10. **Krogh, A.,** Dissolved substances as food of aquatic organisms, *Biol. Rev.*, 6, 412, 1931.

11. **Stephens, G. C. and Schinske, R. A.**, Uptake of amino acids by marine invertebrates, *Limnol. Oceanogr.*, 6, 175, 1961.
12. **Jorgensen, C. B.**, August Pütter, August Krogh, and modern ideas on the use of dissolved organic matter in aquatic environments, *Biol. Rev.*, 51, 291, 1976.
13. **Stephens, G. C.**, Uptake of organic material by aquatic invertebrates. I. Uptake of glucose by the solitary coral, *Fungia scutaria*, *Biol. Bull.*, 123, 648, 1962.
14. **Chien, P. K., Stephens, G. C., and Healey, P. L.**, The role of ultrastructure and physiological differentiation of epithelia in amino acid uptake by the bloodworm, *Glycera dibranchiata*, *Biol. Bull.*, 142, 219, 1972.
15. **Pearse, J. S. and Pearse, V. B.**, Removal of glycine from solution by the sea urchin *Strongylocentrotus purpuratus*, *Mar. Biol.*, 19, 281, 1973.
16. **Anderson, J. W. and Stephens, G. C.**, Uptake of organic material by aquatic invertebrates. VI. Role of epiflora in apparent uptake of glycine by marine crustaceans, *Mar. Biol.*, 4, 243, 1969.
17. **Castille, F. L., Jr. and Lawrence, A. L.**, The role of bacteria in the uptake of hexoses from seawater by postlarval penaeid shrimp, *Comp. Biochem. Physiol.*, 64A, 41, 1979.
18. **Chien, P. K. and Rice, M. A.**, Autoradiographic localization of exogenously supplied amino acids after uptake by the polychaete *Glycera dibranchiata* Ehlers, *Wassman J. Biol.*, 43, 60, 1985.
19. **Péquignat, E.**, A kinetic and autoradiographic study of the direct assimilation of amino acids and glucose by organs of the mussel *Mytilis edulis*, *Mar. Biol.*, 19, 227, 1973.
20. **Wright, S. H.**, Alanine and taurine transport by the gill epithelium of a marine bivalve: effect of sodium on influx, *J. Membr. Biol.*, 95, 37, 1987.
21. **Stevens, B. R. and Preston, R. L.**, The transport of L-alanine by the integument of the marine polychaete, *Glycera dibranchiata*, *J. Exp. Zool.*, 212, 119, 1980.
22. **Davis, J. P., Keenan, C. L., and Stephens, G. C.**, Na^+-dependent amino acid transport in bacteria-free sea urchin larvae, *J. Comp. Physiol. B*, 156, 121, 1985.
23. **Wright, S. H.**, Multiple pathways for amino acid transport in *Mytilus* gill, *J. Comp. Physiol. B*, 156, 259, 1985.
24. **Pajor, A. M., Moon, D. A., and Wright, S. H.**, Sodium D-glucose cotransport in the gill of marine mussels: studies with intact tissues and brush-border membrane vesicles, *J. Membr. Biol.*, 107, 77, 1989.
25. **Manahan, D. T., Wright, S. H., and Stephens, G. C.**, Simultaneous determination of net uptake of 16 amino acids by a marine bivalve, *Am. J. Physiol.*, 244, R832, 1983.
26. **Pajor, A. M. and Wright, S. H.**, L-Alanine uptake in brush border membrane vesicle from the gill of a marine bivalve, *J. Membr. Biol.*, 96, 209, 1987.
27. **Buck, M. and Schlichter, D.**, Driving forces for the uphill transport of amino acids into epidermal brush border membrane vesicles of the sea anemone, *Anemonia sulcata* (Cnidaria, Anthozoa), *Comp. Biochem. Physiol.*, 88A, 273, 1987.
28. **Pajor, A. M. and Wright, S. H.**, Uptake of lysine and proline *via* separate a-neutral amino acid transport pathways in *Mytilus* gill brush border membranes, *J. Membr. Biol.*, 107, 237, 1989.
29. **Preston, R. L.**, Sodium amino acid cotransport systems in marine invertebrates, in *Comparative Aspects of Sodium Cotransport Systems*, Kinne, R. K. H., Ed., Karger, Basel, 1990, 1.
30. **Wright, S. H. and Secomb, T. W.**, Epidermal taurine transport in marine mussels, *Am. J. Physiol.*, 247, R346, 1984.
31. **Jorgensen, C. B.**, Patterns of uptake of dissolved amino acids in mussels (*Mytilus edulis*), *Mar. Biol.*, 73, 177, 1983.
32. **Wright, S. H., Wunz, T. M., and Silva, A. L.**, Betaine transport in the gill of a marine mussel, *Mytilus californianus*, *Am. J. Physiol.*, 263, R226, 1992.
33. **Neufeld, D. S. and Wright, S. H.**, Basolateral transport of taurine in epithelial cells of isolated, perfused *Mytilus* gill, *J. Exp. Biol.*, 198, 465, 1995.
34. **Silva, A. L. and Wright, S. H.**, Integumental taurine transport in *Mytilus* gill: short-term adaptation to reduced salinity, *J. Exp. Biol.*, 162, 265, 1992.
35. **Benyajati, S. and Johnson, J. L.**, Characteristics of renal transport of taurine in reptilian brush-border membranes, *Am. J. Physiol.*, 260, R879, 1991.
36. **Huxtable, R. J.**, Physiological actions of taurine, *Physiol. Rev.*, 72, 101, 1992.
37. **Turner, R. J.**, B-Amino acid transport across the renal brush-border membrane is coupled to both Na and Cl, *J. Biol. Chem.*, 261, 16060, 1986.
38. **Munck, L. K. and Munck, B. G.**, Distinction between chloride-dependent transport systems for taurine and β-alanine in rabbit ileum, *Am. J. Physiol.*, 262, G609, 1992.
39. **Thoroed, S. M. and Fugelli, K.**, Characterization of the Na^+-dependent taurine influx in flounder erythrocytes, *J. Comp. Physiol. B*, 163, 307, 1993.
40. **Wright, S. H. and Secomb, T. W.**, Epithelial amino acid transport in marine mussels: role in net exchange of taurine between gills and sea water, *J. Exp. Biol.*, 121, 251, 1986.
41. **Crane, R. K.**, Na-dependent transport in the intestine and other animal tissues, *Fed. Proc.*, 24, 1000, 1965.
42. **Murakami, A. and Takahashi, K.**, Correlation of electrical and mechanical responses in nervous control of cilia, *Nature*, 257, 48, 1975.
43. **Segel, I. H.**, *Enzyme Kinetics*, John Wiley & Sons, New York, 1975, 1.

44. **Turner, R. J.**, Quantitative studies of cotransport systems: models and vesicles, *J. Membr. Biol.*, 76, 1, 1983.
45. **Zelkovic, I., Stejskal-Lorenz, E., Lohstroh, P., Budreau, A., and Chesney, R. W.**, Anion dependence of taurine transport by rat renal brush-border membrane vesicles, *Am. J. Physiol.*, 256, F646, 1989.
46. **Benyajati, S. and Bay, S. M.**, Sodium-taurine cotransport in reptilian renal brush-border membrane vesicles, *Pflugers Arch.*, 421, 168, 1992.
47. **Chesney, R. W., Gusowski, N., Dabbagh, S., Theissen, M., Padilla, M., and Diehl, A.**, Factors affecting the transport of β-amino acids in rat renal brush-border membrane vesicles. The role of external chloride, *Biochim. Biophys. Acta*, 812, 702, 1985.
48. **Bogé, G. and Pérès, G.**, Chloride requirements of sodium cotransport systems, *Comp. Physiol.*, 7, 186, 1990.
49. **Gerencser, G. A.**, Na^+ absorption in *Aplysia* intestine: Na^+ fluxes and intracellular Na^+ and K^+ activities, *Am. J. Physiol.*, 244, R412, 1983.
50. **Barrow, G. A.**, Ed., *Physical Chemistry*, 2 McGraw-Hill, New York, 1966, 1.
51. **Almeida, E. A. C., Ueda, R., Benjeloun, Z., and Chien, P. K.**, Amino acid uptake and epithelial ultrastructure in the black snail *Tegula funebralis* Adams, *Wassman J. Biol.*, 47, 86, 1989.
52. **Bricteux-Gregoire, S., Duchâteau-Bosson, G., Jeuniaux, C., and Florkin, M.** Constituants osmotiqument actifs des muscles adducteurs d'*Ostrea edulis* adaptée a l'eau de mer ou a l'eau saumâtre, *Arch. Int. Physiol. Biochim.*, 72, 267, 1964.
53. **Bricteux-Gregoire, S., Duchâteau-Bosson, G., Jeuniaux, C., and Florkin, M.**, Constituants osmotiquement actifs des muscles adducteurs de *Mytilus edulis* adaptée a l'eau de mer ou a l'eau saumâtre, *Arch. Int. Physiol. Biochim.*, 72, 116, 1964.
54. **Silva, A. L. and Wright, S. H.**, Short-term cell volume regulation in *Mytilus* gill californianus, *J. Exp. Biol.*, 194, 47, 1994.
55. **Pierce, S. K. and Rowland. L. M.**, Proline betaine and amino acid accumulation in sea slugs (*Elysia chlorotica*) exposed to extreme hyperosmotic conditions, *Physiol. Zool.*, 61, 205, 1988.
56. **Wunz, T. M. and Wright, S. H.**, Betaine transport in rabbit renal brush-border membrane vesicles, *Am. J. Physiol.*, 264, F948, 1993.
57. **Qafaiti, M. and Stephens, G. C.**, Distribution of amino acids to internal tissues after uptake in the annelid *Glycera dibranchiata*, *J. Exp. Biol.*, 136, 177, 1988.
58. **Rice, M. A. and Stephens, G. C.**, Influx and transepithelial flux of amino acids in the mussel, *Mytilus edulis*, *J. Exp. Biol.*, 135, 275, 1988.
59. **White, K. M.**, Mytilus, in *L.M.B.C. Memoirs on the Typical British Marine Plants & Animals*, Daniel, R. J., Ed., University Press of Liverpool, 1937, 1.
60. **Manahan, D. T.**, The uptake and metabolism of dissolved amino acids by bivalve larvae, *Biol. Bull.*, 164, 236, 1983.
61. **Jaeckle, W. B. and Manahan, D. T.**, Amino acid uptake and metabolism by larvae of the marine worm *Urechis caupo* (Echiura), a new species in axenic culture, *Biol. Bull.*, 176, 317, 1989.
62. **Wright, S. H.**, Nutrient transport across the integument of marine invertebrates, in *Advances in Comparative and Environmental Physiology*, Vol 2, Gilles, R., Ed., Springer-Verlag, New York, 1988, 173.
63. **Braven, J., Evens, R., and Butler, E. I.**, Amino acids in sea water, *Chem. Ecol.*, 2, 11, 1984.
64. **Siebers, D. and Winkler, A.**, Amino-acid uptake by mussels, *Mytilus edulis*, from natural sea water in a flow-through system, *Helgol. Meeresunters*, 38, 189, 1984.
65. **Mopper, K. and Lindroth, P.**, Diel and depth variations in dissolved free amino acids and ammonium in the Baltic Sea determined by shipboard HPLC analysis, *Limnol. Oceanogr.*, 27, 336, 1982.
66. **Jorgensen, N. O. G.**, Uptake of glycine and release of primary amines by the polychaete *Nereis virens* (Sars) and the mud snail *Hydrobia neglecta* Muus, *J. Exp. Mar. Biol. Ecol.*, 47, 281, 1980.
67. **Welborn, J. R. and Manahan, D. T.**, Seasonal changes in concentrations of amino acids and sugars in seawaters of McMurdo Sound, Antarctica: uptake of amino acids by asteroid larvae, *Antarct. J. U.S.*, 26, 1993.
68. **Manahan, D. T., Shilling, F. M., Welborn, J. R., and Colwell, S. J.**, Dissolved organic material in seawater as a source of nutrition for invertebrate larvae from McMurdo Sound, Antarctica, *Antarct. J. U.S.*, 25, 206, 1990.
69. **Shilling, F. M. and Manahan, D. T.**, Nutrient transport capacities and metabolic rates scale differently between larvae of an antarctic and temperate echinoderm, *Antarct. J. U.S.*, 26, 1993.
70. **Smith, D. F.**, Small-scale spatial heterogeneity in dissolved nutrient concentrations, *Limnol. Oceanogr.*, 31, 167, 1986.
71. **Jaeckle, W. B. and Manahan, D. T.**, Growth and energy imbalance during the development of a lecithotrophic molluscan larva (*Haliotis rufescens*), *Biol. Bull.*, 177, 237, 1989.
72. **Jaeckle, W. B. and Manahan, D. T.**, Experimental manipulations of the organic composition of seawater: implications for studies of energy budgets in marine invertebrate larvae, *J. Exp. Mar. Biol. Ecol.*, 156, 273, 1992.
73. **Zurburg, W. and De Zwaan, A.**, The role of amino acids in anaerobiosis and osmoregulation in bivalves, *J. Exp. Zool.*, 215, 315, 1981.
74. **Johannes, R. E. and Webb, K. L.**, Release of dissolved organic compounds by marine and freshwater invertebrates, in *Organic Matter in Natural Waters*, Hood, D., Ed., University of Alaska, 1970, 257.

75. **Johannes, R. E., Coward, S. J., and Webb, K. L.,** Are amino acids an energy source for marine invertebrates? *Comp. Biochem. Physiol.*, 29, 283, 1969.
76. **Bamford, D. R. and McCrea, R.,** Active absorption of neutral and basic amino acids by the gill of the common cockle, *Cerastoderma edule*, *Comp. Biochem. Physiol.*, 50A, 811, 1975.
77. **Gomme, J.,** D-Glucose transport across the apical membrane of the surface epithelium in *Nereis diversicolor*, *J. Membr. Biol.*, 62, 29, 1981.
78. **Gomme, J.,** Recycling of D-glucose in collagenous cuticle: a means of nutrient conservation? *J. Membr. Biol.*, 62, 47, 1981.
79. **Gomme, J.,** Laminar water flow, amino acid absorption, and amino acid recycling in the mussel gill, *Am. Zool.*, 22, 1982. (Abstract)
80. **Pierce, S. K., Jr.,** Volume regulation and valve movements by marine mussels, *Comp. Biochem. Physiol.*, 39A, 103, 1971.
81. **Gilles, R.,** Volume regulation in cells of euryhaline invertebrates, *Curr. Top. Membr. Transp.*, 30, 205, 1987.
82. **Yancey, P. H., Clark, M. E., Hand, S. C., Bowlus, R. D., and Somero, G. N.,** Living with water stress: evolution of osmolyte systems, *Science*, 217, 1214, 1982.
83. **Chamberlin, M. E. and Strange, K.,** Anisosmotic cell volume regulation: a comparative view, *Am. J. Physiol.*, 257, C159, 1989.
84. **Pierce, S. K. and Greenberg, M. J.,** The nature of cellular volume regulation in marine bivalves, *J. Exp. Biol.*, 57, 681, 1972.
85. **Baginski, R. M. and Pierce, S. K.,** Anaerobiosis: a possible source of osmotic solute for high-salinity acclimation in marine molluscs, *J. Exp. Biol.*, 62, 589, 1975.
86. **Pierce, S. K. and Greenberg, M. J.,** The initiation and control of free amino acid regulation of cell volume in salinity-stressed marine bivalves, *J. Exp. Biol.*, 59, 435, 1973.
87. **Amende, L. M. and Pierce, S. K.,** Free amino acid mediated volume regulation of isolated *Noetia ponderosa* red blood cells: control by Ca^{2+} and ATP, *J. Comp. Physiol. B*, 138, 291, 1980.
88. **Pierce, S. K. and Rowland-Faux, L. M.,** Ionomycin produces an improved volume recovery by an increased efflux of taurine from hypoosmotically stressed molluscan red blood cells, *Cell Calcium*, 13, 321, 1992.
89. **Smith, L. H., Jr. and Pierce, S. K.,** Cell volume regulation by molluscan erythrocytes during hypoosmotic stress: Ca^{2+} effects on ionic and organic osmolyte effluxes, *Biol. Bull.*, 173, 407, 1987.
90. **Pierce, S. K., Politis, A. D., Smith, L. H., Jr., and Rowland, L. M.,** A Ca^{2} influx in response to hypo-osmotic stress may alter osmolyte permeability by a phenothiazine-sensitive mechanism, *Cell Calcium*, 9, 129, 1988.
91. **Pierce, S. K., Politis, A. D., Cronkite, D. H., Rowland, L. M., and Smith, L. H., Jr.,** Evidence of calmodulin involvement in cell volume recovery following hypo-osmotic stress, *Cell Calcium*, 10, 159, 1989.
92. **Deaton, L. E.,** Potentiation of hypoosmotic cellular volume regulation in the quahog, *Mercenaria mercenaria*, by 5-hydroxytryptamine, FMRFamide, and phorbol esters, *Biol. Bull.*, 178, 260, 1990.
93. **Amende, L. M. and Pierce, S. K.,** Cellular volume regulation in salinity stressed molluscs: the response of *Noetia ponderosa* (Arcidae) red blood cells to osmotic variation, *J. Comp. Physiol. B*, 138, 283, 1980.
94. **Deaton, L. E.,** Hypoosmotic volume regulation in bivalves: protein kinase C and amino acid release, *J. Exp. Zool.*, 268, 145, 1994.
95. **Bartberger, C. A. and Pierce, S. K.,** Relationship between ammonia excretion rates and hemolymph nitrogenous compounds of a euryhaline bivalve during low salinity acclimation, *Biol. Bull.*, 150, 1, 1976.
96. **Strange, K. B. and Crowe, J. H.,** Acclimation to successive short term salinity changes by the bivalve *Modiolus demissus*. II. Nitrogen metabolism, *J. Exp. Zool.*, 210, 227, 1979.
97. **Gainey, L. F., Jr.,** The response of the Corbiculidae (Mollusca: Bivalvia) to osmotic stress: the cellular response, *Physiol. Zool.*, 51, 79, 1978.
98. **Livingstone, D. R., Widdows, J., and Fieth, P.,** Aspects of nitrogen metabolism of the common mussel *Mytilus edulis*: adaptation to abrupt and fluctuating changes in salinity, *Mar. Biol.*, 53, 41, 1979.
99. **Vooys, C. G. N. de,** Anaerobic metabolism in sublittoral living *Mytilus galloprovincialis* in the Mediterranean-IV. Role of amino acids in adapatation to low salinities during anaerobiosis and aerobiosis, *Comp. Biochem. Physiol.*, 100A, 423, 1991.
100. **Willmer, P. G.,** Volume regulation and solute balance in the nervous tissue of an osmoconforming bivalve (*Mytilus edulis*), *J. Exp. Biol.*, 77, 157, 1978.
101. **Banderali, U. and Roy, G.,** Anion channels for amino acids in MDCK cells, *Am. J. Physiol.*, 263, C1200, 1992.

Chapter 6

Nitrogen Metabolism, Excretion, Osmoregulation, and Cell Volume Regulation in Elasmobranchs

Leon Goldstein and Deborah F. Perlman

CONTENTS

INTRODUCTION

The hallmark strategy of ureosmoregulation by elasmobranch fish (Chondrichthyes: sharks, skates, rays, and chimaeras) has been of interest to scientists for a long time. Smith's 1936 review of the discovery and examination of the high plasma and tissue urea concentrations in elasmobranchs entrenched the nitrogen end product, urea, as an intra- and extracellular osmolyte functioning in osmoregulation.[1] He correctly suggested that this strategy of ureosmoregulation "should be of interest to the palaeontologist and the ichthyologist, as well as to the biochemist and physiologist."[1] Various questions from diverse disciplines have been subsequently addressed concerning the use of urea by elasmobranchs. These questions include the nature and routes of its metabolism and excretion, problems and solutions of urea toxicity, environmental adaptations, and evolutionary patterns of ureosmoregulation. Work in these areas has been comprehensively reviewed[2,3] and will only be briefly reiterated here.

Recent interest (especially by the paleontologists and ichthyologists) has focused on the strategy and distribution of ureosmoregulation, urea excretion (ureotelism), and urea biosynthesis as employed by elasmobranchs and a few other fish and vertebrates. Clues to answer basic evolutionary questions (e.g., the freshwater or marine origin of vertebrates, the vertebrate ancestor of tetrapods) may come from these studies. The elasmobranchs have proven to be an excellent model for biochemists and physiologists to study cell volume regulation. Examination of the transport, accumulation, and metabolism of nitrogenous compounds other than urea (e.g., taurine, β-alanine, betaine, sarcosine) in various elasmobranch tissues has yielded an understanding of cell volume regulation that may be quite applicable to many vertebrates.

In this chapter we will review some of the evolutionary links provided by ureosmoregulation in elasmobranchs and other species, and address current studies on the mechanisms employed in the use of nitrogen compounds in osmoregulation and cell volume regulation as they are exemplified in elasmobranchs. The application of this understanding to other classes will be noted.

0-8493-8411-7/95/$0.00+$.50

EVOLUTIONARY LINKS: UREOSMOREGULATION, UREOTELISM, AND UREA BIOSYNTHESIS

There still continues to be debate as to whether vertebrates originated in a marine or freshwater environment, whether tetrapods evolved from marine or freshwater fish, and the origin of the use of urea as an osmotic and/or excretory compound.[4-7] These debates remain unresolved because of the conflict between location of fossil remains (predominantly marine for protovertebrates and both freshwater and estuarine for tetrapods) and the osmotic regulatory patterns of extant examples of the various classes. Interestingly, a better understanding of ureosmoregulation, ureotelism, and urea biosynthesis by elasmobranchs and some exceptional fish and other vertebrates may provide important information to help resolve these questions.

The classic answers for the location of origin for both vertebrates and tetrapods had been freshwater.[4-6] This answer was supported by the fact that both living fish and tetrapods have a plasma and tissue electrolyte composition and kidney function compatible with a freshwater ancestry. Regardless of their phylogenetic relationship, freshwater fish maintain the ionic and osmotic content of their plasma and cells at approximately 300 mOsm. The lampreys (members of the Agnatha or jawless fish), the most primitive class of fish, like the teleosts (members of the Actinopterygii or ray-finned fish), a large subclass of the bony fish (the Osteichthyes) clearly show this pattern. When found in freshwater (both permanently or with the euryhaline habit of moving back and forth), these fish, hypertonic to their environment, offset osmotic water gain by producing a copious urine with their glomerular kidney and balance ion loss by actively transporting salts. When in seawater they maintain the same ionic condition (300 mOsm salts) but now are hypoosmotic and hypoionic to their environment (approximately 1000 mOsm). They achieve this hypoosmotic regulation by reducing their urine flow, drinking saltwater, retaining the water, and actively secreting the ions.

The elasmobranch fish (class Chrondrichthyes) have a plasma ionic pattern of approximately 400 to 600 mOsm, somewhat above the teleosts but well below that of seawater (1000 mOsm). The elasmobranchs employ an osmoregulatory strategy different from the bony fish. Rather than being hypoosmotic in the marine environment these cartilaginous fish are slightly hyperosmotic by the significant additional presence of plasma and tissue urea (approximately 400 mOsm) and trimethylamine oxide (TMAO) (50 to 100 mOsm). Although primarily represented today by marine forms, euryhaline elasmobranchs osmoconform to their environment through alteration of urea excretion and/or synthesis[2] in addition to hypoionic regulation.[8] Those few truly freshwater elasmobranchs (*Potomotrygon sp.*) have negligible blood urea and appear to have lost the ability for ureosmoregulation in spite of the presence (albeit at low levels) of the complete urea biosynthetic enzymes.[2] Interestingly "higher" vertebrates (e.g., amphibians, reptiles, birds, and mammals) also maintain the pattern of approximately 300 mOsm plasma and tissue ionic concentrations although these various classes cope with the terrestrial problem of water availability in different ways.

The traditional argument had been that the osmoregulatory strategies of both bony fish and elasmobranchs supported the early freshwater derivation of vertebrates with each employing alternative strategies for invasion of the marine environment.[6] However it has also been suggested that a 300 to 600 mOsm plasma ionic concentration is obligatory for any *active* vertebrate living in either fresh- or saltwater for reasons of cell physiology (e.g., enzyme and neuronal activity) and therefore cannot be used to argue for or against a freshwater vertebrate origin.[5] Thomson suggested that ureosmoregulation was added to ionic regulation as "a fundamental adaptation of this early marine vertebrate stock" and "is retained as a primitive condition in chondrichthyan fishes, in lungfishes, in the coelacanths and in tetrapods.... The loss of urea synthesis and retention mechanisms in actinopterygians, is, however, probably an adaptive response to freshwater conditions, but it probably occurred well after the first radiations of the Chondrostei had evolved."[5]

Griffith added an intermediate option by suggesting that the earliest vertebrate may have been anadromous (i.e., breeding in fresh water and moving into the sea as an adult).[6] This actually was alluded to earlier by Thomson[5] and appears consistent with both paleontological data and the life history and osmoregulatory patterns of the living primitive and more recently evolved classes of fish. Griffith argues that the marine protochordate ancestor, a filter feeder, entered the estuaries in the Cambrian era to exploit the suspended food resource found there. The exposure to dilute salinities inherent in the estuarine environment promoted the evolution of glomerular kidneys and development of renal and gill ion pumps for sodium and chloride absorption. As Griffith states,

> Reproduction of the ancestral vertebrate initially occurred in the sea, next within the estuary, and finally moved into freshwater rivers, placing the vulnerable filter-feeding larvae in a habitat free of competitors and predators and without the environmental instability of the estuaries. Breeding in the very restricted area of freshwater rivers would also enhance the probability of the adults finding mates.... An anadromous life syle — with reproduction and larval development in fresh water and the adults feeding in the productive estuaries and coastal marine area, would be a reasonable strategy for the protovertebrate during the Cambrian as it is today for lampreys, sturgeons and a host of teleosts.[6]

He also provided reasonable arguments for why such an anadromous lifestyle could account for the evolution of such vertebrate characteristics as specialized senses of olfaction, vision, current detection by the lateral line, acceleration by the semicircular canals, wide and efficient metabolic scope, and bone deposition for both mineral storage and protection. Regardless of where the first vertebrates evolved, the extant vertebrates have maintained the low ionic concentration pattern with the elasmobranchs additionally using the nitrogen compound urea as a critical osmolyte.

The problem of why and how the alternative ureosmotic strategy evolved in the elasmobranchs has been addressed since Smith clearly defined the situation in 1936. A topic related and contributing to this issue is why a few exceptional nonelasmobranch fish and representatives from other unrelated groups use urea for osmoregulation and/or for ammonia detoxification in instances of limited water availability. Has ureotelism evolved several times independently or can it be explained as a monophyletic trait with minor biochemical and function changes? Can the inability for urea synthesis be explained by gene deletion of key enzymes? Can this issue provide clues for the evolutionary relationships among the coelacanth, the lungfish, and the direct ancestor of tetrapods?

Griffith and Pang had suggested that the surprising incidence of ureosmoregulation in the bony fish coelacanth (*Latimeria*), the sole surviving species of the subclass, Crossopterygii, must have evolved independently from that of the elasmobranchs.[9] This remarkable relic of the Devonian era is of extreme interest both because of its close relation to the group Rhipidistia, the candidate for the direct ancestor of the tetrapods, and its ureosmotic behavior which is similar to the elasmobranchs. The argument was based on the absence of renal urea reabsorption in *Latimeria*, the different modes of internal fertilization between this group and the Chondricthians, the presumed independent origin of ureosmoregulation by the amphibian, *Rana cancrivora*, and the estimated low cost for ureosomoregulation vs. ionic regulation. Recent investigation of this question sheds doubt on their conclusion, however. Griffith and Pang had proposed that, based on very rough estimates, hyperosmotic ureosmoregulation was less expensive energetically than hypoosmotic ionic regulation; the achievement of hyperosmolarity in the marine environment via accumulation of urea eliminated the osmotic problem leaving the fish only to cope with excess salts. This was, therefore, the less demanding strategy adopted by elasmobranchs (in the Silurian era) when atmospheric oxygen was perhaps only 10% of the present day value. The later invasion of the sea by teleosts utilized the more energetically expensive ionic hyporegulation and only occurred when the oxygen atmospheric level was considerably higher.[9] Interesting recent recalculations by Kirschner, utilizing the current, more complete understanding of hypoosmotic regulation and water and ionic fluxes, have suggested that the two modes of osmoregulation were in fact about equally efficient and therefore energetics cannot be used to answer where, why, and in what temporal and phyletic sequence these vertebrates evolved the two modes of osmoregulation.[7]

An examination of the few bony fish that do have urea synthetic ability and, during certain environmental conditions, become ureotelic has contributed information to the questions of why and when urea synthesis and accumulation evolved. Recent studies have provided several exceptions to the earlier survey that had led Brown and Cohen to suggest that teleosts (subclass Actinopterygii) were incapable of urea synthesis via the ornithine urea cycle employed by the elasmobranchs and ureotelic tetrapods.[10] Read reported that the marine toadfish, *Opsanus*, had the surprising ability to synthesize urea (and at significant rates) via the ornithine-urea cycle but did not appear to excrete urea.[11] Since the plasma urea concentrations of the toadfish were at most 4 m*M* and urea excretion negligible, urea did not seem to play a significant role in osmoregulation (as in the elasmobranchs) nor as a nitrogen excretory product (as in mammals).

Two explanations for this curious exception by the toadfish have been proposed. Mommsen and Walsh suggested that urea may serve as a transient nitrogen store similar to that used in mammalian ruminants and hibernators.[12] Griffith noted, however, that the toadfish may in fact be sporadically ureotelic, excreting nitrogen wastes as urea in times of limited water availability.[13] He demonstrated bursts of urea excretion occurring every few days when the fish were maintained in dilute (20%) seawater, consistent with urea synthesis followed by urea sequestration in the bladder and infrequent micturition. Griffith offered two possible explanations for this ureotelism in lieu of the metabolically less expensive and more usual ammonia excretion (ammoniotelism) via the skin and gills. The first related to the toadfish's method of parental care; it lays eggs in restricted areas (e.g., under stones, in tin cans) and the male parent guards the nest until they hatch. The parental ammonia excretion could be toxic to the eggs because they brood these eggs laid in a restricted, intertidal location with limited water flow. Synthesis of non-toxic urea from toxic ammonia eliminates this problem. A second suggestion was that urea facilitated water excretion in the aglomerular toadfish, acting as an osmolyte pulling water into the renal tubule or bladder and increasing urine flow when the fish is in a dilute estuarine environment. The increased urea excretion occurred experimentally only when the toadfish was placed in a dilute environment.

Several bony fish have now been reported to have a complete ability to synthesize urea. Important examples are the marine coelacanth, *Latimeria chalumna*, (subclass Crossopterygii),[14] the freshwater lungfish (subclass Dipnoi),[15] and recent reports of the freshwater air-breathing teleost, *Heteropneustes fossilis* (subclass Actinopterygii).[16,17] Criticial studies of the enzymology for urea biosynthesis by elasmobranchs and these other species provide additional information for tracing the evolutionary pattern of urea metabolism. Anderson clearly defined the characteristics (substrates, cofactors, intracellular location) of the ornithine-urea cycle enzymes.[18-21] He described three different carbamoyl-phosphate synthetase enzymes that occur in different classes and appear to have different physiological roles: (1) CPS I [ammonia as substrate, n-acetyl glutamate (NAG)-dependent, mitochondrial location] used for ammonia detoxification and urea excretion in tetrapods; (2) CPS II (glutamine as substrate, NAG-independent and cytosolic location) involved in pyrimidine nucleotide synthesis in animals and plants; and (3) CPS III (glutamine as substrate, NAG-dependent, mitochondrial location) utilized for urea production and osmoregulation by elasmobranchs. The last enzyme in the urea cycle, arginase, had a mitochondrial location in both the elasmobranchs and the coelacanth, but it was cytosolic in amphibians and mammals. As a small aside, Anderson's group reported the isolation and purification of another enzyme in the urea cycle, ornithine carbamoyl transferase (OTC), from the dogfish, *Squalus acanthias*, never previously isolated from any species of fish.[22] Unlike carbamoyl-phosphate synthetase it did not appear to have any unique properties that distinguished it from mammalian OTC.

Using this information, Mommsen and Walsh developed a cladogram grouping fish and various other higher vertebrates by their ability for urea synthesis, their urea cycle enzyme type (I, III), and location and physiological function of urea metabolism.[12] From this array they hypothesized that urea synthesis is a monophyletic trait in vertebrates with two key changes occurring before the evolution of the lungfish: (1) CPS III to CPS I, and (2) replacement of mitochrondrial arginase with a cytosolic equivalent. The marine coelacanth (like the elasmobranchs) has CPS III, mitochondrial arginase, and uses urea in osmoregulation. In contrast, the biosynthesis of urea in the lungfish occurs via CPS I (like the tetrapods) with cytosolic arginase and is used for ammonia detoxification. The toadfish CPS enzyme is type III with mitochondrial arginase (similar to that in elasmobranchs) even though it may appear to function more for ammonia detoxification than osmoregulation.[13] The freshwater air-breathing catfish, *Heteropneustes fossilis*, has recently been analyzed and appears to have both the CPS I and III enzymes and the mitochondrial arginase.[23] This would align this unusual fish more closely with the elasmobranchs than with the freshwater air breathing lungfish.

The application of current molecular biological techniques to a comparative analysis of the three CPS synthesizing enzymes may provide us with some definitive answers for how, when, where, and for what functional purposes urea synthesis evolved. A close examination of the sequence homology of the urea cycle enzymes and their genetic regulation in response to environmental stress may resolve these issues. Anderson's group reported progress along this line with the recent availability of a cDNA probe for the CPS III gene in the dogfish, *S. acanthias*.[24] Considerable amino acid sequencing work has already been accomplished on CPS II.[25-28] Information gained from such an analysis will also supplement recent molecular studies of the mitochondrial DNA of the lungfish[29] and V_H gene organization in the coelacanth.[30]

OSMOREGULATORY MECHANISMS: ORGANIC OSMOLYTES

The regulation of nitrogenous compounds in the osmoconforming elasmobranchs by excretion, metabolism, and transport in response to changes in environmental salinity was reviewed by Perlman and Goldstein in 1988.[2] As these fish move into waters of varying salinities they function as incomplete osmoconformers, changing their extra- and intracellular osmolytes appropriately while remaining hyperosmotic to the environment. The contribution of the various osmolytes to this process differs between the extra- and intracellular compartments (Figure 1).[31] In the extracellular compartment the organic components (urea, TMAO, amino acids) and inorganic salts (Na^+, K^+ and Cl^-) were all reduced approximately equally when the skates were adapted from full- to half-strength seawater.

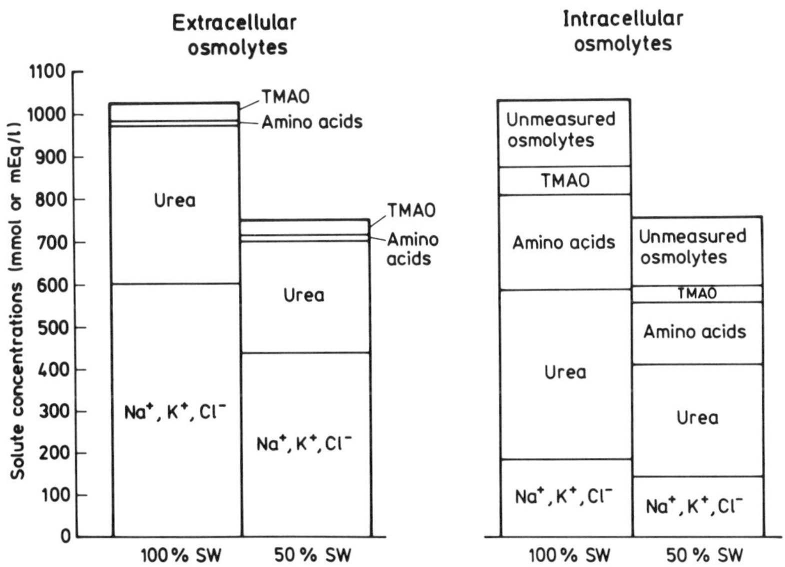

Figure 1 Extracellular (plasma) and intracellular (muscle cell) osmolyte concentrations for the little skate, *Raja erinacea*, acclimated to 100 or 50% seawater (SW). TMAO, trimethylamine oxide. (From King, P. A. and Goldstein, L., *Am. J. Physiol.*, 230, 925, 1976. With permission.)

In the intracellular compartment, however, the organic compounds showed a greater percentage decrease than the salts upon seawater dilution; organic compounds decreased by 36% while the salts decreased only by 16%. In this intracellular compartment the organic compounds form a larger fraction of the total intracellular osmolarity than they do of the extracellular osmolarity (60 vs. 40%), mostly due to the greater concentration of intracellular amino acids (19 vs. 1%). Cell membranes are highly permeable to urea so this organic compound essentially is in equilibrium between the two compartments. Recent work is demonstrating, however, that specific carriers may exert more control over urea distribution (see Chapter 13). Because cell membranes are less permeable to TMAO and amino acids the regulation of these organic compounds is of prime importance for the elasmobranch to maintain cell volume and function upon environmental and extracellular dilution.

There appears to be a different emphasis on the osmoregulatory role played by TMAO and the amino acids among species due to their different intracellular concentrations. The dogfish (*S. acanthias*) and the stingray (*Dasyatis americana*)[32] have higher cellular TMAO concentrations (150 to 200 m*M*) than the skate (*Raja erinacea*) (approximately 60 m*M*) and correspondingly lower intracellular amino acid concentrations (approximately 50 vs. 200 m*M*). Accordingly, research has focused on the skate for understanding the mechanisms for manipulation of amino acids in cell volume regulation (see below).

AMINO ACIDS

Work until 1988 (reviewed by Perlman and Goldstein) has described the important specific responses of two β-amino acids, taurine and β-alanine, and the methylamine, sarcosine, to the cellular osmolarity changes in elasmobranchs.[2] Following adaptation of skates to dilute seawater these intracellular constituents were rapidly released from the cells, reducing their intracellular concentrations and causing a significant increase in their plasma concentrations. These compounds must in turn be eliminated from the whole body via metabolism or excretion so the appropriate plasma concentrations can be maintained. This process was shown to be accomplished by increased oxidation of β-alanine and sarcosine and increased urinary excretion of taurine. Ballantyne et al. have further shown the appropriate stimulation or inhibition of the mitochondrial oxidation of various amino acids when varying the medium osmolarity.[33-35] In addition, mitochondrial oxidation of glutamate increased in both the elasmobranch (*R. erinacia*) and the teleost (*Pseuodopleuronectes americanus*) in a comparable manner when the osmolarity of the incubation medium was varied by omission of TMAO or by omission of the known nonpermeating sucrose/KCl osmolytes. Decreasing the medium osmolarity by omitting permeant urea did not stimulate mitochondrial oxidation. Ballantyne et al. suggested that the osmotic gradient created by the slowly permeating solute TMAO across the mitochondrial membrane was a signal for the amino acid oxidation changes. Unlike urea, TMAO poorly permeates membranes; equilibration of injected TMAO between plasma and various tissues takes 20 d in dogfish erythrocytes and is only 25% equilibrated in muscle by the same time.[36]

TAURINE

Recent research has focused in greater detail on how the interesting β-amino acid, taurine is excreted by the fish kidney. Following the demonstration that taurine excretion was increased in dogfish and skate placed in dilute seawater (70%) and was accomplished by net tubular secretion,[37] there has been interest in understanding the mechanism for renal taurine secretion. Marine teleosts show this similar net secretion of taurine and have provided an easier experimental model for study. Work with the flounder and killifish teased renal tubule preparations, brush-border membrane vesicles, and isolated perfused flounder renal tubules have characterized the basolateral uptake step of this secretory process as a Na^+-cotransport system specific for β-amino acids (taurine and β-alanine).[38] The primary cell culture technique for the flounder (*P. americanus*) kidney developed by Dickman and Renfro[39] provided an opportunity to examine both sides (basolateral and luminal) of the fish renal tubule, measure the unidirectional fluxes in parallel (in an Ussing chamber), and to address more completely the renal transport sytem for taurine in the kidney.[40] Presumably this knowledge

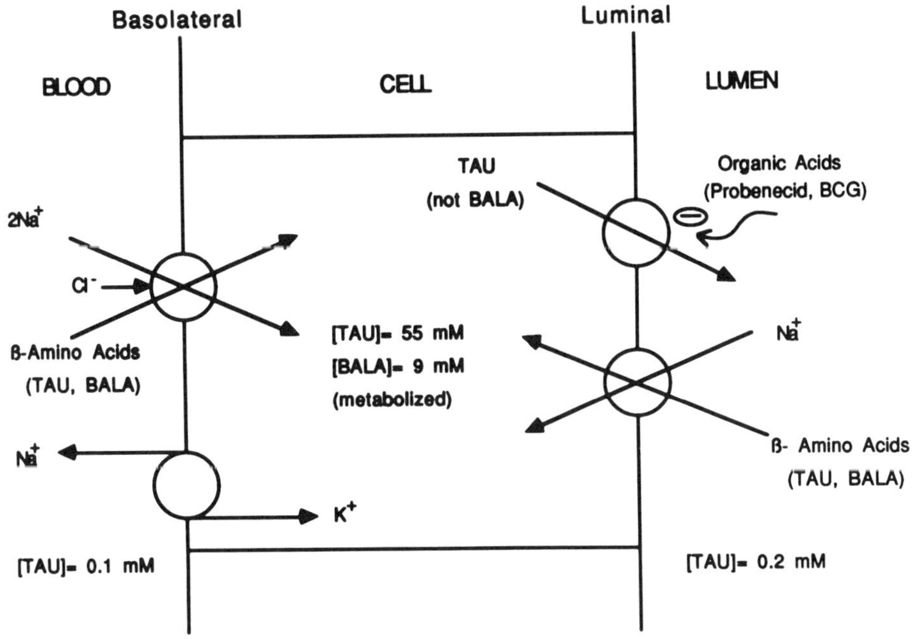

Figure 2 Hypothetical scheme for secretion of taurine by the flounder renal epithelium. Measured intracellular amino acid concentrations in brackets. TAU, taurine; BALA, β-alanine; BCG, bromocresol green. (From Perlman, D. F., Renfro, J. L., and Goldstein, L., *Am. J. Physiol.*, 261, R1155, 1991. With permission.)

can be applied to the similar system in elasmobranchs. The experimental manipulation of the intracellular β-alanine concentration by the inhibition of its oxidation allowed β-alanine to accumulate an intracellular concentration that equals that of taurine. This did not result in ß-alanine secretion, however. Therefore, the renal secretion of taurine, and not β-alanine, was shown to be due not to differences in the basolateral uptake nor intracellular concentration, but appeared to be accomplished by a carrier specific for taurine and not β-alanine in the luminal membrane (Figure 2). The taurine luminal carrier was shown to be competitively inhibited by the organic anions — bromocresol green and probenecid — but not paraaminohippurate (PAH). This work distinguished the taurine carriers as different on the two sides of the renal tubule and contributed to the understanding of the vectorial transport of taurine. In mammals there is a net reabsorptive transport of taurine and the characteristics of the luminal transport system are similar to the basolateral side in the flounder. It would be interesting to determine if the mammalian basolateral system corresponds to the flounder's luminal system.

CELL VOLUME REGULATION

A great deal of attention has recently been paid to taurine as a significant intracellular osmolyte for the regulatory volume response in various elasmobranch tissues. The knowledge gained from such studies is important because taurine itself is utilized in a similar manner by numerous vertebrate (including mammalian) cells and the elasmobranch cells provide a useful experimental model for understanding the cellular responses and signalling pathways involved in cell volume regulatory mechanisms.

Excellent reviews by Chamberlin and Strange[41] and Gilles[42] covered the various approaches used by cells, organisms, and scientists to anisosmotic cell volume regulation with a specific attention to the comparative aspects. An anisosmotic medium causes an immediate increase or decrease in cell size that is generally corrected by regulatory volume decrease (RVD) or regulatory volume increase (RVI) achieved by modulating membrane transport and/or metabolic pathways to change the concentration of intracellular osmolytes and entrained water. The osmolytes may be inorganic ions, or small organic molecules (e.g., amino acids, sugars, polyols, methylamines) each requiring a transport and/or metabolic pathway. The β-amino acid, taurine, was shown to be specifically released in response to cell swelling as a cell volume regulatory decrease (RVD) response in the elasmobranchs,[43,44] various tissues of euryhaline teleosts,[45,46] and Ehrlich ascites tumor cells.[47] The regulatory volume increase (RVI) response involving taurine has not been as thoroughly examined. The cells all have a Na^+-cotransport uptake system specific for taurine and other β-amino acids[38,47] but the mechanisms for increasing intracellular concentrations following exposure to hyperosmotic conditions have not been well studied.

SKATE ERYTHROCYTES

The skate (*R. erinacea*) erythrocytes have proven to be an excellent model system for analyzing the RVD mechanism; under normal physiology they are directly exposed to varying osmolarities, and they are easily collected, observed, assayed, and manipulated. Studies have been directed toward understanding the membrane transport system and the cell signaling pathways involved in the taurine RVD response. Various lines of evidence suggested that a channel, in some way associated with the anion exchanger, Band 3, was involved. Early studies showed that the RVD taurine efflux by skate erythrocytes was inhibited by various stilbene disulfonates — known blockers of the anion exchanger known as Band 3.[48,49] Skate erythrocytes demonstrated increased binding of such an inhibitor, 4,4,-diisothiocyanostilbene-2,2'-disulfonic acid (DIDS), after exposure to hypotonic medium. This DIDS binding was almost exclusively to a protein of molecular mass of approximately 100 kDa.[50] By immunoprecipitation with antibody specific for human Band 3 this skate protein was shown to be the equivalent of human Band 3. Phosphorylation of the Band 3 protein also correlated with hypotonically stimulated taurine efflux.[50] In contrast to the skate, the primitive agathans, hagfish and lamprey, known to have a limited Band 3 component, showed little taurine RVD response, and very reduced DIDS binding to the erythrocyte membranes.[51] Thus comparative evidence provides additional support of the involvement of Band 3 in the taurine RVD response.

RVD taurine transport may be through a channel associated with Band 3 rather than directly through the anion carrier.[52] Although blocked by a wide variety of anion exchange (Band 3) inhibitors, this RVD response of Na^+-independent taurine transport did not saturate well above plasma concentrations; it was not competitively inhibited by structural analogs (β-alanine), was activated in both the influx and efflux directions and was not transtimulated by taurine nor β-alanine. This evidence was consistent with a channel rather than a carrier being the taurine transporter. The channel has been shown to be size selective, estimated to be between 5.7 and 6.3 A in diameter. In addition to taurine a variety of other amino acids in this size range exhibited volume activated transport, inhibitable by DIDS, while larger amino acids did not.[53] Representatives of other small organic osmolytes (e.g., the polyol, inositol, and the trimethylamine, betaine), were also transported in response to hypoosmotic media in skate erythrocytes and were inhibited by Band 3 blockers.[54] These results suggested that there may be a common pathway for the three groups of organic solutes during RVD. A recent experiment tested neutral organic molecules of similar size (but not zwitterions like the amino acids) and showed their increased transport with hypotonic exposure; their transport was not affected, however, by Band 3 inhibitors.[56] This suggested that not only size but charge may be important for passage through this channel.

A model has been proposed whereby a volume sensitive channel is present in (or located near) the Band 3 protein in the cell membrane (Figure 3).[52] In the nonstimulated cell the channel is restricted and DIDS-inhibitable amino acid transport is low. In the volume-activated state phosphorylation of Band 3 may cause the channel to change conformation, open, and facilititate the RVD response with the efflux of taurine or other similarly sized organic compounds. The volume activated but non-DIDS-inhibitable compounds may pass through this same channel, but not interact with the sites blocked by the Band 3 inhibitors, or perhaps pass through a separate volume activated channel of similar size.

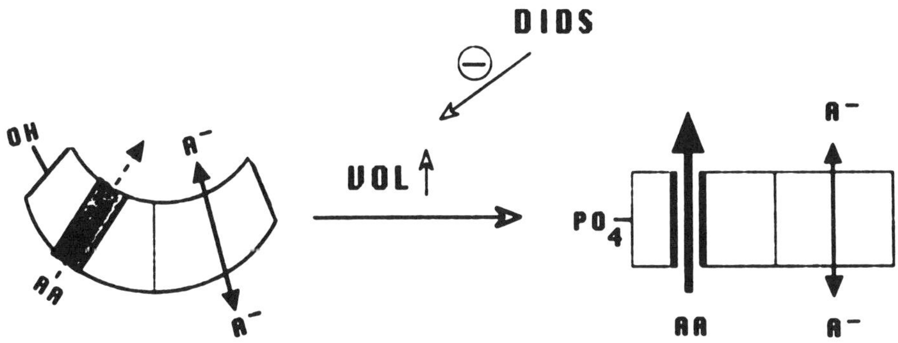

Figure 3 Hypothetical scheme for volume-activated transport of amino acids via a channel opened in or near Band 3 following phosphorylation of the protein. (From Goldstein, L. and Davis, E. M., *Am. J. Physiol.*, 1994.)

Much attention has focused on the signalıng pathways that may initiate the volume activated response. By what mechanism do the cells sense the change in volume, pressure, or extra- or intracellular composition and then transduce the signal via a secondary messenger system(s) to achieve the RVD taurine efflux response (via the proposed model of Band 3 modification)? Cell swelling appears to be the signal in skate erythrocytes as isosomotic expansion of these cells with permeant solutes elicited the same RVD response as hypotonic expansion.[57]

Various studies have looked at what second messengers may be involved in this RVD response. In skate erythrocytes evidence supports a system involving phospholipase D activation of the membrane component diacylglycerol (DAG) and resultant phosphorylation of Band 3 possibly through protein kinase C (PKC) but the exact details await further clarification.[55] The calcium ionophore, A23187, stimulated taurine efflux under isosomotic conditions.[44] However, cytoplasmic Ca^{2+} was not elevated during hypotonic stress in these skate erythrocytes.[58] The membrane component, DAG, increased on isosmotic and hypotonic stimulation.[57,59,60] While two different phospholipases may generate DAG (phospholipase C and phospholipase D), Musch and Goldstein demonstrated that it is phospholipase D that is activated in both hypotonic and isosmotic conditions.[55] As mentioned above Band 3 is phosphorylated during the RVD response; however, studies have not provided a firm understanding of this step.[50] PKC activators and inhibitors gave equivocal results. The PKC activator, phorbol ester, significantly (180%) increased taurine efflux. However, the PKC inhibitor, staurosporin, had only a small inhibitory effect; the protein phosphatase inhibitor, okadaic acid, had no effect and the protein kinase A activator, bromoadenosine cylic-monophosphate, had a slight inhibitory effect. A tyrosine kinase inhibitor, genistein, had no effect. How these signaling steps, phospholipase D, DAG, PKC, and Ca^{2+} are ultimately linked in this RVD response still requires further examination.

VOLUME RESPONSE IN OTHER FISH CELLS

Studies on teleost erythrocytes have shown a similar DIDS sensitive taurine efflux response on osmotic swelling.[61-63] Both flounder and trout erythrocytes appear to have a volume-activated channel that transports different classes of osmolytes but the exact identity of these transporters and their relation to the skate transporter is uncertain. Kirk et al.[62] suggested that the flounder efflux may occur through a volume-activated anion (Cl^-) channel. However, recent studies on skate erthyrocytes with pyridoxal phosphate (which does not block anion channels in several mammalian systems) gave supportive evidence that the skate taurine RVD channel differed from the anion channel.[54] Pyridoxal phosphate inhibited taurine transport twice as effectively as sulfate exchange, suggesting that the mechanism of inhibition of the two transport processes may be different. It may also affect the Band 3 channel via allosteric effects affecting taurine transport in addition to competing for sulfate at the exchange site.

Other nonerythroid elasmobranch cells have been examined for taurine transport systems and have a similar pattern with a few interesting exceptions. Ballatori and Boyer published a detailed study on the uptake and efflux mechanisms of taurine in isolated skate hepatocytes in isosmotic and hyposmotic media.[64,65] Their focus on metabolic inhibitors and sulfhydryl-reactive reagents emphasized the energy dependence of the RVD and the protein transporter involved but have not provided detailed information on an actual mechanism. Interestingly, like the erythrocytes, the RVD response was DIDS sensitive but unlike the erythrocytes was not stimulated by Ca^{2+} and phorbol esters. Further studies need to be done to establish whether there is a different second messenger system involved in the hepatocytes or whether the difference is due to varying experimental conditions.

Ziyadeh et al.[66] have examined the RVD response of dogfish rectal gland tissue slices. The rectal gland cells had a similar taurine efflux response to that demonstrated in skate erthyrocytes but the signal appeared to be a reduced Na^+ concentration rather than cell swelling. Taurine efflux was abolished on exposure to isosmotic medium when either the permeant Li^+ ion or impermeant N-methyl-D-glucamine ion was substituted for the impermeant Na^+ ion. Taurine efflux also did not occur when the medium was made hypoosmolar with urea deletion but normal Na^+ concentration. How this RVD response interacts with the cell cytoskeleton remains an important question. In shark rectal gland cells the cytoskeletal component F-actin disappeared, rapidly but transiently, along the basolateral cell face (as measured by fluorescence changes) when exposed to hypotonic media.[67] Within 5 to 10 min a gradual reconstitution of cell F-actin was seen. Since Band 3 is linked with the erythrocyte cytoskeleton via ankyrin, the RVD-cytoskeleton connection demands further examination in skate erythrocytes.

Two interesting questions relating to taurine accumulation and release have been addressed in skate embryonic heart preparations. Elasmobranch hearts contain high concentrations of taurine and use it for volume regulation in a manner similar to other elasmobranch tissues[68] and other vertebrate hearts.[69] Apart from a recently published exception,[70] fish were understood to be unable to synthesize taurine, acquiring it totally via their diet. The availability of freshly laid skate eggs from a breeding colony at Mote Marine Laboratory, Sarasota, FL provided the opportunity to examine the source and mechanism for taurine accumulation at an early developmental stage.[71] By approximately 25 d (one-third of development), skate embryo hearts had taurine intracellular concentrations and taurine uptake ratios (tissue/medium) equivalent to those in adults. Calculations suggested that there was sufficient taurine stored in the yolk to supply the embryo until hatching. Further studies with the embryonic heart have shown a RVD taurine efflux response similar to that demonstrated in the skate erythrocytes with the notable exception of insensitivity to DIDS inhibition.[72] DIDS did not bind to the embryonic heart membranes as it did to erythrocytes membranes suggesting the absence of the Band 3 component.

Recent studies have proven that taurine is an important intracellular osmolyte in the volume regulatory responses of various mammalian cells. It is interesting to make comparisons between mammalian tissues and elasmobranch tissues to look for a evolutionary conserved mechanism. Both cultured kidney (MDCK) cells[73,74] and cultured astrocytes from mouse cerebral cortex[75] had a remarkedly similar RVD response to that in elasmobranch erythrocytes. These cells contain high intracellular concentrations of taurine (and other amino acids: glutamate, glycine). The amino acid efflux rates increased markedly on exposure to hypotonic medium and were inhibited by DIDS. Strange et al.[76] have shown a similar pattern for the release of taurine and inositol from rat C6 brain glioma cells. Like the skate erthryocytes the signaling pathway may involve PKC and protein kinase A. Recent patch-clamp studies on these C6 glioma cells have suggested that the taurine and inositol efflux may be through a volume-sensitive anion channel[77] similar to that suggested for the flounder erthyrocytes.[62] Further studies will clarify whether or not the RVD response in all these cell types and species involves a common size-specific volume regulated mechanism controlled via a common signaling pathway.

Burg et al.[78,79] have examined the accumulation of organic osmolytes [taurine, inositol, betaine, sorbitol, glycerophosphocholine (GPC)] by the mammalian renal medullary cells. Like elasmobranch erythrocytes, these renal cells face very high blood and interstitial osmolarity due to the operation of the renal concentrating mechanism. Burg et al. have described the mechanisms for the increased biosynthesis (sorbitol), the reduced degradation (GPC), and increased mRNA transcription for transport proteins (taurine, betaine, and inositol) necessary for the accumulation of these osmolytes in hypertonic conditions. Similar isolation and expression of the mRNA and cDNA for the putative RVD efflux system would give an exciting way to compare this mechanism among different cells and species.

CONCLUSION

Early investigations of nitrogen metabolism in elasmobranchs were concerned with their interesting adaptive strategy of ureosmoregulation. The mechanisms for modulating urea and TMAO extracellular concentrations in response to salinity changes have now been reasonably defined. Recent attention to nonelasmobranch fish that have the capacity for ureosmoregulation and closer examination of urea biosynthetic enzymes that differ among species now provide the opportunity to look for more detailed biochemical evidence of the evolutionary pattern of urea metabolism. The availability of cDNA libraries, amino acid sequencing methods, and the understanding of gene regulation in response to environmental changes can now be applied to further interpret these evolutionary relationships.

Previous attention had been focused on the regulation of extracellular osmolytes (urea and TMAO) by elasmobranchs in response to varying salinities. Later reviews[2] described the utilization of other nitrogen compounds as intracellular osmoltyes, available for use in cell volume regulation when the tissues are exposed to varying osmolarity. Important notice was made of several amino acids (i.e., taurine, β-alanine) that were released in response to anisosmotic swelling and their important roles in adjusting cell volume. Recent investigations of the taurine efflux from skate erythrocytes exposed to anisosmotic media have suggested a model in which a channel may be formed in or near the cell membrane anion exchanger, Band 3, upon volume activation that handles not only taurine efflux but other osmolytes as well. Attention is now being focused on the signaling pathways that activate such a response. Modern molecular tools can now be applied to these studies to further understand this response.

REFERENCES

1. **Smith, H. W.,** The retention and physiological role of urea in the Elasmobranchii, *Biol. Rev.*, 11, 49, 1936.
2. **Perlman, D. F. and Goldstein L.,** Nitrogen metabolism, in *Physiology of Elasmobranch Fishes*, Shuttleworth, T. J., Ed., Springer-Verlag, Berlin, 1988, chap. 9.
3. **Yancey, P. H.,** Organic osmotic effectors in cartilaginous fishes, in *Transport Processes, Iono- and Osmoregulation*, Gilles, R. and Gilles-Baillien, M., Eds., Springer-Verlag, Berlin, 1985, 424.
4. **Bray, A. A.,** The evolution of the terrestrial vertebrates: environmental and physiological considerations, *Philos. Trans. R. Soc. London Ser. B*, 309, 289, 1985.
5. **Thomson, K. S.,** The ecology of devonian lobe-finned fishes, in *The Terrestrial Environment and the Origin of Land Vertebrates*, Pachin, A. L., Ed., Academic Press, London, 1980, 187.
6. **Griffith, R. W.,** Freshwater or marine origin of the vertebrates?, *Comp. Biochem. Physiol.*, 87A, 523, 1987.
7. **Kirschner, L. B.,** The energetics of osmotic regulation in ureotelic and hypoosmotic fishes, *J. Exp. Zool.*, 267, 19, 1993.
8. **Shuttleworth, T. J.,** Salt and water balance — extrarenal mechanisms, in *Physiology of Elasmobranch Fishes*, Shuttleworth, T. J., Ed., Springer-Verlag, Berlin, 1988, chap. 6.
9. **Griffith, R. W. and Pang, P. K. T.,** Mechanisms of osmoregulation in the coelacanth: evolutionary implications, *Occas. Pap. Cal. Acad. Sci.*, 134, 79, 1979.
10. **Brown, G. W., Jr. and Cohen, P. P.,** Comparative biochemistry of urea synthesis. 3. Activities of urea-cycle enzymes in various higher and lower vertebrates, *Biochem. J.*, 75, 1960.

11. **Read, L. J.,** The presence of high ornithine-urea cycle enzyme activity in the teleost *Opsanus tau., Comp. Biochem. Physiol.*, 39B, 409, 1971.
12. **Mommsen, T. P. and Walsh, P. J.,** Evolution of urea synthesis in vertebrates: the piscine connection, *Science*, 243, 72, 1989.
13. **Griffith, R. W.,** Guppies, toadfish, lungfish, coelacanths and frogs: a scenario for the evolution of urea retention in fishes, *Environ. Biol. Fish.*, 32, 199, 1991.
14. **Goldstein, L., Harley-DeWitt, S., and Forster, R. P.,** Activities of ornithine-urea cycle enzymes and of trimethylamine oxidase in the coelacanth, *Latimeria chalumnae, Comp. Biochem. Physiol.*, 44B, 357, 1973.
15. **Funkhouser, D., Goldstein, L., and Forster, R. P.,** Urea biosynthesis in the South American lungfish, *Lepidosiren paradoxa*: relation to its ecology, *Comp. Biochem. Physiol.*, 41A, 439, 1972.
16. **Saha, N. and Ratha, B. K.,** Active ureogenesis in a freshwater air-breathing teleost, *Heteropneustes fossilis, J. Exp. Zool.*, 241, 137, 1987.
17. **Saha, N. and Ratha, B. K.,** Comparative study of ureogenesis in freshwater, air-breathing teleosts, *J. Exp. Zool.*, 252, 1, 1989.
18. **Anderson, P. M.,** A glutamine and N-acetyl-L-glutamate-dependent carbamyl phosphate synthestase activity in the teleost *Micropterus salmoides, Comp. Biochem. Physiol.*, 54B, 261, 1976.
19. **Anderson, P. M.,** Glutamine- and N-acetylglutamate-dependent carbamoyl phosphate synthetase in elasmobranchs, *Science*, 208, 291, 1980.
20. **Anderson, P. M.,** Purification and properties of the glutamine- and N-acetyl-L-glutamate-dependent carbamoyl phosphate synthetase from liver of *Squalus acanthias, J. Biol. Chem.*, 256, 12228, 1981.
21. **Anderson, P. M.,** Glutamine-dependent carbamoyl-phosphate synthetase and other enzyme activities related to the pyrimidine pathway in spleen of *Squalus acanthias* (spiny dogfish), *Biochem. J.*, 261, 523, 1989.
22. **Xiong, X. and Anderson, P. M.,** Purification and properties of ornithine carbamoyl transferase from liver of *Squalus acanthias, Arch. Biochem. Biophys.*, 270, 198, 1989.
23. **Dkhar, J., Saha, N., and Ratha, B. K.,** Ureogenesis in a freshwater teleost: an unusual subcellular localization of ornithine-urea cycle enzymes in the freshwater air-breathing teleost *Heteropneustes fossilis, Biochem. Int.*, 25, 1061, 1991.
24. **Cao, X., Kemp, J. R., and Anderson, P. M.,** Subcellular localization of two glutamine-dependent carbamoyl-phosphate synthetases and related enzymes in liver of *Micropterus salmoides* (largemouth bass) and properties of isolated liver mitochondria: comparative relationships with elasmobranchs, *J. Exp. Zool.*, 258, 24, 1991.
25. **Simmer, J. P., Kelly, R. E., Rinker, A. G., Jr., Scully, J. L., and Evans, D. R.,** Mammalian carbamyl phosphate synthetase (CPS), *J. Biol. Chem.*, 265, 10395, 1990.
26. **Davidson, J., Gadiparthi, N., Rao, N., Niswander, L., Andreano, C., Tamer, C., and Chen, K.-C.,** Organization and nucleotide sequence of the 3′ end of the human CAD gene, *DNA Cell Biol.*, 9, 667, 1990.
27. **Evans, D. R., Bein, K., Guy, H. I., Liu, X., Molina, J. A., and Zimmermann, B. H.,** CAD gene sequence and the domain structure of the mammalian multifunctional protein CAD, *Biochem. Soc. Trans.*, 21, 186, 1993.
28. **Herve, G., Nagy, M., Le Gouar, M., Penverne, B., and Ladjimi, M.,** The carbamoyl phosphate synthetase-aspartate transcarbamoylase complex of *Saccharomyces cerevisiae*: molecular and cellular aspects, *Biochem. Soc. Trans.*, 21, 195, 1993.
29. **Meyer, A. and Wilson, A. C.,** Origin of tetrapods inferred from their mitochondrial DNA affiliation to lungfish, *J. Mol. Evol.*, 31, 359, 1990.
30. **Amemiya, C. T., Ohta, Y., Litman, R. T., Rast, J. P., Haire, R. N., and Litman, G. W.,** V_H gene organization in a relict species, the coelacanth *Latimeria chalumnae*: evolutionary implications, *Proc. Natl. Acad. Sci. U.S.A.*, 90, 6661, 1993.
31. **King, P. A. and Goldstein, L.,** Organic osmolytes and cell volume regulation in fish, *Mol. Physiol.*, 4, 53, 1983.
32. **Forster, R. P. and Goldstein, L.,** Intracellular osmoregulatory role of amino acids and urea in marine elasmobranchs, *Am. J. Physiol.*, 230, 925, 1976.
33. **Ballantyne, J. S. and Moon, T. W.,** Solute effects on mitochondria from an elasmobranch (*Raja erinacea*) and a teleost (*Pseudopleuronectes americanus*), *J. Exp. Zool.*, 239, 319, 1986.
34. **Ballantyne, J. S., Moyes, C., and Moon, T. W.,** Osmolarity affects oxidation of sarcosine by isolated hepatocytes and mitochondria from a euryhaline elasmobranch, *J. Exp. Zool.*, 238, 267, 1986.
35. **Moyes, C. D., Moon, T. W., and Ballantyne, J. S.,** Osmotic effects on amino acid oxidation in skate liver mitochondria, *J. Exp. Biol.*, 125, 181, 1986.
36. **Goldstein, L., Hartman, S. C., and Forster, R. P.,** On the origin of trimethylamine oxide in the spiny dogfish, *Squalus acanthias, Comp. Biochem. Physiol.*, 21, 719, 1967.
37. **Schrock, H., Forster, R. P., and Goldstein, L.,** Renal handling of taurine in marine fish, *Am. J. Physiol.*, 242, R64, 1982.
38. **Wolff, N. A., Perlman, D. F., and Goldstein, L.,** Ionic requirements of peritubular taurine transport in *Fundulus* kidney, *Am. J. Physiol.*, 250, R984, 1986.
39. **Dickman, K. G. and Renfro, J. L.,** Primary culture of flounder renal tubule cells: transepithelial transport, *Am. J. Physiol.*, 251, F424, 1986.

40. **Perlman, D. F., Renfro, J. L., and Goldstein, L.**, Taurine secretion in cultured winter flounder renal epithelium, *Am. J. Physiol.*, 261, R1155, 1991.
41. **Chamberlin, M. E. and Strange, K.**, Anisosmotic cell volume regulation: a comparative view, *Am. J. Physiol.*, 257, C159, 1989.
42. **Gilles, R.**, Comparative aspects of cell osmoregulation and volume control, *Ren. Physiol. Biochem.*, 5, 277, 1988.
43. **Forster, R. P. and Goldstein, L.**, Amino acids and cell volume regulation, *Yale J. Biol. Med.*, 52, 497, 1979.
44. **Leite, M. V. and Goldstein, L.**, Ca^{2+} ionophore and phorbol ester stimulate taurine efflux from skate erythrocytes, *J. Exp. Zool.*, 242, 95, 1987.
45. **Fugelli, K. and Zachariassen, K. E.**, The distribution of taurine, gamma-aminobutyric acid and inorganic ions between plasma and erythrocytes in flounder (*Platichthys flesus*) at different plasma osmolalities, *Comp. Biochem. Physiol.*, 55A, 173, 1976.
46. **Fincham, D. A., Wolowyk, M. W., and Young, J. D.**, Volume-sensitive taurine transport in fish erythrocytes, *J. Membr. Biol.*, 96, 45, 1987.
47. **Hoffmann, E. K. and Lambert, I. H.**, Amino acid transport and cell volume regulation in Ehrlich ascites tumour cells, *J. Physiol.*, 338, 613, 1983.
48. **Goldstein, L., Brill, S. R., and Freund, E. V.**, Activation of taurine efflux in hypotonically stressed elasmobranch cells: inhibition by stilbene disulfonates, *J. Exp. Zool.*, 254, 114, 1990.
49. **Goldstein, L. and Brill, S. R.**, Volume-activated taurine efflux from skate erythrocytes: possible Band 3 involvement, *Am. J. Physiol.*, 260, R1014, 1991.
50. **Musch, M. W., Leffingwell, T. R., and Goldstein, L.**, Band 3 modulation and hypotonic-stimulated taurine efflux in skate erythrocytes, *Am. J. Physiol.*, 266, R65, 1994.
51. **Brill, S. R., Musch, M. W., and Goldstein, L.**, Taurine efflux, band 3, and erythrocyte volume of the hagfish (*Myxine glutinosa*) and lamprey (*Petromyzon marinus*), *J. Exp. Zool.*, 264, 19, 1992.
52. **Goldstein, L. and Musch, M. W.**, Volume-activated amino acid transport and cell signaling in skate erythrocytes, *J. Exp. Zool.*, 268, 133, 1994.
53. **Haynes, J. K. and Goldstein, L.**, Volume-regulatory amino acid transport in erythrocytes of the little skate, *Raja erinacea*, *Am. J. Physiol.*, 265, R173, 1993.
54. **Goldstein, L. and Davis, E. M.**, Taurine, betaine and inositol share a volume-sensitive transporter in skate erythrocyte cell membrane, *Am. J. Physiol.*, 267, R426, 1994.
55. **Musch, M. W. and Goldstein, L.**, Hypotonicity stimulates phosphatidylcholine hydrolysis and generates diacylglycerol in erythrocytes, *J. Biol. Chem.*, 265, 13055, 1990.
56. **Russell, R. L., Cauble, G. D., and Goldstein, L.**, Organic solute permeability in erythrocytes of the little skate, *Raja erinacea*, *J. Exp. Zool.*, 267, 628, 1993.
57. **Goldstein, L. and Brill, S.**, Isosmotic swelling of skate (*Raja erinacea*) red blood cells causes a volume regulatory release of intracellular taurine, *J. Exp. Zool.*, 253, 132, 1990.
58. **Ashley, C. C., Goldstein, L., McConnell, F. M., and Simpson, A. W. H.**, Measurement of free calcium ($[Ca^{2+}]$) in skate (*Raja erinacea*) red cells during osmotic challenge, *J. Physiol.*, 410, 34P, 1989.
59. **McConnell, F. M. and Goldstein, L.**, Intracellular signals and volume regulatory response in skate erythrocytes, *Am. J. Physiol.*, 255, R982, 1988.
60. **Goldstein, L.**, Volume regulation in the erythrocyte of the little skate, *Raja erinacea*, *J. Exp. Zool. Suppl.*, 2, 136, 1989.
61. **Nonnotte G. and Truchot, J.-P.**, Cell volume regulation by erythrocytes of the euryhaline fish, *Platichthys flesus*, after hyposmotic stress in bicarbonate/carbon-dioxide-buffered medium, *Cell Physiol. Biochem.*, 2, 336, 1992.
62. **Kirk, K., Ellory, J. C., and Young, J. D.**, Transport of organic substrates via a volume-activated channel, *J. Biol. Chem.*, 267, 23475, 1992.
63. **Garcia-Romeu, F., Cossins, A. R., and Motais, R.**, Cell volume regulation by trout erythrocytes: characteristics of the transport systems activated by hypotonic swelling, *J. Physiol.*, 440, 547, 1991.
64. **Ballatori, N. and Boyer, J. L.**, Taurine transport in skate hepatocytes. I. Uptake and efflux, *Am. J. Physiol.*, 262, G445, 1992.
65. **Ballatori, N. and Boyer, J. L.**, Taurine transport in skate hepatocytes. II. Volume activation, energy and sulfhydryl dependence, *Am. J. Physiol.*, 262, G451, 1992.
66. **Ziyadeh, F. N., Feldman, G. M., Booz, G. W., and Kleinzeller, A.**, Taurine and cell volume maintenance in the shark rectal gland: cellular fluxes and kinetics, *Biochim. Biophys. Acta*, 943, 43, 1988.
67. **Ziyadeh, F. N., Mills, J. W., and Kleinzeller, A.**, Hypotonicity and cell volume regulation in shark rectal gland: role of organic osmolytes and F-actin, *Am. J. Physiol.*, 262, F468, 1992.
68. **Boyd, T. A., Cha, C.-J., Forster, R. P., and Goldstein, L.**, Free amino acids in tissues of the skate *Raja erinacea* and the stingray *Dasyatis sabina*: effects of environmental dilution, *J. Exp. Zool.*, 199, 435, 1977.
69. **Huxtable, R. J.**, Physiological actions of taurine, *Physiol. Rev.*, 72, 101, 1992.
70. **Divakaran, S., Ramanathan, S., and Ostrowski, A. C.**, Endogenous production of taurine in two teleost fish: *Coryphaena hippurus* and Red hybrid tilapia, *Comp. Biochem. Physiol.*, 101B, 321, 1992.

71. **Goldstein, L., Luer, C. A., and Blum, P. C.,** Taurine accumulation by the heart of embryonic skates, *Raja eglanteria*, *J. Exp. Biol.*, 150, 449, 1990.
72. **Goldstein, L., Luer, C. A., and Blum, P. C.,** Taurine transport characterisitcs of the embryonic skate (*Raja eglanteria)* heart, *J. Exp. Biol.*, 182, 291, 1993.
73. **Sanchez-Olea, R., Pasantes-Morales, H., Lazaro, A., and Cereijido, M.,** Osmolarity-sensitive release of free amino acids from cultured kidney cells (MDCK), *J. Membr. Biol.*, 121, 1, 1991.
74. **Roy G. and Banderali, U.,** Channels for ions and amino acids in kidney cultured cells (MDCK) during volume regulation, *J. Exp. Zool.*, 268, 121, 1994.
75. **Sanchez-Olea R., Moran, J., Schousboe, A., and Pasantes-Morales, H.,** Hyposmolarity-activated fluxes of taurine in astrocytes are mediated by diffusion, *Neurosci. Lett.*, 130, 233, 1991.
76. **Strange, K., Morrison, R., Shrode, L., and Putnam, R.,** Mechanism and regulation of swelling-activated inositol efflux in brain glial cells, *Am. J. Physiol.*, 265, C244, 1993.
77. **Jackson, P. S. and Strange, K.,** Volume-sensitive anion channels mediate swelling activated inositol and taurine efflux, *Am. J. Physiol.*, 265, C1489, 1993.
78. **Garcia-Perez, A. and Burg, M. B.,** Role of organic osmolytes in adaptation of renal cells to high osmolality, *J. Membr. Biol.*, 119, 1, 1991.
79. **Burg, M. B.,** Molecular basis for osmoregulation of organic osmolytes in renal medullary cells, *J. Exp. Zool.*, 268, 171, 1994.

Chapter 7

Nitrogen Excretion in Terrestrial Crustaceans

Michael J. O'Donnell and Jonathan C. Wright

CONTENTS

INTRODUCTION

The Crustacea, while remaining predominantly an aquatic group, includes several independent terrestrial radiations.[1,2] Groups in which nitrogenous excretion has been studied comprise the brachyuran (Grapsidae, Gecarcinidae, Ocypodidae) and anomuran (*Birgus*, *Coenobita*) decapods, oniscidean isopods, and talitrid amphipods. Aquatic crustaceans exploit the large branchial surface area for elimination of nitrogenous waste as ammonia (NH_4^+ and/or NH_3).[2,3] Surprisingly, the majority of terrestrial crustacean groups have retained branchial ammonia excretion (as opposed to using alternative end products), thus presenting the problem of concomitant water losses. In contrast to the continuous release of ammonia in aquatic animals, all terrestrial crustaceans studied display intermittent excretory patterns. In so doing, they restrict periods of fluid release and associated water loss. However, the high toxicity of ammonia requires that it be stored in a functional bladder or sequestered as a nontoxic form between excretory bouts.

Following accumulation, sequestered ammonia can be mobilized and excreted directly into the branchial fluid. This fluid may be secreted *in situ*, or may originate as periodic secretion of primary urine from antennal or maxillary glands. Ammonia elimination can then follow one of two pathways: (1) branchial fluid may be periodically eliminated from the branchial chamber, or (2) elevated ammonia levels in branchial fluid can be used to generate a high P_{NH_3}, allowing ammonia to be

0-8493-8411-7/95/$0.00+$.50

released as gaseous NH_3 while retaining branchial fluid. Volatilization can be facilitated by an increase in branchial fluid pH, thereby elevating the NH_3/NH_4^+ ratio. When ammonia has been released, and P_{NH_3} decreases, branchial fluid may then be reingested orally.

Alternatives to ammonotely include ureotely and uricotely. Although uricotely has been clearly demonstrated in one species (*Birgus latro*), there is no evidence of a functional urea cycle in any Crustacea, although uricolytic production of urea may be possible in some species. By eliminating nitrogenous waste in an insoluble form, uricotely minimizes both excretory water loss and nitrogenous waste toxicity.

In the following sections, we will discuss decapods, isopods, and amphipods separately. Subsequent sections will then consider nonexcretory functions of uric acid and mechanisms by which excretory pathways may be modulated. Finally, some brief comparisons will be drawn between excretory patterns in the terrestrial Crustacea and other land arthropods.

NITROGEN METABOLISM AND EXCRETION IN SEMITERRESTRIAL DECAPODS

MODES OF EXCRETION

Ammonium Ion Excretion in Reprocessed Urine (P) by Gecarcinids

In gecarcinids, the primary urine is isoosmotic with the hemolymph, and ammonium concentrations are low in both fluids (0.36 to 1.46 m*M*).[4,6] Primary urine is excreted into the branchial chamber where it is reprocessed by the gills; ammonium is added and salts are reabsorbed. In hypoosmotic conditions, for example, the concentration of NH_4^+ in the final excretory product (P) is elevated more than tenfold in *Gecarcinus lateralis* and *Cardisoma guanhumi*, while Na^+ and Cl^- concentrations decline by as much as 90%.[4] Urine reprocessing permits these species to excrete dilute urine when water loaded.[5] Even greater increases in NH_4^+ concentration have been reported for *Gecarcinus natalis*.[6] Mean values of $[NH_4^+]$ of 10.8 m*M* in P exceed hemolymph and urine levels more than 30-fold, and individual values as high as 73 m*M* have been measured (Figure 1A). A plausible mechanism for branchial ammonium excretion involves Na^+/NH_4^+ exchange across the apical membranes, as proposed for the grapsid *Geograpsus grayi* (see below) and several marine species.[7]

Although branchial ammonia release accounts for the majority of nitrogenous waste excretion in the gecarcinids, other excretory pathways may be exploited. In *G. natalis*, 12% of ammonia is volatilized from P and/or feces, and soluble ammonia in feces comprises a further 14.5%. Excretion of urea is negligible in *G. natalis*[6] but contributes 5 to 10% of total N excretion in the mangrove crab *Cardisoma carnifex* when freshwater is freely available.[8] In *Cardisoma guanhumi*, urea concentrations in P increase substantially when dietary N is supplemented.[9] The metabolic significance of ureotely is unclear since the urea cycle remains to be demonstrated in any crustacean. However, significant activities of some enzymes of uricolysis have been reported for hepatopancreas of *G. lateralis* and *C. guanhumi*.[57] Uricotely constitutes an additional, though labile, mode of excretion in gecarcinids. In some populations of *C. carnifex* and *C. guanhumi*, substantial internal deposits of uric acid may be accumulated (e.g., 55% of nonshell dry weight in *C. guanhumi*).[10] In other populations these accumulations are wanting, as noted in *C. carnifex* in Moorea.[8] When present, urate is apparently accumulated within the hemolymph bound as crystalline precipitates within amorphous tissue (D. Wolcott, personal communication); possible functions are discussed below.

Gaseous Ammonia Excretion by *Geograpsus grayi*

Geograpsus grayi is a carnivore with high nitrogen intake and N excretion rates (107 μmol N kg^{-1} h^{-1}) approximately three times those of a comparably sized herbivorous species, *G. natalis* (Figure 1B). The bulk of nitrogenous excretion (approximately 80%[6,11]) involves cyclic volatilization of NH_3 over periods of 1 to 2 d alternating with similar periods of substantially reduced NH_3 release. An additional 12.3% of nitrogenous waste is excreted as NH_4^+ and amino acids in the feces. Increased urine flow under conditions of high water intake elevates the proportion of NH_4^+ eliminated in reprocessed urine to as much as 22% of total ammoniacal excretion.

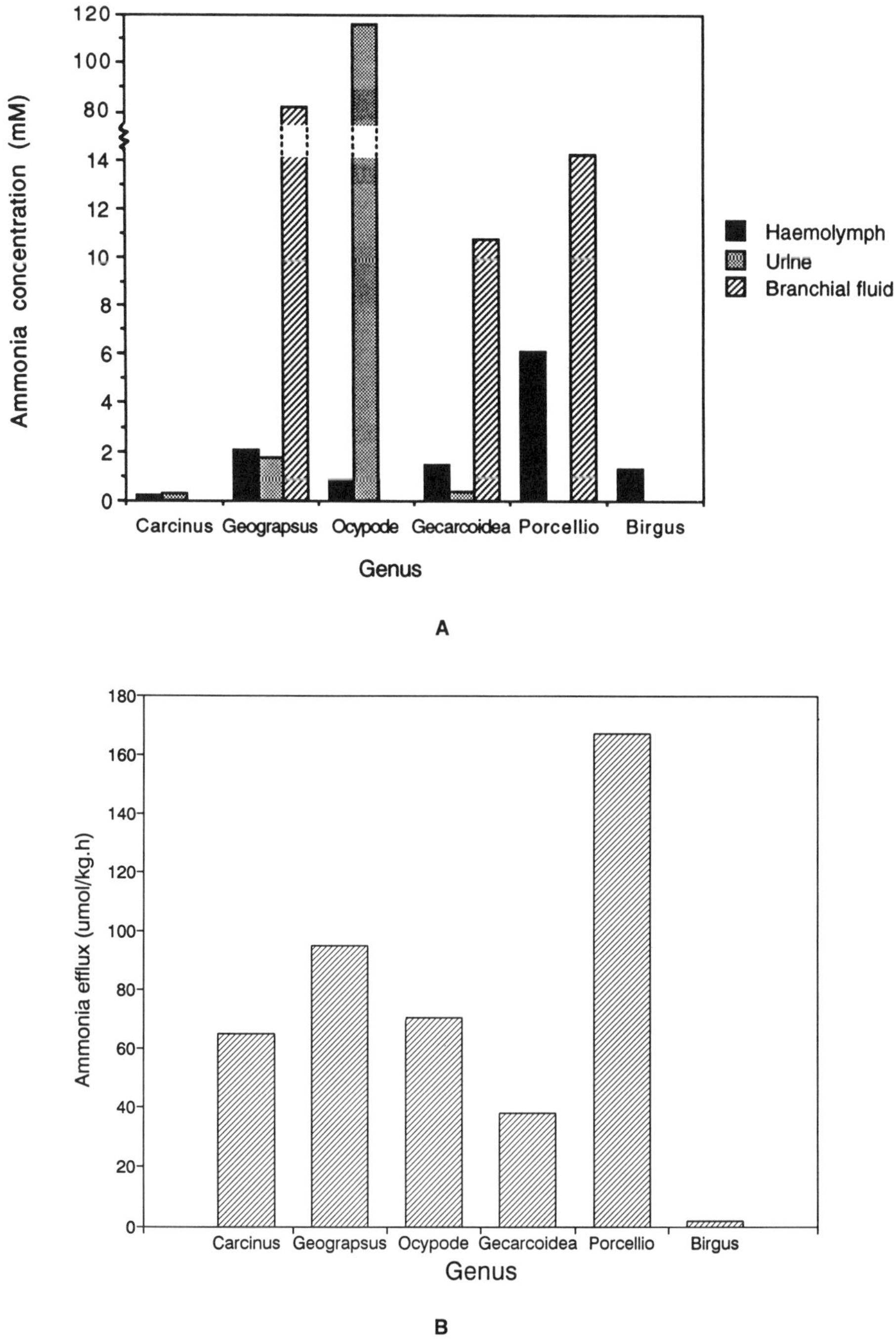

Figure 1 (A) Millimolar concentration of total ammonia (NH_3 + NH_4^+) in body fluids of terrestrial Crustacea and a marine genus, *Carcinus*. The greatest concentrations are seen in the excretory fluids of *Geograpsus*,[11] *Ocypode*,[13] and *Porcellio*[24] (in which branchial fluid concentrations may exceptionally exceed 100 m*M*). Such elevated concentrations serve to augment the P_{NH_3} to permit NH_3 volatilization. Aquatic crustaceans such as *Carcinus* excrete ammonia continuously across the branchial epithelium, maintaining low hemolymph levels.[61] *Birgus* is primarily uricotelic;[15] (B) cumulative rates of ammonia excretion (μmol kg^{-1} h^{-1}) averaged over several excretory bouts in Crustacea; data from references noted in (A). The high excretory rates of the carnivorous *G. grayi* reflect its high-nitrogen diet; by contrast, the vegetarian gecarcinids such as *Gecarcoidea* have low rates of nitrogenous excretion. The uricotelic anomuran, *B. latro*, excretes only about 15% of total nitrogenous wastes as ammonia.

Gaseous ammonia is released from the branchial chamber. Urine from the antennal glands is passed to the branchial chamber, where total ammonia concentrations increases 40- to 50-fold during excretory bursts.[11] Mean ammonia concentrations in primary urine, hemolymph and branchial fluid are 1.72, 2.07, and 80.6 m*M*, respectively, and branchial fluid ammonia concentrations as high as 219 m*M* have been recorded (Figure 1A).[11] The increase in branchial fluid total ammonia concentration apparently involves amiloride-sensitive Na^+/NH_4^+ exchange across the apical membranes of the branchial epithelium. Passive diffusion of NH_3 can clearly be ruled out since concentrations in hemolymph and branchial fluid would favor diffusion in the opposite direction. Similarly, it is highly unlikely that the electrochemical gradient would favor outward diffusion of NH_4^+: the calculated equilibrium potential for NH_4^+ across the epithelium is 94 mV (hemolymph positive), while typical transepithelial potentials in Crustacea are negative and much smaller.

Ammonia volatilization is a two-stage process in *G. grayi*. High branchial NH_4^+ concentrations produced by reprocessing allow a significant P_{NH_3} for volatilization. A concomitant elevation of pH from 7.6 in hemolymph to 8.1 in branchial fluid increases the proportion of NH_3 approximately fourfold to 8% of total ammonia, producing a further proportional increase in P_{NH_3}.[11] The increase in pH appears to involve SITS-sensitive Cl^-/HCO_3^- exchange across the apical membranes of the branchial epithelium. Beating of the scaphognathite (gill bailer) circulates air through the branchial chamber, maintaining the P_{NH_3} gradient during volatilization.[11]

Elimination of gaseous NH_3 results in the accumulation of H^+ in the branchial fluid following dissociation of NH_4^+. In the absence of simultaneous acid excretion or reabsorption, the branchial fluid pH decreases, reducing the NH_3/NH_4^+ ratio and depressing the P_{NH_3} for volatilization. Cl^-/HCO_3^- exchange during reprocessing could maintain the buffer capacity of the branchial fluid. In the presence of apical carbonic anhydrase (CA), outward transport of bicarbonate could be followed by volatilization of CO_2 along with NH_3, resulting in simultaneous excretion of acid–base[11]:

$$NH_4^+ + HCO_3^- \longrightarrow NH_3 + CO_2 + H_2O \qquad (1)$$

An important advantage of ammonia volatilization is that the amount of nitrogen excreted per unit volume of water lost can be varied. Indeed, under dry conditions; *G. grayi* may avoid liquid water loss entirely by reingesting P.[11] This capacity for water conservation may permit exploitation of relatively dry habitats as well as N-rich diets. The resultant adaptive flexibility may contribute to the terrestrial success of the Grapsidae.

The osmotic consequence of the two apical exchange processes, following volatilization of NH_3 and CO_2, is a reduction of mean urine osmolality from 837 mOsm kg^{-1} to 158 mOsm kg^{-1}.[6] Ion conservation is thus directly coupled to elimination of both nitrogenous waste and CO_2; the large osmotic gradient between branchial fluid and hemolymph presumably serves an additional role in water recovery. The proposal of apical Cl^-/HCO_3^- and Na^+/NH_4^+ exchangers is intriguing because such a design is normally associated with NaCl uptake by gills of freshwater crustaceans.[7] Because the grapsids appear to have followed an intertidal route from sea to land, these apical exchangers may have arisen *de novo* following the adoption of a terrestrial habit.

Ammonium Ion Excretion in Primary Urine of *Ocypode quadrata*

Whereas the gecarcinids and grapsids excrete ammonia at the gill, the ghost crab, *Ocypode quadrata*, excretes ammonia primarily across the antennal gland epithelium.[12,13] While hemolymph ammonia concentrations are low (averaging 0.9 m*M*), concentrations of ammonia in the primary urine may exceed 200 m*M*, and average 116 m*M* (Figure 1A).[13] The total concentration of other nitrogenous wastes in the urine is < 12 m*M*.

The urine contains much less Na^+ (255 m*M*) than Cl^- (415 m*M*), raising the possibility of a functional coupling of Na^+ resorption and ammonia enrichment by the antennal glands. The finding of elevated Na^+/K^+ ATPase activity in the antennal glands, exceeding the corresponding activity in the posterior gills, is consistent with this suggestion.[13] The authors note that their data are consistent with basolateral substitution of hemolymph NH_4^+ for K^+ by the ATPase. The low urine pH (5.4 to 5.5) and inverse relationship between urine pH and urine ammonia suggest that ammonia concentration is augmented by luminal acid trapping, possibly involving apical Na^+/H^+ exchange.

Subsequent urine reprocessing by the gill of *O. quadrata* leads to evolution of gaseous ammonia, a decline in Cl^- concentration, and increases in pH and CO_2 concentration.[12] Collectively, these data suggest simultaneous excretion of acid–base, as proposed for *G. grayi* (see above). Active transport is required to maintain the large osmotic gradient between hemolymph and branchial fluid, and Cl^-/HCO_3^- exchange is likely involved. Overall, Na^+ and Cl^- loss is minimized, and NH_3 is excreted. Comparison of ammonia output rates in the urine with those in the P indicate that 60 to 90% of urinary ammonia is volatilized. Occlusion of the nephropores leads to a sevenfold decline in net ammonia release, indicating that additional excretion of ammonia across the branchial epithelium is of minor significance.

Excretion of high ammonia concentrations in the primary urine may be a general characteristic of the Ocypodidae. Previous findings of high ammonia levels (75 m*M*) in the urine of *Uca pugilator*[14] have been confirmed (92 m*M*; pH 7.1) and extended to another species, *Ucides cordatus*, (80 m*M*; pH 6.0).[12] Green et al.[14] also measured ammonia concentrations of up to 20 mM in the hemolymph; antennal secretion may therefore coincide with a general mobilization of ammonia into the hemolymph from a sequestered precursor.

Uric Acid Excretion by *Birgus latro*

The anomuran *Birgus latro* is one of the most terrestrial of decapods. This species occurs widely within the tropical Indo-Pacific, and is commonly known as the coconut crab or robber crab. These names refer to its habit of climbing coconut palms to feed on the nuts, and its predilection for carrying off almost any portable article. The gills are small, so the heavily vascularized wall of the branchial chamber forms the major respiratory surface.

Both the site and mode of nitrogenous waste excretion in *Birgus* provide a stark contrast to the mechanisms employed by the other terrestrial decapods. The predominant fraction (96.2%) of total excreted nitrogen is lost via the feces, with the remainder comprising gaseous ammonia released across the lung/gills.[15] Fecal excretory nitrogen consists primarily of uric acid (82.6%), along with lower amounts of ammonium (11.9%), free amino acids, and urea. Uric acid is excreted into the midgut intermittently, and is then released as white strings distinct from the remaining fecal material. Synthesis of uric acid apparently occurs in the midgut gland (hepatopancreas), which possesses uniformly high activity of xanthine oxidase, the enzyme which oxidizes xanthine to uric acid. Xanthine oxidase activities in the lung, gill, and hindgut epithelia are undetectable. Since *B. latro* excretes most of its nitrogenous waste as uric acid, it presumably can synthesize purines *de novo*. However, the locations of purine cycle enzymes other than xanthine oxidase are unknown.

Extensive deposits of uric acid and elevated hemolymph urate levels are also found in other semiterrestrial anomurans (*Coenobita brevimanus*, *Coenobita perlatus*[16]) and brachyurans (*C. guanhumi*,[10] *G. natalis*, and *Holthuisana transversa*). In *G. grayi* and *G. natalis*, blood urate levels are as high as 0.23 and 0.08 m*M*, respectively (P. Greenaway, peronal communication). In *H. transversa*, values of 3.5 and 17 μmol l^{-1} have been measured in animals on low- and high-nitrogen diets, respectively (S. Linton and P. Greenaway, personal communication). These findings suggest that *B. latro* is not the only species capable of synthesizing purines. Brachyuran and anomuran lineages separated at least as early as the Jurassic period[17] with the independent radiations of terrestrial groups occurring much later.[1] Purine synthetic pathways remain to be detected in most aquatic Crustacea,[3] although *de novo* purine synthesis has been demonstrated in *Artemia*[18] suggesting that purine synthesis may have been an ancestral crustacean trait. This would make independent reacquisition of purine synthesis in terrestrial groups, proposed by Greenaway and Morris,[15] evolutionarily straightforward. An important caveat is that urate synthesis and accumulation does not necessarily indicate significant levels of urate excretion; uricolytic pathways may degrade the predominant fraction to urea or ammonia as discussed further below.

NITROGEN METABOLISM AND EXCRETION IN ISOPODS

GASEOUS AMMONIA EXCRETION IN ONISCIDEANS

The most fully terrestrial crustaceans are the oniscidean isopods, variously known as woodlice, sowbugs, and potato bugs. Unlike the decapods (which have some aquatic stage), they are viviparous, brooding the juveniles (mancas) in a ventral, fluid-filled marsupium, and thereby overcoming the need to return to water for reproduction. In spite of possessing much higher integumental water permeabilities than found in mesic-xeric insects and arachnids, the oniscideans have radiated since the early Tertiary to become a highly diverse and successful group with over 3500 described species. Like the majority of decapods, however, they remain primarily ammonotelic. Ammonia is liberated in gaseous form, accounting for over 90% of total nitrogen excretion.[19,20] A small proportion of soluble ammonia (<10%) is voided with the feces.

Ammonia excretion shows a pronounced diel rhythm in fasting animals, with most ammonia being released diurnally when animals are inactive.[21,22] Net nitrogenous excretion declines three- to fourfold in feeding animals, presumably because of reduced catabolism of tissue proteins, and the diel rhythm is all but abolished.[23] Significant seasonal variation in N excretion has also been documented; spring-summer excretion exceeds that in fall-winter about fivefold.[23] It is not known whether this difference reflects seasonal changes in metabolic rate or dietary nitrogen.

In the short term, ammonia volatilization is episodic, with excretory bouts of up to 2 h alternating with typically much longer periods (hours to several days) of minimal ammonia release. NH_3 excretion involves rapid mobilization of ammonia into the hemolymph from a sequestered form, followed by excretion into the thin film of fluid which bathes the ventral pleopods (Figure 2); NH_3 then volatilizes from the pleon fluid.[24] Hemolymph total ammonia concentrations in nonexcreting animals are approximately 1.5 m*M*. During excretory bouts, ammonia levels increase abruptly, frequently reaching 20 m*M* and exceptionally exceeding 60 m*M*. Ammonia levels in pleon fluid increase simultaneously. Whether ammonia diffuses as NH_3 or NH_4^+, and whether it follows a

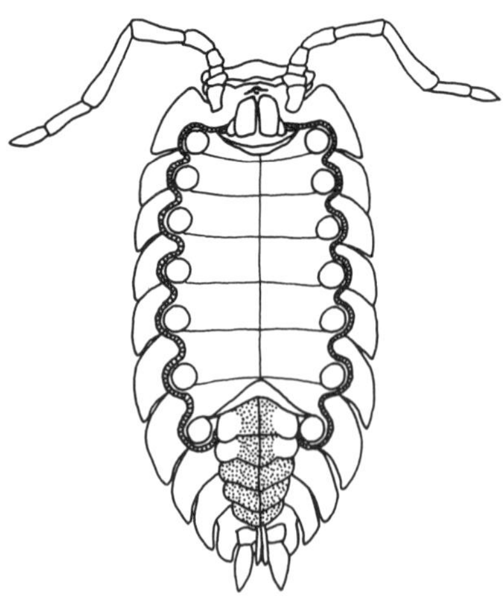

Figure 2 Schematic diagram of the ventral surface of a typical oniscidean isopod *P. scaber* (legs excluded). The imbricate pleopodal exopods (stippled) conceal the smaller endopods in the fluid-bathed pleoventral chamber. Ammonia is secreted across the endopodal epithelium and volatilized from pleon fluid. The pleopods also function in water vapor absorption. Urine is secreted periodically from the maxillary glands and is conveyed posteriorly to the pleon via the water transport system (WS) (hatched lines) for reprocessing.

transcellular or paracellular route, is not known. Pleon fluid is secreted *in situ* and is thus derived independently of maxillary urine.

Augmentation of NH_3 volatilization is apparently achieved by virtue of the very high pleon fluid ammonia levels, rather than by alkalinization. The pH of both hemolymph (7.60 ± 0.08) and pleon fluid (7.60 ± 0.04) is relatively constant, regardless of associated ammonia levels.[24] At this pH, a 20 m*M* total ammonia concentration in the pleon fluid would represent an NH_3 concentration of 0.45 m*M*, thus creating a substantial P_{NH_3} gradient to favor volatilization. Elevated pleon fluid ammonia levels are invariably associated with a characteristic behavior involving metachronal beating of the pleopodal exopods. This presumably serves a ventilatory role, removing NH_3 from the boundary layers overlying the pleon fluid, and thus maintaining the P_{NH_3} gradient.

Ammonotely is generally considered maladaptive for terrestrial animals because of the large volumes of water required to dilute excretory ammonia to subtoxic concentrations.[25] Isopods have exploited two adaptive strategies to overcome the problems of water loss associated with ammonotely. First, ammonia release is confined to periods of high relative humidity, minimizing evaporative water loss from the pleon fluid. Rapid volatilization is facilitated by exceptional ammonia tolerance which permits development of large P_{NH_3} gradients. A second strategy is the evolution in the Crinocheta of a capacity for active water vapor absorption (WVA), which is employed diurnally to replenish water losses sustained during nocturnal foraging.[26-28] Bouts of ammonia volatilization are apparently confined to these periods of WVA,[24] representing a unique example of N excretion actually associated with a net *gain* of water.

WVA involves colligative lowering of water vapor pressure of the pleon fluid by means of molar concentrations of NaCl.[29] This brings the pleon fluid into passive equilibrium with ambient relative humidities ranging from 85.5 to 92% for species studied. In humidities exceeding these *uptake thresholds*, water vapor condenses on the pleon fluid. Uptake thresholds thus define the minimum humidities at which ammonia volatilization can take place without accompanying evaporation of water from the pleon. Hyperosmotic fluid production and ammonia volatilization cease when humidities fall below threshold values.

One adaptive hurdle for the evolution of ammonia volatilization in oniscidean isopods, as in *G. grayi*, would be the net base loss and accumulation of protons in the pleon fluid. Without buffering, declining pH would rapidly reduce NH_4^+ dissociation or NH_3 levels to the point where the P_{NH_3} would be negligible and volatilization would cease. The high chloride concentration in pleon fluid suggests the possibility of Cl^-/HCO_3^- exchange, with buffering of protons being followed by volatilization of CO_2, as proposed for *G. grayi* and *O. quadrata*. Again, this would presumably depend on substantial CA activity in the apical membrane of the endopods.

The intermittent release of high concentrations of ammonia into the hemolymph indicates that ammonia is sequestered in a less toxic form following deamination. When *Porcellio scaber* is experimentally loaded with ammonia by exposure to high ambient P_{NH_3}, glutamine (GLN) and glutamate (GLU) are synthesized in direct proportion to P_{NH_3} and the duration of exposure.[30] On subsequent transfer to an ammonia-free atmosphere, ammonia excretion increases, with concomitant declines in whole animal GLN and GLU. Deamination of GLN and GLU can account for approximately 35% of the increase in ammonia production. Somatic tissues of *P. scaber* show high glutaminase and glutamate dehydrogenase activities.[31] Furthermore, seasonal variations in glutaminase activity concur with seasonal patterns of ammonia production in *P. scaber*. A primary role for glutamine in ammonia sequestration has been noted in other crustaceans. Elucidation of ammoniogenesis pathways in the marine crabs *Carcinus* and *Cancer* indicates glutamine as the major storage product in these species.[32] The elevated glutaminase activities in the gills of these species suggest predominantly branchial excretion.

Additional amino acids accumulated during ammonia loading of *P. scaber* include alanine, arginine, glycine, and proline (Wright, J. C., Caveney, S., Reichert, J., and O'Donnell, M. J., submitted). These studies, using HPLC, reveal glutamine, arginine, and glycine to be the major forms into which ammonia is incorporated, accounting for 34, 25, and 20% of total nitrogen sequestration, respectively. Approximately 40% of sequestration is in the hepatopancreas, with the remaining amino acids accumulating in lower concentrations in the general somatic tissues. The pleopodal endopods do not constitute a significant site of amino acid storage. The large accumulations of arginine observed following ammonia loading are intriguing since the requisite enzyme pathways for synthesis of ornithine and citrulline precursors remain to be demonstrated in crustaceans. Nevertheless, these studies strongly suggest that arginine is not an essential amino acid in oniscideans.

Evolution of ammonia volatilization in terrestrial isopods, as with other terrestrial crustaceans, annelids, and gastropods, probably involved only slight adaptive modifications of existing excretory organs. The thin-walled endopods probably serve as the major excretory surface in both terrestrial and aquatic species. Pleopodal ventilation is a continuous process in aquatic isopods, circulating water for respiratory exchange. Terrestrial species thus inherited the neuromuscular control for ventilation employed in gaseous ammonia release and WVA. Enlargement of the imbricating pleopodal exopods in terrestrial species conferred two advantages. First, apposition of the exopodal margins to the pleon sternites provided an effective sealing of the more permeable endopods, minimizing water loss in active animals. Second, the greater surface area enhanced ventilation and respiratory exchange. Demonstration of water vapor absorption in all Crinocheta studied, as well as in a littoral ligiid,[27] suggests that the coupling of WVA and ammonotely may have evolved in the earliest stages of the littoral-terrestrial transition. A possible scenario for the evolution of water vapor absorption in isopods has been proposed.[33]

Like other crustaceans, oniscidean isopods lack the full complement of enzymes for the urea cycle and urea is undetectable in *Oniscus asellus* using methods sensitive to 2.6 ng. mg $tissue^{-1}$.[20] Uric acid, on the other hand, may be accumulated in significant concentrations in the epidermis, forming visible plaques. These contain approximately 93% of total urate present (0.113 μg mg $tissue^{-1}$). Assuming a 3-week intermoult period, levels of urate in the body wall represent a storage of approximately 0.4 μg $animal^{-1}$ d^{-1}, or 0.13 μg N $animal^{-1}$ d^{-1}.[20] Whether any of this is accumulated from ammonia via the purine ring is unknown, although neither adenase nor xanthine oxidase have measureable activities. By contrast, all the enzymes of uricolysis are present, with maximal activities occurring in the body wall. Urate may therefore be sequestered from dietary sources as a nitrogen store.

The role of the maxillary glands in the oniscidean isopods is enigmatic. Maxillary urine is excreted into the water transport system (WS) which comprises an external capillary system of hydrophilic cutaneous hairs bordering the invaginated pleurae (Figure 2).[34] The WS moves urine via the pleon to the rectum where water is reabsorbed; accumulated excretory materials may then be voided with the feces.[34,35] The evolution of the WS, and the observation that maxillary urine is excreted in oniscideans, albeit irregularly, clearly indicates that maxillary urination is of some physiological significance. Ablation of the maxillary glands results in elevated hemolymph calcium levels.[36] More generally, however, since oniscideans will need to conserve ions, and urine apparently plays a minimal role in nitrogenous excretion, urine may serve primarily in the excretion of nonnitrogenous wastes such as organic acids or toxins of dietary origin such as alkaloids.

Excretion of Ammonia Via the Gut in Tylidae

Although little studied from a physiological standpoint, the Tylidae are strikingly different from other oniscideans and may merit classification within a separate suborder. One indication of their remote affinities is that ammonia volatilization does not serve as a significant mechanism for nitrogenous excretion. The Mediterranean woodland species, *Helleria brevicornis*, excretes 140 to 190 μmol NH_3 kg^{-1} h^{-1} when feeding, similar to rates of N turnover in the Crinocheta (Wright, J. C., unpublished data). Over 90% of this ammonia is present in the feces; fecal uric acid and urea levels are negligible. It is not known whether ammonia is excreted directly across the hindgut or originates externally via branchial or maxillary gland excretion and is then conveyed to the hindgut via the WS.

NITROGENOUS EXCRETION IN AMPHIPODS

Of the over 5500 described species of amphipods only about 50 can be considered to have become emancipated from aquatic habitats. All belong to the family Talitridae and three genera, *Talitrus*, *Orchestia*, and *Talorchestia*, predominate.[37] Ecologically, they may be classified in two categories: the supralittoral and the forest litter species. The supralittoral amphipods show a broad zoogeographic range and typically burrow among drift in or above the strand line. Forest litter species are largely restricted to the subtropical Indo-Pacific.

In the supralittoral genus, *Orchestia*, ammonia comprises the preponderance of fecal nitrogen.[38] The authors attributed low N excretion rates to a suppression of N metabolism in response to terrestriality. However, similar low excretion rates determined for isopods were subsequently shown to underestimate the predominant gaseous NH_3 component.[23] On this basis, and in the absence of subsequent investigations, there is a strong possibility that talitrids, like isopods, excrete significant levels of gaseous ammonia. This could be volatilized across the pereopodal gills (epipodites). Urinary excretion is less plausible given the apparent absence of an external fluid-conducting system to convey urine to the branchial epithelium or an alternative volatilizing surface.

Urate spherules are accumulated in the pericardial septum of the cavernicolous, freshwater amphipod *Niphargus*.[39] These are thought to be incorporated from dietary purines as a nitrogen reserve and may also play a role in ion sequestration. It is not known whether similar incorporation of urate occurs in terrestrial genera.

NONEXCRETORY FUNCTIONS OF URIC ACID

In addition to serving as the primary waste product for elimination of excess nitrogen in *B. latro*, uric acid may be accumulated in significant amounts in a number of primarily ammonotelic species. Since some of these species apparently synthesize urate *de novo*, the considerable energetic cost may be offset by ancillary functions of urate. A number of possible ancillary functions are considered below.

STORAGE EXCRETION AND NITROGEN SEQUESTRATION

In gecarcinid crabs such as *C. guanhumi* and *C. carnifex*, uric acid may represent a form of storage excretion. Wolcott and Wolcott[40] noted that populations of crabs with ready access to water have low or undetectable hemolymph urate,[9] while crabs from drier habitats have high levels.[10,16] Furthermore, Bright[41] reported that urate levels in *C. crassum* varied inversely with groundwater supply. Crabs may depend on burrow water to bathe the gills for ammonia release.[42] If this is true, uricotely in *Cardisoma* spp. may be a response to burrow water depletion, and may consequently vary according to season.[9] A direct test of the relationship between water availability and uricotely in *C. guanhumi*, compared across intermolt period and a wide range of size classes, produced no evidence, however, for increased urate deposition in response to water stress (Wolcott, D. L. and DeVries, M. C., personal communication).

Whether or not storage excretion of urate occurs in brachyurans, the wide variations in hemolymph urate levels in *Cardisoma* spp. do indicate variable exploitation of uricotely according to some physiological need. One possible function is nitrogen sequestration in herbivorous species in which dietary nitrogen levels are frequently low or erratic. *C. guanhumi* fed on a nitrogen-supplemented diet show a tenfold increase in hemolymph urate levels compared to animals sampled directly from the field.[40] Facultative use of uricotely is biochemically plausible given the evidence for purine synthesis in the anomuran *B. latro*, and given that urate deposits in some gecarcinids apparently exceed levels explicable on the basis of dietary purine intake.[3] Most, if not all, Crustacea possess enzymes for catabolism of urates; the amino groups thus produced may be incorporated into amino acids or excreted as urea.[59] Alternatively, urate deposits may accumulate to be eliminated in some way during the molt, as proposed for *C. guanhumi*.[11]

EFFECTS ON O_2 TRANSPORT

In several aquatic decapods, the decrease in hemocyanin oxygen affinity via the Bohr shift during metabolic acidosis is partially counteracted by the opposing effect of the anaerobic end product L-lactate. Urate also enhances hemocyanin oxygen affinity, and may play an important role in aquatic species exposed to environmental hypoxia,[43,44] but apparently not postexercise.[58] In contrast, in *B. latro*, the effect of lactate is minimal, and it appears that urate actually diminishes the response to lactate. In general, the O_2 affinity of hemocyanin in terrestrial decapods appears to be relatively insensitive to modulators such as lactate and urate.[45] For these species, it appears that O_2 delivery is regulated by adjustments of branchial ventilation and perfusion, rather than by modulation of hemocyanin O_2 affinity.

POSSIBLE ION SEQUESTRATION

A further possible role of urate is in the binding of Na^+ during hemolymph osmoregulation.[39,46] A precedent for Na^+ sequestration as urates in the Isopoda is seen in *Asellus aquaticus*. In this freshwater species, substantial quantities of urate (about 1.8 to 2.5 μmol in a 50 mg animal) are present in the Zenker's organ and serve as a long-term sodium store.[46] In terrestrial isopods, however, urate does not appear to play a major role in sequestration of hemolymph Na^+ during dehydration. Inulin space measurements show that water loss during desiccation occurs almost exclusively from the hemolymph, until hemolymph volume is negligible. Given that both *Armadillidium vulgare* and *P. scaber* can maintain near-stable hemolymph osmolality until over 90% of hemolymph is depleted[47] (Wright, J. C., O'Donnell, M., and Sazgar, S., unpublished data), substantial sequestration of Na^+ and Cl^- must occur, since these ions account for over 70% of hemolymph osmolality.[48,60] Na^+/urate ratios in integumental urate plaques increase from 0.41 in hydrated *A. vulgare* to 0.53 in dehydrated specimens. Assuming whole-animal urate levels of 0.24 to 0.74 μmol,[38] the increased Na^+/urate binding during dehydration can account for only 1 to 2% of requisite Na^+ sequestration. Studies on *P. scaber* similarly fail to indicate a significant role of urate in Na^+ storage. This species possesses much lower urate levels (about 0.08 μmol per animal) and typically lacks visible urate plaques in the integument. Comparisons of total sodium content in specific tissues of hydrated and dehydrated *P. scaber* indicate only minor Na^+ sequestration in the gut, hepatopancreas, and somatic tissues ("body wall"). Together with Na^+ crystallized externally in the pleural water transport system, these extra-hemolymph Na^+ stores can explain only about 1% of sequestration. These data, therefore, leave the hemolymph itself as the probable site of Na^+ storage during dehydration. Because Na^+:urate ratios as high as 24 have been demonstrated in birds,[49,50] it is possible that accumulations of urate in isopods other than *A. vulgare* and *P. scaber* may play a role in Na^+ sequestration. Possible mechanisms of Cl^- sequestration during dehydration still need to be investigated. A possible role of urate in cation sequestration has been investigated in only one decapod species. Energy-dispersive x-ray spectroscopic studies of urate deposits from *B. latro* failed to detect any cations, indicating excretion of pure uric acid (Greenaway, P., personal communication).

MODULATION OF EXCRETORY PATHWAYS: FACULTATIVE UREOTELY IN *PARATELPHUSA HYDRODROMUS*

The semiterrestrial crab *Paratelphusa hydrodromus* is common in tropical Asia, inhabiting streambanks and digging burrows which reach down to the water table. It shows a decline in glutaminase activity as ammonotely is substituted by ureotely in increasing salinities.[51] It is not known whether a similar shift from ammonotely to ureotely occurs in response to prolonged aerial exposure and dehydration. Urea synthesis *in vitro* by hepatopancreas incubated in the presence of NH_4^+, ornithine, citrulline, and arginine has been reported,[51] although the enzymes responsible have not been identified. This result does not confirm the presence of a complete ornithine-urea cycle, because the presence of arginase alone could explain the observed finding.

COMPARISONS OF CRUSTACEANS WITH OTHER TERRESTRIAL GROUPS

AMMONOTELY IN INSECTS

Although the majority of insects are uricotelic, the American cockroach *Periplaneta americana* is primarily ammonotelic; between 27 and 49% of total nitrogenous waste voided with the feces comprises ammonia and little or no urate is detectable in the excreta[52] (see also Chapter 8). Blowfly larvae, *Sarcophaga bullata*, excrete nitrogenous waste primarily as ammonium chloride.[53] In both species, ammonium ions are apparently transported across the hindgut, with luminal concentrations exceeding 200 m*M* in *Sarcophaga*.

THE ROLE OF AMMONIUM URATES IN N EXCRETION AND pH REGULATION

The desert locust *Schistocerca gregaria* excretes nitrogenous waste primarily as highly insoluble ammonium urate.[54] The mean ammonia concentration in the excreta is 263 mmol kg wet mass^{-1}, and ammonia is excreted at up to three times the rate of total urate. As the authors note, it is unlikely that dibasic ammonium urate is incorporated because ammonium$_2$urate has a pK of 10.3 and the fecal pellets are highly acidic. Furthermore, ammonium:urate ratios have been shown not to exceed 1:1 at pH 7.[50] Excess ammonium in *S. gregaria* fecal pellets is thought to be precipitated with organic anions. Excretion of NH_4^+ contributes to hemolymph acid–base regulation since fecal NH_4^+:urate ratios increase following acid injection into the hemolymph; fecal NH_4^+ can theoretically account for 15% of acid equivalents removed from the hemolymph during recovery from acid loading.

SUMMARY

The evolution of a terrestrial habit in crustacean families has involved a number of specialized changes in nitrogenous excretory physiology (Figure 3). The most conservative strategy is typified by the gecarcinids, which transport NH_4^+ across the branchial epithelium during urine reprocessing. Urine is subsequently eliminated from the branchial chamber. A modification of this basic pattern is seen in the grapsid, *G. grayi*, in which alkalinization of branchial fluid boosts the P_{NH3} and ammonia is volatilized from the branchial chamber. Subsequently, branchial fluid may be partly or completely reingested depending on the animal's hydration status.

Branchial ammonia volatilization has evolved convergently in the oniscidean isopods. Following deamination, amino groups are largely sequestered as glutamine, glycine, and arginine. Ammonia is then intermittently released into the hemolymph in high concentrations and secreted across the pleopodal endopods into the pleon (branchial) fluid. Elevated ammonia levels in the pleon fluid provide a sufficient P_{NH_3} for volatilization without alkalinization. Ventilatory beating of the pleopods accelerates dissemination of NH_3. Simultaneous evaporative water losses are overcome by restricting ammonia excretion to periods of high ambient humidity when animals exploit colligative water vapor absorption.

A more radical departure from the presumed branchial excretion of marine ancestors is seen in the ocypodid crabs. In this family, ammonia is excreted across the labyrinth epithelium of the antennal gland (continuously?) and is stored in the bladder where acid trapping detoxifies NH_3 to NH_4^+.[13] Primary urine is released periodically and passed into the branchial chamber where ions are reabsorbed and the pH increases. A substantial fraction (60 to 90%) of ammonia is then volatilized. Excretion of ammonia across the branchial epithelium is of minor significance.

In all crustaceans exploiting gaseous ammonia excretion, acid–base balance presents an adaptive hurdle since NH_3 excretion involves net base-loss. Available evidence suggests that accumulated

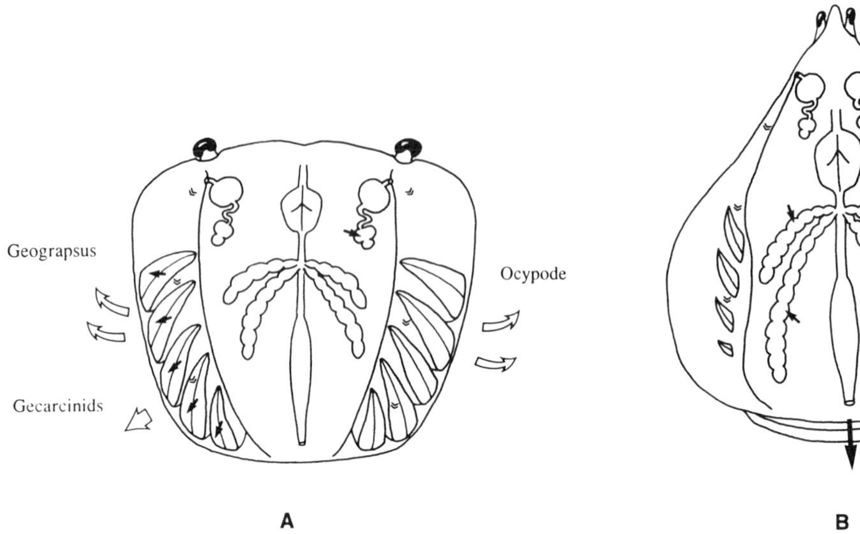

Figure 3 Schematic diagrams summarizing the modes of nitrogenous excretion in semiterrestrial decapods. Small solid arrows indicate the primary site of nitrogenous waste excretion from the hemolymph; paired arrowheads indicate the path of urine flow. (A) Gecarcinids, grapsids, and ocypodids are primarily ammonotelic. Gecarcinids and grapsids excrete ammonia across the branchial epithelium. Gecarcinids then expel ammonia in the reprocessed, liquid urine (P) (straight, open arrow), while *G. grayi* (Grapsidae) volatilizes ammonia from the branchial surface (curved arrows). By contrast, ocypodids excrete ammonia across the labyrinth epithelium of the antennal gland; it is then conveyed to the gills in the primary urine. As with *G. grayi*, branchial reprocessing of urine is accompanied by ammonia volatilization; (B) the anomuran *B. latro* is primarily uricotelic. Uric acid is excreted by the lobes of the very large midgut gland (much more extensive than shown) and expelled with the feces. The gills are greatly reduced with respiration being subsumed by the heavily vascularized wall of the branchial cavity.

protons may be buffered by bicarbonate and then excreted as CO_2. In this way, the opposing acid–base disturbances consequent on nitrogenous and respiratory excretion are mutually eliminated. This may further explain why gaseous ammonia excretion employs the branchial surface in all groups despite the obvious compromises for water conservation. Branchial CO_2 excretion in decapods is generally believed to involve basolateral conversion of HCO_3^- to CO_2 by membrane-bound CA; CO_2 then diffuses outward para- or transcellularly.[55,56] By contrast, buffering of protons by branchial fluid HCO_3^- and subsequent CO_2 excretion would depend on an apical membrane-bound CA. The existence of such an enzyme in the branchial epithelia of crustaceans excreting gaseous ammonia remains to be determined.

B. latro is the only terrestrial crustacean to have largely abandoned branchial ammonotely. It is perhaps significant that it is also the only species to have switched virtually entirely from gill breathing to lung breathing. The predominant fraction of nitrogenous waste comprises uric acid which is excreted by the midgut gland and voided with the feces. Significant levels of urate are also accumulated under certain conditions by brachyurans and probably represent storage excretion. Minor levels of urate in other terrestrial crustaceans are apparently derived from dietary sources. They may serve diverse functions including nitrogen storage, modulation of hemocyanin O_2-binding and ion sequestration.

The semiterrestrial crab, *P. hydrodromus*, shows a progressive switch from ammonotely to ureotely in response to both increasing salinity and desiccation. A complete urea cycle remains to be demonstrated for any crustacean and the pathways of urea synthesis in *P. hydrodromus* are unknown. Nevertheless, it remains the only species known to exploit ureotely to a significant extent, constituting an interesting anomaly in this regard.

ACKNOWLEDGMENTS

The authors are grateful to Dr. Peter Greenaway and Dr. Donna Wolcott for their helpful comments on a draft of this chapter.

REFERENCES

1. **Hartnoll, R. G.,** Evolution, systematics and geographical distribution, in *Biology of the Land Crabs*, Burggren, W. W. and McMahon, B. R., Eds., Cambridge University Press, New York, 1988,
2. **Greenaway, P.,** Nitrogenous excretion in aquatic and terrestrial Crustacea, *Mem. Queensland Mus.*, 31, 215, 1991.
3. **Claybrook, D. L.,** Nitrogen metabolism, in *Biology of Crustacea*, Vol. 5, Mantel, L. H., Ed., Academic Press, New York, 1983, 163.
4. **Wolcott, D. L.,** Nitrogen excretion is enhanced during urine recycling in two species of terrestrial crab, *J. Exp. Zool.*, 259, 181, 1991.
5. **Wolcott, T. G. and Wolcott, D. L.,** Extrarenal modification of urine for ion conservation in ghost crabs, *Ocypode quadrata* Fabricius, *J. Exp. Mar. Biol. Ecol.*, 91, 93, 1985.
6. **Greenaway, P. and Nakamura, T.,** Nitrogenous excretion in two terrestrial crabs *Gecarcoidea natalis* and *Geograpsus grayi*, *Physiol. Zool.*, 64, 767, 1991.
7. **Towle, D. W.,** Sodium transport systems in gills, *Comp. Physiol.*, 7, 241, 1990.
8. **Wood, C. M. and Boutilier, R. G.,** Osmoregulation, ionic exchange, blood chemistry, and nitrogenous waste excretion in the land crab *Cardisoma carnifex*; a field and laboratory study, *Biol. Bull.*, 169, 267, 1985.
9. **Horne, F. R.,** Nitrogen excretion in Crustacea. I. The herbivorous land crab *Cardisoma guanhumi* Latreille, *Comp. Biochem. Physiol.*, 26, 687, 1968.
10. **Gifford, C. A.,** Accumulation of uric acid in the land crab, *Cardisoma guanhumi*, *Am. Zool.*, 8, 521, 1968.
11. **Varley, D. G. and Greenaway, P.,** Nitrogenous excretion in the terrestrial carnivorous crab *Geograpsus grayi*: site and mechanism of excretion, *J. Exp. Biol.*, 190, 179, 1994.
12. **DeVries, M. C. and Wolcott, D. L.,** Gaseous ammonia evolution is coupled to reprocessing of urine at the gills of ghost crabs, *J. Exp. Zool.*, 267, 97, 1993.
13. **DeVries, M. C., Wolcott, D. L., and Holliday, C. W.,** High ammonia and low pH in the urine of the ghost crab *Ocypode quadrata*, *Biol. Bull.*, in press.
14. **Green, J. W., Harsch, M., Barr, L., and Prosser, C. L.,** The regulation of water and salt by the fiddler crabs *Uca pugnax* and *Uca pugilator*, *Biol. Bull.*, 116, 76, 1959.
15. **Greenaway, P. and Morris, S.,** Adaptations to a terrestrial existence by the robber crab, *Birgus latro* L. III. Nitrogenous excretion. *J. Exp. Biol.*, 143, 333, 1989.
16. **Henry, R. P. and Cameron, J. N.,** A survey of blood and tissue nitrogen compounds in terrestrial decapods of Palau, *J. Exp. Zool.*, 218, 83, 1981.
17. **Warner, G. F.,** *The Biology of Crabs*, Elek Science, London, 1977.
18. **Liras, A., Rotllan, P., and Llorente, P.,** De novo purine biosysnthesis in the crustacean *Artemia*: influence of salinity and geographical origin, *J. Comp. Physiol.* B., 162, 263, 1992.
19. **Sloan, W. C.,** Unpublished report cited in Hartenstein, R., Nitrogen metabolism in the terrestrial isopod *Oniscus asellus*, *Am. Zool.*, 8, 507, 1968.
20. **Hartenstein, R.,** Nitrogen metabolism in the terrestrial isopod *Oniscus asellus*, *Am. Zool.*, 8, 507, 1968.
21. **Wieser, W., Schweizer, G., and Hartenstein, R.,** Patterns in the release of ammonia by terrestrial isopods, *Oecologia*, 3, 390, 1969.
22. **Kirby, P. K. and Harbaugh, R. D.,** Diurnal patterns of ammonia release in marine and terrestrial isopods, *Comp. Biochem. Physiol.*, 47A, 1313, 1974.
23. **Wieser, W. and Schweizer, G.,** A re-examination of the excretion of nitrogen by terrestrial isopods, *J. Exp. Biol.*, 52, 267, 1970.
24. **Wright, J. C. and O'Donnell, M. J.,** Total ammonia concentration and pH of haemolymph, pleon fluid and maxillary urine in *Porcellio scaber* Latreille (Isopoda, Oniscidea): Relationship to ambient humidity and water vapour uptake, *J. Exp. Biol.*, 176, 233, 1993.
25. **Needham, J.,** Contributions of chemical physiology to the problems of reversibility in evolution, *Biol. Rev.*, 13, 225, 1938.
26. **Wright, J. C. and Machin, J.,** Water vapour absorption in terrestrial isopods, *J. Exp. Biol.*, 154, 13, 1990.
27. **Wright, J. C. and Machin, J.,** Active water vapor absorption in isopods: evidence for pressure cycling as a supplement to the colligative uptake mechanism, *Physiol. Zool.*, 66, 193, 1993.
28. **Wright, J. C. and Machin, J.,** Active water absorption and the water budget of terrestrial isopods (Crustacea, Isopoda, Oniscidea), *Biol. Bull.*, 184, 243, 1993.

29. **Wright, J. C. and O'Donnell, M. J.,** Osmolality and electrolyte composition of pleon fluid in *Porcellio scaber* (Crustacea, Isopoda, Oniscidea): implications for water vapour absorption, *J. Exp. Biol.*, 164, 189, 1992.
30. **Wright, J. C., O'Donnell, M. J., and Reichert, J.,** Effects of ammonia loading on *Porcellio scaber*: glutamine and glutamate synthesis, ammonia excretion and toxicity, *J. Exp. Biol.*, 188, 143, 1994.
31. **Wieser, W. and Schweizer, G.,** Der Gehalt an Ammoniak und Freien Aminosuren sowie die Eigenschaften einer Glutaminase bei *Porcellio scaber* (Isopoda), *J. Comp. Physiol.*, 81, 73, 1972.
32. **King, F. D., Cucci, T. L., and Bidigare, R. R.,** A pathway of nitrogen metabolism in marine decapod crabs, *Comp. Biochem. Physiol.*, 80B, 401, 1985.
33. **Wright, J. C. and O'Donnell, M. J.,** Water vapour absorption and ammonia volatilization: adaptations for terrestriality in isopods, *Crustacean Issues*, in press.
34. **Hoese, B.,** Morphologie und Funktion des Wasserleitungssystems der terrestrichen Isopoden (Crustacea, Isopoda, Oniscoidea), *Zoomorphologie*, 98, 135, 1981.
35. **Drobne, D. and Fajgelj, A.,** Use of Tc-99m-pertechetate to follow liquid water uptake by *Porcellio scaber*, *J. Exp. Biol.*, 178, 275, 1993.
36. **Maissiat, J.,** Contribution a l'etude du role endocrine de la glande antennaire de l'Oniscoide *Porcellio dilatatus* Brandt. Effet de son ablation sur l'equilibre hydro-mineral, *C.R. Soc. Biol.*, 1972, 916.
37. **Hurley, D.,** Transition to land in amphipods, *Am. Zool.*, 8, 327, 1968.
38. **Dresel, E. B. and Moyle, V.,** Nitrogenous excretion in amphipods and isopods, *J. Exp. Biol.*, 27, 210, 1950.
39. **Graf, F. and Michaut, P.,** Chronologie du developement et evolution du stockage de calcium et des cellules de urates chez *Niphargus schellenbergi* (Karaman). *Int. J. Speleol.*, 7, 247, 1975.
40. **Wolcott, D. L. and Wolcott, T. G.,** Nitrogen limitation in the herbivorous land crab *Cardisoma guanhumi*, *Physiol. Zool.*, 60, 262, 1987.
41. **Bright, D. B.,** Digestion, excretion and metabolism, *Am. Zool.*, 8, 541, 1968.
42. **Kormanik, G. A. and Cameron, J. N.,** Ammonia excretion in animals that breathe water: a review, *Mar. Biol. Lett.*, 2, 11, 1981.
43. **Morris, S., Bridges, C. R., and Grieshaber, M. K.,** A new role for uric acid: modulator of hemocyanin oxygen affinity in crustaceans, *J. Exp. Zool.*, 235, 135, 1985.
44. **Morris, S.,** Organic ions as modulators of respiratory pigment function during stress, *Physiol. Zool.*, 63, 253, 1990.
45. **Morris, S., Greenaway, P., and McMahon, B. R.,** Adaptations to a terrestrial existence in the robber crab, *Birgus latro* L. I. An *in vitro* investigation of blood gas transport, *J. Exp. Biol.*, 140, 477, 1988.
46. **Lockwood, A. P. M.,** The osmotic and ionic regulation of *Asellus aquaticus* (L.), *J. Exp. Biol.*, 36, 546, 1959.
47. **Holdich, D. M. and Mayes, K. R.,** Blood volume and total water content of the woodlouse, *Oniscus asellus*, in conditions of hydration and desiccation. *J. Insect Physiol.*, 27, 547, 1976.
48. **Barrett, R. T.,** unpublished report cited in Little, C., *The Colonisation of Land*, Cambridge University Press, 1983.
49. **Braun, E. J.,** Renal response of the starling (*Sturnus vulgaris*) to an intravenous salt load, *Am. J. Physiol.*, 234, F270, 1978.
50. **McNabb, R. A. and McNabb, F. M. A.,** Physiological chemistry of uric acid: solubility, colloid and ion-binding properties, *Comp. Biochem. Physiol.*, 67A, 27, 1980.
51. **Krishnamoorthy, R. V. and Srihari, K.,** Changes in the excretory patterns of the freshwater field crab *Paratelphusa hydrodromous* upon adaptation to higher salinities, *Mar. Biol.*, 21, 341, 1973.
52. **Mullins, D. E. and Cochran, D. G.,** Nitrogenous excretory material from the American cockroach, *J. Insect Physiol.*, 19, 1007, 1973.
53. **Prusch, R. D.,** Excretion of NH_4Cl by the hindgut of *Sarcophaga bullata* larvae, *Comp. Biochem. Physiol.*, 41A, 215, 1972.
54. **Harrison, J. F. and Phillips, J. E.,** Recovery from acute hemolymph acidosis in unfed locusts. II. The role of ammonium and titratable acid excretion, *J. Exp. Biol.*, 165, 97, 1992.
55. **Henry. R. P.,** Membrane-associated carbonic anhydrase in gills of the blue crab *Callinectes sapidus*, *Am. J. Physiol.*, 252, R966, 1987.
56. **Henry, R. P.,** Branchial and branchiostegite carbonic anhydrase in decapod crustaceans: the aquatic to terrestrial transition, *J. Exp. Zool.*, 259, 294, 1991.
57. **Lallier, F. H. and Walsh, P. J.,** Activities of uricase, xanthine oxidase, and xanthine dehydrogenase in the hepatopancreas of aquatic and terrestrial crabs, *J. Crust. Biol.*, 11, 506, 1991.
58. **Lallier, F. H. and Walsh, P. J.,** Urate does not accumulate in the haemolymph of exercised blue crabs, *Callinectes sapidus*, *J. Exp. Biol.*, 154, 581, 1990.
59. **Regnault, M.,** Nitrogen excretion in marine and fresh-water crustaceans, *Biol. Rev.*, 62, 1, 1987.
60. **Alikhan, M. A.,** Ionic content of the haemolymph in *Porcellio laevis* Latreille (Porcellionidae, Isopoda), *Comp. Biochem. Physiol.*, 44A, 47, 1973.
61. **Harris, R. R. and Andrews, M. B.,** Total NPS pool and ammonia net efflux rate changes in *Carcinus maenas* during acclimation to low environmental salinity, *Comp. Biochem. Physiol.*, 82A, 3301, 1985.

Chapter 8

Nitrogen Metabolism and Excretion in Locusts

Jon F. Harrison

CONTENTS

INTRODUCTION

Since the extensive reviews of Cochran,[1,2] a number of fundamental aspects of nitrogen excretion in insects have been elucidated. Basic advances have been made in understanding (1) the mechanisms of urate and ammonium secretion, (2) the importance of ammonium as a nitrogenous waste in insects, and (3) the relationship between ammonium secretion and acid–base physiology. The bulk of this work has been carried out with locusts. In addition, there has been substantial research related to the feeding behavior and nutritional physiology of locusts, particularly in relation to nitrogen. In this review I have integrated recent information on nitrogen intake, metabolism, and excretion in locusts in an attempt to develop a more coherent view of the physiological handling of the element considered most likely to limit growth in herbivorous insects. In some cases (particularly nonrenal nitrogen metabolism), critical information is not available for locusts, and I have relied on results from other insect groups.

In contrast to vertebrate physiologists, insect physiologists have largely ignored the topics of nitrogen metabolism and excretion; surveys of recent biological databases show about 40 times more published papers on nitrogen metabolism and excretion in lower vertebrates than in insects over the last four years. This is unfortunate, given the importance of insects to ecology, agriculture, and medicine. One of the most striking conclusions to be drawn by this review is that many fundamental aspects of nitrogen metabolism and excretion in insects remain unclear. I have chosen to highlight these gaps in our knowledge in the hope of stimulating further research on this fundamental aspect of insect homeostasis.

0-8493-8411-7/95/$0.00+$.50

For the purpose of this review, I will use the term "total ammonia" to refer to NH_4^+ plus NH_3, ammonia to refer specifically to nonionic NH_3, and ammonium to refer to ionic NH_4^+. Total urate indicates urate plus uric acid, with the term uric acid only being used to refer specifically to the nonionic uric acid or uric acid dihydrate. Since the pK_2 for uric acid is 10.3,[3] and there is no evidence for pHs greater than 7.5 in the hemolymph, cells, or alimentary lumen of locusts,[4-6] I have used the term urate to refer only to the monovalent anion.

NITROGEN AS A NUTRIENT IN LOCUSTS

When considering nitrogen metabolism and excretory mechanisms, it would be useful to be able to define clearly the "goals" of the organism. This is somewhat difficult for nitrogen excretion in locusts. As noted by Cochran[1] "it is well established that most animals consume larger amounts of nitrogen in their diets than is necessary for the support of body functions." This statement is somewhat contradicted, however, by the finding that nitrogen is generally a limiting nutrient for herbivorous insects.[7,8]

A number of studies have attributed poor growth of locusts to low dietary nitrogen levels.[9-11] Locusts may compensate for low levels of nitrogen in grass and leaves by increasing feeding rates.[12] Behavioral mechanisms can not completely compensate for low dietary protein, however.[13] Together, these studies suggest that an important goal of nitrogen metabolism and excretion in locusts should be to conserve nitrogen. This raises the question, "Why do nitrogen-limited animals excrete nitrogen?"

"Nitrogen limitation" in these cases must in fact be a shortage of certain essential amino acids, since the other amino acids can be synthesized. Therefore, we would predict that locusts should catabolize and excrete excess nonessential, but not essential, amino acids. Although the essential amino acids have not been specifically determined for grasshoppers, most insect orders require the same ten essential amino acids[14] (nine are listed in Table 1, the tenth is tryptophan). Grasshopper diets are particularly deficient in phenolic essential amino acids such as phenylalanine under laboratory conditions.[15,16] Most herbivorous insects, including locusts, are generalist feeders in nature; and it has been suggested that this behavior may function to provide a balanced diet.[17] Nonetheless, it would seem that physiological mechanisms that would allow the conservation of essential amino acids would be useful.

NITROGEN METABOLISM

TOTAL URATE METABOLISM

Amino acid catabolism primarily produces carbon dioxide, ammonium, and bicarbonate. Each of these is potentially toxic unless eliminated. The respiratory system excretes carbon dioxide, while ammonium and bicarbonate must be disposed of by other means. In terrestrial insects, including locusts, the primary metabolic fate of ammonium and bicarbonate is the synthesis of total urate in the fat body. Uric acid metabolism has received little attention in locusts, so information from studies of Lepidoptera and cockroaches will be summarized here.

The details of the synthetic pathway for uric acid are not known for any insect, but evidence suggests that the pathway is similar to that of birds[1] (see also Chapter 10). Apparently, as in birds, the precursors of total urate are glutamine, glycine, formate, carbon dioxide, and aspartate.[1] Xanthine dehydrogenase catalyzes the final steps of the uric acid synthesis pathway, and has been used as a marker enzyme to identify the fat body as the primary tissue involved in total urate synthesis.[18] Total urate is stored as protein-containing granules within membrane-bound structures in the fat body tissue.[19-21] These granules contain potassium at 10 to 30% the molarity of total urate,[21,22] suggesting that both uric acid and potassium urate salts exist.

In certain cockroaches, termites, flies, beetles, aphids, and homopterans, there is some evidence for microbially mediated degradation of uric acid.[1] At least in roaches, stored total urate serves as a mobile reserve of nitrogen. In cockroaches, total urate stores increase in response to high nitrogen

Table 1 Hemolymph Concentrations and Urine-to-Hemolymph Ratios (U/H) for Amino Acids in *S. gregaria*

	Hemolymph mmol l^{-1}	U/H		Hemolymph mmol l^{-1}	U/H
Essential amino acids			Nonessential amino acids		
Arginine	1.3	0	Alanine	0.7	1.4
Histidine	1.1	0	Aspartate	0.1	1.9
Isoleucine	0.3	0	Asparagine	0.8	0
Leucine	0.3	0	Glutamate	0.1	5.1
Lysine	1.0	0	Glutamine	5.0	.15
Methionine	0.3	0	Glycine	16.7	.25
Phenylalanine	0.5	0	Proline	12.5	3.2
Threonine	0.4	0	Serine	1.7	.2
Valine	0.5	0	Tyrosine	0.7	0

Note: Hemolymph data for locusts fed on lettuce; urine refers to Malpighian tubule fluid collected *in situ*.

From Chamberlin, M. E. and Phillips, J. E., *Can J. Zool.*, 60, 2745, 1982. With permission.

intake, and decrease during protein deprivation.[23] During food deprivation, total urate stores decrease, presumably serving as a nitrogen store for the insect.[23,24] The metabolic pathway for the catabolism of total urate is unknown, although a variety of evidence suggests involvement of mycetocyte bacteria that are symbiotic within the cockroach fat body.[2] Similarly, microbial degradation of uric acid within the gut serves to conserve nitrogen in termites.[25]

Locusts and most other insects can degrade some fraction of total urate to allantoin or allantoic acid, but can not catabolize these substances any further.[1] Total urate stores in locusts are less than 10% of those found in roaches.[26] Despite the evidence that nitrogen can be a limiting nutrient in locusts, there is no evidence that total urate serves as a nitrogen reserve.

Total urate synthesized in the fat body must be released for transit through the hemolymph for excretion. At least under some conditions in Lepidoptera, 20-hydroxyecdysone controls the release of total urate from the fat body.[27] Total urate released from the fat body moves through the hemolymph, presumably in solution, to the Malpighian tubules for excretion.

METABOLISM OF PROTEIN AND AMINO ACIDS

In locusts, as in most other insects, the hemolymph is a large reservoir of amino acids. Three nonessential amino acids, glycine, proline, and glutamine, are high in concentration relative to all other amino acids in the locust *Schistocerca gregaria* (Table 1[28]). In *Locusta migratoria*, amino acid and protein levels in the hemolymph decrease during starvation, suggesting that these compounds serve as metabolic reserves.[29] Hemolymph proteins also serve as a nitrogen source for vitellogenesis in locusts and many other insects.[30]

Proline is an important metabolic fuel during activity in other insects,[31] but in locusts, locomotion is considered to be fueled by carbohydrate and lipid.[32,33] However, amino acids serve as the primary metabolic substrate for the locust hindgut. Evidence from inhibitor studies suggests that the hindgut catabolizes amino acids via the glutamate dehydrogenase pathway.[34,35] Proline is the primary metabolic fuel for the support of ion transport in the rectum.[35,36] At physiological concentrations, only glucose and five amino acids support ion transport in the ileum.[34] The hindgut epithelium obtains amino acids primarily from the lumen, rather than the hemolymph.[34,36]

Do the mechanisms for amino acid uptake and metabolism by the hindgut conserve essential amino acids for the locust? Amino acids, including essential amino acids, can diffuse passively through Malpighian tubules of the blood feeder *Rhodnius prolixus in vitro*.[37] However, no essential amino acids occur in Malpighian tubule fluid collected *in situ* from *S. gregaria* (Table 1). Some nonessential amino acids that occur in tubular urine have lower concentrations in the hemolymph than many of the essential amino acids (Table 1). It is unlikely that locust tubules are impermeable

Table 2 Amino Acids Which Support Short-Circuit Current in *S. gregaria*

	% of maximum current		% of maximum current
Essential amino acids tested		Nonessential amino acids tested	
Arginine	0	Alanine	46
Histidine	0	Asparagine	100
Lysine	0	Glutamine	77
Valine	0	Proline	43
		Serine	60

Note: Individual amino acids applied bilaterally at hemolymph concentrations; no other exogenous substrates present.

From Peach, J. L. and Phillips, J. E., *J. Insect Physiol.*, 37, 255, 1991. With permission.

to essential amino acids, since they are permeable to inulin.[37] Therefore, it seems likely that locust tubules, or anterior regions of the ileum, reabsorb essential amino acids. At hemolymph concentrations, applied bilaterally, only nonessential amino acids serve as respiratory substrates for the ileum (Table 2). Renal handling of amino acids seems well designed for the conservation of essential amino acids for organismal protein synthesis.

NITROGEN EXCRETION IN LOCUSTS

COMPOSITION OF THE NITROGENOUS WASTES

Numerous studies have confirmed that total urate and its derivatives (allantoin, allantoic acid) are the primary nitrogenous wastes of most terrestrial insects.[18,26,38] However, a wide variety of other nitrogenous compounds occur in the excreta of terrestrial insects, including the purines hypoxanthine and xanthine (precursors of uric acid), total ammonia, urea, amino acids, tryptophan derivatives (i.e., ommocromes), and pteridines.[1]

Careful research by Mullins and Cochran demonstrated that in many cockroach species, total ammonia rather than uric acid is the major nitrogenous waste.[2,23,39] Little ammonia is volatilized from the body surface; the majority is excreted in the feces as ammonium.[40] Mullins and Cochran predicted that "Given the small attention which has been paid to this problem in insects and the rapid volatilization of ammonia from excreta, it would not be surprising if numerous other terrestrial species are shown to be ammonotelic when it is specifically looked for."[1]

In locusts, early measurements showed that total urate is the predominate nitrogenous waste, with some contribution of allantoin, allantoic acid, urea, and amino acids (Table 3). Most, but not all studies reported only trace measures of ammonium (Table 3). These early studies may have underestimated fecal total ammonia levels, however, due to two technical problems. First, some early authors oven dried the excreta prior to analysis. As pointed out by Mullins and Cochran,[39] oven drying can volatilize large fractions of the total ammonia in excreta. In *S. gregaria*, 12 h of drying at 60°C results in loss of 50% of fecal ammonia even from pellets with a pH <5 (Harrison, unpublished data). To lower the partial pressure of NH_3 and conserve total ammonia in the excreta, fecal pellets should be collected rapidly, transferred to an acidic medium, and frozen.[41] A second technical problem is that the total ammonia may be present as relatively insoluble ammonium urate. Ammonium levels measured in the supernatant fractions of 20× dilutions of fecal pellets can be as low as 10% of those measured with dilutions sufficient to solubilize all urate.[41] Because the fraction of ammonium present as a precipitate may vary depending on the pH of the feces and the concentration of urate, assays of fecal total ammonium in insects should begin with a study of successive dilutions on the concentration measured. Recent studies suggest that total ammonia accounts for 10 to 46% of nitrogen excretion in locusts (Table 3). At the pHs commonly occurring in locust fecal pellets (7 to 4.5), virtually all of the total ammonia will be in the form of NH_4^+ (as the pK for NH_4^+ is above 9).[42]

Table 3 Percent Contribution to Nitrogen Excretion in Grasshoppers

Species	Diet	Total urate	Total ammonia	Urea	Amino acids	Arginine	Total allantoin	Tryptophan derivatives	References
Melanoplus bivittatus	lettuce	81	2	8	7	1	NM	NM	75
Schistocerca gregaria	—	72	28	0	NM	NM	NM	NM	76
Locusta migratoria	—	89	trace	trace	NM	NM	11	NM	26
Schistocerca gregaria	—	86	trace	trace	NM	NM	14	NM	26
Schistocerca gregaria	unfed	54	46	NM	NM	NM	NM	NM	41
Taeniopoda eques	wheat	90	10	NM	NM	NM	NM	NM	66
Taeniopoda eques	unfed	80	20	NM	NM	NM	NM	NM	66

Note: NM, not measured.

CHEMISTRY OF TOTAL URATE

Although the total urate in insect fecal pellets is commonly referred to as uric acid, the proportion of total urate that is uric acid is unknown. This is because (1) other urate salts, particularly ammonium urate, are also poorly soluble, (2) the colloidal chemistry of urate solutions is very complex, (3) cations (i.e., ammonium, calcium, potassium) in solution can exchange with protons in precipitated uric acid crystals, and (4) the effects of protein and other organic components in the feces on urate chemistry are uncertain.

The pK_1 of uric acid is 5.75, and the pK_2 is 10.3.[3] The K_{sp} of uric acid in imidazole buffer is about 1.2×10^{-9}.[43] "True solubilities" of 0.4 to 6 $mmol^{-1}$ depending on pH have been reported for uric acid in salines.[43-46] Because the pH of the rectal lumen is acidic (pH 5 to 6 in unfed and fed locusts[4,6]), and total urate concentrations are usually over 100 mmol kg^{-1}, it is clear that in the absence of other cations, over 95% of the total urate would be in the form of precipitated uric acid.

Ammonium, potassium, and sodium urate salts are also poorly soluble, and the activities and concentrations of these cations are much higher than the hydrogen ion activity (Table 4). Urate enters the hindgut from the Malpighian tubules at concentrations of 1 to 5 mmol l^{-1}.[47,48] As water is reabsorbed in the hindgut, concentrations increase to over 100 mmol kg^{-1}. Based on the solubility products, precipitation of uric acid should occur when the hindgut pH drops below 6. However, potassium urate salts should form at potassium activities above 100 mmol l^{-1}, similar to potassium levels in locust Malpighian tubule fluid.[49] Ammonium urate deposition should occur at total ammonia concentrations above 3 mmol l^{-1}, which occur in tubular fluid *in vivo*.[47] Sodium urate precipitation should also occur (Table 4).

Table 4 Solubility of Urate Salts, Cation Concentrations, and Activities in *S. gregaria*

	Urate salt Ksp[43]	Concentration in feces, mmol l^{-1} [41,66]	Activity in rectal fluid, mmol l^{-1} [52]	Concentration in Malpighian tubule fluid[47,49]
Ammonium	8.5×10^{-6}	25–325	—	5–14
Potassium	1.5×10^{-4}	258	7	140
Sodium	4.9×10^{-5}	128	75	30
Hydrogen	1.2×10^{-9}	–	10^{-5}	10^{-7}

Note: Total urate concentrations in locust feces range from 70 to 300 mmol kg^{-1};[41,66] while concentrations of total urate in the Malpighian tubule fluid entering the hindgut range from 1 to 5 mmol l^{-1}.[47]

Unfortunately, the chemical properties of urates are extremely complex and not well explained by simple solubility. Uric acid and urate salts form both lyophilic and lyophobic colloids in physiological solutions, allowing supersaturation to be achieved for considerable periods of time.[43,50] O'Donnell et al.[51] showed that concentrations of total urate in lower *Rhodnius* tubules may reach 20 mmol l^{-1} for hours regardless of urine pH, with these concentrations falling by over 20-fold in 2 d. These results suggest that the properties of these supersaturated or colloidal solutions might determine whether urate or uric acid is formed.

McNabb and McNabb[43] showed that cations (particularly ammonium, calcium, and magnesium) at concentrations less than those observed in locust feces could quickly reduce the solubility of uric acid from 6 to under 1 mmol l^{-1} (pH 6.7), and replace protons in uric acid precipitates. Low pH (below 6.1) prevented the inclusion of sodium into uric acid precipitates[43]; however, the effect of pH on incorporation of other cations into precipitates has not been investigated. A further complication is the lack of any information on how proteins or other organics in the insect hindgut might affect total urate precipitation. In summary, the chemical nature of total urate in locust feces (and most insects) is difficult to predict.

O'Donnell et al.[51] have made the only measurements of cations in relatively pure total urate precipitates of insects. In *Rhodnius* lower tubules, sodium and potassium urate salts account for about 20% of the total urate in precipitates (assuming monovalent salts); ammonium content of the precipitates was not measured. In unfed *S. gregaria*, more than twice as much ammonium as total urate is excreted, and over 90% of the ammonium is precipitated (Table 4[41]). These results suggest that a large fraction of the total urate must be present as ammonium urate, but are not conclusive because urate granules were not purified. Given the complexities of urate chemistry and the large variability in the concentrations of cations observed in the excreta of different insect species, it seems likely that the fraction of uric acid in total urate precipitates varies widely. For quantification of net acid excretion, the total urate in the feces can be solubilized and titrated directly,[41] but this does not resolve the chemical nature of the total urate because other compounds account for more than 60% of fecal buffering.

NITROGEN EXCRETION AND WATER CONSERVATION

A basic paradigm of comparative insect physiology is that aquatic insects excrete ammonia while terrestrial insects excrete uric acid.[1,18] Ammonia excretion is simple and inexpensive for aquatic organisms; while uric acid excretion serves to eliminate nitrogen in a precipitate form without losing water. The high surface-to-volume ratio of small insects exacerbates problems of water conservation for terrestrial insects, perhaps driving the evolution of uricotely. Ammonia excretion is thought to be incompatible with water conservation because of the high solubility of ammonium salts (the solubility of ammonium chloride is about 6 mol l^{-1}). As noted in sections above, however, in a uricotelic organism, ammonium may be excreted as precipitated ammonium urate. Unfed *S. gregaria* locusts excreted more ammonium than total urate in fecal pellets containing less than 10% water, demonstrating that ammonium excretion is clearly compatible with water conservation in locusts.[41] In contrast, in the American cockroach, increasing nitrogen intake is strongly correlated with ammonium excretion and *ad lib* water intake.[23] Thus in a nonurate excreting insect, ammonium excretion does interfere with water conservation.

LOCATION AND MECHANISM OF TOTAL URATE SECRETION

The locust renal system consists of approximately 200 Malpighian tubules, which secrete the primary urine, and a hindgut (primarily the ileum and rectum) which performs selective reabsorption and secretion. Although the epithelial mechanisms of urate transport have not been investigated in detail in any grasshopper, data to date suggest that the most important mechanism is active transport by the Malpighian tubules.

O'Donnell et al.[51] have investigated the mechanism of urate transport by the Malpighian tubules of the blood-sucking bug *Rhodnius prolixus*. They demonstrated that active transport is limited to

Table 5 Rates of Total Ammonia and Total Urate Secretion in *S. gregaria*

	Malpighian tubules	Ileum	Rectum	Fecal
J_{amm}, unstimulated (unfed)	.06	0.6	0.4	0.15
J_{amm}, cAMP stimulated	0.1	1.4	1.9	NM
J_{ur}, unfed	.03	NM	NM	0.06
J_{ur}, fed	.06	NM	NM	NM

Note: Rates presented in μmol h^{-1}. Rates for the ileum and rectum are *in vitro* rates with or without cAMP;[52,57] rates for the Malpighian tubules are from *in situ* cannulation of fed or unfed locusts;[47,48] fecal rates refer to fecal excretion in unfed animals.[41]

the lower tubule, and is activated by feeding and elevations of hemolymph total urate. There is some evidence that the transport mechanism is located on the basolateral membrane. In *Rhodnius*, the rate of total urate secretion by the Malpighian tubules is sufficient to account for whole animal excretion rates.

Active transport of urate also occurs in the Malpighian tubules of locusts.[51] Excretion rates of urate by Malpighian tubules (J_{ur}, μmol h^{-1}) cannulated *in situ* were about half of the J_{ur} measured for unfed *S. gregaria* (Table 5). It is likely that this difference simply reflects variation in total urate excretion due to dietary or ontogenetic effects. It is also possible that the *in situ* preparation affected animal health and reduced tubular secretion rate. However, this comparison does suggest that it would be useful to test the capacity of other sections of the hindgut for total urate secretion to confirm that the Malpighian tubules are the only site of secretion. If this is the case, comparison of urate levels in the tubular fluid and hindgut may be used for analysis of water reabsorption *in vivo*.[52]

LOCATION AND MECHANISM OF TOTAL AMMONIA SECRETION

Considerable information is now available concerning the nature of total ammonia secretion (J_{amm}, μmol h^{-1}) by various regions of the locust renal system. Ammonia is secreted throughout the locust Malpighian tubule-hindgut complex.[52] It is also possible that total ammonia is secreted by the salivary glands or midgut tissue of the locust. Total ammonia concentrations are approximately 10 mmol l^{-1} in the crop and midgut fluid of unfed *S. gregaria*, about 10 times lower than in the lumen of the hindgut.[6] Given 95% water removal by the hindgut of unfed locusts,[52] it is possible that total ammonia derived from the crop or midgut contributes considerably to total ammonia excretion. However, total ammonia secretion has not been examined for the salivary glands or midgut of any insect, so it is possible that some of the total ammonia measured in the crop and midgut is derived from digestive or microbial processes. The summed *in vitro* J_{amm} for the Malpighian tubules, ileum, and rectum are more than sufficient to account for total ammonia excretion in locusts,[52] supporting the primary role of the Malpighian tubule-hindgut complex of locusts in total ammonia excretion (Table 5).

Malpighian Tubules

The Malpighian tubules form the primary urine as an isosmotic secretion, driven by a H^+ ATPase and H^+/K^+ exchange.[53,54] Chamberlin and Phillips[28,36] showed that the Malpighian tubules actively secrete amino acids which serve as a metabolic substrate for the rectum. In addition, the tubules secrete total ammonia, achieving concentrations in the tubular fluid of 5 to 13 mmol l^{-1}.[47,48] Total ammonia secretion by the tubules (0.06 mmol h^{-1}) is about 5% of the total ammonia secretion by the Malpighian tubule-hindgut complex in *S. gregaria*.[47,48, 52]

Ileum

The ileum is in many ways analogous to the vertebrate proximal tubule, absorbing ions, nutrients, and water isosmotically.[55,56] The primary metabolic substrates for the ATP production necessary to drive fluid and ion transport are amino acids, mostly derived from the lumen.[34] Metabolism of these amino acids produces ammonium, over 80% of which is transported to the luminal side.[52] The mechanism by which the ileum transports ammonium to the lumen is not known. J_{amm} is insensitive to luminal pH, suggesting that J_{amm} is not accomplished by diffusion trapping.[52] J_{amm} is also not affected by perfusing the lumen with nominally Na^+-free saline, or by amiloride, suggesting that transport does not occur by a Na^+/NH_4^+ exchange mechanism.[52]

Rectum

The rectum performs final adjustment of the ionic content of the feces, absorbing ions, nutrients, and water, and secreting acid and total ammonia. Because the rectum is the only region of the renal system that is capable of absorbing water against a concentration gradient, it is likely that most concentration and precipitation of nitrogenous wastes occurs here.

Thomson et al.[57] have provided evidence suggesting that ammonium is secreted by the rectum via an apical Na^+/NH_4^+ exchange mechanism. As in the ileum, the primary source of ammonium for rectal secretion is derived from metabolism of amino acids (particularly proline), reabsorbed from the lumen.[57] Chamberlin has shown that rectal mitochondria oxidize proline, apparently via a glutamate dehydrogenase mediated pathway.[35,36] *In vitro*, over 95% of the total ammonia secreted goes to the luminal side.[57] J_{amm} is unaffected by the transepithelial pH, indicating that, as for the ileum, diffusion trapping is not a major mechanism of total ammonia secretion.[57] J_{amm} is inhibited by removal of luminal Na^+, or by amiloride.[57]

Regulation of Renal J_{amm}

Two hormones have been partially purified that stimulate ion and water transport by the hindgut, chloride transport stimulating hormone (CTSH), and ion transport peptide (ITP).[52] Many, but not all of the effects of CTSH and ITP appear to be mediated by rises in cAMP.[52,58-60] Ileal, but not rectal J_{amm} has been reported to be stimulated by cAMP.[52,61] However, neither ITP, nor extracts of the corpora cardiaca or ventral ganglia stimulate ileal J_{amm}.[62] These results suggest that strong neural or hormonal regulation of J_{amm} occurs by an as yet unidentified mechanism.

NITROGEN EXCRETION AND ACID–BASE REGULATION

An understanding of the relationship between renal nitrogen excretion, hepatic nitrogen metabolism, and organismal acid–base regulation is critical to understanding the function and evolution of each of these "separate" systems. Regulation of total urate synthesis could potentially participate in locust pH regulation as total urate formation will consume bicarbonate.[63-65] The system has the potential to be a particularly effective regulator of organismal pH since both uric acid and potassium urate are formed in the insect fat body.[21,22] Unfortunately, the sensitivity of the insect uric acid synthetic pathway to pH is unknown. Changes in locust acid–base status are not correlated with changes in total urate excretion,[41,66] but effects of pH on total urate synthesis of the fat body have not been examined.

Ammonium secretion is often considered to be equivalent to net acid excretion in animals. A model based on current cellular data for locusts in which ammonium secretion by the hindgut is functionally equivalent to net acid secretion is diagrammed in Figure 1. Amino acids (primarily proline) are actively absorbed by the rectal epithelia, primarily from the lumen via a putative H+/proline symport.[34,36,67] Neutral amino acids are the primary metabolic substrate of both the ileum and rectum.[34,36] Enzymatic, isolated mitochondrial, and inhibitor studies indicate that both transamination and complete oxidation of proline occur, but that complete oxidation predominates in the intact rectal tissue.[35] Complete oxidation of a neutral amino acid produces NH_4^+ and HCO_3^- in equal quantities.[68] The CO_2 produced will leave the cell via diffusion (primarily to the more alkaline hemolymph *in vivo*) and be eliminated by the respiratory system. The ammonium produced is secreted sidedly to the lumen, primarily via a Na^+/NH_4^+ exchange mechanism (in the rectum).[57] The bicarbonate produced leaves the cell to the hemolymph via a SITS- and DIDS-sensitive transporter (i.e., a Cl^-/HCO_3^- exchanger).[69] Assuming constancy of intracellular pH, secretion of NH_4^+ to the

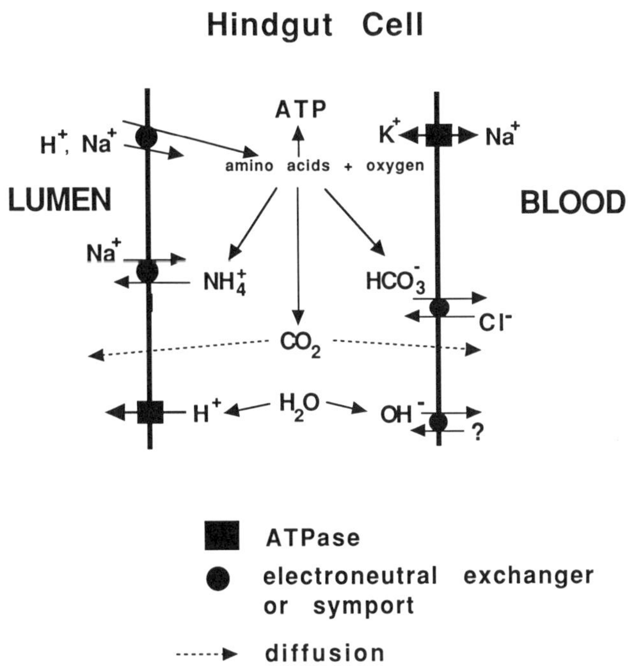

Figure 1 A model of ammonium secretion by the rectum of locust, in which ammonium secretion is functionally equivalent to bicarbonate reabsorption (or acid secretion). See text for explanation.

lumen is accompanied by an equivalent rate of transport of bicarbonate to the hemolymph. The net result is that ammonium secretion is functionally equivalent to acid secretion (or base reabsorption).

Not all of the empirical data available support this model. Most significantly, rectal acid secretion to the lumen has been reported to be equivalent to serosal base excretion (about 1.5 $\mu eq\ cm^{-2}\ h^{-1}$), when the recta are bathed bilaterally in pH 7 saline.[70] Under the same conditions, rectal ammonium secretion is approximately 0.6 $\mu eq\ cm^{-2}\ h^{-1}$.[57] If ammonium secretion to the lumen is accompanied by serosal bicarbonate transfer, serosal titratable base secretion would be expected to exceed luminal acid secretion by 40%. While ammonium and acid–base transport were not measured simultaneously, these data do suggest that under these experimental conditions either (1) most ammonium is produced via transamination (without complete oxidation of the amino acid and consequent bicarbonate production, contrary to the findings of Chamberlin and Phillips[35]), or (2) the bicarbonate produced in association with ammonium production is transported equally to the lumen and the hemolymph, contrary to current models of rectal acid–base transport.[52]

At the moment, it can not be definitively determined whether ammonium excretion is functionally equivalent to acid excretion from the hemolymph in locusts. However, results from five separate studies on three grasshopper species are consistent with participation of regulated J_{amm} in acid–base regulation (Figure 2). In case A1, unfed *S. gregaria* were injected with 10 μeq of HCl, causing a 0.5 fall in blood pH from which the locusts recovered in 8 h.[6] The rise in ammonium excretion potentially accounted for 15% of the acid equivalents removed from the hemolymph during recovery.[41] In case A2, *ad lib* fed *Taenipoda eques* were injected with 10 μeq of HCl, causing a 0.1 fall in blood pH from which the locusts recovered in 3 h.[66] In this case, increases in J_{amm} contributed about 50% of the removal of the injected acid load, assuming J_{amm} is equivalent to net acid excretion.

In two of three studies, reductions in J_{amm} also apparently contributed to recovery from base challenges in locusts. In case B1, *ad lib* fed *Melanoplus differentialis* were injected with 3 μeq of NaOH, causing a 0.1 rise in blood pH from which the locusts recovered in 1 h. There was a significant rise in titratable base excretion which quantitatively accounted for the acid–base recov-

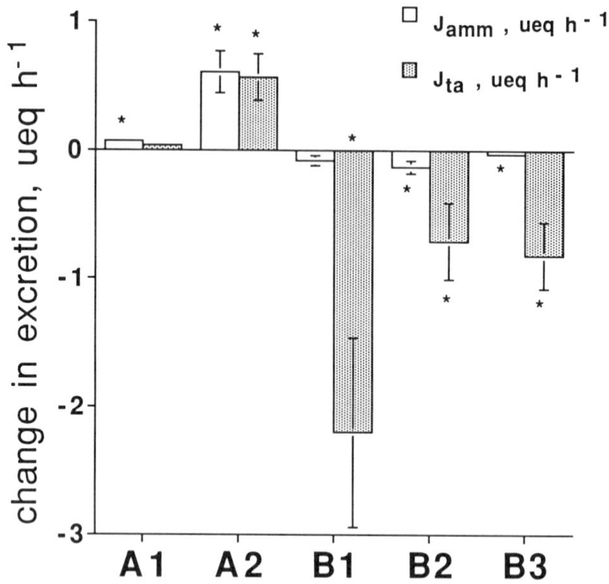

Figure 2 Change in total ammonia (J_{amm}) and titratable acid (J_{ta}) excretion rate (with standard error) after an acid (A1 and A2) or base (B1, B2, B3) challenge. A1, HCl injection into unfed *S. gregaria*; A2, HCl injection into fed *T. eques*; B1, NaOH injection into fed *M. differentialis*; B2, feeding on wheat sprouts vs. starvation for *T. eques*; B3, terpene ingestion for *M. differentialis*. See text for further description of the experiments.

ery; J_{amm} decreased, accounting for 12% of the recovery (Harrison et al., in preparation). In case B2, *T. eques* were feed wheat sprouts, which induced a large increase in titratable base excretion, without a significant change in J_{amm}.[66] In case B3, *M. differentialis* were fed lettuce spiked with the terpene menthol, which induced a change from titratable acid to titratable base excretion, accompanied by a small but significant drop in J_{amm} (Stewart and Harrison, in preparation). In all five of these studies, control groups injected with saline showed no changes in acid or nitrogen excretion, supporting the conclusion that the changes in J_{amm} and acid secretion were in response to the acid–base challenge.

It should be noted that all five of these studies involved relatively acute acid–base challenges, and that studies with vertebrates often show that increases in ammonium excretion are more important components of the pH regulatory response when the stimulus is chronic.[71,72] This is apparently because chronic acidosis allows more time for induction of enzymes which catabolize glutamate.[73] Thus it seems reasonable to suggest that studies to date have underestimated the importance of regulated J_{amm} to long-term pH homeostasis in locusts.

The regulatory mechanism for the control of variation in J_{amm} with acid–base status is unknown. J_{amm} of both the ilea and recta is insensitive to hemolymph or luminal pH, to ITP, or to locust neurohaemal gland extracts.[52,57] Three studies have documented a stimulation of ileal J_{amm} by cAMP, however, suggesting that hormonal or neural control of ileal J_{amm} could be a major mechanism of regulating body pH.[52]

SUMMARY

The nitrogen metabolism and excretory system of locusts serves several functions. First, the system is designed to conserve essential amino acids. A variety of evidence suggests that essential amino acids can be a limiting nutrient in locusts. Locusts apparently lack the capacity to catabolize total urate once formed, and thus use hemolymph protein and amino acid pools as reserves. The renal system prevents loss of essential amino acids, probably by reabsorbing essential amino acids from the urine. Second, the system sequesters and excretes potentially toxic ammonium and bicarbonate derived from catabolism of amino acids. Most nitrogen is excreted as total urate, but 10 to 46% is excreted as ammonium. Excretion of nitrogen as precipitated ammonium urate allows for maximal nitrogen excretion with minimal water loss, which may be important at times for these animals. Third, the data suggest that regulation of ammonium excretion is an integral component of locust acid–base regulation, contributing up to 50% of the regulatory response to an acute acid–base challenge. The convergence between insects and terrestrial vertebrates in the use of regulated J_{amm} for acid–base regulation is particularly striking since, in both cases, the excretory organs are virtually unique in their metabolic specialization on amino acids, and the process is dependent on specialized functions in a diversity of tissues (e.g., amino acid secretion by the Malpighian tubules). In both insects and terrestrial vertebrates, J_{amm} is easily regulated and apparently serves to excrete acid in excess of that possible by titratable acid excretion.[74]

ACKNOWLEDGMENTS

I thank J. Fewell, S. Roberts, and K. Krolikowski for comments on the manuscript, and N. Gonzalez, J. Stewart, M. Kennedy, and L. Neinaber for research assistance. This research was partially funded by NSF awards DCB 90-20284 and IBN 9317784.

REFERENCES

1. **Cochran, D. G.,** Nitrogenous Excretion, *Comprehensive Insect Physiology, Biochemistry and Pharmacology,* 1985.
2. **Cochran, D. G.,** Nitrogen excretion in cockroaches, *Annu. Rev. Entomol.,* 30, 29, 1985.
3. **Hitchings, G. H.,** Uric acid: Chemistry and synthesis, in *Handbook of Experimental Pharmacology,* Kelley, W. N. and Weiner, I. M., Eds., Springer-Verlag, New York, 1978, 1.
4. **Thomson, R. B., Speight, J. D., and Phillips, J. E.,** Rectal acid secretion in the desert locust, *Schistocerca gregaria, J. Insect Physiol.,* 34, 829, 1988.
5. **Harrison, J. M.,** Temperature effects on intra- and extracellular acid-base status in the American locust, *Schistocerca nitens, J. Comp. Physiol. B,* 158, 763, 1988.
6. **Harrison, J. F., Wong, C. J. H., and Phillips, J. E.,** Recovery from acute haemolymph acidosis in unfed locusts I. Acid transfer to the alimentary lumen is the dominant mechanism, *J. Exp. Biol.,* 165, 85, 1992.
7. **Slansky, F. J. and Scriber, J. M.,** Food consumption and utilization, in *Comprehensive Insect Physiology, Biochemistry and Pharmacology,* Kerkut, G. A. and Gilbert, L. I., Eds., Pergamon Press, Oxford, 1985, 87.
8. **Bernays, E. A. and Barbehenn, R.,** Nutritional ecology of grass foliage-chewing insects, in *Nutritional Ecology of Insects, Mites, Spiders and Related Invertebrates,* Slansky, F., Jr. and Rodriguez, J. G., Eds., John Wiley and Sons, New York, 1987, 147.
9. **Uvarov, B.,** *Grasshoppers and Locusts, a Handbook of General Acridology,* Vol. 2, Centre for Overseas Pest Research, London, 1977.
10. **Lee, S. S. and Wong, I. M.,** The relationship between food plants, haemolymph protein, and ovarian development in *Oxya japonica* Willemse (Orthoptera:Acrididae), *Acrida,* 8, 1, 1979.
11. **Scriber, J. M. and Slansky, F.,** The nutritional ecology of immature insects, *Annu. Rev. Entomol.,* 26, 183, 1981.
12. **Abisgold, J. D. and Simpson, S. J.,** The physiology of compensation by locusts for changes in dietary protein, *J. Exp. Biol.,* 129, 329, 1987.
13. **Simpson, S. J. and Abisgold, J. D.,** Compensation by locusts for changes in dietary nutrients: behavioural mechanisms, *Physiol. Entomol.,* 10, 443, 1985.
14. **Dadd, R. H.,** Arthropod nutrition, in *Chemical Zoology,* Florkin, M. and Scheer, B. T., Eds., Academic Press, New York, 1970, 35.
15. **Bernays, E. A. and Woodhead, S.,** Incorporation of dietary phenols into the cuticle in the tree locust *Anacridium melanorhodon, J. Insect Physiol.,* 28, 601, 1982.
16. **Bernays, E. A. and Woodhead, S.,** The need for high levels of phenylalanine in the diet of *Schistocerca gregaria* nymphs, *J. Insect Physiol.,* 30, 489, 1984.
17. **Bernays, E. A., Bright, K., Howard, J. J., Raubenheimer, D., and Champagne, D.,** Variety is the spice of life: frequent switching between foods in the polyphagous grasshopper *Taeniopoda eques* burmeister (Orthoptera:Acrididae), *Anim. Behav.,* 44, 721, 1992.
18. **Cochran, D. G.,** Excretion in insects, in *Insect Biochemistry and Function,* Candy, D. J. and Kilby, B. A., Eds., Chapman and Hall, London, 1975, 177.

19. **Dean, R. L., Lock, M., and Collins, J. V.,** Structure of the fat body, in *Comprehensive Insect Physiology, Biochemistry, and Pharmacology*, Kerkut, G. A. and Gilbert, L. J., Eds., Pergamon Press, New York, 1985, 155.
20. **Buckner, J. S.,** Subcellular localization of uric acid storage in the fat body of *Manduca sexta* during the larval-pupal transformation, *J. Insect Physiol.*, 31, 741, 1985.
21. **Buckner, J. S., Henderson, T. A., Ehresmann, D. D., and Graf, G.,** Structure and composition of urate storage granules from the fat body of *Manduca sexta, Insect Biochem.*, 20, 203, 1990.
22. **Mullins, D. E.,** Isolation and partial characterization of uric acid spherules obtained from cockroach tissues (Dictyoptera), *J. Insect Physiol.*, 62A, 699, 1979.
23. **Mullins, D. E.,** Nitrogen metabolism in the American cockroach: an examination of whole body ammonium and other cations excreted in relation to water requirements, *J. Exp. Biol.*, 61, 541, 1974.
24. **Mullins, D. E. and Cochran, D. G.,** Nitrogen metabolism in the American cockroach. II. An examination of negative nitrogen balance with respect to mobilization of uric acid stores, *Comp. Biochem. Physiol. A*, 50, 501, 1975.
25. **Potrikus, C. J. and Breznak, J. A.,** Gut bacteria recycle uric acid nitrogen in termites-a strategy for nutrient conservation, *Proc. Natl. Acad. Sci.*, 78, 4601, 1981.
26. **Razet, P.,** Recherches sur l'Uricolyse chez les insectes, *Bull. Soc. Sci. Bret.*, 36, 5, 1961.
27. **Ehresmann, D. D., Buckner, J. S., and Graf, G.,** Uric acid translocation from the fat body of *Manduca sexta* during the pupal-adult transformation: effects of 20-hydroxyecdysone, *J. Insect Physiol.*, 36, 173, 1990.
28. **Chamberlin, M. E. and Phillips, J. E.,** Regulation of hemolymph amino acid levels and active secretion of proline by Malpighian tubules of locusts, *Can. J. Zool.*, 60, 2745, 1982.
29. **Hill, L. and Goldsworthy, G. J.,** The utilization of reserves during starvation of larvae of the migratory locust, *Comp. Biochem. Physiol.*, 36, 61, 1970.
30. **Hill, L.,** Neurosecretory control of haemolymph protein concentration during ovarian development in the desert locust, *Insect Physiol.*, 8, 609, 1962.
31. **Bursell, E.,** The role of proline in energy metabolism, *Energy Metabolism in Insects*, 135, 1981.
32. **Weis-Fogh, T.,** Fat combustion and metabolic rate of flying locusts (*Schistocerca gregaria* Forskal), *Phil. Trans. R. Soc. London*, 237, 1, 1952.
33. **Jutsum, A. R. and Goldsworthy, G. J.,** Fuels for flight in *Locusta*, *J. Insect Physiol.*, 22, 243, 1976.
34. **Peach, J. L. and Phillips, J. E.,** Metabolic support of chloride-dependent short-circuit current across the locust (*Schistocerca gregaria*) ileum, *J. Insect Physiol.*, 37, 255, 1991.
35. **Chamberlin, M. E. and Phillips, J. E.,** Oxidative metabolism in the locust rectum, *J. Comp. Physiol. B*, 151, 191, 1983.
36. **Chamberlin, M. and Phillips, J. E.,** Metabolic support of chloride-dependent short-circuit current across locust rectum, *J. Exp. Biol.*, 99, 349, 1982.
37. **Maddrell, S. H. P. and Gardiner, B. O. C.,** The passive permeability of insect Malpighian tubules to organic solutes, *J. Exp. Biol.*, 60, 641, 1974.
38. **Bursell, E.,** The excretion of nitrogen in insects, in *Advances in Insect Physiology*, Beament, J. W. L., Treherne, J. E., and Wigglesworth, V. B., Eds., Academic Press, New York, 1967, 33.
39. **Mullins, D. E. and Cochran, D. G.,** Nitrogen excretion in cockroaches: uric acid is not a major product, *Science*, 177, 699, 1972.
40. **Mullins, D. E. and Cochran, D. G.,** Nitrogen metabolism in the American cockroach: an examination of whole body and fat body regulation of cations in response to nitrogen balance, *J. Exp. Biol.*, 61, 557, 1974.
41. **Harrison, J. F. and Phillips, J. E.,** Recovery from acute haemolymph acidosis in unfed locusts II. Role of ammonium and titratable acid excretion, *J. Exp. Biol.*, 165, 97, 1992.
42. **Cameron, J. N.,** *Principles of Physiological Measurement*, Academic Press, New York, 1986, 278.
43. **McNabb, R. A. and McNabb, F. M.,** Physiological chemistry of uric acid: solubility, colloid and ion-binding properties, *Comp. Biochem. Physiol.*, 67A, 27, 1980.
44. **Peters, J. P. and Van Slyke, D. D.,** *Quantitative Clinical Chemistry*, Williams & Wilkens, Baltimore, MD, 1946.
45. **Porter, P.,** Physio-chemical factors involved in urate calculus formation. I. Solubility, *Res. Vet. Sci.*, 4, 580, 1963.
46. **Seegmiller, J. E., Lancaster, L., and Howell, R. R.,** Biochemistry of uric acid and its relation to gout, *N. Eng. J. Med.*, 268, 712, 1963.
47. **Stagg, A. P., Harrison, J. F., and Phillips, J. E.,** Acid-base variables in Malpighian tubule secretion and response to acidosis, *J. Exp. Biol.*, 159, 433, 1991.
48. **Stagg, A. P.,** Role of the Malpighian tubules in acid-base regulation in the desert locust *Schistocerca gregaria,* Ph.D. thesis, University of British Columbia, 1991.
49. **Anstee, J. H., Bell, D. M., and Fathpour, H.,** Fluid and cation secretion by the Malpighian tubules of *Locusta*, *J. Insect Physiol.*, 25, 373, 1979.
50. **Porter, P.,** Physio-chemical factors involved in urate calculus formation. II. Colloidal flocculation, *Res. Vet. Sci.*, 4, 592, 1963.
51. **O'Donnell, M. J., Maddrell, S. H. P., and Gardiner, B. O. C.,** Transport of uric acid by the Malpighian tubules of *Rhodnius prolixus* and other insects, *J. Exp. Biol.*, 103, 169, 1983.

52. **Phillips, J. E., Thomson, R. B., Audsley, N., Peach, J. L., and Stagg, A. P.,** Mechanisms of acid-base transport and control in locust excretory system, *Physiol. Zool.*, 67, 95, 1994.
53. **Harvey, W. R.,** Physiology of V-ATPases, *J. Exp. Biol.*, 172, 1, 1992.
54. **Wieczorek, H.,** The insect V-ATPase, a plasma membrane proton pump energizing secondary active transport: molecular analysis of electrogenic potassium transport in the tobacco hornworm midgut, *J. Exp. Biol.*, 172, 335, 1992.
55. **Irvine, B., Audsley, N., Lechleitner, R., Meredith, J., Thomson, B., and Phillips, J.,** Transport properties of locust ileum *in vitro*: effects of cyclic AMP, *J. Exp. Biol.*, 137, 361, 1988.
56. **Lechleitner, R. A., Audsley, N., and Phillips, J. E.,** Composition of fluid transported by locust ileum: influence of natural stimulants and luminal ion ratios, *Can. J. Zool.*, 67, 2662, 1989.
57. **Thomson, R. B., Thomson, J. M., and Phillips, J. E.,** NH_4^+ transport in acid-secreting insect epithelium, *Am. J. Physiol.*, 254, R348, 1988.
58. **Chamberlin, M. E. and Phillips, J. E.,** Effects of stimulants of electrogenic ion transport on cyclic AMP and cyclic GMP levels in locust rectum, *J. Exp. Zool.*, 245, 9, 1988.
59. **Audsley, N., McIntosh, C., and Phillips, J. E.,** Isolation of a neuropeptide from locust corpus cadiacum which influences ileal transport, *J. Exp. Biol.*, 173, 261, 1992.
60. **Audsley, N., McIntosh, C., and Phillips, J. E.,** Actions of ion-transport peptide from locust corpus cardiacum on several hindgut transport processes, *J. Exp. Biol.*, 173, 275, 1992.
61. **Lechleitner, R. A. and Phillips, J. E.,** Effects of corpus cardiacum, ventral ganglia, and proline on absorbate composition and fluid transport by locust hindgut, *Can. J. Zool.*, 67, 2669, 1989.
62. **Audsley, N.,** Purification of a neuropeptide from the corpus cardiacum of the desert locust which influences ileal transport, Ph.D. thesis, University of British Columbia, 1991.
63. **Atkinson, D. E. and Camien, M. N.,** The role of urea synthesis in the removal of metabolic bicarbonate and the regulation of blood pH, *Curr. Top. Cell. Regul.*, 21, 261, 1982.
64. **Atkinson, D. E.,** Functional roles of urea synthesis in vertebrates, *Physiol. Zool.*, 65, 243, 1992.
65. **Haussinger, D., Ed.,** *pH Homeostasis: Mechanisms and Control*, Academic Press, New York, 1988, 445.
66. **Harrison, J. F. and Kennedy, M. J.,** *In vivo* studies of the acid-base physiology of grasshoppers: the effect of feeding state on acid-base and nitrogen excretion, *Physiol. Zool.*, 67, 120, 1994.
67. **Meredith, J. and Phillips, J. E.,** Sodium-independent proline transport in the locust rectum, *J. Exp. Biol.*, 137, 341, 1988.
68. **Walser, M.,** Roles of urea production, ammonium excretion, and amino acid oxidation in acid-base balance, *Am. J. Physiol.*, 250, F181, 1986.
69. **Thomson, R. B. and Phillips, J. E.,** Characterization of acid/base transport in an insect epithelium, *Fed. Proc.*, 44, 643, 1985.
70. **Thomson, R. B., Audsley, N., and Phillips, J. E.,** Acid-base transport and control in locust hindgut: artifacts caused by short-circuit current, *J. Exp. Biol.*, 155, 455, 1991.
71. **Sullivan, L. P.,** Renal mechanisms involved in acid-base regulation, in *Acid-base Regulation in Animals*, Heisler, N., Ed., Elsevier Science Publishers B.V., New York, 1986, 83.
72. **Schoolwerth, A. C.,** Regulation of renal ammoniagenesis in metabolic acidosis, *Kidney Int.*, 40, 961, 1991.
73. **Curthoys, N. P.,** Cellular distribution and induction of the enzymes of renal ammoniagenesis and gluconeogenesis, in *pH Homeostasis: Mechanisms and Control*, Haussinger, D., Ed., Academic Press, London, 1988, 323.
74. **Harrison, J. F.,** Respiratory and ionic aspects of acid-base regulation in insects: an introduction, *Physiol. Zool.*, 67, 1, 1994.
75. **Brown, A. W. A.,** The excretion of ammonia and uric acid during the larval life of certain muscoid flies, *J. Exp. Biol.*, 13, 131, 1936.
76. **Chauvin, R.,** Contribution a l'etude physiologique du criquet pelerin et du determinisme des phenomenes gregaires, *Ann. Soc. Entomol. Fr.*, 110, 133, 1947.

Chapter 9

Molecular Aspects of Ureogenesis in Amphibians

Burr G. Atkinson

CONTENTS

INTRODUCTION

The nature of an animal's nitrogenous excretory product is believed to be a function of its phylogenetic position and its environment.[1-4] The principle end product of protein metabolism of aquatic animals is ammonia which is easily lost by diffusion. Even in small quantities, however, ammonia is toxic to land-dwelling animals. Thus, terrestrial animals have developed enzyme systems for the conversion of ammonia to the less toxic compounds, urea, and/or uric acid. Birds and reptiles primarily excrete uric acid whereas mammals are ureotelic, urea being the end product of protein metabolism. The realization that ammonotelism is incompatible with a terrestrial existence led to the recognition of the importance of the role of ureotelism in the evolutionary emergence of life from an aquatic to a terrestrial environment. Between these permanently aquatic and strictly terrestrial animals are the amphibians, vertebrates which during their life cycle make a transition from an aquatic to a fully or partially terrestrial environment. In fact, the Amphibia are the first class of Vertebrata which are thought to have successfully colonized land. Although not all amphibians are completely adapted to terrestrial life independent of water, the origins of the adaptations to terrestrial environment in the whole vertebrate phylum may, in some cases, be found within the molecular/genetic features of the class Amphibia. Thus, the unique position of the Amphibia in the evolutionary tree[4] and the striking evolutionary mimicry of the postembryonic development of an ammonotelic aquatic tadpole into a ureotelic terrestrial frog provide a model system for assessing the molecular mechanisms which (1) govern the expression of genes or gene products responsible for the synthesis of urea, and (2) are instrumental in permitting the invasion of land by an aquatic animal.

0-8493-8411-7/95/$0.00+$.50

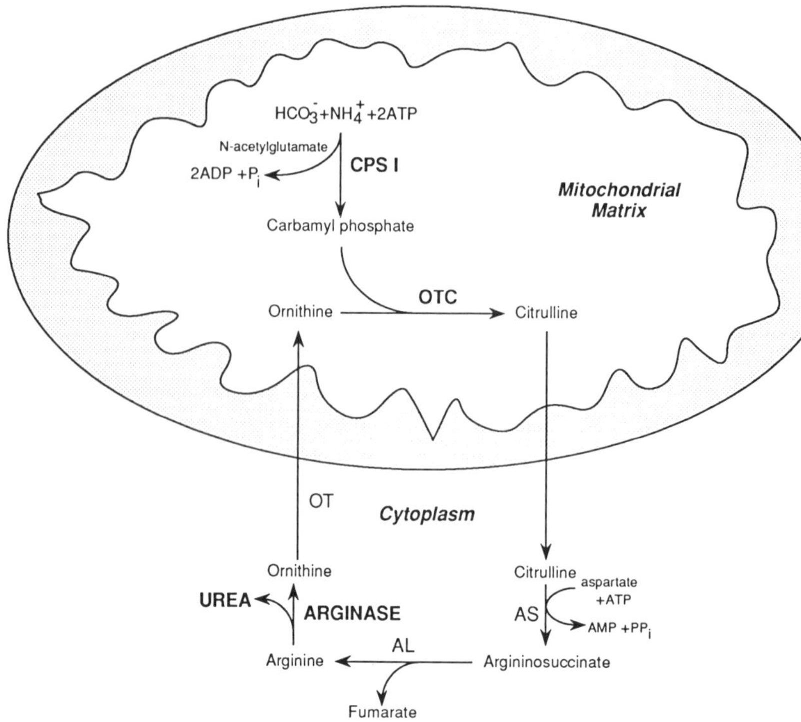

Figure 1 The ornithine urea-cycle enzymes responsible for the production of urea in the liver cells of *Rana catesbeiana*. Two of the urea cycle enzymes, carbamyl-phosphate synthetase-I (CPS-I) and ornithine transcarbamylase (OTC) are localized in the mitochondrial matrix and the other three enzymes — argininosuccinate synthetase (AS), argininosuccinate lyase (AL), and arginase — are cytoplasmic. The enzymes shown in bold represent the products of genes whose expressions are emphasized in this review. OT, ornithine transporter.

BIOCHEMICAL ASPECTS OF UREOGENESIS IN AMPHIBIANS: A BRIEF HISTORICAL REVIEW

Postembryonic development in most amphibians involves biochemical, physiological, and morphological changes which prepare an aquatic tadpole for its transition to a full or partially terrestrial adult — a process termed metamorphosis.[5-9] One of the most significant metabolic alterations that occur during the metamorphosis of most amphibians is the shift from ammonotelism to ureotelism.[5,6,10-13]

UREA BIOSYNTHESIS AND THE POSTEMBRYONIC DEVELOPMENTAL TRANSITION OF SOME AMPHIBIANS FROM AMMONOTELIC TADPOLES TO UREOTELIC ADULTS

The enzymes responsible for the synthesis of urea in the metamorphosing tadpole and adult, ureotelic amphibian are shown in Figure 1. The first two enzymes in the ornithine-urea cycle, carbamyl-phosphate synthetase-I (CPS-I or CPSase I) and ornithine transcarbamylase (OTC), are located in the mitochondrion, while the other three enzymes — argininosuccinate synthetase (AS), argininosuccinate lyase (AL), and arginase — are extramitochondrial. These enzymes are found in the liver of most amphibians[14] and have been characterized and extensively studied in the Amphibia by Cohen et al.[6] The identification of the amphibian CPSase as CPS-I is based on its mitochondrial localization and its dependency on ammonia as a nitrogen-donating substrate and N-acetyl-L-glutamate as an allosteric activator.[13] Cohen et al. measured the activity of the enzymes in the ornithine-urea cycle in tadpoles at different stages of metamorphosis and correlated the activities of these enzymes with the levels of ammonia and urea excreted by the metamorphosing tadpole.[15] The data, summarized in several reviews,[5,6,13] demonstrate that both the activities of the enzymes involved in urea biosynthesis and urea excretion increase at a time in development (Stages XVIII to XX) when gross morphological changes in the tadpole (e.g., loss of the tadpole's tail) are only just beginning to be evident. Indeed, these biochemical changes in the liver of this organism occur in *anticipation* of new demands and, as discussed in subsequent sections, their initiation and completion are dependent on and tightly controlled by thyroid hormone.

UREA BIOSYNTHESIS AND THE TRANSITION OF SOME AMPHIBIAN AMMONOTELIC TADPOLES TO UREOTELIC ADULTS IS PRECOCIOUSLY INDUCED BY ADMINISTRATION OF THYROID HORMONE TO PREMETAMORPHIC TADPOLES

Although the obligatory control of amphibian metamorphosis by thyroid hormones was established by Gudernatsch, Allen and others from 1912 to 1930,[16-19] the fact that this hormone can also induce premetamorphic tadpoles to excrete urea precociously was not demonstrated until 1953.[11] Shortly thereafter, suitable assays were developed for measuring the activities of the different enzymes involved in urea biosynthesis, and a number of studies showed that the activities of all of the enzymes of the ornithine-urea cycle are significantly increased during thyroid hormone-induced metamorphosis.[20-22] The striking increase (>30-fold) in the level of CPS-I activity during thyroid hormone-induced metamorphosis and its correlation with urea excretion[15] prompted studies focused on elucidating the means by which thyroid hormone effects the activity of this enzyme. The availability of a specific antibody to CPS-I permitted Cohen and colleagues to demonstrate that the increase in CPS-I activity in the liver of tadpoles exposed to thyroxine is the result of synthesis *de novo*.[23] These latter observations have been confirmed by others[24-26] and have clearly implicated an early, if not direct, role for thyroid hormone in urea biosynthesis.

MOLECULAR ASPECTS OF UREOGENESIS IN AMPHIBIANS

It is obvious from the biochemical results briefly reviewed in the preceding sections that the obligatory change from ammonotelism to ureotelism, which occurs in some amphibians in anticipation of a terrestrial lifestyle, is a result of a dramatic, *concerted* increase in the levels of urea cycle enzyme activities in the tadpole liver. The fact that these same changes can be precociously induced by exogenous administration of the thyroid hormones, thyroxine (T4) or 3,5,3′-triiodothyronine (T3), to the free-living, premetamorphic tadpole[6,13,27] and the observation that thyroid hormone causes an enhanced synthesis and accumulation of at least one of the enzymes necessary for urea synthesis, CPS-I,[23] promoted this metamorphosing system as a model for studying the *molecular* effects of thyroid hormone on the expression of the genes encoding the ornithine-urea cycle enzymes. Studies aimed at elucidating the molecular means by which this hormone influences the enhanced synthesis and accumulation of CPS-I have revealed that the hormone-induced occurrence of these events in the tadpole liver is preceded by an accumulation of translatable CPS-I mRNAs.[24-26,28,29] The hormone-induced increase in the level of CPS-I mRNAs is a permanent change and is thought to be sufficient to account for the increased amount of CPS-I protein. These observations, albeit restricted to the gene(s) encoding CPS-I, clearly support the conjecture that thyroid hormone regulates the expression of the ornithine-urea cycle enzymes at the level of gene transcription.[24,26,28,29] However, the realization that the data supporting this hypothesis relied primarily on studies with CPS-I sequences, and on results that were obtained *without* the use of amphibian CPS-I cDNA or genomic probes, caused us to reassess the hypothesis and extend previous studies. As a first step, we prepared cDNA libraries of *Rana catesbeiana* liver poly(A)+ mRNAs and isolated cDNA clones from them which contained nucleotide sequences encoding CPS-I,[30] OTC,[26] and arginase.[31] Subsequently, as summarized in this review, we characterized the sequences in these cDNAs and used them to assess the expression of genes encoding these particular *R. catesbeiana* ornithine-urea cycle enzymes.

CPS-1

```
   1 MTRILSVFKT AKTGVLNAAA HRYRGFSKAG VRLMSVKAQT ANLVLEDGTK
  51 IKGYSFGHPA SVAGEVIFNT GLGGYVEAVT DPSYHGQILT LTNPIIGNGG
 101 APDTKARDAY GLMKYIESEN IQASGLLVLD YSHEYSHWGA VKSLSEWLHE
 151 EKVPALCGID TRMLAKKIRD NKGAVLGKIE FEGQPVEFID PNKRNLIAEV
 201 STKETKVFGK GNPVRIVAVD CGVKHNIIRQ LVKRGAEVHL VPWNHDFSQM
 251 EYDGLLITSG PGNPELAKPL IQNLKKVFQS DRPEPIFGIC KGNEIAALAA
 301 GGKTYRLPMA NRGQNQPVMI TLNGQAFITA QNHAYAVDNN SLPAGWKPLF
 351 VNINDQSNEG IMHETKPIFT SQFHPEANPG PVDTEFLFDV YMSLIKKGKG
 401 TTLTSVMPKP ALQSKRIDVA KVLILGSGGL SIGQAGEFDY SGSQAVKAMK
 451 EENVKTVLMN PNIASVQTNE VGLKQADTVY FLPITPQFVT EVIKAEKTDG
 501 IILGMGGQTA LNCGVELFKR GVLKEYGVRV LGTSVESIMF TEDRQLFSDK
 551 LNEIKEPIAP SFAVESVKDA LEAADKIGYP VMIRSAYALG GLGSGLCPDK
 601 ETLTDLATKA LAMTNQILVE RSVVGWKEIE YEVVRDAADN CVTVCNMENV
 651 DAMGVHTGDS IVVAPCQTLS NEECQMLRAV SIKVVRHLGI VGECNIQFAL
 701 HPTSLEYVII EVNARLSRSS ALASKATGYP LAFIAAKIAL GIPLPEIKNV
 751 VSGKTTACFE PSLDYMVTKI PRWDLDRFHG ASGLIGSSMK SVGEVMAIGR
 801 TFEESFQKAL RMCHPSVDGF TSNLPMNKAW SSDVNLRKEM AEPTSTRMYS
 851 MAKAIQSGIS LDEINKLTAI DKWFLYKMQG ILNMEKTLKG SRSESVPEET
 901 LRRAKQIGFS DRYIGKCLGL SETQTRELRL NKNVKPWVKQ IDTLAAEYPA
 951 ITNYLYLTYN GQEHDIKFDD HGMMVLGCGP YHIGSSVEFD WCAVSSIRTL
1001 RHVGKKTVVV NCNPETVSTD FDECDKLYFE ELSQERIMDV FQLEQCDGCI
1051 ISVGGQIPNN LAVPLYKNGV KIMGTSPMQI DRAEDRSIFS AVLDELQIAQ
1101 APWKAVNSLD DALQFTKTVG YPCLLRPSYV LSGSAMNVVY GEEELKTFLA
1151 EATRVSQEHP VVITKFIEGA REVEMDAVGK EGRVISHAIS EHVEDAGVHS
1201 GDATLMIPTQ SISQGAIEKV KIATKKIATA FAISGPFNVQ FLVRGNDVLV
1251 IECNLRASRS FPFVSKTLGV DFIDVATKVM IGEKIDESSL PTLERPVIPA
1301 DYVGIKAPMF SWPRLRGADP VLKCEMASTG EVACFGQNVY SAFLKAMIST
1351 GFKLPQKGIL IGIQHSFRPH FLGTAQTLKD EGFKLYATEA TADWLNANDI
1401 TATPVAWPSQ EGQSGPSSIY KLIKEGNIDM VINLPNNNTK YVRDNFAIRR
1451 TAVDTGTALL TNFQVVKMFA EAIKYSGDLD AKSLFHYRQF GGAKPS*
```

Figure 2 The cDNA-derived amino acid sequence of a *R. catesbeiana* CPS-I precursor protein. The site at which the leader sequence is cleaved from the mammalian CPS-I precursor to produce the mature enzyme[33] is indicated by double arrows. The highly conserved ATP binding sites are boxed and shaded. A solid circle denotes the position of a cysteine which has been shown to be essential for amidotransferase activity in *Escherichia coli* and yeast.[67] An arrowhead denotes the point at which an additional amino acid, a serine, exists in the CAD protein.[35]

CHARACTERIZATION OF THE NUCLEOTIDE AND DERIVED AMINO ACID SEQUENCES OF cDNAS ENCODING *R. catesbeiana* CPS-I, OTC, AND ARGINASE

A composite cDNA sequence (4716 nucleotides) for CPS-I was derived from 18 different overlapping clones. It contains a 4488-nucleotide open reading frame (ORF) which encodes for a putative CPS-I precursor protein of 1496 amino acids with a molecular mass of 163,579.[30] Comparison of the nucleotide sequence in the ORF of this *Rana* CPS-I reveals that it shares 74% identity with the nucleotide sequence in the ORF of the rat CPS-I,[32,33] and only 53% identity with the nucleotide sequence in the glutamine aminotransferase and CPS domains of the hamster CAD (see Chapter 2 for definition of CAD).[34] The amino acid sequence derived from the putative ORF of this *Rana* CPS I is shown in Figure 2. Comparative studies disclose that it shares an 81% identity with the rat CPS I sequence[32,33] but only 51 and 37% identity with the hamster CAD[34] and *Escherichia coli* CPS[32] sequences, respectively.[30] *Rana* CPS-I, like other mitochondria-targeted proteins, possesses a 33-amino acid leader, or signal, sequence at its N-terminus which contains eight basic- and no acidic-amino acid residues. The *Rana* sequence, like the rat CPS-I sequence, is missing a serine (at position 1365) which has been shown to be an important phosphorylation site in the cytosolic CAD multifunctional enzyme.[35] Several amino acids important in carbamate intermediate formation (glycine 591 and 595, and histidine 656) and in carbamyl-phosphate generation (histidine 1192

OTC

```
  1 MLHHMRTIIN ASWRYGNKCI VRQFGFSQTY SQLKGRDLLT LKNYSAEEIK
 51 YLLWVAADLK YRIKEKGEYL PLLQGKSLAM IFEKRSTRTR LSTETGFALL
101 GGHPSFLTTQ DIHLGVNESL KDTARVLSGM TDAVLARVYH QSDLEVLAEE
151 ASIPIVNGLS DDYHPIQILA DYLTIQEHYG HLKGLTISWI GDGNNVLHSI
201 MMSAAKFGMH LHIATPKGYE PNSSLTEAAK QFSKECGTKL LMTNDPLEAA
251 NGANVLVTDT WVSMGQEEEK KKRLLDFKGY QITMKTAKLA APNWIFLHCL
301 PRKPEEVDDE VFYCPKSLVF QEAENRKWTI MGVMVSLLTD YSPQLLRPTF
351 *
```

Figure 3 The cDNA-derived amino acid sequence of a *R. catesbeiana* OTC precursor protein. The site at which the leader sequence is cleaved from the mammalian OTC precursor to produce the mature enzyme[38-40] is indicated by an arrowhead. The catalytic binding sites for carbamyl phosphate (amino acids 86 to 90)[39] and ornithine (amino acids 296 to 304)[42] are boxed and shaded.

and 1199, and glutamate 1252) are completely conserved in the *Rana* sequence.[36,37] While the rat and *Rana* CPS-I amino acid sequences appear to be highly conserved, the cysteine residue that reacts with glutamine in glutamine-dependent CPSases[33] is retained in the *Rana* CPS-I (amino acid 290) even though this amphibian's CPS-I cannot utilize glutamine as a nitrogen-donating substrate (Anderson, P. M. and Atkinson, B. G., unpublished results). These observations suggest that the determining factor in the loss of capability for the *Rana* CPS-I to utilize glutamine as a nitrogen-donating substrate must involve amino acid replacements other than the active site cysteine residue.

Characterization of *Rana* cDNA clones encoding OTC (1360 nucleotides) revealed that they contained an ORF of 1050 nucleotides which encodes for a putative OTC precursor protein of 350 amino acids with a molecular mass of 39,591.[26] Comparison of the nucleotide sequence in the ORF of this *Rana* OTC reveals that it shares 74, 72, and 71% identity with the nucleotide sequence in the ORF of cDNAs reported to encode human,[38] rat,[39] and mouse[40] OTCs, respectively.[26] The amino acid sequence derived from the putative ORF of this *Rana* OTC is shown in Figure 3. Comparative studies disclose that it shares 75, 77, and 76% identity with the rat,[39] mouse,[40] and human[38] OTC sequences, respectively.[26] Like the rat OTC,[41] the amino-terminal end of the *Rana* OTC possesses a leader or signal sequence (30-nucleotides long) which contains 6 basic amino acid residues. The catalytic binding sites for carbamyl phosphate (amino acids 86 to 90) and ornithine (amino acids 296 to 304) are completely conserved.[39,42] The use of antibodies directed against OTC on Western blots of the protein translated *in vitro* from transcripts of the *Rana* OTC cDNAs confirmed that these cDNAs encoded an OTC protein and Southern blot hybridization analyses of enzyme-digested *R. catesbeiana* genomic DNA suggest that OTC arises from a single-copy gene in this organism.[26]

Screening of our *Rana* liver libraries with a fragment from a *Xenopus laevis* cDNA reported to encode arginase[43] resulted in the isolation of a number of identical overlapping cDNAs of which two (1516 and 1520 nucleotides long) contained complete ORFs.[31] In each case the ORF is 969 nucleotides in length and encodes for a protein of 323 amino acids with a molecular mass of 35,253. Comparison of the nucleotide sequence in the ORF of this *Rana* arginase reveals that it shares 70, 66, and 65% identity with the nucleotide sequence in the ORF of cDNAs reported to encode *Xenopus*,[43] human,[44] and rat[45] arginase, respectively.[31] The amino acid sequence derived from the putative ORF of this *Rana* arginase is shown in Figure 4. Comparative studies reveal that it shares 65, 71, and 68% identity with the rat[45], *Xenopus*[43] and human[44] arginase sequences, respectively.[31] The fact that no mitochondrial-targeting leader/signal sequence is evident in the amino-terminal end of the *Rana* arginase supports the biochemical evidence that this enzyme, unlike *Rana* CPS-I and OTC, is a constituent of the cytoplasm. Southern blot hybridization analyses of *R. catesbeiana* genomic DNA digested with various restriction enzymes suggest that arginase, like the *Rana* OTC, originates from a single-copy gene in this organism.[31]

Arginase

```
  1 MSERTKRSVG VLGAPFSKGQ ARGGVEEGPI YIRRAGLIEK LEELEYEVRD
 51 YGDLHFPELP CDEPFQNVKN PRTVGQAAEK VANAVSEVKR SGRVCLTLGG
101 DHSLAVGTIT GHAKVHPDLC VVWVDAHADI NTPITSPSGN LHGQPVSFLI
151 RELQTKVPAI PGFSWVQPSL SAKDIVYIGL RDVDPGEHYI LKTLGIKSYS
201 MSDVDRLTIN KVMEETIEFL VGKKKRPIHL SFDIDGLDPS VAPATGTPVP
251 GGLTYREGMY ITEQLYNTGL LSAVDMMEVN PSRGETERES KLTVNTSLNM
301 ILSCFGKARE GFHASSLRVP DLI*
```

Figure 4 The cDNA-derived amino acid sequence of *R. catesbeiana* arginase.

EXPRESSION OF GENES ENCODING CPS-I, OTC, AND ARGINASE IS UPREGULATED IN THE LIVER OF THYROID HORMONE-INDUCED AND SPONTANEOUSLY METAMORPHOSING *R. catesbeiana* TADPOLES

The isolation and characterization of the nucleotide sequences encoding two of the *R. catesbeiana* mitochondrial ornithine-urea cycle enzymes, CPS-I and OTC, and one of the cytosolic enzymes involved in this cycle, arginase, enabled assessment of the intracellular levels of the respective mRNAs in liver of *R. catesbeiana* tadpoles undergoing thyroid hormone-induced or spontaneous metamorphosis. To adequately demonstrate differences in the level of these mRNAs during the metamorphosis of this amphibian, several different methods have been used to substantiate any observed changes. For example, mRNA was isolated from the liver of thyroid hormone-induced and spontaneously metamorphosing tadpoles and was assessed by reverse transcription-polymerase chain reactions (RT-PCR) using PCR primers specific for each of the mRNAs,[46] and by Northern and dot-blot hybridization analyses using nucleotide sequences corresponding to the 3′-untranslated regions (UTRs) of the mRNAs as probes. Finally, using mRNA-specific 3′-UTR sequences subcloned into pBluescript II SK- as probes, the relative level of specific mRNAs in the liver of thyroid hormone-induced and spontaneously metamorphosing tadpoles was analyzed by *in situ* hybridization.

The accumulation of mRNAs encoding CPS-I, OTC, and arginase in the liver of spontaneously and thyroid hormone-induced metamorphosing *Rana* tadpoles has been the subject of several recent reports.[26,30,31,47] A summary of the results and their relationship to urea excretion are shown in Figures 5 and 6. It is obvious from the data depicted in these figures that mRNAs encoding these ornithine-urea cycle enzymes accumulate before urea excretion is elevated. Moreover, the correlation between the accumulation of these mRNAs and increasing levels of endogenous thyroid hormone in a spontaneously metamorphosing tadpole (Figure 5B) support the contention that the precocious thyroid hormone-induced accumulation of these mRNAs (Figure 6A) is a reflection of what occurs normally during this organism's postembryonic development.

The observation (see Figure 6A) that little detectable accumulation of these ornithine-urea cycle mRNAs occurs until approximately 24 h after a tadpole has been treated with thyroid hormone (a so-called "lag" period) has been difficult to interpret. The concept that this lag period may represent a period of time required either for the replication and production of cells capable of responding to this hormone or for the synthesis and/or interaction of other extrahepatic factors necessary for this response was addressed. Results from *in situ* hybridizations using CPS-I, OTC, and arginase-specific probes indicate that most, if not all, of the liver cells concurrently accumulate these mRNAs and that this response is the result of a "reprogramming" of cells present in the liver and *not* due to the expression of these genes exclusively in a new/replicating population of cells.[48,49] A role for the requirement of extrahepatic factors, other than thyroid hormone, was ruled out by demonstrating that mRNAs encoding CPS-I, OTC, and arginase accumulate, in a similar fashion, in serum-free cultures of tadpole liver treated with thyroid hormone, whereas they do not accumulate in cultures lacking thyroid hormone.[26,30,47] The demonstration that neither the generation of a new, responding population of cells nor the synthesis and/or interaction of extrahepatic factors other than

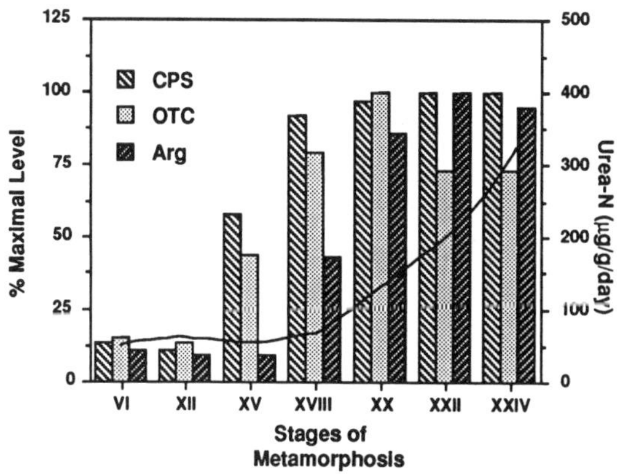

A

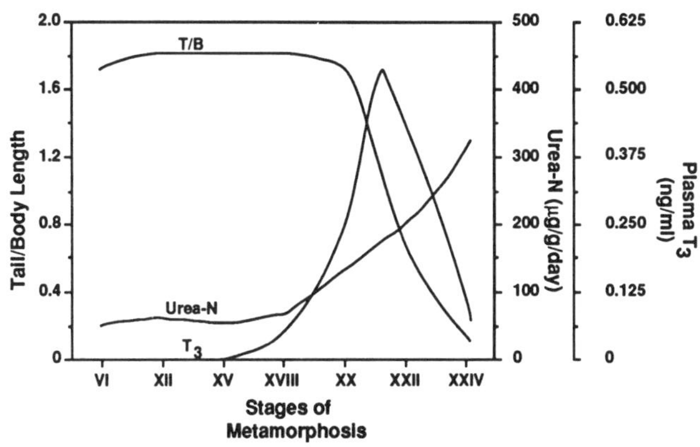

B

Figure 5 Relationship of CPS-I, OTC, and arginase mRNA levels in the liver of spontaneously metamorphosing *R. catesbeiana* tadpoles to the tadpole's level of urea excretion, degree of morphological change, and level of endogenous thyroid hormone. (A) Summary bar graphs comparing the accumulation of CPS-I, OTC, and arginase (Arg) mRNAs in the liver of spontaneously metamorphosing tadpoles to the amount of urea excreted by the tadpoles. The average relative amounts of the mRNA transcripts were determined by laser densitometric quantitation of dot-blot hybridizations and RT-PCR products, and the plotted values were corrected based on the abundance of an mRNA which has been shown not to change in abundance during metamorphosis[26,47]; (B) comparison of the levels of urea-nitrogen excreted by spontaneously metamorphosing *R. catesbeiana* tadpoles with a morphological change in the tadpoles (tail loss) and with the relative amount of endogenous thyroid hormone reported in the plasma[71] of staged-tadpoles. The tadpoles were staged (Stage of Metamorphosis) according to Taylor and Kollros.[72]

thyroid hormone are required for the upregulation of these urea cycle mRNAs and urea biosynthesis eliminates them as a means to account for the lag period.

In the last few years, several reports concerned with the response of other amphibian tadpole tissues to thyroid hormone, have also demonstrated the presence of a lag period between the administration of the hormone and the response of genes whose expression is characteristic of a particular adult tissue phenotype.[50-54] These observations have resulted in the conjecture that the lag period may represent a time in which certain molecular events occur that initiate, in a cell-specific manner, the terminal differentiation of cells which, in turn, causes them to express genes characteristic of the adult cell/tissue phenotype. This hypothesis implies that thyroid hormone may not

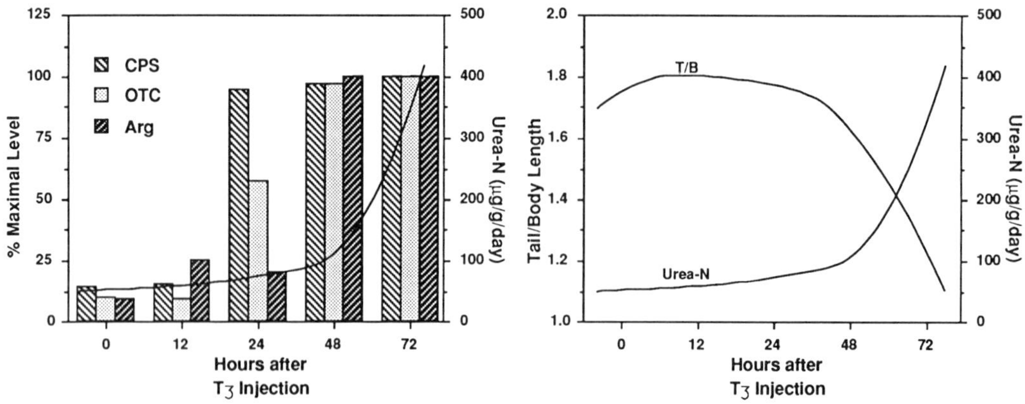

Figure 6 Relationship of CPS-I, OTC, and arginase mRNA levels in the liver of thyroid hormone-induced metamorphosing *R. catesbeiana* tadpoles to the tadpole's level of urea-nitrogen excretion and degree of morphological change. (A) Summary bar graphs comparing the accumulation of CPS-I, OTC, and arginase (Arg) mRNAs in the liver of thyroid hormone (T3)-induced metamorphosing tadpoles to the amount of urea excreted by the tadpoles. The average relative amounts of the mRNAs transcripts were determined as in Figure 5 and the plotted values were corrected based on the abundance of mRNAs encoding either actin or ferritin[47]; (B) comparison of the amount of urea-nitrogen excreted by thyroid hormone (T3)-induced metamorphosing tadpoles with a morphological change in the tadpoles (tail loss). These data, as well as the urea-nitrogen and tail loss data shown in Figure 5, are from results obtained (unpublished) in the author's laboratory and compare favorably with those published by Ashley et al.[73] Stage VI to XIII tadpoles were injected either with T3 (3×10^{10} mol/g body wt) or a vehicle solution (200 μM NaOH) lacking T3, and sacrificed at the indicated times.

directly affect the expression of the ornithine-urea cycle enzyme genes but, instead, initiates a terminal differentiation of the amphibian's liver cells which, in turn, results in the expression of the urea cycle enzyme genes. Attention has therefore shifted towards testing this hypothesis and elucidating this hormone's involvement in molecular events occurring in the amphibian's liver cells during the lag period.

MOLECULAR MECHANISMS BY WHICH THYROID HORMONE INFLUENCES THE EXPRESSION OF GENES ENCODING THE ORNITHINE-UREA CYCLE ENZYMES AND INITIATES THE BIOSYNTHESIS OF UREA

The interaction of thyroid hormone with a gene is mediated through the binding of this hormone to a nuclear-localized receptor protein. Unliganded thyroid hormone receptor proteins (TRs) are normally bound to thyroid hormone response elements (TREs) in the promoter regions of target genes. Once a thyroid hormone-TR complex is formed, it can act as a transcription factor and may initiate and/or upregulate the expression of the target gene. While the means by which thyroid hormones interact with genes is reasonably well understood, the actual mechanism by which it regulates the expression of specific genes is poorly understood.[9,55] The recent characterization of TRs as products of the erb A protooncogene[56,57] has enabled investigators, however, to focus on the mechanism(s) by which liganded TRs act as transcription factors. Moreover, the use of sequences from C-erb A cDNA clones has resulted in the isolation and characterization of cDNA and genomic clones encoding the TR sequences from a number of organisms,[55] including two different amphibians, *X. laevis*[58,59] and *R. catesbeiana*.[26,60,61] Although sequences from two different TR-encoding genes, TRα and TRβ, have been characterized in both of these amphibians,[58-61] alternative promoter usage and alternative splicing of the primary transcripts from these genes is thought to result in a diversity of receptor isoforms.[59,62] The use of cDNA and genomic clones encoding the TRs has enabled investigators to assess the response of TRα and TRβ genes to thyroid hormone in a variety of amphibian tadpole tissues.

Studies with the liver of the *R. catesbeiana* tadpole demonstrate that mRNAs from both of the TR-encoding genes are concurrently upregulated by administration of thyroid hormone to the tadpole, albeit to quite different extents (TRα ~3-fold and TRβ >11-fold).[26,31,61] Moreover, the thyroid hormone-induced upregulation of tadpole liver TR mRNAs (an autoinduction) was found to precede the upregulation of the urea cycle enzyme mRNAs by at least 12 h.[26,30,31,47] This latter observation places the response time of the TR genes in the so-called lag phase and categorizes this response in the liver as one of the earliest molecular events influenced by thyroid hormone.

The temporal difference in the thyroid hormone-induced accumulation of the TR and urea cycle enzyme mRNAs suggests a direct role for thyroid hormone on the expression of the genes encoding one or more of the TRs and a less direct role for thyroid hormone on the expression of the genes encoding the liver-specific, urea cycle enzymes. In order to further substantiate, albeit circumstantially, the notion that thyroid hormone, complexed with its receptor protein, does not directly influence the expression of the CPS-I or OTC genes, the promoter regions of the genes encoding *Rana* CPS-I and OTC were isolated and sequenced. The results confirmed our suspicions that sequences characteristic of TREs were not found in the promoter regions of these genes.[63] Thus, these results support the contention that thyroid hormone receptor protein complexes *do not directly interact* with the genes encoding the ornithine-urea cycle enzymes CPS-I and OTC.

The absence of TREs in the promoter regions of the CPS-I and OTC genes prompted a search in the promoter regions of these genes for regulatory elements corresponding to other known transcription factors; the promoters of both genes were found to contain CCAAT/enhancer core binding protein (C/EBP) elements.[63] The fact that C/EBPs are transcription factors, which have been implicated in both the terminal differentiation of mammalian hepatocytes and liver-specific gene expression,[64-66] raised the possibility that an amphibian C/EBP-like transcription factor might be involved in terminally differentiating these cells and/or might be required for upregulating the expression of the ornithine-urea cycle enzyme genes. Guided by this notion, we isolated and characterized clones from our *R. catesbeiana* cDNA liver libraries which contained information encoding two different C/EBP-like DNA-binding proteins.[48] Characterization of the amino acid sequence in the bZIP DNA-binding domains of these proteins suggests that RcC/EBP-1 and RcC/EBP-2 encode *Rana* homologues of the mammalian C/EBPα[67,68] and δ,[69] respectively. Northern and dot-blot hybridization analyses revealed that although the amount of RcC/EBP-2 mRNAs in tadpole liver remains constant throughout metamorphosis, the RcC/EBP-1 mRNAs are upregulated during both spontaneous and thyroid hormone-induced metamorphosis.

The thyroid hormone-induced upregulation of RcC/EBP-1 mRNAs was found to occur during the lag period, prior to or at approximately the same time as the upregulation of the TR mRNAs.[48] These results suggest that thyroid hormone may be directly affecting the expression of the gene encoding RcC/EBP-1, as well as the genes encoding one or both of its receptors, TRα and TRβ. The fact that RcC/EBP-1 mRNAs accumulate during the lag period is compatible with the notion that the protein product from these mRNAs, a putative transcription factor, may be important for the activation and/or transcription of genes expressed after the lag period. The presence of C/EBP binding elements in the promoter regions of the CPS-I and OTC genes of *R. catesbeiana* and the expression of these genes after the lag period lends support to the contention that RcC/EBP-I, like its mammalian homologue,[64-66] may help initiate the expression of liver-specific genes, such as those encoding these urea cycle enzymes, and, perhaps, those responsible for the terminal differentiation of this amphibian's liver cells.

The results from the studies described in this review provide a foundation for describing a putative molecular mechanism to account for the thyroid hormone-induced expression of the genes encoding the ornithine-urea cycle enzymes, CPS-1 and OTC, in the liver of *R. catesbeiana*. The molecular mechanism proposed, as shown in Figure 7, places heavy emphasis on the expression of TRβ; however, it should be recognized that TRα may play a cooperative or independent role in the proposed scheme. The model suggests that thyroid hormone, in combination with its receptor, initially upregulates genes encoding its own receptor and genes encoding other, perhaps liver-specific (e.g., RcC/EBP-1), transcription factors. Although, as suggested by Tata,[8] the activation of the genes encoding these transcription factors may be subject to different thresholds of TR concentration, all of

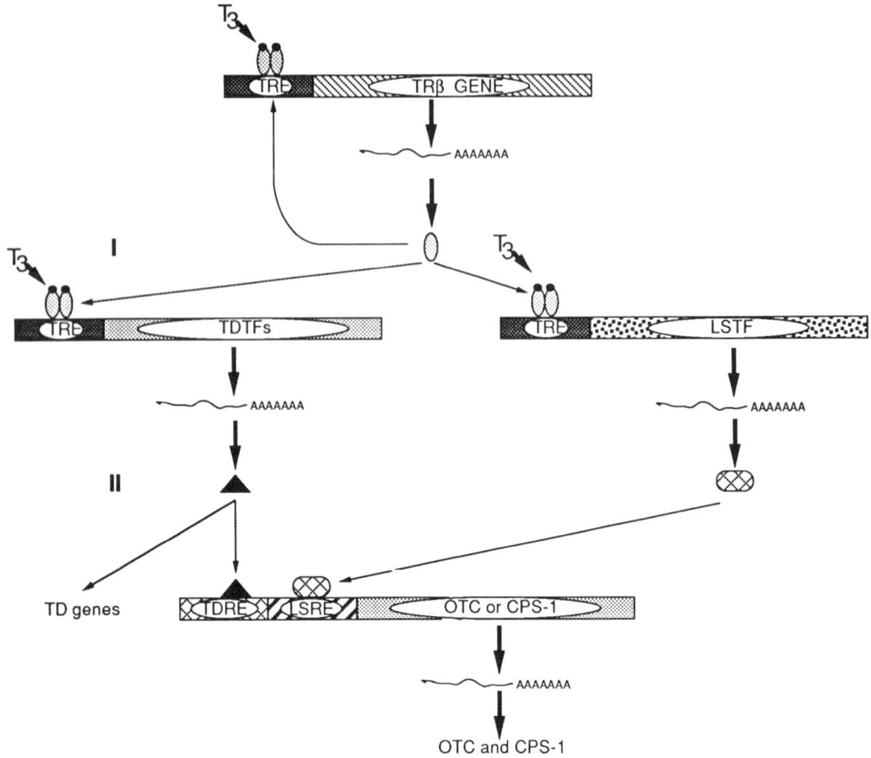

Figure 7 Putative molecular mechanism(s) to account for the thyroid hormone-induced expression of the genes encoding the ornithine-urea cycle enzymes, CPS-1 and OTC, in the liver of *R. catesbeiana* tadpoles. The molecular mechanism(s) proposed consists of two phases: the first phase (I) depicts a scheme for the molecular events thought to occur during the lag period which is the time between the administration of thyroid hormone and the expression of the ornithine-urea cycle enzyme genes CPS-1 and OTC; the second phase (II) depicts a scheme for initiating the molecular events, the expression of the CPS-1 and OTC genes, occurring after the lag period. Phase I depends on thyroid hormone (T_3) binding to thyroid hormone receptor proteins (TRβs) which are bound, as dimers, to thyroid hormone response elements (TREs) located in the promoter regions of genes proposed to be *directly* regulated by this hormone. Phase II is thought to be initiated by the products of the genes directly regulated by this hormone-TRβ complex. The products and transcription factors (TDTF, terminal differentiation transcription factor; LSTF, liver-specific transcription factor) are envisioned to bind to their appropriate response elements (TDRE and LSRE) in the promoter regions of the CPS-1 and OTC genes, and, subsequently, activate and sustain the expression of these ornithine-urea cycle enzyme encoding genes.

these molecular events are proposed to occur during the lag period. The molecular events occurring after the lag period (i.e., the upregulation of the genes encoding CPS-1 and OTC), are envisioned to be initiated by the products (transcription factors) of the genes which were upregulated during the lag period. Finally, the simplistic, molecular mechanism(s) outlined in this scheme raises the possibility that some of the genes initially activated by thyroid hormone encode transcription factors which lead to or direct the terminal differentiation of this amphibian's liver cells.

CONCLUSION

The bulk of the nitrogen excreted by Amphibia is either urea or ammonia. The relative amounts of these liver compounds vary from species to species. Urea is the most important end product in those species which spend at least part of their time out of water. Ammonia predominates only in the aquatic tadpole stages and in the completely aquatic adult forms, such as *X. laevis*. *Xenopus*, unlike most adult Amphibia, excretes more ammonia than urea, but under unfavorable conditions, such as osmotic stress (e.g., dehydration or a saline environment), is capable of synthesizing abnormal amounts of urea and retaining it in its tissues.[74,76] The molecular mechanisms responsible for the transitory, stress-induced synthesis of urea in this organism may depend on factors similar to those described for heat shock or other types of stress.[77] The studies presented in this review have focused on results from investigations with an anuran, *R. catesbeiana*, which is, at least in terms of urea biosynthesis, more typical of the Amphibia. Thus, these investigations were undertaken based on the assumptions (1) that the expression of the genes encoding the ornithine-urea cycle enzymes is not a characteristic of the tadpole liver phenotype whereas it is a characteristic of the adult liver phenotype in this organism, and (2) that thyroid hormone is responsible, directly or indirectly, for initiating and orchestrating the concerted, permanent expression of these genes in the liver of *R. catesbeiana.*

The results presented highlight the obligatory requirement of thyroid hormone to initiate the permanent expression of the genes encoding the ornithine-urea cycle enzymes in the liver of *R. catesbeiana.* It also focuses on the observation that thyroid hormone does not appear to be *directly* influencing the expression of two, if not more, of the genes encoding these urea cycle enzymes. Indeed, the absence of thyroid hormone response elements in the promoter regions of the CPS-1 and OTC genes of *R. catesbeiana*, and the 24-h lag period observed between the administration of this hormone and any detectable accumulation of the mRNAs encoding these urea cycle enzymes, clearly promotes the thought that the response of these genes to thyroid hormone is an indirect or secondary response. Attention has therefore been placed on the lag period, and on uncovering the genes whose expression is being directly effected by thyroid hormone during this period.

The hypothetical mechanism(s) suggested for influencing thyroid hormone to initiate the expression of the ornithine-urea cycle enzyme genes (see Figure 7) proposes that this hormone initially upregulates the expression of genes encoding transcription factors which, in turn, upregulates the expression of the genes encoding the ornithine-urea cycle enzymes. While it appears possible that an amphibian homologue of the mammalian C/EBP may be one of the transcription factors responsible for the liver-specific expression of the genes encoding these enzymes and/or for the terminal differentiation of the liver cells of *R. catesbeiana*, no functional assessments have been completed for C/EBP or any other transcription factor. Although studies to establish the cis-elements and transacting factors involved in the thyroid hormone-induced upregulation of these amphibian genes are still in their infancy, it is apparent that the use of molecular techniques has made a great impact on our progress. Indeed, resolution of the transcription factors required for the permanent expression of the ornithine-urea cycle enzyme genes in the liver of an adult, ureotelic amphibian should be forthcoming, and their (its) identity may provide some molecular-based insight for understanding how the evolution of land vertebrates took place.

ACKNOWLEDGMENTS

The author wishes to express his gratitude to Dr. Rob Dean and Dr. Cheryl Ketola for their helpful suggestions in preparing this review. The work from the author's laboratory was supported by funds from the Natural Sciences and Engineering Research Council of Canada.

REFERENCES

1. **Needham, J.,** *Chemical Embryology*, Vol. 2, Cambridge University Press, Cambridge, England, 1931, 1136.
2. **Needham, J.,** *Biochemistry and Morphogenesis*, Cambridge University Press, Cambridge, England, 1942, 51.
3. **Baldwin, E.,** *An Introduction To Comparative Biochemistry*, 2nd ed., Cambridge University Press, Cambridge, England, 1940.
4. **Brown, G. W., Jr. and Cohen, P. P.,** Comparative biochemistry of urea synthesis. III. Activities of urea-cycle enzymes in various higher and lower vertebrates, *Biochem. J.*, 75, 82, 1960.
5. **Cohen, P. P.,** Biochemical differentiation during amphibian metamorphosis, *Science*, 168, 533, 1970.
6. **Cohen, P. P., Brucker, R. F., and Morris, S. M.,** Cellular and molecular aspects of thyroid hormone action during amphibian metamorphosis, in *Hormonal Proteins and Peptides*, Li, C. H., Ed., Academic Press, New York, 1978, 273.

7. **Gilbert, L. I. and Frieden, E.,** *Metamorphosis: A Problem in Developmental Biology*, Plenum Publishing, New York, 1981.
8. **Tata, J. R.,** Gene expression during metamorphosis: An ideal model for postembryonic development, *Bioessays*, 15, 239, 1993.
9. **Atkinson, B. G.,** Metamorphosis: Model system for studying gene expression in postembryonic development, *Dev. Genet.*, 15, 313, 1994.
10. **Munro, A. F.,** Nitrogen excretion and arginase activity during amphibian development, *Biochem. J.*, 33, 1957, 1939.
11. **Munro, A. F.,** The ammonia and urea excretion of different species of amphibian during their development and metamorphosis, *Biochem. J.*, 54, 29, 1953.
12. **Frieden, E.,** Biochemical adaptation and anuran metamorphosis, *Am. Zool.*, 1, 115, 1961.
13. **Cohen, P. P.,** Biochemical aspects of metamorphosis: transition from ammonotelism to ureotelism, The Harvey Lectures, Series 60, 119, 1966.
14. **Balinsky, J. B.,** Nitrogen metabolism in amphibians, in *Comparative Biochemistry of Nitrogen Metabolism*, Vol. 2, Campbell, J. W., Ed., Academic Press, New York, 1970, chap. 11.
15. **Paik, W. K. and Cohen, P. P.,** Biochemical studies on amphibian metamorphosis, I. The effect of thyroxine on protein synthesis in the tadpole, *J. Gen. Physiol.*, 43, 683, 1960.
16. **Gudernatsch, J. F.,** Feeding experiments on tadpoles. I. The influence of specific organs given as food on growth and differentiation: a contribution to the knowledge of organs with internal secretion, *Wilhelm Roux Arch Entwicklungsmech Organismen*, 35, 457, 1912.
17. **Allen, B. M.,** The results of extirpation of the anterior lobe of the hypophysis and of the thyroid of *Rana pipiens* larvae, *Science*, 44, 755, 1916.
18. **Allen, B. M.,** The development of the thyroid glands of *Bufo* and their normal relation to metamorphosis, *J. Morphol.*, 32, 489, 1919.
19. **Allen, B. M.,** The influence of the thyroid gland and hypophysis upon growth and development of amphibian larvae, *Q. Rev. Biol.*, 4, 325, 1929.
20. **Dolphin, J. and Frieden, E.,** Biochemistry of amphibian metamorphosis, II. Arginase activity, *J. Biol. Chem.*, 217, 735, 1955.
21. **Brown, G. W., Jr. and Cohen, P. P.,** in *The Chemical Basis of Development*, McElroy, W. and Glass, B., Eds., John Hopkins Press, Baltimore, 1958, 495.
22. **Brown, G. W., Jr., Brown, W. R., and Cohen, P. P.,** Comparative biochemistry of urea synthesis, *J. Biol. Chem.*, 234, 1775, 1959.
23. **Metzenberg, R. L., Marshall, M., Paik, W. K., and Cohen, P. P.,** The synthesis of carbamyl phosphate synthetase in thyroxin-treated tadpoles, *J. Biol. Chem.*, 236, 162, 1961.
24. **Kim, K. H. and Cohen, P. P.,** Actinomycin D inhibition of thyroxine-induced synthesis of carbamyl phosphate synthetase, *Biochim. Biophys. Acta*, 166, 574, 1968.
25. **Pouchelet, M. and Shore, G.,** Chloramphenicol inhibits hormone-dependent induction of cytoplasmic mRNA coding for the mitochondrial enzyme, carbamyl phosphate synthetase, in *Rana catesbeiana* tadpoles, *Biochim. Biophys. Acta*, 654, 67, 1981.
26. **Helbing, C., Gergely, G., and Atkinson, B. G.,** Sequential up-regulation of thyroid hormone β receptor, ornithine transcarbamylase, and carbamyl phosphate synthetase mRNAs in the liver of *Rana catesbeiana* tadpoles during spontaneous and thyroid hormone-induced metamorphosis, *Dev. Genet.*, 13, 289, 1992.
27. **Wixom, R. L., Reddy, M. K., and Cohen, P. P.,** A concerted response of the enzymes of urea biosynthesis during thyroxine-induced metamorphosis of *Rana catesbeiana*, *J. Biol. Chem.*, 247, 3684, 1972.
28. **Morris, S. M.,** Thyroxine elicits divergent changes in mRNA levels for two urea cycle enzymes and one gluconeogenic enzyme in tadpole liver, *Arch. Biochem. Biophys.*, 259, 144, 1987.
29. **Galton, V. A., Morganelli, C. M., Schneider, M. J., and Yee, K.,** The role of thyroid hormone in the regulation of hepatic carbamyl phosphate synthetase activity in *Rana catesbeiana*, *Endocrinology*, 129, 2298, 1991.
30. **Helbing, C. and Atkinson, B. G.,** 3,5,3′-triiodothyronine-induced carbamyl phosphate synthetase gene expression is stabilized in the liver of *Rana catesbeiana* tadpoles during heat shock, *J. Biol. Chem.*, 269, 11743, 1994.
31. **Atkinson, B. G., Helbing, C., and Chen, Y.,** Reprogramming of gene expression in the liver of *Rana catesbeiana* tadpoles during spontaneous and thyroid hormone-induced metamorphosis, in *Perspectives in Comparative Endocrinology*, Davey, K. S., Tobe, S. S., and Peter, R. G., Eds., National Research Council of Canada, Ottawa, 1994, 416.
32. **Nyunoya, H., Broglie, K. E., and Lusty, C. J.,** The gene coding for carbamyl-phosphate synthetase I was formed by fusion of an ancestral glutaminase gene and a synthetase gene, *Proc. Natl. Acad. Sci. U.S.A.*, 82, 2244, 1985.
33. **Nyunoya, H., Broglie, K. E., Widgren, E. E., and Lusty, C. J.,** Characterization and derivation of the gene coding for mitochondrial carbamyl phosphate synthetase I of rat, *J. Biol. Chem.*, 260, 9346, 1985.
34. **Simmer, J. P., Kelly, R. E., Rinker, A. G., Jr., Scully, J. L., and Evans, D. R.,** Mammalian carbamyl phophate synthetase (CPS): cDNA sequence and evolution of the CPS domain of the Syrian hamster multifunctional protein CAD, *J. Biol. Chem.*, 265, 10395, 1990.

35. **Carrey, E. A., Campbell, D. G., and Hardie, D. G.,** Phosphorylation and activation of hamster carbamyl phosphate synthetase II by cAMP-dependent protein kinase: a novel mechanism for regulation of pyrimidine nucleotide biosynthesis, *EMBO J.*, 4, 3735, 1985.
36. **Post, L. E., Post, D. J., and Raushel, F. M.,** Dissection of the functional domains of *Escherichia coli* carbamoyl phosphate synthetase by site-directed mutagenesis, *J. Biol. Chem.*, 265, 7742, 1990.
37. **Lusty, C. J. and Liao, M.,** Substitution of Glu841 by lysine in the carbamate domain of carbamyl phosphate synthetase alters the catalytic properties of the glutaminase subunit, *Biochemistry*, 32, 1278, 1993.
38. **Horwich, A. L., Fenton, W. A., Williams, K. R., Kalousek, F., Kraus, J. P., Doolittle, R. F., Konigsberg, W., and Rosenberg, L. E.,** Structure and expression of a complementary DNA for the nuclear coded precursor of human mitochondrial ornithine transcarbamylase, *Science*, 224, 1068, 1984.
39. **Kraus, J. P., Hodges, P. E., Williamson, C. L., Horwich, A. L., Kalousek, F., Williams, K. R., and Rosenberg, L. E.,** A cDNA clone for the precursor of rat mitochondrial ornithine transcarbamylase: comparison of rat and human leader sequences and conservation of catalytic sites, *Nucl. Acids Res.*, 13, 943, 1985.
40. **Veres, G., Gibbs, R. A., Scherer, S. E., and Caskey, C. T.,** The molecular basis of sparse fur mutations, *Science*, 237, 415, 1987.
41. **Mori, M., Miura, S., Tatibana, M., and Cohen, P. P.,** Processing of a putative precursor of rat liver ornithine transcarbamylase, a mitochondrial matrix enzyme, *J. Biochem.* (Tokyo), 88, 1829, 1980.
42. **Marshall, M. and Cohen, P. P.,** Evidence for an exceptionally reactive arginyl residue at the binding site for carbamyl phosphate in bovine ornithine transcarbamylase, *J. Biol. Chem.*, 255, 7287, 1980.
43. **Xu, Q., Baker, B. S., and Tata, J. R.,** Developmental and hormonal regulation of the *Xenopus* liver-type arginase gene, *Eur. J. Biochem.*, 211, 891, 1993.
44. **Haraguchi, Y., Takiguchi, M., Amaya, Y., Kawamoto, S., Matsuda, I., and Mori, M.,** Molecular cloning and nucleotide sequence of cDNA for human liver arginase, *Proc. Natl. Acad. Sci. U.S.A.*, 84, 412, 1987.
45. **Kawamoto, S., Amaya, Y., Murakami, K., Tokunaga, F., Iwanaga, D., Kobahashi, K., Saheki, T., Kimura, S., and Mori, M.,** Complete nucleotide sequence of cDNA and deduced amino acid sequence of rat liver arginase, *J. Biol. Chem.*, 262, 6280, 1987.
46. **Innis, M. A., Gelfand, D. H., Sninsky, J. J., and White, T. J.,** *PCR Protocols: a Guide to Methods and Applications*, Academic Press, San Diego, 1990.
47. **Helbing, C.,** Thyroid Hormone-Induced Changes in Gene Expression in the Bullfrog, *Rana catesbeiana*, Tadpole Liver, Ph.D. thesis, The University of Western Ontario, London, Ontario, Canada, 1993.
48. **Chen, Y., Hu, H., and Atkinson, B. G.,** Characterization and expression of C/EBP-like genes in the liver of *Rana catesbeiana* tadpoles during spontaneous and thyroid hormone-induced metamorphosis, *Dev. Genet.*, 15, 366, 1994.
49. **Hu, H., Chen, Y., and Atkinson, B. G.,** Expression of C/EBP genes in the liver of *Rana catesbeiana* tadpoles undergoing spontaneous and thyroid hormone-induced metamorphosis, presented at Canadian Federation of Biological Sciences, Montreal, June 15–18, 1994.
50. **Mathisen, P. and Miller, L.,** Thyroid hormone induces constitutive keratin gene expression during *Xenopus laevis* development, *Mol. Cell. Biol.*, 9, 1823, 1989.
51. **Buckbinder, L. and Brown, D. D.,** Thyroid hormone-induced gene expression changes in the developing frog limb, *J. Biol. Chem.*, 267, 25786, 1992.
52. **Shi, Y.-B. and Brown, D. D.,** The earliest changes in gene expression in tadpole intestine induced by thyroid hormone, *J. Biol. Chem.*, 278, 20312, 1993.
53. **Wang, Z. and Brown, D. D.,** Thyroid hormone-induced gene expression program for amphibian tail resorption, *J. Biol. Chem.*, 268, 16270, 1993.
54. **French, R. P., Warshawsky, D., Tybor, L., Mylniczenko, N. D., and Miller, L.,** Upregulation of AP-2 in the skin of *Xenopus laevis* during thyroid hormone induced metamorphosis, *Dev. Genet.*, 15, in press, 1994.
55. **Lazar, M. A.,** Thyroid hormone receptors: multiple forms, multiple possibilities, *Endocrinol. Rev.*, 14, 184, 1993.
56. **Weinberger, C., Thompson, C. C., Ong, E. S., Lebo, R., Gruol, D. J., and Evans, R. M.,** The *c-erb-A* gene encodes a thyroid hormone receptor, *Nature*, 324, 641, 1986.
57. **Sap, J., Munoz, A., Damm, K., Goldberg, Y., Ghysdael, J., Leutz, A., Berg, H., and Vennstrom, B.,** The c-erb-A protein is a high affinity receptor for thyroid hormone, *Nature*, 324, 635, 1986.
58. **Brooks, A. R., Sweeney, G., and Old, R. W.,** Structure and functional expression of a cloned *Xenopus* thyroid hormone receptor, *Nucl. Acids Res.*, 17, 9395, 1989.
59. **Yaoita, Y. and Brown, D. D.,** A correlation of thyroid hormone receptor gene expression with amphibian metamorphosis, *Genes Dev.*, 4, 1917, 1990.
60. **Schneider, M. J. and Galton, V. A.,** Regulation of c-erbA-α messenger RNA species in tadpole erythrocytes by thyroid hormone, *Mol. Endocrinol.*, 5, 201, 1991.
61. **Davey, J. C., Schneider, M. J., and Galton, V. A.,** Cloning of a thyroid hormone responsive *Rana catesbeiana* c-erbA-β gene, *Dev. Genet.*, 15, 339, 1994.
62. **Shi, Y-B., Yaoita, Y., and Brown, D. D.,** Genomic organization and alternative promoter usage of the two thyroid hormone receptor β genes in *Xenopus laevis*, *J. Biol. Chem.*, 267, 733, 1992.

63. **Chen, Y. and Atkinson, B. G.,** Characterization of the promoter regions of the carbamyl phosphate synthetase (CPS-1) and ornithine transcarbamylase (OTC) genes from *Rana catesbeiana*, presented at Canadian Federation of Biological Sciences, Montreal, June 15–18, 1994.
64. **Umek, R. M., Friedman, A. D., and McKnight, S. L.,** CCAAT-enhancer binding protein: a component of a differentiation switch, *Science*, 251, 288, 1991.
65. **Friedman, A. D., Landschulz, W. H., and McKnight, S. L.,** CCAAT/enhancer binding protein activates the promoter of the serum albumin gene in cultured hepatoma cells, *Genes Dev.*, 3, 1314, 1989.
66. **Sladek, F. M. and Darnell, J. E.,** Mechanisms of liver-specific gene expression, *Curr. Opin. Gen. Dev.*, 2, 256, 1992.
67. **Landschulz, W. H., Johnson, P. F., Adashi, E. Y., Graves, B. J., and McKnight, S. T.,** Isolation of a recombinant copy of the gene encoding C/EBP, *Genes Dev.*, 2, 786, 1988.
68. **Lincoln, A. J., Williams, S. C., and Johnson, P. F.,** A revised sequence of rat c/ebp gene, *Genes Dev.*, 8, 1131, 1994.
69. **Williams, S. C., Cantwell, C. A., and Johnson, P. F.,** A family of C/EBP-related proteins capable of forming covalently linked leucine zipper dimers *in vitro*, *Genes Dev.*, 5, 1553, 1991.
70. **Nyunoya, H. and Lusty, C. J.,** Sequence of the small subunit of yeast carbamyl phosphate synthetase and identification of its catalytic domain, *J. Biol. Chem.*, 259, 9790, 1984.
71. **Galton, V. A. and St. Germain, D. L.,** Putative nuclear triiodothyronine receptors in tadpole liver during metamorphic climax, *Endocrinology*, 117, 912, 1985.
72. **Taylor, A. C. and Kollros, J. J.,** Stages in the normal development of *Rana pipiens* larvae, *Anat. Rec.*, 94, 7, 1946.
73. **Ashley, H., Katti, P., and Frieden, E.,** Urea excretion in the bullfrog tadpole: effect of temperature, metamorphosis, and thyroid hormones, *Develop. Biol.*, 17, 293, 1968.
74. **Balinsky, J. B., Choritz, E. L., Coe, C. G. L., and Van Der Schans, G. S.,** Urea cycle enzymes and urea excretion during the development and metamorphosis of *Xenopus laevis*, *Comp. Biochem. Physiol.*, 22, 53, 1967.
75. **Balinsky, J. B., Choritz, E. L., Coe, C. G. L., and Van Der Schans, G. S.,** Amino acid metabolism and urea synthesis in naturally aestivating *Xenopus laevis*, *Comp. Biochem. Physiol.*, 22, 59, 1967.
76. **Janssens, P. A. and Cohen P. P.,** Biosynthesis of urea in the estivating African lungfish and in *Xenopus laevis* under conditions of water-shortage, *Comp. Biochem. Physiol.*, 24, 887, 1968.
77. **Atkinson, B. G. and Walden, D. B.,** *Changes in Eukaryotic Gene Expression in Response to Environmental Stress*, Academic Press, Orlando, Florida, 1985.

Chapter 10

Excretory Nitrogen Metabolism in Reptiles and Birds

James W. Campbell

CONTENTS

INTRODUCTION

Terrestrial vertebrates use two mechanisms for detoxication of ammonia generated during hepatic amino acid catabolism: (1) its conversion to urea for excretion, ureoteley; and (2) its conversion to uric acid, uricoteley. In both mechanisms, the primary ammonia-detoxifying reaction is located within the mitochondrial matrix of liver cells. In ureotelic tetrapods, the combined actions of carbamyl-phosphate synthetase-I (CPS-I), the primary ammonia-fixing enzyme, and ornithine transcarbamylase (OTC) result in the formation of citrulline. Citrulline then exits to the cytosol for conversion to urea for excretion. In uricotelic tetrapods, the primary ammonia-fixing enzyme in the matrix is glutamine synthetase (GS). Glutamine formed by the action of this enzyme then exits to the cytosol for conversion to urate for excretion. Both hepatic mechanisms are present in extant tortoises[1] and possibly the tuatara[2] which exhibit facultative uricoteley dependent upon the availability of water.[2,3] Because of the ancient origin of chelonians,[4] this may well represent the primitive cotylosaurian ("stem reptile") condition. During subsequent divergence of mammalian and archosaurian lines, one or the other hepatic ammonia detoxication system became predominant. The nature, distribution, and possible evolution of these two ammonia-detoxifying mechanisms among the higher tetrapod vertebrates will be considered in this chapter.

TETRAPOD PHYLOGENY

The amniote tetrapods may be divided into the mammalian and reptilian-avian lines (Figure 1). These lines of descent arose from the Cotylosauria and were represented primitively by the

0-8493-8411-7/95/$0.00+$.50

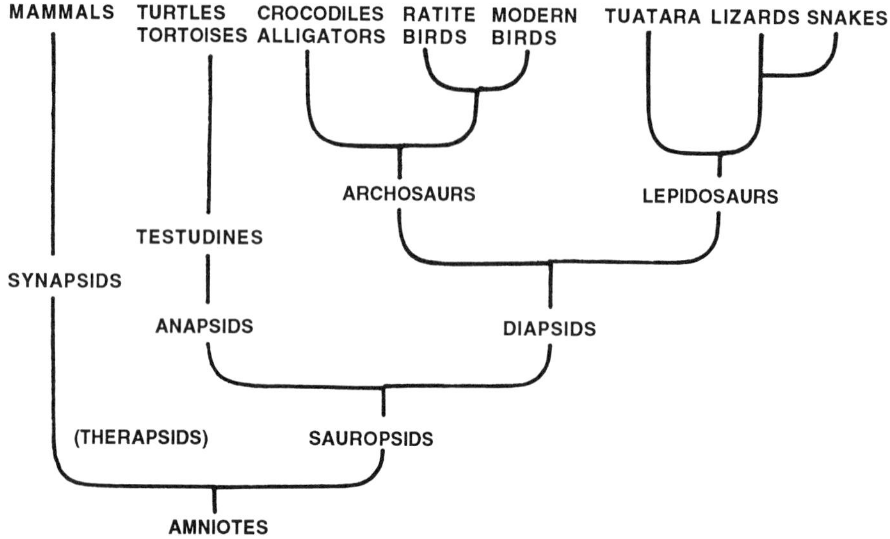

Figure 1 Phylogeny of the amniote tetrapods. (From Benton, M. J., *J. Mol. Evol.*, 30, 409, 1990. With permission.)

mammalian-like therapsid reptiles on the one hand, and the sauropsid reptiles, on the other.[5] The latter included the reptilian ancestors of the dinosaurs and their kin (archosaurs), the primitive turtles (testudines), and ancestors of modern reptiles (lepidosaurs). The mammalian and reptilian-avian lines are generally thought to have diverged around 300 million years ago (mya). The sauropsid line now contains the extant turtles and tortoises, lizards, snakes, the tuatara (Rhyncocephalia), alligators, crocodiles, and birds. Although the possibility of a bird-mammal relationship exists,[6] the preponderance of morphological and molecular evidence supports the grouping of birds and crocodilian separately from mammals as sister groups in the Archosauria.[7-10] Extant birds and crocodilians therefore represent the sole survivors of the dinosaurs.

EVOLUTIONARY DIVERGENCE OF UREOTELIC AND URICOTELIC TETRAPODS

The dinosaurs were the most successful land vertebrates, having been the dominant species for the 120 to 130 my between the late Triassic period and the end of the Cretaceous period. During the earlier part of the Triassic, the mammalian-like therapsids were dominant.[11] During the latter part of this period, however, the therapsids virtually disappeared, to be replaced by the archosaurs.[12-15] Although mammals arose during the late Triassic, they remained as small innocuous insectivores until the demise of the dinosaurs at the end of the Cretaceous. Even today, however, the known number of avian-reptilian species outnumber the known mammalian species three to one.[16]

The development of increased locomotory ability and endothermy most likely contributed to the marked radiation of the archosaurs during the late Triassic.[12] A third major factor may have been development of the ability to adapt to the increasing aridity of this period.[17] Most of the large continental mass of Pangea was distributed along the equator during the Triassic, making for predominantly arid tropical or subtropical environments.[18] The development of uricoteley as a means of detoxifying ammonia, along with a cloaca with a high postrenal reabsorptive capacity as a water-conserving mechanism, may have been a critical part of the physiological adaptation to these environments.[13] The ability of primitive fish to synthesize urea as an osmolyte and the presence of ureotely in terrestrial amphibians suggests this was the primitive mode of nitrogen excretion in early amniote tetrapods. Not only did ureoteley persist as the sole mechanism for hepatic ammonia-detoxification in mammals, but urea was also retained as a major osmolyte in the counter current concentrating mechanism used to produce a hypertonic urine in the mammalian kidney (see also Chapter 13).[19] The avian kidney, on the other hand, utilizes mainly sodium chloride to create an osmotic gradient.[20] Whereas the concentrating ability of the archosaurian kidney is less than that of certain desert mammals,[21] the postrenal reabsorption of water and electrolytes by the lower intestine and cloaca in reptiles, birds, and crocodilians results in the final excretion of a semisolid paste of urate containing a minimal amount of water.[22-25] Few, if any, mammals can minimize excretory water loss to this extent. There is, however, the interesting observation on water-deprived cricetid rodents indicating they excrete up to 47% of their total excretory nitrogen as crystalline allantoin, normally the product of the action of uricase on uric acid.[26] Knowing the metabolic source of this uric acid would be interesting since the excretion of waste nitrogen as purine has not heretofore been observed in mammals. In any event, in those tetrapod vertebrates capable of "facultative uricoteley", as defined by a shift toward an increased percentage urate nitrogen excreted, this transition always occurs during conditions of water restriction. Examples include treefrogs,[27-29] tortoises,[30] certain lizards,[31,32] the tuatara,[33] and crocodilians.[22] Water-loaded hummingbirds normally excrete a high percentage of their excretory nitrogen as ammonia. Decreasing the water load causes a return to urate excretion,[34] reaffirming the physiological advantage of uricoteley for water conservation. The development of this mode of excretion may have been especially advantageous to herd dinosaurs that had to migrate long distances to find vegetation and water.[35]

In addition to water conservation, the development of uricoteley may also have had at least one additional selective advantage. This was the antioxidant properties of urate. These properties of uric acid are well known,[36] and it has been suggested that they are the reason for the loss of the enzyme uricase during primate evolution.[37] The transition to plasma urate as an antioxidant was especially critical following the loss of the ability to synthesize ascorbate by primates.[38] There is some evidence that during the initial period of the rapid radiation of the archosaurs some 300 mya, Earth's atmosphere was much more highly oxidizing than it is now. Based on calculations of reduced and oxidized carbon and sulfur[39] and direct analyses of amber-entrapped air bubbles,[40] Earth's atmosphere may have contained as much as 50% more oxygen at this time (see, however, Ref. 41). Uricoteles also lack the enzyme uricase, so presumably their elevated plasma urate levels would have been of selective advantage under these conditions. Fruit flies lacking the ability to produce urate (xanthine dehydrogenase mutants; see Figure 5) show increased sensitivity to environmental oxygen in further support of this idea.[42]

AMMONIA-DETOXIFYING SYSTEMS

The formation of excretory urea and urate both involve metabolic pathways that were primitively biosynthetic ("nutritional") pathways for either protein (arginine) or nucleic acid (purine nucleotides) components, which during evolution were usurped for an excretory function.[43] In their original biosynthetic mode, they were mainly regulated by the availability and amount of their end products, arginine or purine mononucleotides, relative to the need for these products. To function for ammonia detoxication, however, end-product inhibition had to be eliminated making them "open" pathways (i.e., substrate limited). This conversion required several modifications, not only in the catalytic and regulatory properties of the enzymes, but also in their compartmentation within liver cells. The latter was especially important because ammonia is released within the mitochondrial matrix during hepatic amino acid catabolism. While there is considerable information on the cellular and molecular aspects of amino acid catabolism and end-product formation in mammals, there is relatively little such information available for reptiles, crocodilians, and birds. Much of what is available comes from studies on domestic birds, especially chickens. Therefore, at this time, "generalizations" with respect to the cellular and molecular aspects of amino acid catabolism and ammonia detoxication in uricoteles are at best, hypotheses, and, at worst, mere speculation.

SOURCES OF SUBSTRATE: AMMONIA FORMATION

Unlike carbohydrates and lipids, excess dietary amino acids cannot be stored as complex molecules.[44] Rather, they are catabolized in liver and their carbon skeletons are converted to glucose for either storage as glycogen or for use as a metabolic fuel by extrahepatic tissues. During fasting, or other glucocorticoid-induced protein catabolic states, amino acids are derived mainly from the breakdown of muscle protein and are returned to the liver for gluconeogenesis. Normal working muscle also supplies amino acids to the liver, especially alanine and glutamine.[45] Alanine is quite gluconeogenic and the interchange between liver and muscle has been referred to as the glucose-alanine cycle,[46] analogous to the Cori or glucose-lactate cycle between the two organs. Liver thus receives amino acids from the gut via the hepatic portal vein and from the extrahepatic tissues via the systemic circulation. The supply of amino acids to the liver is dependent on complex interactions involving rates of proteolysis, rates of membrane transport, acid–base balance, etc.[47] Because of its central role in modulating circulating levels of glucose, the liver has been referred to as the "glucostat" of the body.[48] Again, much of the information available regarding the function of the liver in this capacity comes from mammals, especially the laboratory rat, an omnivore. Reptiles and crocodilians are predominantly sporadically feeding carnivores. Many bird species are also carnivorous; others, including the chicken, are omnivores, and still others are strictly herbivorous so data obtained from laboratory rodents and domestic chickens may have limited applicability to many reptilian and avian species.

The first step in amino acid gluconeogenesis for most amino acids is their transamination in the cytosol of hepatocytes. Some amino acids, such as asparagine, glycine, serine, and threonine, may also undergo direct deamination in the cytosol, at least in mammals. It is unlikely that serine is deaminated in uricoteles because of its role in urate formation. The most common acceptor for cytosolic transamination is α-ketoglutarate (Figure 2). Cytosolic transamination of most amino acids therefore results in glutamate formation; hepatic glutamate levels increase twofold during the feeding period in rats.[49] Alanine increases even more dramatically during this time, probably reflecting an exchange between it and glutamate via alanine aminotransferase.[50] The exact mechanism for the increase in alanine is not clear, however, but it may reflect the buildup of a pool of alanine that can later be converted to pyruvate when the concentration of the latter starts decreasing as it is utilized for gluconeogenesis. Aminotransferases are at or near equilibrium so the rate and direction of their reactions are substrate concentration dependent. In any event, the amino function of cytosolic glutamate represents a major pool of "waste nitrogen" and flux even though this pool is determined by the availability of amino acids.

In mammalian liver mitochondria, there are two transport systems for glutamate in the inner membrane (Figure 2): (1) an H^+ symporter, and (2) an aspartate-glutamate antiporter.[51,52] Uptake of glutamate by the former results in its deamination via glutamate dehydrogenase (GDH), located exclusively in the mitochondrial matrix[53] and, by the latter, transamination with oxaloacetate to form aspartate via the mitochondrial isozyme of aspartate aminotransferase. Although GDH is also localized exclusively in the mitochondrial matrix in chicken liver,[54] there is no information on the mitochondrial transport systems for glutamate in either birds or reptiles. There are, however, clear differences between the fate of intramitochondrial glutamate in some mammals (e.g., rat) and birds (e.g., chicken). For example, aspartate aminotransferase, which is required for the intramitochondrial transamination of glutamate to aspartate in the matrix of mammalian liver mitochondria, is located on the outer surface of the inner membrane in chicken liver mitochondria.[55] Aspartate formation from glutamate therefore effectively takes place in the cytosol in chicken liver rather than in the mitochondrial matrix.[56,57] However, whether aspartate is formed in the matrix or in the cytosol, it is utilized for end-product (urate[56] or urea[58]) synthesis in the cytosol. Another unique aspect of chicken hepatic amino acid metabolism is the exclusive localization of alanine aminotransferase within the mitochondrion.[59] The response of this enzyme to starvation-induced gluconeogenesis suggests that it acts mainly to form pyruvate from alanine for carboxylation to oxaloacetate which is then converted to phospho*enol*pyruvate (PEP) via PEP carboxykinase (PEPCK). The latter enzyme is also localized exclusively in the mitochondrial compartment in chicken liver (Figure 2).[57] In carnivorous birds such as the murre, a cytosolic isozyme of PEPCK and both mitochondrial and

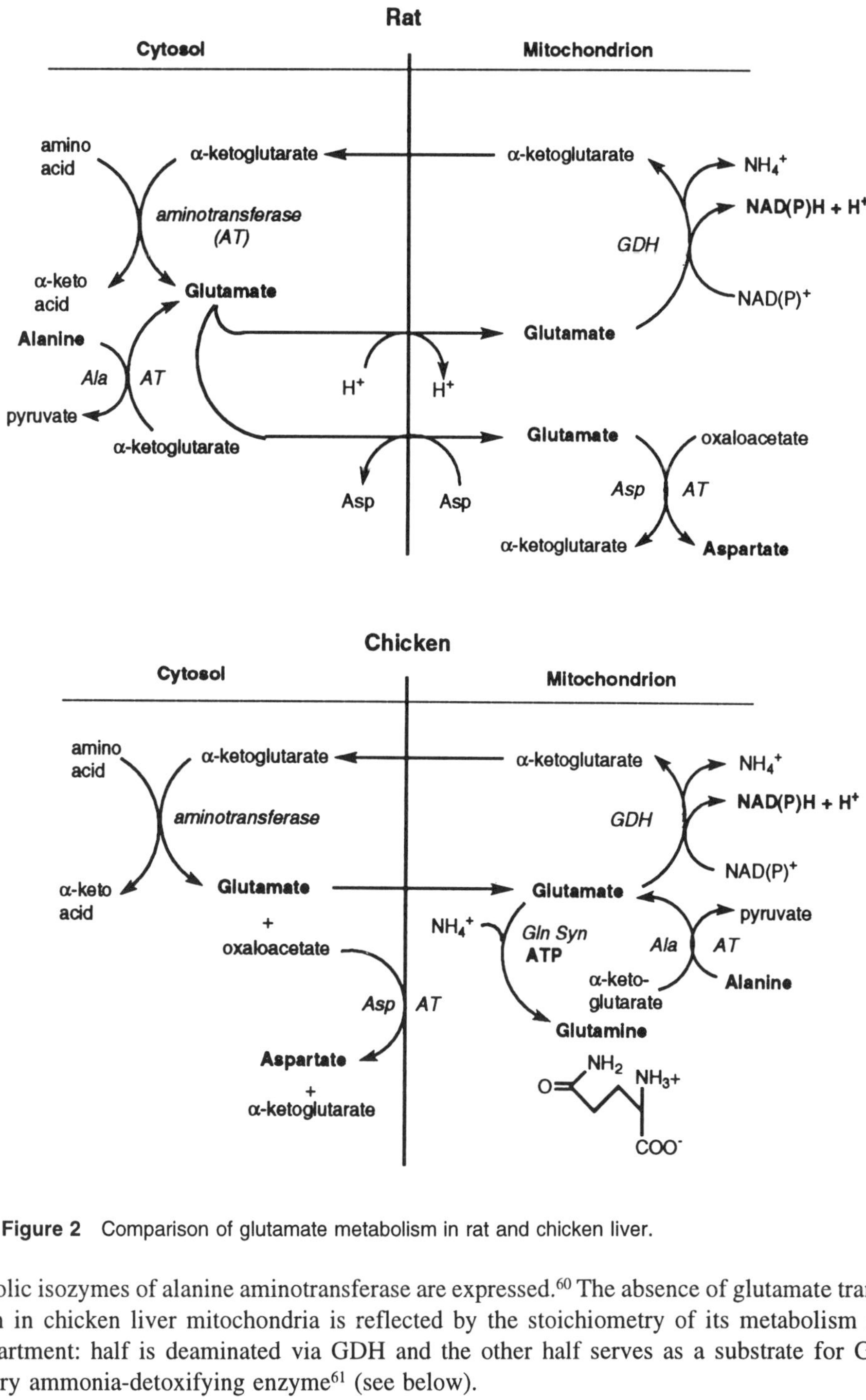

Figure 2 Comparison of glutamate metabolism in rat and chicken liver.

cytosolic isozymes of alanine aminotransferase are expressed.[60] The absence of glutamate transamination in chicken liver mitochondria is reflected by the stoichiometry of its metabolism in this compartment: half is deaminated via GDH and the other half serves as a substrate for GS, the primary ammonia-detoxifying enzyme[61] (see below).

Another major source of ammonia for hepatic urea synthesis in mammals is the amide function of glutamine.[50] As indicated above, glutamine is an end product of working muscle and is also formed in other extrahepatic tissues such as brain and intestine in mammals and birds and presumably also in reptiles and other higher tetrapods. However, the mitochondrial localization of glutamine synthetase in uricotelic liver, demonstrated both by cellular fractionation[62] and immunocytochemistry,[63] results in modification of the hepatic metabolism of glutamine in these species (Figure 2). In mammals and presumably other ureoteles, there are high levels of phosphate-activated glutaminase in the matrix of liver mitochondria which result in rapid uptake and breakdown of glutamine.[64,65] If high levels of glutaminase were also present in the matrix compartment of uricotelic liver mitochondria along with high levels glutamine synthetase, a futile cycle of glutamine synthesis and breakdown would result. A glutaminase can be detected in chicken liver mitochondria,[54] but its activity is only 1/40th of that in rat liver mitochondria and it is localized mainly in the outer mitochondrial membrane. Because of this localization, it has been suggested that there may be a "vectorial" metabolism of glutamine in avian liver mitochondria based on whether it is exiting or entering.

Ammonia arriving via the portal and systemic circulation also serves as a substrate for urea formation in ureoteles. Ammonia absorbed by the avian intestine[66] stimulates both hepatic glutamine and urate production[67,68] indicating that this is also true in uricoteles. Because of its alkaline pKa (>9), extrahepatic ammonia exists mainly as NH_4^+. To be detoxified and converted to either urea or urate, it must traverse the hepatocyte and mitochondrial membranes. Specific transport mechanisms for NH_4^+ exist, but not in higher vertebrates, so NH_3 is generally considered to be the form of ammonia that is permeable to membranes,[57,69] although NH_4^+ can also substitute for K^+ on the Na^+, K^+ ATPase.[56] Although NH_4^+ is taken up by hepatocytes, the specific nature of this process does not appear to have been characterized. There is little ammonia uptake by liver during metabolic acidosis[70] which could be due to proportionally lower NH_3 under these conditions. This would then account for the diversion of NH_4^+ to the kidney for excretion during metabolic acidosis.[69] GS is cytosolic in mammalian hepatocytes and presumably detoxifies NH_3 entering or formed in that compartment. However, its expression is restricted to only a few perivenule cells in mammalian liver and it therefore presumably functions only in a "failsafe" manner if the capacity for urea synthesis is exceeded in the periportal and other hepatocytes which express CPS-I and the remainder of the urea cycle.[71,72] GS is thus not normally involved in the detoxification of cytosolic ammonia in mammalian liver. In avian and crocodilian liver, GS is mitchondrial and is expressed in all hepatocytes,[73,74] similar to the expression pattern of CPS-I in mammalian liver. Therefore, in both ureotelic and uricotelic ammonia-detoxifying systems, the primary ammonia-binding enzyme in liver is located within the mitochondrial matrix, so that ammonia taken up by hepatocytes or formed in the cytosol must eventually enter this compartment for detoxication. NH_3 is the actual substrate for both CPS-I[75] and GS.[76] There is evidence in mammals for the direct channeling of the NH_3 taken up by hepatic mitochondria to CPS-I which results in its conversion to the carbamoyl function of citrulline via an inner membrane–CPS-I–OTC complex.[77,78] It is not known whether a similar channeling of NH_3 to GS takes place in uricotelic liver mitochondria.

UREAGENESIS

The "ornithine-urea" cycle consists of five enzymes: CPS-I, OTC, argininosuccinate (ASA) synthetase, ASA lyase, and arginase. The reactions catalyzed and their location in mammalian hepatocytes are shown in Figure 3. CPS-I and OTC are located in the matrix of these mitochondria and may occur as a complex on the inner mitochondrial membrane.[78] The complex may also include glutamate dehydrogenase (GDH) and specific transaminases,[79,80] (Figure 2) forming an efficient mechanism for channeling the amino function of amino acids into the carbamoyl function of citrulline. The kinetics of GDH are generally unfavorable for the deamination reaction in dilute, homogeneous solution, but this may not be true for the enzyme in such a complex. The primary ammonia-fixing enzyme is CPS-I which is allosterically activated by N-acetylglutamate (NAGA).[81] There are three animal CPSs that are defined by their requirement for this activator, their nitrogenous substrate, and their subcellualr localization (see also Chapter 2). CPS-II is an NAGA-independent enzyme that utilizes glutamine as its substrate[82] and occurs universally in the cytosol of eukaryotic cells as part of a multienzyme complex involved in pyrimidine biosynthesis.[83] CPS-III, the probable

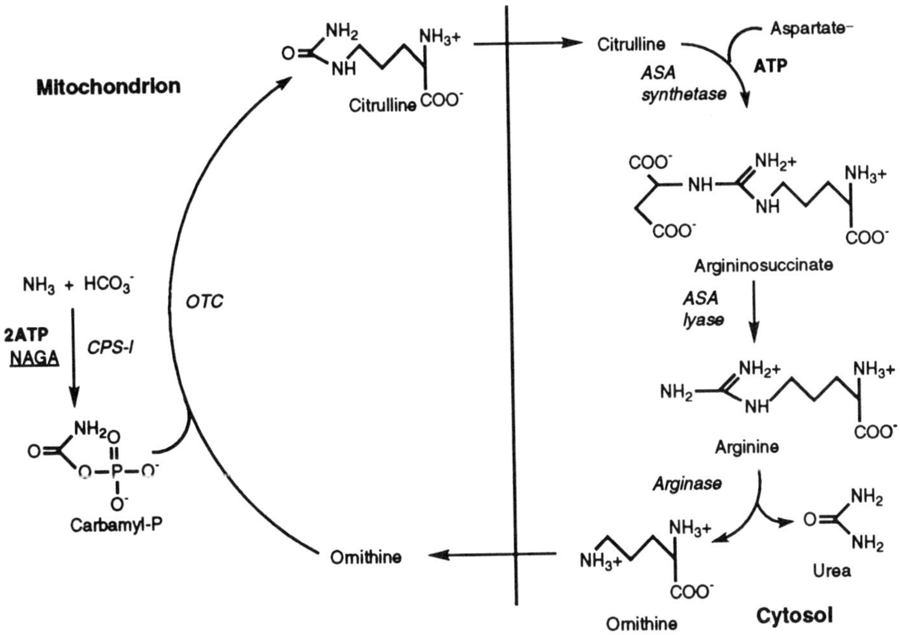

Figure 3 Compartmentation of the urea cycle in mammalian liver.

ancestor of CPS-I, is an NAGA-dependent enzyme that also utilizes glutamine,[84] occurring in mitochondria of representatives of at least three major invertebrate phyla.[85] This invertebrate enzyme is also present in elasmobranchs[86] and some teleosts.[87] The sporadic occurrence of CPS-III in fish plus the presence of an ammonia- and NAGA-dependent CPS, presumably CPS-I, in the coelacanth[88] and lungfish[89] and the definite presence of CPS-I in amphibia[90] suggest that CPS-III did not persist long after primitive fish and was replaced early in the pretetrapod line by CPS-I.

Carbamyl phosphate formed by CPS-I in mammalian liver mitochondria is converted to citrulline in the presence of ornithine and OTC. Citrulline then exits to the cytosol via a citrulline-ornithine antiporter.[91] In the cytosol, citrulline is converted to urea by the sequential actions of ASA synthetase, ASA lyase, and arginase (Figure 3). Mammalian liver arginase is a basic protein occurring in multiple molecular forms[92] which will interact with acidic macromolecules, including those in membranes. Although there is as yet no evidence for a multifunctional complex including arginase and the ASA synthetase system in the cytosol, these three enzymes may be closely associated with the mitochondrion in intact cells.[93]

As indicated in Figure 3, the biosynthesis of urea directly utilizes four pyrophosphate bonds from three ATPs (ATP is converted to AMP + PP_i in the ASA synthetase reaction). Fumarate is one of the products of ASA lyase and the sequential actions of fumarate and malate dehydrogenase (MDH) in the cytosol could generate NADH + H^+ which is energetically equivalent to three ATPs. These cytosolic reducing equivalents are important in the energetics of gluconeogenesis so the coupling of this process with ureagenesis may reduce the "cost" of urea formation.

Urea is a neutral compound with the functionality of an amide.[95] As pointed out several years ago,[96] the overall stoichiometry for urea synthesis, as well as for urate synthesis, results in the generation of H^+ since the substrate is $-NH_3^+$ and the product is $-NH_2$ or, in the case of urate, =NH. In terms of the actual generation of H^+ that could contribute to general acid–base homeostasis, this would probably only occur if all of the enzymes involved acted in dilute, homogeneous solution. Because of metabolic compartmentation in hepatocytes and the exchange of H^+ across the mitochondrial inner membrane via various translocation systems, the final fate of the protons generated during ammonia detoxication in the mitochondrial matrix is difficult to acertain. The NH_3 generated in this compartment exits as the *neutral* carbamoyl function of citrulline or as the *neutral* amide function of glutamine. In ureotelic liver, the protons could exit via the citrulline-ornithine antiporter to reenter the matrix on the glutamate symporter (Figure 2), or they could be pumped out by the electron transport system to reenter via the ATPase, etc. As discussed above, a main substrate for ureagenesis, NH_4^+, is shunted to the kidney for excretion during metabolic acidosis. This "shunting" may be due to the decreased percentage of ammonia present as NH_3 under these conditions. Glutamine, another major substrate for ureagenesis, is also shunted to the kidney during metabolic acidosis, due, apparently, to decreased glutamine uptake by both hepatocytes and their mitochondria.[65] While ureagenesis is responsive to acid–base balance and may serve to maintain hepatic HCO_3^- levels under certain physiological conditions, its primary function is ammonia detoxification.[47] This is also probably true for uricogenesis.

In its primitive nutritional capacity of providing arginine for protein synthesis, the "ornithine-urea" cycle was regulated by the availability of arginine. This can be seen today in primitive eukaryotes such as yeast and fungi in which arginine limitation markedly stimulates transcription of subunits of the arginine-specific carbamyl-phosphate synthetase, CPS-A; high arginine, on the other hand, represses transcription.[97,98] To function as an ammonia-detoxifying pathway, the basic arginine pathway had to be integrated with arginase and its regulation modified so that, end-product effects, if any, had to be "feedforward." In ureotelic tetrapods, there is both short- and long-term regulation of this type.

CPS-I may be the main point of short-term regulation of urea synthesis in mammals when NH_4^+ and HCO_3^- are not limiting (see also Chapter 12).[47] Ureoteles have also evolved a mechanism in which arginine has a regulatory function. This process is via the allosteric activation of CPS by NAGA. NAGA activates both CPS-III, the presumed ancestoral protein, and CPS-I. The synthesis of NAGA via the mitochondrial synthetase is markedly stimulated by arginine.[99] There are, however, some questions remaining with respect to the role of intramitochondrial NAGA in both short- and long-term regulation of the urea cycle.[47] Ornithine availability is also a controlling factor in urea synthesis. A requirement for exogenous ornithine to sustain high rates of ureagenesis is reflected by the dietary arginine requirement of the cat.[100] Because of the cat's almost exclusively protein diet, a high rate of ureagenesis is dictated by a high rate of amino acid gluconeogenesis.[101,102] Ornithine may activate CPS-I, either directly or indirectly,[47,103] as well as serve as a substrate for ureagenesis. Based on levels of enzyme activity, the urea cycle may be limited by ASA metabolism under conditions when NH_3, ornithine, and aspartate are saturating.[47] Physiologically, the short-term regulation of ureagenesis is probably most important postprandially, especially in carnivores.

The long-term regulation of the urea cycle occurs during adaptation to chronic increases in dietary protein intake or to other protein catabolic states, such as starvation. This generally involves increased transcription of the genes for the enzymes of the cycle, especially for CPS-I.[104,105] The increased transcription of the urea cycle genes is hormonally mediated by both cAMP(glucagon)- and glucocorticoid(cortisol)-dependent mechanisms. There may also be increased stabilization of enzyme mRNA as well as possible stabilization of the enzymes themselves.[106] Such mechanisms are also involved in the perinatal development of the capacity for ureagenesis in mammals.[107-109] Again, this simply reflects an adaptation to alimentation of food, a *de facto* increase in dietary protein. In summary, the response of the urea cycle in ureoteles is to "feedforward" under conditions of increased ammonia concentrations requiring an increased flux through the cycle. For the short term, this involves mainly activation of CPS-I when all substrates are saturating. For the long term, the levels of the enzymes are increased through transcriptional activation.

URICOGENESIS

The structure and numbering system for uric acid is shown in Figure 4. The formation of the purine ring involves the sequential additions of atoms from the precursors glutamine, glycine, HCO_3^-, and

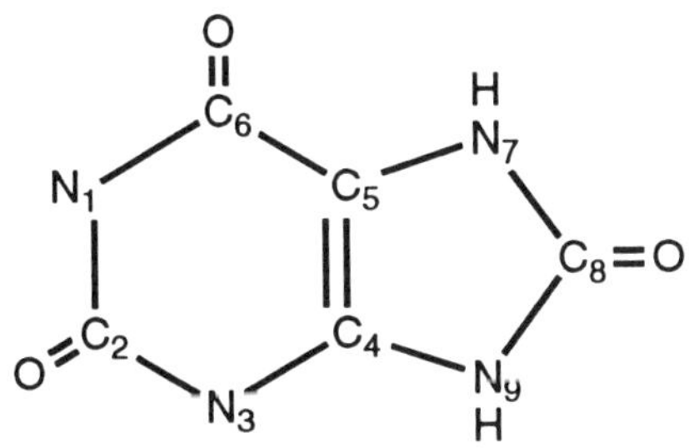

Figure 4 Structure and numbering system for uric acid.

of one-carbon units as 5,10-methylene tetrahydrofolate (5,10-$MeFH_4$) (Figure 5). If there is direct recycling of ribose or ribose phosphates formed during the conversion of the nucleoside inosine to the purine hypoxanthine, then the purine pathway can be depicted as a "urate cycle" in uricoteles with 5-P-ribose acting as the carrier for the eventual end product, urate. This is analogous to one of the ways ornithine acts in the urea cycle. 5-P-Ribose for purine nucleotide biosynthesis is generally thought to come from the pentose pathway,[111] but this has not been critically investigated in uricoteles. It is important in terms of the energetics of ATP utilization by the urate pathway. Two reactions are shown for P-ribose recovery in Figure 5: The first reaction (rx. 14), is the conversion of 1-P-ribose, formed by the action of nucleoside phosphorylase on inosine, to 5-P-ribose by phosphopentomutase[112]; in the second reaction (rx. 15), 5-P-ribose is converted to 5-P-ribose-PP by pyrophosphokinase. An additional reaction is catalyzed by "5′-IMP phosphohydrolase" in which IMP is converted directly to 5-P-ribose and hypoxanthine.[113] It has also been suggested that the purine salvage enzyme, hypoxanthine-guanine phosphoribosyltransferase (HGPRT), acting in the reverse direction, converts IMP directly to 5-P-ribose-PP and hypoxanthine in the presence of pyrophosphate.[114] Avian liver is, however, deficient in the salvage enzymes,[115] a condition which may be important in the regulation of the urate pathway (see below). Unless the HGPRT in uricoteles is kinetically modified so that it has little IMP-synthesizing activity, it seems unlikely that it is part of a urate cycle.

The primary ammonia-detoxifying enzyme in the urate pathway is GS which is localized in mitochondria in uricotelic liver.[62] Hepatic GS in uricoteles is therefore equivalent to CPS-I in ureotelic liver. Glutamine formed intramitochondrially by GS during amino acid catabolism (Figure 2) contributes N3 and N9 of the purine ring via its amide function (Figure 5). The K_m of ammonia (NH_3 + NH_4^+) for chicken liver GS is 0.5 m*M*[116] which is two to four times lower than the K_m of ammonia of CPS-I.[47] Early studies on purine biosynthesis were performed using the soluble cellular fraction of pigeon liver. With this system, incorporation of all nitrogenous substrates into the purine ring was obtained *except* for ammonia.[117] It was shown later that the inclusion of liver mitochondria greatly stimulated glycine incorporation into purine by the cytosolic system.[118] GS, theretofore known only as a cytosolic enzyme, was then shown to be mitochondrial in both avian and reptilian liver.[62] Mitochondria may also be a major source of one-carbon units (as 5,10-$MeFH_4$)[118] (rxs. 3 and 9; Figure 5). Serine hydroxymethyltransferase is the enzyme catalyzing the formation of 5,10-$MeFH_4$ from serine. Whereas this methyltransferase may be predominately mitochondrial in avian liver, it occurs in both the cytosolic and mitochondrial compartments in mammalian liver.[119] The remainder of the enzymes of the urate pathway are in the cytosol. Most of these enzymes occur as part of multifunctional proteins in both chicken and humans.[120] The multifunctional enzymes that have been identified include those that catalyze Reactions 2, 3, and 5; Reactions 6 and 7; and Reactions 9 and 10. None of these reactions utilize glutamine as a substrate. Because of the instability of P-ribosylamine (rx. 1), it would appear advantageous for P-ribose-PP amidotransferase

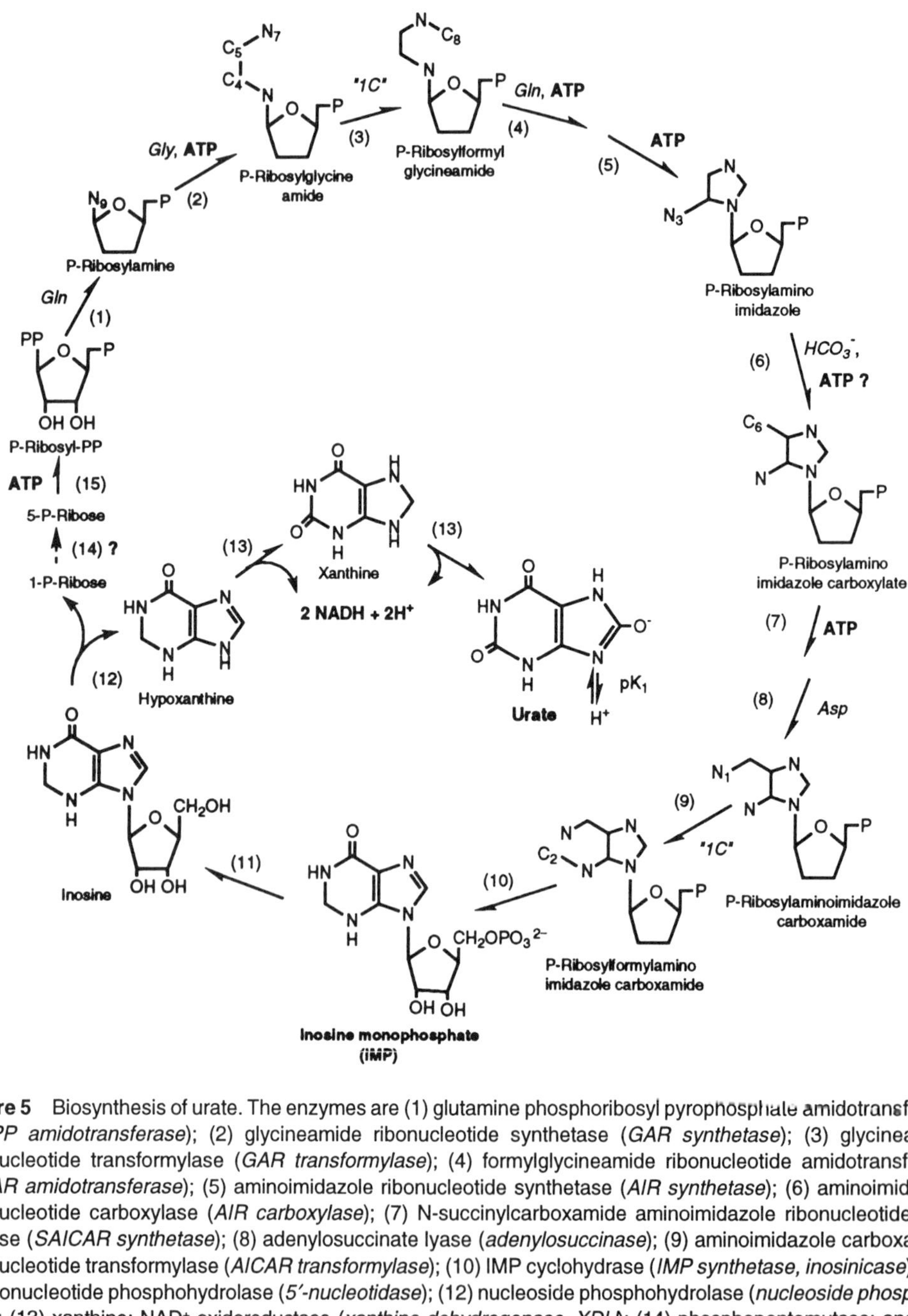

Figure 5 Biosynthesis of urate. The enzymes are (1) glutamine phosphoribosyl pyrophosphate amidotransferase (*PRPP amidotransferase*); (2) glycineamide ribonucleotide synthetase (*GAR synthetase*); (3) glycineamide ribonucleotide transformylase (*GAR transformylase*); (4) formylglycineamide ribonucleotide amidotransferase (*FGAR amidotransferase*); (5) aminoimidazole ribonucleotide synthetase (*AIR synthetase*); (6) aminoimidazole ribonucleotide carboxylase (*AIR carboxylase*); (7) N-succinylcarboxamide aminoimidazole ribonucleotide synthetase (*SAICAR synthetase*); (8) adenylosuccinate lyase (*adenylosuccinase*); (9) aminoimidazole carboxamide ribonucleotide transformylase (*AICAR transformylase*); (10) IMP cyclohydrase (*IMP synthetase, inosinicase*); (11) 5′-ribonucleotide phosphohydrolase (*5′-nucleotidase*); (12) nucleoside phosphohydrolase (*nucleoside phosphorylase*); (13) xanthine: NAD^+ oxidoreductase (*xanthine dehydrogenase, XDH*); (14) phosphopentomutase; and (15) ATP: ribose-5-P pyrophosphotransferase (*5′phosphoribosyl pyrophosphokinase*). Abbreviated structures are used except for the last five compounds of the pathway. The pK_a (5.4) for uric acid ionization is from Reference 110.

to be associated with the first multifunctional protein in the pathway so its unstable product could be channeled directly to the next enzyme, GAR synthetase (rx. 2). This would be analogous to the association of CPS-I, whose product is also labile, with OTC in mammalian liver mitochondria.

The synthesis of urate for excretion has generally been considered to be an energetically expensive process.[114] Because of the uncertainty with respect to the metabolism of ribose and the P-riboses in uricotelic liver, it is not possible to estimate this energetic cost exactly with respect to

ATP utilization. As shown in Figure 5, six pyrophosphate bonds, as five ATPs, are utilized (ATP is converted to 5-ribosyl-PP in rx.15). Two additional ATP-requiring steps have been demonstrated in the prokaryote, *E. coli*, where Reaction 6 consists of two subreactions, one of which requires ATP.[121] In *E. coli*, Reaction 3 also utilizes formate and ATP rather than 5,10-MeFH_4.[122] In calculating the energetics of urate formation, generation of NADH + H^+ via xanthine dehydrogenase (XDH; rx. 13) must also be considered. This is especially critical because of the high energetic cost of gluconeogenesis (2 mol NADH; 4 mol ATP; and 2 mol GTP/mole glucose synthesized).[123] As shown in Figure 5, 2 mol of reduced NAD are formed in the cytosol per mole urate synthesized; energetically equivalent to *six ATPs*. A mole of fumarate is also formed in the cytosol during urate synthesis (rx.8) which can be utilized to generate an additional mole of reduced NAD via fumarase and MDH. Thus there is the potential for *generating* the equivalent of *nine ATPs* during urate synthesis. As first pointed out by Coolbear et al.,[124] the XDH reaction may be a major source of reducing equivalents for cytosolic amino acid gluconeogenesis in birds, especially when PEPCK is exclusively mitochondrial.[57] XDH is present in all uricotelic species but, unlike its counterpart in mammals, it cannot be converted to the oxidase which utilizes O_2 rather than NAD^+ as an electron acceptor.[125]

Unlike urea, which is a neutral end product, uric acid is a moderately strong organic acid with a pK_a of 5.4. Its anion, urate, is therefore present mainly as the K^+, Na^+, or NH_4^+ urate salt under normal physiological circumstances.[111] It was therefore necessary for uricoteles to develop efficient mechanisms for recovering any essential cations that might otherwise be lost during urate excretion.[126] The limited solubility of uric acid also poses some unique problems with respect to its vascular transport and renal handling.[127] The recent demonstration of a severe hyperuricemia and renal pathology in uricase "knock-out" mice[128] in which the uricase gene has been inactivated illustrates how critical the evolution of special physiological mechanisms for handling uric acid as an end product was to uricoteles.

Little is known regarding the short-term regulation of uricogenesis in uricoteles. Unlike the mammalian enzyme, avian GS is not activated appreciably by α-ketoglutarate but does show a small percentage activation by its product, glutamine, *in vitro*.[116] Glutamine also stimulates urate synthesis *in vivo*,[129] but this may simply be due to its acting as a substrate for Reactions 1 and 4 in the cytosol. Avian GS, like its mammalian counterpart CPS-I, is inhibited by physiological levels of Ca^{2+},[116] but it is not known if this is part of a physiologically significant regulatory mechanism. The octameric vertebrate GSs do not undergo an adenylation-deadenylation cycle which is the main mechanism for end-product regulation of the dodecameric prokaryotic enzymes.[130] PRPP synthesis (rx. 15) is a key regulatory step in the purine pathway in prokaryotes and mammals. The enzyme catalyzing this reaction, PRPP synthetase, occurs as an oligomeric complex in which some of the subunits are inhibitory.[131] In its capacity of supplying purine nucleotides, the most critical enzyme in the short-term regulation of the *de novo* purine biosynthetic pathway is PRPP amidotransferase (rx. 1). This enzyme is subject to feedback inhibition by AMP and GMP and also by IMP.[132] Although the iron-containing mammalian and avian enzymes are similar in structure, there are differences between them in their sensitivities to inhibition by these nucleotides. Partially purified pigeon PRPP amidotransferase is sensitive to AMP and GMP inhibition, but loses this sensitivity with further purification.[133] The chicken enzyme appears to be far less sensitive to purine nucleotide inhibition than the rat enzyme.[134] There have also been reports that AMP and GMP stimulate urate synthesis in intact chick heptocytes.[135] A general deficiency in the salvage enzymes, HGPRT and adenine PRT (APRT), in avian liver[115] may be critical in this differential sensitivity of the avian PRPP amidotransferse to purine nucleotide inhibition. First of all, the salvage enzymes utilize PRPP to convert hypoxanthine, guanine, and adenine to their respective nucleotide monophosphates and therefore compete with PRPP amidotranferase for substrate. The levels of PRPP are some 200 times greater in chicken hepatocytes than those in rat hepatocytes[115] which may reflect the decreased levels of salvage enzymes in the former. Second, high levels of PRPP cause disaggregation of the

Table 1 Response of Chicken Liver Enzymes to Fasting and High Dietary Protein Intake[137]

Enzyme	% Change: Fast (48 h)	% Change: High protein diet (8 d)
Alanine aminotransferase	122	1167
	(9900[a], 179[b])	—
Aspartate aminotransferase	15	88
Glutamate dehydrogenase	11	5
Glutamine synthetase	−10	295
	(200[c])	(120)[c]
PRPP amidotransferase (rx. 1)	100	350
	(67)[c]	(237)[c]
Adenylosuccinase (rx. 8)	(114)[b]	—
5′-Nucleotidase (rx. 11)	14	100
Phosphorylase (rx. 12)	31	331
Xanthine dehydrogenase (rxs. 13)	100	445
	(793)[b]	—

Note: Rounded and recalculated from original.[137] For reactions catalyzed, see Figures 2 and 5.

[a] Seven-d fast.[59]

[b] Four-d fast.[138]

[c] Seven-d fast and 5 vs. 50% dietary protein.[134]

amidotransferase, rendering it less sensitive to inhibition by the nucleotides,[136] most likely due to the dissociation of an inhibitory subunit. Finally, a deficiency in the salvage enzymes simply means smaller amounts of AMP and GMP are formed were they to be inhibitory to PRPP amidotransferase. Relieving the purine nucleotide monophosphate inhibition of PRPP amidotransferase thus appears to be the most important means for the short-term regulation of uricogenesis, assuring that the pathway operates in an "open," substrate (amino acid/ammonia)-limited manner.

In birds, there is increased urate excretion with increased dietary protein intake and fasting, so long-term adaptation of the urate pathway probably involves some sort of concerted increase in the levels of enzymes of the pathway as in the urea pathway. However, only a few enzymes of amino acid catabolism and uricogenesis have been measured under conditions of fasting or high dietary protein intake (Table 1). The response to increased dietary protein intake is the most straightforward. Under these conditions there is a very marked increase in alanine aminotransferase which, in chicken liver, is exclusively mitochondrial.[59] This increase is therefore consistent with the role of the aminotransferase in pyruvate formation for gluconeogenesis (Figure 2). In fasting murres, which are carnivorous, there is a 21% decrease in hepatic alanine aminotransferase indicating they, like carnivorous mammals,[101] probably maintain high rates of gluconeogenesis irrespective of dietary protein intake. There is also a fairly clear positive response by the other enzymes for amino acid catabolism, ammonia detoxification, and urate synthesis during protein-catabolic states (Table 1). We have found chicken liver GS activity to increase about threefold after a week on a high protein (60%) diet, consistent with the values reported by others. As shown in Figure 6, the increase in chicken liver GS on high protein diets is at the pretranslational level in that there is an increase in GS mRNA levels. Of the urate pathway enzymes measured during high dietary protein intake, XDH shows the largest increase. XDH also shows the most dramatic increase during the neonatal period in which there is an increase in the capacity for uricogenesis.[57,141] The increase in XDH is glucocorticoid dependent, unlike the increase in GS during this period.[141,142] The increase in PRPP amidotransferase and nucleoside phosphorylase during high protein intake is about 25% less than that of XDH. 5'Nucleotidase also increases, which suggests that long-term adaptation of the urate pathway does involve increases in most, if not all, of

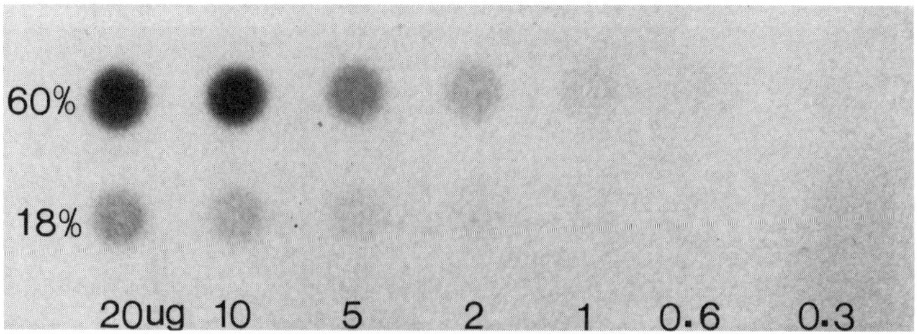

Figure 6 Induction of chicken liver glutamine synthetase mRNA by high dietary protein.[140] White Leghorn hens were fed standard Purina chicken food which contains 18% protein. The high protein diet was supplemented with casein to a final protein content of 60%. After 1 week on these diets, total liver RNA was isolated for serial dilution "dot blotting" using a chicken glutamine synthetase cDNA probe (pCGS11).[130] Numbers along the horizontal axis refer to μg total liver RNA. Hepatic glutamine synthetase activity increased threefold on the high protein diet compared with that on the standard diet. [Northern blotting (not shown) established the hybridizing mRNA to be the correct size for glutamine synthetase mRNA.]

the enzymes for uricogenesis in a concerted manner. The neonatal increase in the capacity for uricogenesis in the chicken also involves increases in the enzymes for hepatic amino acid catabolism as well as in GS and coincides with the development of the capacity for gluconeogenesis which is absent in the prenate.[57] Again, as in the development of the capacity for ureagenesis in the mammalian neonate, the development of the capacity for uricogenesis in the avian neonate and presumably other uricoteles, is directed toward the alimentation of a protein-containing diet and the capacity for hepatic amino acid gluconeogenesis.

THE UREOTELIC AND URICOTELIC SYSTEMS COMPARED

There are basic similarities as well as differences between the two hepatic ammonia-detoxifying systems. CPS-I is the primary ammonia-fixing enzyme in ureotelic liver, but this enzyme is not expressed in avian liver. It is, however, expressed in faculative ureouricoteles such as the tortoise.[1] CPS-I is present in most mammalian hepatocytes but there is an especially high level of CPS-I expression in the periportal ones.[71] CPS-I is not, however, expressed except in trace amounts in the perivenule hepatocytes in mammals,[144] where a cytosolic GS is expressed in adult mammals[145] (Figure 7). This cytosolic GS appears not to normally be involved in ammonia detoxification since it decreases with glucocorticoid administration rather than increasing like CPS-I.[146] GS is, on the other hand, the primary ammonia-fixing enzyme in the uricotelic system and, as can be seen in Figure 7, is expressed rather uniformly in both avian and crocodilian liver cells.[63,73,74] This almost[147] mutually exclusive expression of either CPS-I or GS in adult mammalian and avian hepatocytes must represent the final evolutionary divergence of the two detoxication systems. The ontogenetic recapitulation of this can be seen in mammals. Most fetal (Days 1 to 15) mouse hepatocytes express GS but do not show the perivenular concentration of the enzyme.[148] CPS-I mRNA only appears after Day 13 of the gestation period of the mouse and can be detected in low levels at Day 15.[149] Subsequent increases in CPS-I mRNA coincide with the repression of GS in most hepatocytes and its final restriction to the perivenular cells[148] (Figure 7). Treefrogs of the genera *Chiromantis* and *Phyllomedusa* have a functional urea cycle[150] but excrete mainly urate when water deprived.[27-29] Unlike most amphibians, in which a hepatic GS is very low or undetectable, these treefrogs have significant levels of the enzyme as well as high levels of the urate pathway enzymes PRPP amidotransferase[150] and XDH.[151] The hepatic GS in *Phyllomedusa* is almost exclusively in the cytosolic compartment,[150] but is, however, expressed in all hepatocytes along with CPS-I and OTC.[73] The tortoise, which also excretes both urea and urate and is facultatively uricotelic when water deprived, has the regular uricotelic system in addition to the ureotelic system since it expresses GS

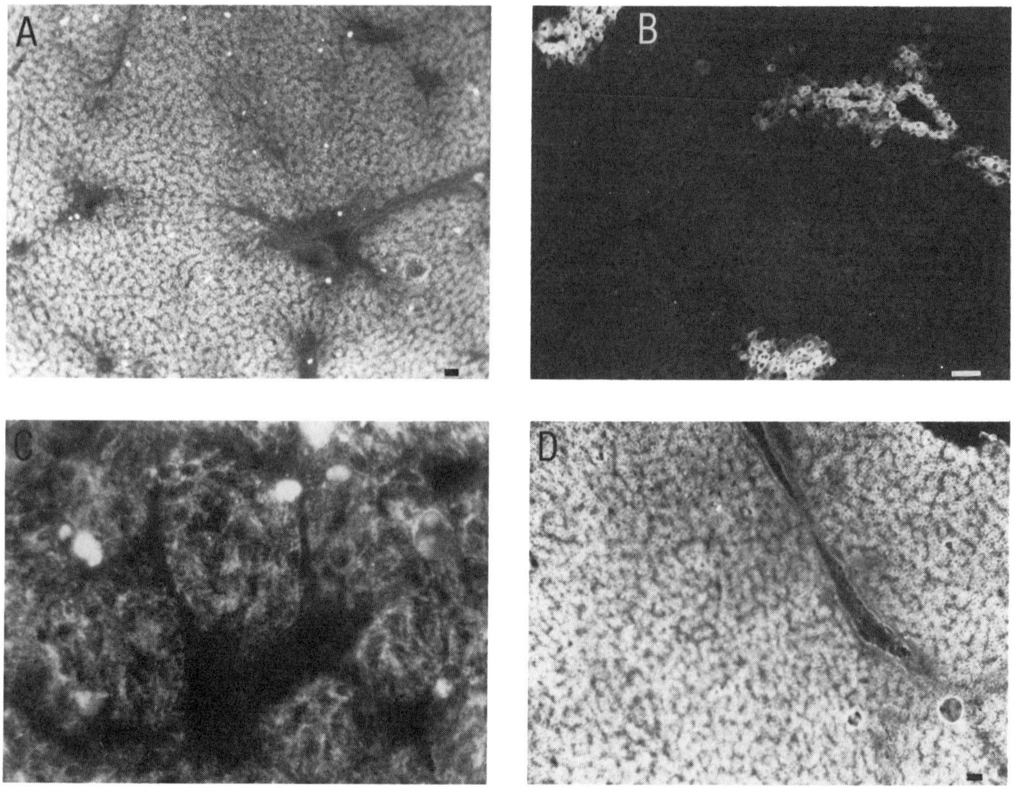

Figure 7 Expression of glutamine synthetase in vertebrate liver. Immunofluorescent staining (white) of the enzyme was as previously described[73,74] using rabbit antichicken glutamine synthetase antibody. A, duck liver; B, hamster liver; C, alligator liver; D, chicken liver. The bars in A, B, and D = 40 μm. (For C, magnification 400X.)

along with CPS-I and OTC in liver cells.[73] This dual expression of CPS-I and GS suggests that the mutual exclusion of one or the other of these enzymes in hepatocytes occurs only in adult mammals. One may speculate that the treefrog condition represents the first "experiment" in utilizing GS, in this case a cytosolic enzyme, as an ammonia-detoxifying enzyme for uricogenesis. The treefrog condition of expressing a cytosolic GS and a mitochondrial CPS-I is thus somewhat recapitulated during early development of the mammal. The tortoise, in which GS has become localized in liver mitochondria along with CPS-I, would then represent the next stage in the evolution of the uricotelic system. The final stage is represented by uricotelic archosaurs and squamate reptiles in which CPS-I is completely repressed in liver so that the urea pathway is no longer functional as an ammonia-detoxifying system.

A major similarity between the two ammonia-detoxifying systems is the reactivity of the mitochondrial product formed for translocation to the cytosol. Ammonia released in the matrix of hepatic mitochondria in ureoteles is converted to the carbamoyl function of citrulline by the combined actions of CPS-I and OTC; in uricoteles, it is converted to the amide function of glutamine by GS. Neither of these functions consume H^+ at physiological pHs. Citrulline and glutamine cannot be separated by routine electrophoresis and, during acid–base titration, both titrate like glycine which does not possess an R chain.[56] Therefore, ammonia released intramitochondrially in either ureoteles or uricoteles exits in a neutral form that will not interfere with normal H^+ flux across the inner mitochondrial membrane. The latter is required to maintain coupling of ATP synthesis to O_2 uptake in oxidative phosphorylation. The experimental evidence for this has been discussed.[56]

The synthesis of both urea and urate consumes a HCO_3^-; however, the compartment in which this occurs differs: in ureoteles, it is in the mitochondrial compartment (Figure 3) whereas in uricoteles, it is in the cytosolic compartment (Figure 5). The two systems also utilize the amino function of aspartate. In ureoteles, aspartate is formed from the intramitochondrial transamination of glutamate, whereas in uricoteles (at least in the chicken) it is effectively formed in the cytosol via glutamate transamination (Figure 2). At high flux rates, an external (dietary) source of an additional amino acid intermediate is required in both pathways. This is ornithine in ureoteles[100] and glycine in uricoteles.[152] The function of these two amino acids is, however, quite different. In addition to its possible regulatory role, ornithine acts as a "carrier" molecule for the end product in ureagenesis. 5-P-Ribose would appear to provide this function in uricogenesis. The protective effect of dietary glucose during ammonia intoxication in birds[152] may reflect a requirement for an external source of ribose during high rates of uricogenesis. This requirement would then be analogous to the requirement for ornithine in ureoteles. On the other hand, glycine actually provides two carbon atoms and one nitrogen atom of the end product in uricogenesis. Another unique aspect of uricogenesis is the involvement of folate derivatives and the requirement for serine and/or glycine to provide one carbon units. Finally, the two end products differ in their reactivity: urea is a neutral, soluble compound whereas uric acid reacts as an acid and is very water insoluble.

EVOLUTIONARY MODIFICATIONS OF THE *DE NOVO* PURINE BIOSYNTHETIC PATHWAY: A SUMMARY

An important modification in the purine biosynthetic pathway that allowed it to function more efficiently in ammonia detoxication was the translocation of GS into the mitochondrial matrix in hepatocytes. This served to locate the enzyme at the site of ammonia formation and also to provide for its conversion to a neutral compound whose exit would not uncouple oxidative phosphorylation during high rates of flux through the pathway. The latter was obviously quite important to carnivorous archosaurs that presumably maintained high rates of amino acid gluconeogenesis due to high dietary protein intake. Based on the situation in treefrogs, the first experiment in uricoteley may have occurred in early amniotes containing a cytosolic GS in liver. In primitive cartilagenous fish, GS is targeted to liver mitochondria by an N-terminal signal sequence on its mRNA translation product.[153] Again, based on hepatic GS being cytosolic in the treefrogs, coding information for this targeting signal was presumably lost in early tetrapod ancestors. Retargeting to hepatic mitochondria in the archosaurs did not involve reacquisition of a mitochondrial targeting presequence but rather, specific amino acid replacements at the N-terminus of mature GS that served to convert this domain to an internal targeting signal.[139]

Release of end-product inhibition of PRPP amidotransferase by purine nucleotides, especially IMP, was fundamental to the use of the urate pathway for ammonia detoxification. This appears to be via the suppression of the purine salvage enzymes, HGPRT and APRT, in uricoteles. These enzymes utilize PRPP as substrate and, when nearly absent, PRPP levels become elevated, causing disaggregation of PRPP amidotransferase, rendering it insensitive to inhibition by the nucleotides. It is likely that other enzymes of the basic biosynthetic pathway also have modified properties but these are not known at present. The two IMP-catabolzing enzymes expressed in birds serve to divert this substrate toward uric acid formation rather than to other purine nucleotide monophosphates. The cytosolic 5′-nucleotidase in avian liver has a high affinity for IMP and a very low affinity for AMP.[154] This enzyme is also present in other vertebrate tissues in which nucleotide catabolism is important. In uricoteles, its kinetic properties decouple IMP flux from the adenine nucleotide pool and shunt IMP to inosine. In chicken liver, purine nucleoside phosphorylase, which converts inosine to hypoxanthine, is a multimeric enzyme that is activated by its substrate, inosine, due to subunit cooperativity,[155,156] diverting inosine to hypoxanthine. The pigeon liver enzyme, on the other hand, is a monomeric enzyme that does not exhibit substrate activation,[157] but this may be related to the special situation in the pigeon which lacks an hepatic XDH (see below). In most other uricoteles, the hypoxanthine formed via the phosphorylase is shunted to xanthine by XDH due to the low levels of salvage enzymes. Uricoteles utilize NAD^+ as an electron acceptor for XDH rather than O_2,[158] providing cytosolic reducing equivalents for gluconeogenesis, thus tightly coupling this process with uricogenesis. Finally, there was a "silencing" of the uricase gene to prevent allantoin formation and therefore allowing for the direct excretion of the sparingly soluble urate. Neither uricase nor an inactive mutant form of the enzyme can be detected in liver of turtle (*Pseudomys picta*)(the scientific names as given in the original references are retained here), pigeon (*Columbia livia*), or chicken (*Gallus gallus*).[159] While these may represent some of the metabolic modifications involved in the detoxication of ammonia via the uric acid pathway, the excretion of the end-product urate also required several critical physiological modifications, especially for its renal handling.[127]

AMMONIA DETOXICATION AND END-PRODUCT EXCRETION IN ADVANCED TETRAPODS

TURTLES

As can be seen in Table 2, aquatic turtles are, in general, primarily ureotelic in terms of their partition of excretory-N. This appears true for both adults and embryos. All species that have been examined also have a complete complement of hepatic urea cycle enzymes (Table 3). The CPS activity that has been measured in turtle liver is ammonia- and NAGA-dependent so is presumed to be CPS-I; the activity in tortoise liver has definitely been shown to be CPS-I.[1] No GS activity can be detected in liver tissue of either aquatic or terrestrial turtles (Table 4), so CPS-I appears to be the sole ammonia-detoxifying enzyme in these species. The levels of most urea cycle enzymes in turtle liver are comparable to those in rat liver, especially if assay temperature differences and overall metabolic rate differences between poikilotherms and homeotherms are taken into account.

Turtles also excrete a significant amount of ammonia-N which, in some cases, is the predominant form of excretory-N. Aquatic turtles should therefore be considered to be facultative ammonoureoteles. A significant amount of urate-N excretion has been reported for the pond slider, *Pseudomys picta*, and solid deposits, presumably urates, are found in its cloaca.[160] Urate-N never accounts for more than 25% of the total N excreted, however, and, in the absence of detectable levels of hepatic GS, it seems unlikely that this urate represents an end product of ammonia detoxification. It more likely represents the end-product of nucleic acid catabolism which, in the absence of uricase,[159] would accumulate in the cloaca.

TORTOISES

Tortoises excrete about equal amounts of urea- and urate-N with most containing relatively large amounts of solid urate deposits in their cloacal urine; there is, however, considerable variability in the partitioning of excreted-N. This is best illustrated by the Texas tortoise, *Gopherus berlandieri* (Table 2), which, depending upon its history, can be primarily ureotelic or primarily uricotelic in its excretion pattern. Both the desert tortoise, *G. agassizii*, and the Texas tortoise express CPS-I and GS in liver mitochondria.[1] Hepatic mitochondria from both species form and release both citrulline and glutamine to the medium *in vitro*. Both the ureotelic and uricotelic hepatic ammonia-detoxifying systems are therefore present and functional in tortoise liver tissue. GDH levels are high in hepatic mitochondria of these two species of tortoises (48.7 to 59.8 mmol/min/g tissue) and glutamate is an effective precursor of both citrulline and glutamine in mitochondria *in vitro*.[1] Ammonia, on the other hand, is mainly converted to citrulline under these conditions. The physiological significance of this apparent differential metabolism of glutamate and ammonia is not clear. One would expect that there are regulatory mechanisms dictating which of the two ammonia-detoxifying systems are utilized under a particular physiological situation. However, the considerable variation in the partitioning of excretory-N between urea- and urate-N in the Texas tortoise is not reflected by changes in the levels of hepatic urea cycle enzymes (Table 3), which show relatively small differences in animals from

Table 2 Nitrogenous End-Product Excretion by Reptiles and Birds

Species	% Total nitrogen[a]			Reference
	Ammonia-N	Urea-N	Urate-N	
TESTUDINES				
Turtles				
Pseudomys scripta (FW)	4–44	45–95	1–24	3
	26–75	16–53	9–21	160
Kinosternon subrubrum (FW)	24	23	<1	161
Pelusios derbianus (FW)	19	24	5	161
Trionyx spiniferus (FW)				162
in ovum day 36	30	70	0	
hatch		24	69	7
Chelydra sperpentina (FW)	11	80	10	160
Chelodina longicollis (FW)				163
hydrated	15	78	4	
dehydrated (20 d)	28–52	27–52	12–16	
Emys orbicularis (amphibFW)	14	47	3	161
Kinixys erosa (T)	6	61	4	161
K. youngii (T)	6	44	5	161
Tortoises				
Testudo sulcata (T)	3	20	55	30
T. leithii (T)	4	49	34	30
T. denticulata (T)	6	29	7	161
T. gracea (T)	4	22	52	161
T. elegans (T)	6	9	56	161
T. ornata (T)	23	47	30	160
T. carolina (T)	16	79	6	160
Gopherus agassizii (T)	3–18	15–50	20–50	3
G. berlandieri (T)				160
freshly collected	9	4	86	
commercial source	5	84	11	
dehydrated (29–37 d)	1	92	7	
dehydrated (60 d)	0	72	28	
RHYNCHOCEPHALIA				
Tuatara				
Sphenodon punctatum (T)				2
low protein diet	3–4	10–12	80–82	
intermediate	3–4	18–20	71–77	
high protein diet	4–5	25–31	57–69	
SQUAMATA				
Snakes				
Coluber constrictor (T)				164
in ovum day 34	0	61	3	
hatch		0	58	20
Zamensis diadema (T)	4	2	69	165
Eryx thebacious (T)	6	0	63	166
Lizards				
Anolis carolinensis (T)	13	13	73	31
Scincus officinalis (T)	3	0	93	167
Chalcides ocellatus (T)	0	0	94	167
Iguanas				
Amblyrhyncus cristatus				168
(amphibSW)	14	0	86	
Cyclura nubilia (T)	<1	1	98–99	169
CROCODILIA				
Crocodiles				
Crocodylus niloticus				170
(amphibFW)	4–5	0	62–73	
fluid urine	67	13	2	171

Table 2 (Continued)

	% Total nitrogen^a			
Species	**Ammonia-N**	**Urea-N**	**Urate-N**	**Reference**
solid urine	6	0	89	
total	66	5	21	172
C. porosus (E)				173
liquid urine FW	97	3	<<1	
liquid urine SW	77	21	2	
solids FW 2.7% by vol				
solids SW 30.5% by vol				
Caimen				
Caiman crocodilus (T)	53	6	27	172
AVES				
Ratites				
Dromaius vovaehollandiae (H)	6	10	84	174
Modern birds				
Mergus merganser (H)	11–15	5	51–76	175
Coturnix coturnix (H)				176
in ovum near hatch	12	15	73	
Calypte anna (H)				34
water loaded 10°C	53	13	35	
dehydrated 40°C	27	8	65	
Gallus domesticus (O)				177
0% dietary protein	29	3	59	
6% dietary protein	16	1	77	
12% dietary protein	15	1	77	
G. domesticus (male)				178
11% dietary protein				
water *ad lib*	17	7	55	
water restricted	15	2	60	
33% dietary protein				
water *ad lib*	11	9	72	
water restricted	21	11	62	
Cathartes aura (C)	16	8	76	179
fed meat	9	4	87	
Anas platyrhynchos (O)				180
FW for drinking	30	1	52	
saline for drinking	28	2	57	

Note: FW, freshwater; SW, seawater; E, estuarine; T, terrestrial; O, omnivore; H, herbivore; and C, carnivore.

[a] Calculated as % of the sum of ammonia-, urea-, and urate-N when total-N data were not available. Molecular weights of 18, 60.1, and 168.1 and percentages N of 83, 46.7, and 33.3% were used, respectively, in calculations.

different backgrounds. These differences (Table 3) probably reflect individual variation rather than physiological adaptation. During true physiological adaptation, enzyme levels usually increase or decrease several fold. The main physiological stimulus for a switch to increased urate-N excretion by tortoises appears to be water restriction or dehydration.[3] Khalil and Haggag[191] suggested that it is the ratio of arginase to XDH (which they measured as the "oxidase" in the absence of added NAD^+) that dictates this switch. While the levels of urea cycle enzymes are generally lower in tortoise liver than in turtle and other ureotelic liver tissue, the level of arginase activity approaches that of a strict ureotele and, as is typical for such, occurs predominantly in the cytosolic compartment (95.2%).[189] Although arginase is usually not considered to be rate limiting in ureagenesis, it might act to control the availability of ornithine, which might be especially critical in the presence of low levels of CPS-I, such as in tortoise liver. On the other hand, it has been suggested that XDH is limiting in uricogenesis, especially in birds.[114] It is not clear, however, how the state of tissue hydration might affect the level of XDH activity. In any event, tortoises are apppropriately classified

Table 3 Urea Cycle Enzymes in Reptiles and Birds vs. Mammals

Species	µmol/min/g tissue[a] CPS-I	OTC	ASA synthetase[b]	ASA lyase	Arginase	Reference
TESTUDINES						
Turtles						
Trionyx muticus (FW)	—	7.1	1.1	0.9	18.6	160
Chrysemys picta (FW)	10.7	213.3	0.8	—	101.0	181
Kinosternon hirtipes (FW)	2.7	174.0	0.9	—	960.0	182
Pseudomys scripta (FW)	2.1	123.6	0.7	—	877.5	182
Chelydra serpentina (FW)	0.5	10.8	0.3	0.9	30.2	160
Clemmys insculpta (T)	0.3	8.9	0.3	0.5	14.6	160
Terrapene ornata (T)	0.1	7.6	0.4	1.0	200.0	160
T. carolina (T)	0.5	10.1	—	0.8	18.4	160
Tortoises						
Gopherus flavomarginatus (T)	0.2	4.6	0.2	—	46.7	182
G. berlandieri (T)	0.5	15.0	0.3	—	493.3	182
	0.4	24.2	—	—	—	1[c]
freshly collected	0.2	7.3	0	0.2	9.4	160
commercial source	0.6	7.2	0.2	0.6	14.1	
dehydrated (29–37 d)	0.5	9.1	0.6	0.5	15.0	
dehydrated (60 d)	0.7	8.6	0.7	0.6	13.7	
G. agassizii (T)	0.5	64.3	—	—	135.3	1[c]
RHYNCOCEPHALIA						
Tuatara						
Spenodon punctatus (T)	—	129.8	0.8	—	587.3	2
SQUAMATA						
Snakes						
Crotalis horridus (T)	0	—	0	0	—	181
C. molossus (T)	0	0	0	—	1.5	182
Thamnophis sirtalis (T)	—	—	—	—	1.3	181
Lizards						
Egnernia cunninghami (T)	—	0	0	—	1.8	163
Gerrhonotus multicarinatus (T)	0	0	0	—	0.5	181
Ctenosaura pectinata (T)	0	0.6	<0.1	—	30.0	182
Calotes versicolor (T)	—	0	0	—	45.7	183
kidney	—	0	0	—	1.8	
CROCODILIA						
Alligators						
Alligator mississippiensis (amphibFW)	0	0	—	—	0.6	74
kidney	—	—	—	—	>>100.0	c
AVES						
Modern birds						
Gallus domesticus (O)	0	0	0	—	2.2[d]	182
3–10 week chick	0	0	0	0	4.0	184
kidney	0	0.2	0.2	0.2	114.0	
embryo	0	0	0	—	0.2–0.3	185
kidney	—	—	0.02–0.05	—	1.4–4.2	
MAMMALIA						
Rodents						
Rattus norvegicus (O)	10.0	380.0	1.4	2.5	497.0	181
	5.7	162.0	2.1	3.0	1125.0	186

Table 3 (Continued)

Note: FW, freshwater; O, omnivore; T, terrestrial.

[a] For liver tissue except as noted; not corrected for different assay temperatures.

[b] Refers to the combined synthetase-ASA lyase system, except for reference 184, in which ASA synthetase was measured directly.

[c] Unpublished data.

[d] Brown[187] has complied the hepatic arginase levels for sixteen species of birds. The levels range from <<0.1 in pigeon to 3.7 units in kingfisher.

Table 4 Glutamine Synthetase in Reptiles and Birds

Species	µmol/min/g tissue[a]	Reference
TESTUDINES		
Turtles		
Chrysemys picta (FW)	0	188
Pseudomys picta (FW)	0	189
Terrepene carolina (T)	0	189
Tortoises		
Gopherus agassizii (T)	1.6	1
G. berlandieri (T)	3.7	1
SQUAMATA		
Snakes		
Thamnophis sirtalis (T)	12.1	188
T. elegans (T)	10.0	62
CROCODILIA		
Alligators		
Alligator mississippiensis (amphibFW)	7.5–11.2	74
	26.0	190
kidney	0	
muscle	0	
AVES		
Modern birds		
Gallus domesticus (O)	0.7	188
	2.1	137
	25.8	62
Columbia livia (O)	1.2	188
	15.5	62
Anas platyrhynchus (O)	3.5	188
Anser anser (O)	4.8	188

Note: FW, freshwater; O, omnivore; T, terrestrial.

[a] Liver tissue unless otherwise noted; not corrected for different assay temperatures.

as facultative ureouricoteles on the basis not only of their partitioning of excretory-N, but also for the duality in their capacity to detoxify ammonia in liver tissue.

THE TUATARA

The tuatara has a typical reptilian excretory system and is predominantly uricotelic with solid urate deposits being present in its cloacal urine.[192] Like tortoises, it also excretes a significant amount of urea-N. The percentage of urea-N excreted appears to increase with increase dietary protein intake whereas the percentage urate-N decreases (Table 2). Although a CPS activity has not been reported in the tuatara, the remainder of urea cycle enzyme activities are present in liver in levels comparable to those of strict ureoteles[2] (Table 3). Because tortoises possess both major hepatic detoxifying systems, it has been suggested that they represent the primitive cotylosaurian condition prior to the divergence of the mammalian and reptilian-avian lines in which only one or the other system was ultimately expressed.[1] It would appear this primitive duality of function with respect to hepatic ammonia detoxication has also persisted through to the Rhyncocephalia.

THE SQUAMATE REPTILES

With the possible exception of the embryo, exemplified by that of the black snake, *Coluber constrictor*,[164] neither snakes nor lizards excrete a significant amount of excretory-N as urea nor do they, in general, excrete much ammonia-N. All attempts to demonstrate a complete set of urea cycle enzymes in squamate reptiles have yielded negative results (Table 3). CPS-I activity is especially undetectable as is, in most cases, ASA synthetase activity. Low levels of OTC, the ASA synthetase system, and arginase have been found in liver of the lizard, *Ctenosaura pectinata*,[182] but the occurrence of these enzymes in isolated species may reflect a situation similar to that discussed below in birds. Neither OTC nor the ASA synthetase system could be detected in other snakes and lizards, and hepatic arginase levels are generally low, typical of uricoteles. A large number of squamate reptiles are adapted to very arid environments and the excretion of urate as the major nitrogenous end product reflects this adaptation. In their transition to uricotelism, the urea cycle appears to have been rendered nonfunctional by "silencing" the CPS-I gene. A heptic GS is expressed in garter snakes of the genus *Thamnophis* in levels higher than those in the facultatively uricotelic tortoises but comparable to those in birds. This GS is localized in hepatic mitochondria[62] and presumably functions as the sole ammonia-detoxifying enzyme for uricogenesis. There is very little ammonia-N excreted by squamate reptiles, in contrast to the uricotelic archosaurs.

CROCODILES AND ALLIGATORS

Crocodiles, alligators, and the caimen also excrete very little urea-N and are either ammonotelic or uricotelic, depending on water load. The American alligator, *Alligator mississippiensis*, excretes near equal parts ammonia- and urate-N in its normal habitat. When forcibly hydrated, ammonia becomes the main end product; when dehydrated, urate excretion predominates.[193] The presence of solid urate in the cloaca of most uricoteles complicates analyses for the partition of N in the excreta and this is especially true for crocodilians. The fluid urine of the Nile crocodile, *Crocodylus niloticus*, contains mainly ammonia whereas the solid deposits are around 90% urate (Table 2). When only the liquid urine is analyzed, as was the case for the saltwater crocodile, *Crocodylus porosus*, the amount of urate-N is greatly underestimated. As shown in Table 2, in the transition from freshwater to seawater, the percentage solids increases from about 3% of the urine volume to over 30%, again indicating that these crocodiles are facultative uricoteles under conditions of water restriction. CPS-I is not detectable in liver of the American alligator[74] which does contain relatively high levels of GS activity. GS activity is not detectable in either muscle or kidney of the American alligator whereas it is present in these tissues in ureoteles.[190] GS is mitochondrial in alligator liver[74,190] and is expressed in all hepatocytes (Figure 7). Thus, although crocodilians are predominantly ammonotelic when water loaded, they are constitutively uricotelic with respect to their hepatic ammonia metabolism. In keeping with a high rate of renal ammonia excretion and a uricotelic liver metabolism, there are high levels of glutaminase-I (6 to 8 μmol/min/g tissue) in kidney but no detectable activity in liver.[190] Surprisingly, alanine aminotransferase activity is very low in alligator liver relative to its activity in kidney. This suggests the kidney may be the main gluconeogenic organ in this species.[190]

Alligator kidney tissue also has very high arginase activity relative to that in liver (Table 3). Khalil and Haggag[194] first pointed out this relationship between liver and kidney in ureoteles and uricoteles. In ureotelic tortoises (*Testudo leithii*), arginase is high in liver but nondetectable in kidney, whereas in the Nile crocodile, two species of lizards (*Uromastyx aegyptia* and *Chaleides ocellatus*), and two species of snakes (*Psammophis sibilans* and *Zamenis diadema*), the enzyme is present in kidney but undetectable in liver. Arginase does, however, occur rather consistently in avian liver so it is not universally absent from uricotelic liver tissue.[187] In ureoteles, hepatic arginase is predominantly cytosolic. In birds, however, the enzyme is mitochondrial in both liver and kidney[195] although there appears to be some discrepancy with respect to its submitochondrial localization.[195,196] The level of renal arginase in chicks is increased when arginine is added to the diet,[197] so presumably the enzyme functions for arginine catabolism in renal amino acid gluconeogenesis in this species. Renal arginase may therefore be responsible for most, if not all, of the urea present in the excreta of some uricoteles simply because of its catabolic role in gluconeogenesis.

BIRDS

The large, flightless ratites are the most primitive of the extant bird species and include the emus, ostriches, rheas, etc. They are generally found in desert or semidesert environments to which they are well adapted. The emu, *Dromaius novaehortlandiae*, for example, excretes mainly urate as a nitrogenous end product and has extremely well-developed postrenal water- and salt-reabsorptive mechanisms.[174] With the exception of the demonstration of low levels of arginase activity in liver of the ostrich, *Struthis camelus*,[187] however, there appear to have been no studies on the enzymes of hepatic amino acid catabolism or of ureagenesis or uricogenesis in ratites despite their unique evolutionary position.

Most modern birds excrete a considerable amount of ammonia-N, much like their archosaurian relatives, the ammonouricotelic crocodiles and alligators (Table 2). Recent observations on Anna's hummingbird, *Calypte anna*, suggest that at least some bird species should also be classified as facultative ammonouricoteles.[34] At low ambient temperatures, hummingbirds increase their nectar intake due to increased metabolic demands. Because evaporative water rates are low at these temperatures, this increased nectar intake results in water loading. Under these condtions, more than 53% of the N excreted by *C. anna* is as ammonia-N. At higher temperatures, nectar intake is decreased, but evaporative water loss is increased, so the birds are relatively dehydrated. Urate-N then becomes the predominant form excreted. It remains a question as to whether this facultative ammonoteley of hummingbirds represents a general response to water loading by birds or is simply one more of hummingbirds' several unique metabolic adaptations. Water restriction in other birds either has little effect on ammonia-N excretion or else it has the opposite effect to that in hummingbirds. For example, it increases, rather than decreases, ammonia-N excretion by chickens on a high protein diet[178] (Table 2). Water restriction also increases ammonia-N excretion in the pigeon, *C. livia*.[198] The Pekin duck, *Anas platyrhyncus*, shows little difference in its partition of excretory-N when maintained on freshwater or on saline (284 m*M* NaCl, 6 m*M* KCl).[180]

The most consistent physiological stimulus for increasing urate-N excretion by modern birds is an increased dietary protein intake. Increases in urate excretion rates under this condition occur in both omnivorous species such as the domestic chicken[177,178] and in carnivorous species such as the turkey vulture, *Cathartes aura*.[179] When fed, turkey vultures increase the urate-N in their excreta from 76 to 87% of the total-N and decrease ammonia-N from 16 to 9%. After 72 h on a high protein diet lacking carbohydrate, there is an increase in plasma urate levels in chickens of from 50 to 200 μ*M*[199] and presumably, a corresponding increase in urate-N excretion. As discussed above (see Table 1), these increases in rates of urate-N excretion on high protein diets are also reflected by increases in levels of activity of hepatic amino acid catabolic enzymes; of the primary hepatic ammonia-detoxifying enzyme, GS; and of several hepatic enzymes in the urate pathway. In the case of GS, the increased activity in the chicken has been shown to coincide with increased hepatic levels of GS mRNA (Figure 6). The capacity for hepatic amino acid gluconeogenesis develops mainly posthatch in birds.[57] This is also true of the capacity for hepatic uricogenesis which, at hatch, is essentially zero.[57] The ability to respond to increased dietary protein is nevertheless present by 3 d posthatch and continues to develop.[200] Transfer of 3- and 5-d chicks to high protein diets results in increases of from 23 to 44% in hepatic aspartate aminotransferase activity, 751% in alanine aminotransferase activity, 6 to 24% in GDH, 174 to 236% in GS, and 124 to 180% in XDH. In the pigeon, which lacks hepatic XDH[201] and probably utilizes the kidney as the main gluconeogenic organ, high protein diets

(50 vs. 5% control) cause an approximate sixfold increase in renal XDH,[202] coincident with about a fourfold increase in plasma hypoxanthine levels, most likely of hepatic origin. Increased ammonia detoxication by GS present in pigeon liver[62] during high protein catabolic states would result in an increased availability of glutamine for *de novo* purine synthesis. Nucleoside phosphorylase activity increases in pigeon liver under these conditions[202] suggesting that the purine pathway to hypoxanthine is intact in this tissue (Figure 5). Hypoxanthine formed via the purine pathway in liver would then be released into the systemic circulation to be taken up by the kidney for conversion to uric acid for excretion.[203] During the latter process, renal XDH provides reducing eqivalents required for gluconeogenesis. XDH is also absent from the liver of prehatch chickens but is present in the kidney.[204] The prehatch chicken therefore recapitulates the interorgan relationship involved in amino acid gluconeogenesis and uricogenesis in the adult pigeon. As shown in Table 5, the pigeon appears to be unique among uricoteles in its deficiency of hepatic XDH. With the possible exception of the chameleon, *Anolis carolinensis*, most uricoteles have high levels of hepatic, renal, and intestinal XDH.

Although starvation or fasting also represent protein catabolic states, these would appear to be more complex states than those resulting from a simple increase in the availability of exogenous protein. At the enzyme level, the enzymes required for amino acid gluconeogenesis and ammonia detoxification also represent endogenous proteins — now the main source of amino acids. There is generally some increase in the enzymes involved in amino acid catabolism and nitrogen end-product formation during starvation, but the magnitude of change is usually less than that resulting from increased dietary protein intake. Some key enzymes of N metabolism may actually decrease in activity during fasting. This is true, for example, of arginase in mammalian liver. The decrease in arginase activity during starvation involves both its rate of synthesis and its rate of degradation.[206] GS, the key ammonia-detoxifying enzyme in avian liver, also decreases, at least in some instances,

Table 5 Xanthine Dehydrogenase in Reptiles and Birds vs. Mammals

	nmol/min/g tissue[a]		
Species	**Liver**	**Kidney**	**Small intestine**
TESTUDINES			
Turtles			
Pseudomys scripta (FW)	<5	<4	8
Chrysemys picta (FW)	6	14	62
Terrapene carolina (T)	9	34	<8
SQUAMATA			
Lizards			
Anolis carolinensis (T)	<7	—	72
Snakes			
Ancistrodon contortix (T)	540	95	580
Thamnophis ordinatus (T)	320	300	78
Natrix erythrogaster (FW)	790	280	410
AVES			
Modern birds			
Gallus domesticus (O)	390	200	180
Columbo columbo (H)	<20	380	38
MAMMALIA			
Rodents			
Rattus norvegicus (O)	120	59	110

Note: FW, freshwater; H, herbivore; O, omnivore; T, terrestrial.

[a] As NAD^+ reduction.

From Krenitsky, T. A., Tuttle, J. V., Cattau, E. L., Jr., and Wang, P., *Comp. Biochem. Physiol.*, 49B, 687, 1974. With permission.

during fasting (Table 1). In other cases, there may be no change in the level of the enzyme. In the carnivorous murre, *Uria aalge*, alanine and aspartate aminotransferases show little change after a 3-d fast.[60] Although XDH has been reported to increase in one case during a short fast,[137] this has not been found in other cases. For example, total XDH levels were found not to change in either liver or kidney of White Leghorn chicks after an 8-d fast but to decrease in liver of Indian River chicks after a 3-day fast,[207] so there is considerable variation in the response of XDH to starvation. Protein sparing is an important part of starvation and this is illustrated by fasting geese in which the total daily N excretion drops from around 5 g/d to less than 1 g during prolonged fasts in which whole body mass decreases.[175] There is, however, some change in partition of excretory-N during this time in that percentage urate-N increases and that of ammonia-N decreases.

AVIAN AMMONIA DETOXICATION *IN VIVO*

Ammonia infused into the portal vein of chickens is excreted both as ammonia and as urate.[68] Ammonia (as ammonium acetate) infused into the systemic circulation increases hepatic glutamine synthesis[67] as well as urate synthesis and excretion[208] by chickens. This ammonia-stimulated increase in urate excretion is inhibited by the irreversible inhibitor of GS, methionine sulfoxime.[209] While these observations generally substantiate the role of hepatic glutamine synthetase and the urate pathway in ammonia detoxification in the intact bird, quantitative interpretations of the data obtained under these experimental conditions are not possible. As pointed out for the urea cycle,[47] a major complication in this experimental approach is the metabolic acidosis that is imposed by high concentrations of ammonium salts. Under these conditions in mammals, ammonia and other precursors are shunted away from the liver to the kidney for excretion. This process also occurs in birds. For example, during the metabolic acidosis imposed on laying birds during eggshell formation, there is a very marked increase in ammonia excretion along with an acidification of the urine.[210] In addition, massive ammonia loads also cause substrates which are not normally limiting to end-product formation, to become so.

URICOGENESIS IN AVIAN HEPATOCYTES

The operation of the urate pathway *in situ* has been studied in chicken hepatocytes and also other chicken cells. Spleen, thymus, Bursica Fabricius, and kidney cells, like hepatocytes, all incorporate [1-^{14}C]glycine into urate indicating there is an extensive extrahepatic capacity for purine synthesis *de novo* and for the conversion of IMP to urate.[135] Hepatocytes have also been used to show the shunting of IMP to urate in birds.[211] Only about 14% of the IMP formed in chicken hepatocytes enters the adenine nucleotide pool whereas 85% is shunted to urate. Incorporation of glycine into urate by hepatocytes is stimulated by AMP and GMP and slightly inhibited by free adenine and guanine.[135] Inhibition by the latter appears not to be due to PRPP depletion since the levels of this precursor do not decrease appreciably in the presence of the free purines. It has been reported that uricogenesis is not stimulated by amino acids or NH_4Cl in hepatocytes taken from fed chicks but is stimulated by these precursors in hepatocytes from 72-h starved animals. Alanine and asparagine were especially stimulatory.[114] In this study, low NH_4Cl concentrations stimulated uricogenesis but high concentrations were inhibitory. Even with low concentrations of the ammonium salt, the inclusion of alanine resulted in inhibition. These results were interpreted to show that added ammonium salts divert gluconeogenic precursors into glutamate or glutamine.[114] Other studies have shown that the inhibition by ammonium salts can be alleviated by the inclusion of glucose.[129] In these studies, there was little difference in the rate of uricogenesis in hepatocytes from fed chicks vs. those from 48-h starved chicks. If anything, the rate was slightly decreased. Several amino acids were also found to stimulate uricogenesis. These were: glutamine > asparagine > alanine > glycine > histidine, in order of their effectiveness. In yet another study,[137] both alanine and glutamine were found to stimulate uricogenesis with alanine being the most effective in hepatocytes from chicks on high protein diets. Such cells are adapted to a high rate of amino acid gluconeogenesis including marked increases in alanine aminotransferase (Table 1).

AVIAN UREA CYCLE ENZYMES

As in squamate reptiles, CPS-I silencing appears to have been a main consequence of the transition to uricotelism by birds. CPS-II, the cytosolic CPS for pyrimidine biosynthesis, is present in pigeon and chick liver[212,213] but CPS-I activity is not detectable.[184] The functional absence of this enzyme from all tissues has been confirmed by demonstrating the inability of live chicks to incorporate [^{14}C]sodium carbonate into the guanidino-C of arginine.[214] The ASA system is present in chick kidney tissue and intact animals convert [carbamoyl-^{14}C]citrulline to arginine, explaining why citrulline can substitute for the dietary requirement of arginine by chicks.[214] This nutritional function would be a positive selection factor for retention of the ASA synthesis system. ASA lyase also functions as a structural lens protein in birds,[215] another possible reason for its retention. Arginase, which is present in both avian liver and kidney tissues, converts arginine to proline or glutamate for energy metabolism and also provides ornithine for polyamine formation, especially during development.[216] What is somewhat surprising from an evolutionary standpoint is the retention of the OTC gene by birds. The only known function of this enzyme in animal tissues is in conjunction with CPS-I for citrulline biosynthesis. The enzyme is present in kidney mitochondria of chickens[184,217] and, although it is not expressed in liver, pigeon liver mitochondria can take up mammalian pre-OTC and process it to the mature form.[218] The chicken enzyme is very similar to the mammalian enzyme[219] and its expression in kidney tissue is genetically regulated.[220] The selective advantage of retaining high levels of OTC in kidney of certain bird species or strains remains to be established.

REFERENCES

1. **Campbell, J. W., Smith, D. D., Jr., and Vorhaben, J. E.,** Avian and mammalian mitochondrial ammonia-detoxifying systems in tortoise liver, *Science*, 228, 349, 1985.
2. **Hill, L. and Dawbin, W. H.,** Nitrogen excretion in the tuatara, *Sphenodon punctatus*, *Comp. Biochem. Physiol.*, 31, 453, 1969.
3. **Dantzler, W. and Schmidt-Nielsen, B.,** Excretion in fresh-water turtle (*Pseudomys scripta*) and desert tortoise (*Gopherus agassizii*), *Am. J. Physiol.*, 210, 198, 1966.
4. **Lee, M. S. Y.,** The origin of the turtle body plan: bridging a famous morphological gap, *Science*, 261, 1716, 1993.
5. **Benton, M. J.,** Phylogeny of the major tetrapod groups: morphological data and divergence dates, *J. Mol. Evol.*, 30, 409, 1990.
6. **Hedges, S. B., Moberg, K. D., and Maxson, L. R.,** Tetrapod phylogeny inferred from 18S and 28S ribosomal sequences and a review of the evidence for amniote relationships, *Mol. Biol. Evol.*, 7, 607, 1990.
7. **deJong, W. W., Zweers, A., Versteeg, M., Dessauer, H. C., and Goodman, M.,** α-Crystallin A sequence of *Alligator mississippiensis* and the lizard *Tupinambis teguixin*: molecular evolution and reptilian phylogeny, *Mol. Biol. Evol.*, 2, 484, 1985.
8. **Larhammar, D. and Milner, R. J.,** Phylogenetic relationship of birds with crocodiles and mammals as deduced from protein sequences, *Mol. Biol. Evol.*, 6, 693, 1989.
9. **Hedges, S. B.,** Molecular evidence for the origin of birds, *Proc. Natl. Acad. Sci. U.S.A.*, 91, 2621, 1994.
10. **Barreto, C., Albrecht, R. M., Bjorling, D. E., Harner, J. R., and Wilsman, N. J.,** Evidence of the growth plate and the growth of long bones in junivile dinosaurs, *Science*, 262, 2020, 1993.
11. **Hotton, N. H., III, MacLean, P. D., Roth, J. J., and Roth, E. C., Eds.,** *The Ecology and Biology of Mammal-like Reptiles*, Smithsonian Institution Press, Washington, D.C., 1986.
12. **Benton, M. J.,** Dinosaur success in the Triassic: a non-competitive ecological model, *Q. Rev. Biol.*, 58, 29, 1983.
13. **Robinson, P. L.,** A problem of faunal replacement on Permo-Triassic continents, *Palaeontology*, 14, 131, 1971.
14. **Charig, A. J.,** Competition between therapsids and archosaurs during the Triassic period: a review and synthesis of current theories, in *The Structure, Development and Evolution of Reptiles*, Ferguson, M. W. J., Ed., Academic Press, New York, 1984, 597.
15. **Benton, M. J.,** Late Triassic extinctions and the origin of dinosaurs, *Science*, 260, 769, 1993.
16. **May, R. M.,** How many species are there on Earth? *Science*, 241, 1441, 1988.
17. **Tucker, M. E. and Benton, M. J.,** Triassic environments, climate, and reptile evolution, *Palaeogeograph. Palaeoclim. Palaeoecol.*, 40, 361, 1982.
18. **Parrish, J. M., Parrish, J. T., and Ziegler, A. M.,** Permian-Triassic paleogeography and paleoclimatology and implications for therapsid distribution, in *The Ecology and Biology of Mammalian-like Reptiles*, Hotton, N. H., III, MacLean, P. D., Roth, J. J., and Roth, E. C., Eds., Smithsonian Press, Washington D.C., 1986, 109.

19. **King, P. A. and Goldstein, L.,** Renal excretion of nitrogenous compounds in vertebrates, *Renal Physiol., Basil*, 8, 261, 1985.
20. **Lien, Y.-H. H., Pacelli, M. M., and Braun, E. J.,** Characterization of organic osmolytes in avian renal medulla: a nonurea osmotic gradient system, *Am. J. Physiol.*, 264, R1045, 1993.
21. **Dantzler, W. H.,** Renal adaptations of desert vertebrates, *Bioscience*, 32, 108, 1982.
22. **Coulson, R. A. and Hernandez, T.,** Alligator metabolism, Studies on chemical reactions *in vivo*, *Comp. Biochem. Physiol.*, 74B, 143, 1983.
23. **Schmidt-Nielsen, B.,** Excretory mechanisms in the animal kingdom: examples of the principle "The whole is greater than the sum of its parts," *Physiol. Zool.*, 61, 312, 1988.
24. **Skadhauge, E.,** *Osmoregulation in Birds*, Springer-Verlag, Berlin, 1981, chap. 5.
25. **Goldstein, D. L. and Braun, E. J.,** Contributions of the kidneys and intestine to water conservation, and plasma levels of antidiuretic hormone, during dehydration in house sparrows (*Passer domesticus*). *J. Comp. Physiol. B.*, 158, 353, 1988.
26. **Buffenstein, R., Campbell, W. E. and Jarris, J. U. M.,** Identification of crystalline allantoin in the urine of African Cricetidae (Rodentia) and its role in their water economy, *J. Comp. Physiol. B.*, 155, 493, 1985.
27. **Loveridge, J. P.,** Observations on nitrogenous excretion and water relations of *Chiromantis xerampelina* (Amphibia, Anura). *Arnoldia*, 5, 1, 1970.
28. **Drewes, R. C., Hillman, S. S., Putnam, R. W., and Sokol, O. M.,** Water, nitrogen and ion balance in the African treefrog *Chiromantis petersi* Boulenger (Anura: Rhacophoridae), *J. Comp. Physiol. B.*, 116, 257, 1977.
29. **Shoemaker, V. H. and Bickler, P. E.,** Kidney and bladder function in a uricotelic treefrog (*Phyllomedusa sauvagei*), *J. Comp. Physiol. B.*, 133, 211, 1979.
30. **Khalil, F. and Haggag, G.,** Ureotelism and uricotelism in tortoises, *J. Exp. Zool.*, 130, 552, 1955.
31. **Dessauer, H. C.,** Biochemical studies on the lizard, *Anolis carolinensis*, *Proc. Soc. Exp. Biol. Med.*, 80, 742, 1952.
32. **Perschmann, C.,** Über die Bedeutung der Nierenpfortader insbesondere für die Ausscheidung von Harnstoff und Harnsäure bei *Testudo Hermanni* Gml. und *Lacerta viridis* Laur. sowie über die Funktion der Harnblase bei *Lacerta viridis* Laur, *Zool. Beitr. Berlin*, 2, 447, 1956.
33. **Schmidt-Nielsen, B. and Schmidt, D.,** Renal function of *Sphenodon punctum*, *Comp. Biochem. Physiol.*, 44A, 121, 1973.
34. **Preest, M. R. and Beuchat, C. A.,** Nitrogen excretion in hummingbirds: ammonioteley in a bird? *The Physiologist*, 37, A-53, 1994.
35. **Bakker, R. T.,** *The Dinosaur Heresies: New Theories Unlocking the Mystery of the Dinosaurs and Their Extinction*, Marrow, New York, 1986.
36. **Becker, B. F.,** Towards the physiological function of uric acid, *Free Rad. Biol. Med.*, 14, 615, 1993.
37. **Ames, B. N., Cathcart, R., Schwiers, E., and Hochstein, P.,** Uric acid provides an antioxidant defense in humans against oxidant- and radical-caused aging and cancer: a hypothesis, *Proc. Natl. Acad. Sci. U.S.A.*, 78, 6858, 1981.
38. **Nishikimi, M., Koshizaka, T., Ozawa, T., and Yagi, K.,** Occurrence in humans and guinea pigs of the gene related to their missing enzyme L-gulono-γ-lactone oxidase, *Arch. Biochem. Biophys.*, 267, 842, 1988.
39. **Berner, R. A.,** Models for carbon and sulfur cycles and atmospheric oxygen: applications to paleozoic geologic history, *Am. J. Sci.*, 287, 177, 1987.
40. **Berner, R. A. and Landis, G. P.,** Gas bubbles in fossil amber as possible indicators of the major gas composition of ancient air, *Science*, 239, 1406, 1988.
41. **Cerling, T. E.,** Does the gas content of amber reveal the composition of palaeoatmospheres? *Nature*, 339, 695, 1989.
42. **Hilliker, A. J., Duyf, B., Evans, D., and Phillips, J. P.,** Urate-null rosy mutants of *Drosophila melanogaster* are hypersensitive to oxygen stress, *Proc. Natl. Acad. Sci. U.S.A.*, 89, 4343, 1992.
43. **Campbell, J. W.,** Arginine and urea biosynthesis in the land planarian: its significance in biochemical evolution, *Nature*, 208, 1299, 1965.
44. **Krebs, H. A.,** Some aspects of the regulation of fuel suply in omnivorous animals, *Adv. Enz. Regul.*, 10, 397, 1972.
45. **Wu, G., Thompson, J. R., Sedgwick, G. W., and Drury, M.,** Formation of alanine and glutamine in chicken (*Gallus domesticus*) skeletal muscle, *Comp. Biochem. Physiol.*, 93B, 609, 1989.
46. **Felig, P.,** The glucose-alanine cycle, *Metabolism*, 22, 179, 1973.
47. **Meijer, A. J., Lamers, W. H., and Chamuleau, R. A. F. M.,** Nitrogen metabolism and ornithine cycle function, *Physiol. Rev.*, 70, 701, 1990.
48. **Jungermann, K. and Katz, N.,** Metabolism of carbohydrates, in *Regulation of Hepatic Metabolism*, Thurman, R. G., Kauffmann, F. C., and Jungermann, K., Eds., Plenum, NY, 1986, 211.
49. **Carey, B., Cheung, C.-W., Cohen, N. S., Brusilow, S., and Raijman, L.,** Regulation of urea and citrulline synthesis under physiological conditions, *Biochem. J.*, 292, 241, 1993.
50. **Nissim, I., Caffano, C., Nissim, I., and Yudkoff, M.,** Relative role of the glutaminase, glutamate dehydrogenase, and AMP-pathways in hepatic ureagenesis: studies with ^{15}N, *Arch. Biochem. Biophys.*, 292, 393, 1992.
51. **LaNoue, K. F. and Schoolwerth, A. C.,** Metabolite transport in mitochondria, *Annu. Rev. Biochem.*, 48, 871, 1979.
52. **Njogu, R. M. and Hoek, J. B.,** The effect of inhibitors of glutamate transport on the pathway of glutamate oxidation in rat liver mitochondria, *FEBS Lett.*, 152, 222, 1983.

53. **McGivan, J. D. and Chappell, J. B.,** On the metabolic function of glutamate dehydrogenase in rat liver, *FEBS Lett.*, 52, 1, 1975.
54. **Vorhaben, J. E. and Campbell, J. W.,** Submitochondrial localization and function of enzymes of glutamine metabolism in avian liver, *J. Cell Biol.*, 73, 300, 1977.
55. **Gil, M., Cascant, M., Cortés, A., and Bozal, J.,** Intramitochondrial location and some characterization of liver aspartate aminotransferase, *Int. J. Biochem.*, 19, 355, 1987.
56. **Campbell, J. W.,** Excretory nitrogen metabolism, in *Environmental and Metabolic Animal Physiology, Comparative Animal Physiology*, 4th ed., Prosser, C. L., Ed., Wiley-Liss, New York, 1991, 277.
57. **Campbell, J. W.,** Excretory nitrogen metabolism and gluconeogenesis during development of higher vertebrates, *Isr. J. Zool.*, in press, 1994.
58. **Meijer, A. J., Gimpel, J. A., Deleeuw, G., Tischler, M. E., Tager, J. M., and Williamson, J. R.,** Interrelationships between gluconeogenesis and ureagenesis in isolated hepatocytes, *J. Biol. Chem.*, 253, 2308, 1978.
59. **Fujiwara, S., Sakuraba, H., and Noguchi, T.,** Identification of chicken liver mitochondrial alanine 2-oxoglutarate aminotransferase and its response to starvation, *Arch. Biochem. Biophys.*, 310, 497, 1994.
60. **Herzberg, G. R.,** Blood amino acids and the distribution of alanine and aspartate aminotransferases in fed and fasted murres (*Uria aalge*), *Comp. Biochem. Physiol.*, 98A, 563, 1991.
61. **Campbell, J. W. and Vorhaben, J. E.,** Avian mitochondrial glutamine metabolism, *J. Biol. Chem.*, 251, 781, 1976.
62. **Vorhaben, J. E. and Campbell, J. W.,** Glutamine synthetase: a mitochondrial enzyme in uricotlic species, *J. Biol. Chem.*, 247, 2763, 1972.
63. **Campbell, J. W., Vorhaben, J. E., and Smith, D. D., Jr.,** Uricoteley: its nature and origin during the evolution of tetrapod vertebrates, *J. Exp. Zool.*, 243, 349, 1987.
64. **McGivan, J. D.,** Metabolism of glutamine and glutamate in the liver-regulation and physiological significance, in *Glutamine and Glutamate in Mammals*, Kvamme, E., Ed., CRC Press, Boca Raton, FL, 1988, 184.
65. **Soboll, S., Lenzen, C., Rettich, D., Gründel, S., and Ziegler, B.,** Characterization of glutamine uptake in rat liver mitochondria, *Eur. J. Biochem.*, 197, 113, 1991.
66. **Karasawa, Y. and Nakata, C.,** Ammonia absorption from different parts of chicken intesine and its quantitative evaluation *in situ*, *Comp. Biochem. Physiol.*, 84A, 747, 1986.
67. **Karasawa, Y., Kamata, K., Kurihara, Y., Onodera, M., Yajima, H., and Aramaki, T.,** Glutamine production in the liver of the chicken intoxicated with ammonia, *Comp. Biochem. Physiol.*, 67A, 505, 1980.
68. **Karasawa, Y.,** Time course of urinary excretion of intraportal ammonia as uric acid and ammonia in chickens fed low- or high-protein diet, *Comp. Biochem. Physiol.*, 83A, 569, 1986.
69. **Flessner, M. F., Wall, S. M., and Knepper, M. A.,** Permeabilities of rat collecting duct segments to NH_3 and NH_4^+, *Am. J. Physiol.*, 260, F264, 1991.
70. **Almond, M. K., Smith, A., Cohen, R. D., Iles, R. A., and Flynn, G.,** Substrate and pH effects on glutamine synthesis in rat liver, Consequences for acid-base regulation, *Biochem. J.*, 278, 709, 1991.
71. **Häussinger, D.,** Nitrogen metabolism in liver: structural and functional organization and physiological relevance, *Biochem. J.*, 267, 281, 1990.
72. **Häussinger, D. and Gerok, W.,** Metabolism of amino acids and ammonia, in *Regulation of Hepatic Metabolism*, Thurman, R. G., Kauffman, F. C., and Jungermann, K., Eds., Plenum, New York, 1986, 253.
73. **Smith, D. D., Jr. and Campbell, J. W.,** Distribution of glutamine synthetase and carbamoyl-phospahte synthetase I in vertebrate liver, *Proc. Natl. Acad. Sci. U.S.A.*, 85, 160, 1988.
74. **Smith, D. D., Jr. and Campbell, J. W.,** Glutamine synthetase in liver of the American alligator *Alligator mississippiensis*, *Comp. Biochem. Physiol.*, 86B, 755, 1987.
75. **Cohen, N. S., Kyan, F. S., Kyan, S. S., Cheung, C.-W., and Raijman, L.,** The apparent K_m of ammonia for carbamoyl phosphate synthetase (ammonia) in situ, *Biochem. J.*, 229, 205, 1985.
76. **Tate, S. S. and Meister, A.,** Glutamine synthetases of mammalian liver and brain, in *The Enzymes of Glutamine Metabolism*, Prusiner, S. and Stadtman, E. R., Eds., Academic Press, New York, 1973, 77.
77. **Powers-Lee, S. G., Mastico, R. A., and Bendayan, M.,** The interaction of rat liver carbamoyl phosphate synthetase and ornithine transcarbamoylase withe inner mitochondrial membranes, *J. Biol. Chem.*, 262, 15683, 1987.
78. **Cohen, N. S., Cheung, C.-W., and Raijman, L.,** Channeling of extramitochondrial ornithine to matrix ornithine transcarbamylase, *J. Biol. Chem.*, 262, 203, 1987.
79. **Fahien, L. A., Kmiotek, E. H., Woldegiorgis, G., Evenso, M., Shargo, E., and Marshall, M.,** Regulation of aminotransferase-glutamate dehydrogenase interactions by carbamyl phosphate synthetase-I, Mg^{2+} plus leucine *versus* citrate and malate, *J. Biol. Chem.*, 260, 6069, 1985.
80. **Salerno, C., Fasella, P., and Fahien, L. A.,** Interaction of aminotransferases with other metabolically linked enzymes, in *Transaminases*, Christen, P. and Metzler, D. E., Eds., Wiley-Interscience, New York, 1985, 192.
81. **Lusty, C. J.,** Catalytically active monomer and dimer forms of rat liver cabamoyl-phosphate synthetase, *Biochemistry*, 20, 3665, 1981.
82. **Levenberg, B.,** Role of L-glutamine as donor of carbamyl nitrogen for the enzymatic synthesis of citrulline in *Agaricus bisporus*, *J. Biol. Chem.*, 237, 2590, 1962.

83. **Evans, D. R., Bein, K., Guy, H. I., Liu, X., Molina, J. A., and Zimmerman, B. H.,** CAD gene sequence and the domain structure of the mammalian multifunctional protein CAD, *Biochem. Soc. Trans.*, 21, 186, 1993.
84. **Tramell, P. R. and Campbell, J. W.,** Carbamyl phosphate synthesis in a land snail, *Strophocheilus oblongus*, *J. Biol. Chem.*, 245, 6634, 1970.
85. **Tramell, P. R. and Campbell, J. W.,** Carbamyl phosphate synthesis in invertebrates, *Comp. Biochem. Physiol.*, 40B, 395, 1971.
86. **Anderson, P. M.,** Glutamine- and N-acetylglutamate-dependent carbamoyl phosphate synthetase in elasmobranchs, *Science*, 208, 291, 1980.
87. **Campbell, J. W. and Anderson, P. M.,** Evolution of mitochondrial enzyme systems in fish: the mitochondrial synthesis of glutamine and citrulline, in *Biochemistry and Molecular Biology of Fishes, Vol. 1*, Hochachka, P. W. and Mommsen, T. P., Eds., Elsevier, Amsterdam, 1991, 43.
88. **Goldstein, L., Harley-Dewitt, S., and Forster, R. P.,** Activities of ornithine-urea cycle enzymes and of trimethylamine oxidase in the coelacanth *Latimeria Chalumnae*, *Comp. Biochem. Physiol.*, 44B, 357, 1973.
89. **Janssens, P. A. and Cohen, P. P.,** Ornithine-urea cycle enzymes in the African lungfish, *Protopterus aethiopicus*, *Science*, 152, 358, 1966.
90. **Marshall, M., Metzenberg, R. L., and Cohen, P. P.,** Purification of carbamyl phosphate synthetase from frog liver, *J. Biol. Chem.*, 233, 102, 1958.
91. **Bradford, N. M. and McGivan, J. D.,** Evidence for the existence of an ornithine/citrulline antiporter in rat liver mitochondria, *FEBS Lett.*, 113, 294, 1980.
92. **Spolarics, Z. and Bond, J. S.,** Multiple molecular forms of mouse liver arginase, *Arch. Biochem. Biophys.*, 260, 469, 1988.
93. **Cheung, C. W., Cohen, N. S., and Raijman, L.,** Channeling of urea cycle intermediates in situ in permeabilized hepatocytes, *J. Biol. Chem.*, 264, 4038, 1989.
94. **Williamson, J. R., Browing, E. T., and Olson, M. S.,** Interrelations between fatty acid oxidation and the control of gluconeogenesis in perfused rat liver, *Adv. Enzyme Regul.*, 6, 67, 1968.
95. **Kennedy, J.,** Properties and chemistry of urea and related intermediates, in *The Urea Cycle*, Grisolia, S., Báguena, R., and Mayor, F., Eds., Wiley, New York, 1976, 39.
96. **Campbell, J. W.,** Nitrogen excretion, in *Comparative Animal Physiology*, 3rd ed., Prosser, C. L., Ed., Saunders, Philadelphia, PA, 1973, 279.
97. **Orbach, M. J., Sachs, M. S., and Yanofsky, C.,** The *Neurospora crassa arg-2* locus, structure and expression of the gene encoding the small subunit of arginine-specific carbamoyl phosphate synthetase, *J. Biol. Chem.*, 265, 10981, 1990.
98. **Davis, R. H.,** Compartmental and regulating mechanisms in the arginine pathways of *Neurospora crassa* and *Saccharomyces cerevisiae*, *Microbiol. Rev.*, 50, 280, 1986.
99. **Meijer, A. J.,** Regulation of carbamoyl-phosphate synthetase (ammonia) in liver in relation to urea cycle activity, in *Metabolic Regulation*, Ochs, R. S., Hanson, R. W., and Hall, J., Eds., Elsevier, Amsterdam, 1985, 171.
100. **Morris, J. G. and Rogers, Q. R.,** Ammonia intoxication in the near-adult cat as a result of dietary deficiency of arginine, *Science*, 199, 431, 1978.
101. **Silva, S. V. P. S. and Mercer, J. R.,** Effect of protein intake on amino acid catabolism and gluconeognesis by isolated hepatocytes from the cat (*Felis domestica*), *Comp. Biochem. Physiol.*, 80B, 603, 1986.
102. **Morris, J. G. and Rogers, Q. R.,** Arginine: an essential amino acid for the cat, *J. Nutr.*, 108, 1944, 1978.
103. **Cohen, N. S., Cheung, C. W., and Raijman, L.,** The effects of ornithine on mitochondrial carbamyl phosphate synthesis, *J. Biol. Chem.*, 255, 10248, 1980.
104. **Morris, S. M., Jr., Moncman, C. L., Rand, K. D., Dizikes, G. J., Cederbaum, S. D., and O'Brien, W. E.,** Regulation of mRNA levels for five urea cycle enzymes in rat liver by diet, cyclic AMP, and glucocorticoids, *Arch. Biochem. Biophys.*, 256, 343, 1987.
105. **Morris, S. M., Jr.,** Regulation of enzymes of urea and arginine synthesis, *Annu. Rev. Nutr.*, 12, 81, 1992.
106. **Ulbright, C. and Snodgrass, P. J.,** Coordinate induction of the urea cycle enzymes by glucagon and dexamethasone is accomplished by three different mechanisms, *Arch. Biochem. Biophys.*, 301, 237, 1993.
107. **Husson, A., Buquet, C., and Vaillant, R.,** Induction of the five urea-cycle enzymes by glucagon in cultured foetal rat hepatocytes, *Differentiation*, 35, 212, 1987.
108. **Renouf, S., Buquet, C., Fairand, A., Benamar, M., and Husson, A.,** Changes in level of argininosuccinate lyase mRNA during induction by glucagon and cyclic AMP in cultured foetal-rat hepatocytes, *Biochem. J.*, 291, 609, 993.
109. **Ryall, J. C., Quantz, M. A., and Shore, G. C.,** Rat liver and intestinal mucosa differ in the developmental pattern and hormonal regulation of carbamoyl-phosphate synthetase I and ornithine carbamoyl transferase gene expression, *Eur. J. Biochem.*, 156, 453, 1986.
110. **Simic, M. G. and Jovanovic, S. V.,** Antioxidation mechanism of uric acid, *J. Am. Chem. Soc.*, 111, 5778, 1989.
111. **Henderson, J. F. and Paterson, A. R. P.,** *Nucleotide Metabolism, An Introduction*, Academic Press, New York, 1973.
112. **Barankiewicz, J. and Henderson, J. F.,** Ribose-1-phosphate metabolism in Ehrlich ascites tumor cells *in vitro*, *Biochim. Biophys. Acta*, 479, 371, 1977.

113. **Kuninaka, A.,** *Kosa Kagku Shinpojiumu*, in *Enzyme Nomenclature*, 12, 65, 1957, Elsevier, Amsterdam, 1973, 226.
114. **Mapes, J. P. and Krebs, H. A.,** Rate-limiting factors in urate synthesis and gluconeogenesis in avian liver, *Biochem. J.*, 172, 193, 1978.
115. **Lipstein, B., Boer, P., and Sperling, O.,** Regulation of *de novo* purine synthesis in chick liver. Role of phosphoribosylpyrophosphate availability and of salvage purine nucleotide synthesis, *Biochem. Biophys. Acta*, 521, 45, 1978.
116. **Vorhaben, J. E., Smith, D. D., and Campbell, J. W.,** Characterization of glutamine synthetase from avian liver mitochondria, *Int. J. Biochem.*, 14, 747, 1982.
117. **Sonne, J. C., Lin, I., and Buchanan, J. M.,** Biosynthesis of the purine: precursors of the nitrogen atoms of the purine ring, *J. Biol. Chem.*, 220, 369, 1956.
118. **Yoshida, T. and Kikuchi, G.,** Significance of glycine cleavage system in glycine and serine catabolism in avian liver, *Arch. Biochem. Biophys.*, 145, 658, 1971.
119. **Usha, R., Savithri, H. S., and Rao, N. A.,** The primary structure of sheep liver cytosolic serine hydroxymethyltransferase and an analysis of the evolutionary relationships among serine hydroxymethyltgransferases, *Biochim. Biophys. Acta*, 1204, 75, 1994.
120. **Zalkin, H.,** Overview of multienzyme systems in biosynthetic pathways, *Biochem. Soc. Trans.*, 21, 203, 1993.
121. **Mueller, E. J., Meyer, E., Rudolph, J., Davisson, V. J., and Stubbe, J.,** N^5-Carboxyaminoimidazole ribonucleotide: evidence for a new intermediate and two new enzymatic activities in the de novo purine biosynthetic pathway of *Escherichia coli*, *Biochemistry*, 33, 2269, 1994.
122. **Marolewski, A., Smith, J. M., and Benkovic, S. J.,** Cloning and characterization of a new purine biosynthetic enzyme: a non-folate glycinamide ribonucleotide transformylase from *E. coli*, *Biochemistry*, 33, 2531, 1994.
123. **Linder, M. J. and Weigel, P. H.,** A balanced equation for gluconeognesis. Are the textbooks all wet? *Trends Biochem. Sci.*, 11, 200, 1986.
124. **Coolbear, K. P., Herzberg, G. R., and Brosan, J. T.,** Subcellular localization of chicken xanthine dehydrogenase, A possible source of reducing equivalents, *Biochem. J.*, 202, 555, 1982.
125. **Zakrezewska, B. and Jezewska, M. M.,** Comparison of xanthine: NAD^+ oxidoreductase from liver of two uricotelic species: hen *Gallus gallus* and snake *Natrix natrix*, *Comp. Biochem. Physiol.*, 97B, 135, 1990.
126. **Long, S. and Skadhauge, E.,** The role of urinary precipitates in the excretion of electrolytes and urate in the domestic fowl, *J. Exp. Biol.*, 104, 41, 1983.
127. **Dantzler, W. H.,** Excretion in reptiles, in *Nitrogen Metabolism and Excretion*, Walsh, P. J. and Wright, P. A., Eds., CRC Press, Boca Raton, FL, 1995, .
128. **Wu, X., Wakamiya, M., Vaishnav, S., Geske, R., Montgomery, C., Jr., Jones, P., Bradley, A., and Caskey, C. T.,** Hyperuricemia and urate nephropathy in urate oxidase-deficient mice, *Proc. Natl. Acad. Sci. U.S.A.*, 91, 742, 1994.
129. **Burns, R. A. and Buttery, P. J.,** Effect of ammonia and amino acids on urate synthesis by chicken hepatocytes, *Arch. Biochem. Biophys.*, 208, 468, 1981.
130. **Rhee, S. G., Choek, P. B., and Stadtman, E. R.,** Regulation of Escherichia coli glutamine synthetase, *Adv. Enzymol. Rel. Areas Mol. Biol.*, 62, 37, 1989.
131. **Kita, K., Ishizuka, T., Ishijima, S., Sonoda, T., and Tatibana, M.,** A novel 39-kDa phosphoribosylpyrophosphate synthetase-associated protein of rat liver. Cloning, high sequence similarity to the catalytic subunits, and a negative regulatory role, *J. Biol. Chem.*, 269, 8334, 1994.
132. **Zalkin, H.,** The amidotransferases, *Adv. Enzymol. Rel. Areas Mol. Biol.*, 66, 203, 1993.
133. **Rowe, P. B. and Wyngaarden, J. B.,** Glutamine phosphoribosylpyrophosphate amidotransferase. Purification, substructure, amino acid composition, and absorption spectra, *J. Biol. Chem.*, 243, 6373, 1968.
134. **Katunuma, N., Matsuda, Y., and Kuroda, Y.,** Phylogenic aspects of different regulatory mechanisms of glutamine metabolism, *Adv. Enzyme Regul.*, 8, 73, 1970.
135. **Badenoch-Jones, P. and Buttery, P. J.,** The effect of added purines on urate and purine synthesis *de novo* by isolated chicken liver, kidney and lymphoid cells, *Biochem. J.*, 158, 549, 1976.
136. **Itoh, R., Holmes, E. W., and Wyngaarden, J. B.,** Pigeon liver amidophosphoribosyltransferase. Ligand-induced alterations in molecular and kinetic properties, *J. Biol. Chem.*, 251, 2234, 1976.
137. **Wiggins, D., Lund, P., and Krebs, H. A.,** Adaptation of urate synthesis in chicken liver, *Comp. Biochem. Physiol.*, 72B, 565, 1982.
138. **Brand, L. M. and Lowenstein, J. M.,** Effect of diet on adenylosuccinase activity in various organs of rat and chicken, *J. Biol. Chem.*, 253, 6872, 1978.
139. **Campbell, J. W. and Smith, D. D., Jr.,** Metabolic compartmentation of vertebrate glutamine synthetase: putative mitochondrial targeting signal in avian liver glutamine synthetase, *Mol. Biol. Evol.*, 9, 787, 1992.
140. **Smith, D. D., Jr. and Campbell, J. W.,** unpublished, 1988.
141. **Pons, A., García, F. J., Palore, A., and Alemany, M.,** Amino-acid metabolism enzyme activities in liver, intestine and yolk sac membrane of developing domestic fowl, *Arch. Int. Physiol. Biochem.*, 94, 219, 1986.
142. **Whitman, J., Mengi, A., and Goldberg, M.,** Activation and inactivation of renal xanthine dehydrogenase activity by dexamethasone during the prenatal period in chick eggs, *Enzyme*, 35, 70, 1986.

143. **Patejunas, G. and Young, A. P.,** Tissue-specific regulation of avian glutamine synthetase expression during development and in response to glucocorticoid hormones, *Mol. Cell. Biol.*, 7, 1070, 1987.
144. **Gaasbeek-Janzen, J. W., Lamar, W. H., Moorman, A. F. M., DeGraaf, A., Los, A. J., and Charles, R.,** Immunohistochemical localization of carbamoylphosphate sythetase (ammonia) in adult rat liver. Evidence for a heterogeneous distribution, *J. Histochem. Cytochem.*, 32, 557, 1984.
145. **Gebhardt, R. and Mecke, D.,** Heterogeneous distribution of glutamine synthetase among rat parenchymal cells *in situ* and in primary culture, *EMBO J.*, 2, 567, 1983.
146. **de Groot, C. J., ten Voorde, G. H. J., van Andel, R. E., te Kortschot, A., Gassbeck-Janen, J. W., Wilson, R. H., Moorman, A. F. M., Charles, R., and Lamers, W. H.,** Reciprocal regulation of glutamine synthetase and carbamoylphosphate synthetase levels in rat liver. *Biochim. Biophys. Acta*, 908, 231, 1987.
147. **Gebhardt, R., Lindros, K., Lamers, W. H., and Moorman, A. F. M.,** Hepatocellular heterogeneity in ammonia metabolism: demonstration of limited colocalization of carbamoylphosphate synthetase and glutamine synthetase, *Eur. J. Cell Biol.*, 56, 464, 1991.
148. **Kuo, C. F., Paulson, K. E., and Darnell, J. W., Jr.,** Positional and developmental regulation of glutamine synthetase expression in mouse liver, *Mol. Cell. Biol.*, 8, 4966, 1988.
149. **Morris, S. M., Jr., Kepka, D. M., Sweeney, W. E., Jr., and Avner, E. D.,** Abundance of mRNAs encoding urea cycle enzymes in fetal and neonatal mouse liver, *Arch. Biochem. Biophys.*, 269, 175, 1989.
150. **Campbell, J. W., Vorhaben, J. E., and Smith, D. D., Jr.,** Heptic ammonia metabolism in a uricotelic treefrog *Phyllomedusa sauvagei*, *Am. J. Physiol.*, 246, R805, 1984.
151. **Shoemaker, V. H. and McClanahan, L. L.,** Enzymatic correlates and ontogeny of uricotelism in tree frogs of the genus *Phyllomedusa*, *J. Exp. Zool.*, 220, 163, 1983.
152. **Bloomfield, R. A., Letter, A. A., and Wilson, R. P.,** The effect of glycine on ammonia intoxication and uric acid biosynthesis in the avian species, *Arch. Biochem. Biophys.*, 129, 196, 1969.
153. **Laud, P. and Campbell, J. W.,** Genetic basis for tissue isozymes of glutamine synthetase in elasmobranchs, *J. Mol. Evol.*, 39, 93, 1994.
154. **Zimmermann, H.,** 5′-Nucleotidase: molecular structure and functional aspects, *Biochem. J.*, 285, 345, 1992
155. **Mora, M., Manzanero, J. C., and Bozl, J.,** Chicken live purine nucleoside phosphorylase activity dependency of the redox state of sulfhydryl groups, *Comp. Biochem. Physiol.*, 87B, 403, 1987.
156. **Ropp, P. A. and Trant, T. W.,** Purine nucleoside phosphorylase. Allosteric regulation of a dissociating enzyme, *J. Biol. Chem.*, 266, 7682, 1991.
157. **Mora, M., Manzanero, J. C., and Bozal, J.,** Pigeon liver purine nucleoside phosphorylase characterization: kinetic behavior and sulfhydryl groups, *Comp. Biochem. Physiol.*, 88B, 1143, 1987.
158. **Nishino, T., Nishino, T., Schopfer, L. M., and Massey, V.,** The reactivity of chicken liver xanthine dehydrogenase with molecular oxygen, *J. Biol. Chem.*, 264, 2518, 1989.
159. **Varela-Echavarría, A., Montes de Oca-Luna, R., and Barrera-Saldaña, H. A.,** Uricase protein sequence: conserved during vertebrate evolution but absent in humans, *FASEB J.*, 2, 3093, 1988.
160. **Baze, W. B. and Horne, F. R.,** Ureogenesis in Chelonia, *Comp. Biochem. Physiol.*, 34, 91, 1970.
161. **Moyle, V.,** Nitrogenous excretion in chelonia reptiles, *Biochem. J.*, 44, 581, 1949.
162. **Packard, G. C. and Packard, M. J.,** Patterns of nitrogen excretion by embryonic softshell turtles (*Trionys spiniferus*) developing in cleidoic eggs, *Science*, 221, 1049, 1983.
163. **Rogers, L. J.,** The nitrogen excretion of *Chelodina longicollis* under conditions of hydration and dehydration, *Comp. Biochem. Physiol.*, 18, 249, 1966.
164. **Clark, H.,** Metabolism of the black snake embryo. I. Nitrogen excretion, *J. Exp. Biol.*, 30, 492, 1953.
165. **Khalil, F.,** Excretion in reptiles. II. Nitrogen constituents of urinary concretions of the oviparous snake *Zamensis diadema*, *J. Biol. Chem.*, 172, 101, 1948.
166. **Khalil, F.,** Excretion in reptiles. III. Nitrogen constituents of urinary concretions of the viviparous snake *Eryx thebacious*, *J. Biol. Chem.*, 172, 105, 1948.
167. **Khalil, F.,** Excretion in reptiles. IV. Nitrogenous constituents in excreta of lizards, *J. Biol. Chem.*, 189, 443, 1951.
168. **Shoemaker, V. H. and Nagy, K. A.,** Osmoregulation in the Galápagos marine iguana, Amblyrhynchus cristatus, *Physiol. Zool.*, 57, 291, 1984.
169. **Christian, K. A. and Torregrosa, D.,** Effect of diet on nitrogenous wastes of the iguana, *Cyclura nubila*, *Comp. Biochem. Physiol.*, 85A, 761, 1986.
170. **Lawrence, P. and Loveridge, J. P.,** Carbon, nitrogen, and energy balance of young crocodiles (Crocodylus niloticus) fed meat, *Physiol. Zool.*, 61, 351, 1988.
171. **Khalil, F. and Haggag, G.,** Nitrogenous excretion in crocodiles, *J. Exp. Biol.*, 35, 552, 1958.
172. **Cragg, M. M., Balinsky, J. B. and Baldwin, E.,** A comparative study of nitrogen excretion in some amphibia and reptiles, *Comp. Biochem. Physiol.*, 3, 227, 961.
173. **Grigg, G. C.,** Plasma homeostasis and cloacal urine composition in *Crocodylus porosus* caught along a salinity gradient, *J. Comp. Physiol. B*, 144, 261, 1981.

174. **Dawson, T. J., Maloney, S. K., and Skadhauge, E.,** The role of the kidney in electrolyte and nitrogen excretion in a large flightless bird, the emu, during different osmotic regmines, including dehydration and nesting, *J. Comp. Physiol. B*, 161, 165, 1991.
175. **Robin, J.-P., Cherel, Y., Girard, H., Géloen, A., and Maho, Y. L.,** Uric acid and urea in relation to protein catabolism in long-term fasting geese, *J. Comp. Physiol. B*, 157, 491, 1987.
176. **Packard, G. C. and Packard, M. J.,** Nitrogen excretion by embryos of a gallinaceous bird and a reconsideration of the evolutionary origin of uricoteley, *Can. J. Zool.*, 64, 691, 1986.
177. **Tasaki, I. and Okumura, J.,** Effect of protein level of diet on nitrogen excretion in fowls, *J. Nutr.*, 83, 34, 1964.
178. **McNabb, F. M. A. and McNabb, R. A.,** Proportions of ammonia, urea, urate and total nitrogen in avian urine and quantitative methods for their analysis on a single urine sample, *Poult. Sci.*, 54, 1498, 1975.
179. **McNabb, F. M. A., McNabb, R. A., Prather, I. D., Conner, R. N., and Adkisson, C. S.,** Nitrogen excretion by turkey vultures, *Condor*, 82, 219, 1980.
180. **Stewart, D. J., Holmes, W. H., and Fletcher, G.,** The renal excretion of nitrogenous compounds by the duck (*Anas platyrhynchos*) maintained on freshwater and hypertonic saline, *J. Exp. Biol.*, 50, 527, 1969.
181. **Brown, G. W., Jun. and Cohen, P. P.,** Comparative biochemistry of urea synthesis. 3. Activities of urea-cycle enzymes in various higher and lower vertebrates, *Biochem. J.*, 75, 82, 1960.
182. **Mora, J., Martuscelli, J., Ortiz-Pineada, J., and Soberón, G.,** The regulation of urea-biosynthesis enzymes in vertebrates, *Biochem. J.*, 96, 28, 1965.
183. **Baby, T. G. and Reddy, S. R. R.,** Arginine metabolism in purinotelic animals. I. Characterization of arginase and absence of other urea cycle enzymes in the lizard, *Calotes versicolog*, *J. Comp. Physiol.*, 140, 261, 1980.
184. **Tamir, H. and Ratner, S.,** Enzymes of arginine metabolism in chicks, *Arch. Biochem. Biophys.*, 102, 249, 1963.
185. **Emmanuel, B. and Gilanpour, H.,** Studies on enzymes of nitrogen metabolism, and nitrogen end products in the developing chick (*Gallus domesticus*) embryo, *Comp. Biochem. Physiol.*, 61B, 287, 1978.
186. **Snodgrass, P. J. and Lin, R. C.,** Induction of urea cycle enzymes of rat liver by amino acids, *J. Nutr.*, 111, 586, 1981.
187. **Brown, G. W., Jr.,** Studies in comparative biochemistry and evolution. 1. Avian liver arginase, *Arch. Biochem. Biophys.*, 114, 184, 1966.
188. **Wu, C.,** Glutamine synthetase. I. A comparative study of its distribution in animals and its inhibition by DL-allo-δ-hydroxylysine, *Comp. Biochem. Physiol.*, 8, 335, 1963.
189. **Smith, D. D., Jr. and Campbell, J. W.,** unpublished, 1986.
190. **Lemieux, G., Craan, A. G., Quenneville, A., Lemieux, C., Berkofsky, J., and Lewis, V. S.,** Metabolic machinery of the alligator kidney, *Am. J. Physiol.*, 247, F686, 1984.
191. **Khalil, F. and Haggag, G.,** Ureotelism and uricotelism in tortoise in relation to arginase and xanthine oxidase activities, *Z. Vergl. Physiol.*, 43, 278, 1960.
192. **Schmidt-Nielsen, B. and Schmidt, D.,** Renal function of *Sphenodon punctatum*, *Comp. Biochem. Physiol.*, 44A, 121, 1973.
193. **Coulson, R. A. and Hernandez, T.,** Nitrogen metabolism and excretion in the living reptile, in *Comparative Biochemistry of Nitrogen Metabolism, Vol. 2, The Vertebrates*, Campbell, J. W., Ed., Academic Press, London, 639, 1970.
194. **Khalil, F. and Haggag, G.,** Xanthine and arginase in the liver and kideny of reptiles, *Z. Vergl. Physiol.*, 43, 269, 1960.
195. **Grazi, E., Magri, E., and Balboni, G.,** On the control of arginine metabolism in chick kidney and liver, *Eur. J. Biochem.*, 60, 431, 1975.
196. **Kadowaki, H., Israel, H. W., and Neshim, M. C.,** Intracellular localization of arginase in chick kidney, *Biochim. Biophys. Acta*, 437, 158, 1976.
197. **Kadowaki, H. and Neishim, M. C.,** An assay for arginase in chicken kidney, *Comp. Biochem. Physiol.*, 61B, 281, 1978.
198. **McNabb, F. M. A., McNabb, R. A., and Ward, J. M., Jr.,** The effects of dietary protein content on water requirements and ammonia excretion in pigeons, *Columbia livia*, *Comp. Biochem. Physiol.*, 43A, 181–185.
199. **Evans, R. M. and Scholz, R. W.,** Metabolic responses of chicks during adaptation to a high protein, "carbohydrate-free" diet, *J. Nutr.*, 101, 1127, 1971.
200. **Pons, A., García, F. J., Palou, A., and Alemany, M.,** Effect of starvation and a protein diet on the amino acid metabolism enzyme activities of the organs of domestic fowl hatchlings, *Comp. Biochem. Physiol.*, 85B, 275, 1986.
201. **Benzinger, T. and Krebs, H. A.,** Über die Harnsäuresynthese in Vogelorganismus, *Klin. Wschr.*, 31, 1206, 1933.
202. **Nagahara, N., Nishino, T., Kanisawa, M., and Tsushima, K.,** Effect of dietary protein on purine nucleoside phosphorylase and xanthine dehydrogenase activities of liver and kidney in chicken and pigeon, *Comp. Biochem. Physiol.*, 88B, 589, 1987.
203. **Tinker, D. A., Brosan, J. T., and Hertzberg, G. R.,** Interorgan metabolism of amino acids, glucose, lactate, glycerol and uric acid in the domestic fowl (*Gallus domesticus*), *Biochem. J.*, 240, 829, 1986.
204. **Strittmatter, C. F.,** Studies on avian xanthine dehydrogenase. Properties and patterns of appearance during development, *J. Biol. Chem.*, 240, 2557, 1965.

205. **Krenitsky, T. A., Tuttle, J. V., Cattau, E. L., Jr., and Wang, P.,** A comparison of the distribution and electron acceptor specificities of xanthine oxidase and aldehyde oxidase, *Comp. Biochem. Physiol.*, 49B, 687, 1974.
206. **Schimke, R. R., Brown, M. B., and Smallman, E. T.,** Turnover of rat liver arginase, *Ann. NY Acad. Sci.*, 102, 587, 1963.
207. **Scholz, R. W.,** Comparative studies on liver and kidney xanthine dehydrogenase in two breeds of domestic chicks (*Gallus domesticus*) during prolonged starvation, *Comp. Biochem. Physiol.*, 36, 503, 1970.
208. **Karasawa, Y. and Tasaki, I.,** Effect on uric acid synhtesis of the infusion of various levels of ammonia acetate or glutamine in chickens, *J. Nutr.*, 103, 1727, 1973.
209. **Karasawa, Y. and Tasaki, I.,** Stimulatory effect of ammonia and glutamine on uric acid synthesis in the chicken (*Gallus domesticus*) treated with methionine sulfoxime, *Comp. Biochem. Physiol.*, 53B, 307, 1976.
210. **Simkiss, K.,** Sex differences in the acid-base balance of adult and immature fowl, *Comp. Biochem. Physiol.*, 34, 777, 1970.
211. **Spychala, J. and van den Berge, G.,** Adenine nucleotide metabolism in isolated chicken hepatocytes, *Biochem. J.*, 242, 551, 1987.
212. **Hager, S. E. and Jones, M. E.,** A glutamine-dependent enzyme for the synthesis of carbamyl phosphate for pyrimidine biosynthesis in fetal rat liver, *J. Biol. Chem.*, 242, 5674, 1967.
213. **Rochovansky, O.,** Pyrimidine biosynthesis in the chick, *Arch Biochem. Biophys.*, 138, 574, 1970.
214. **Tamir, H. and Ratner, S.,** A study of ornithine, citrulline and arginine synthesis in growing chickes, *Arch. Biochem. Biophys.*, 102, 259, 1963.
215. **Piatigorsky, J. and Wistow, G.,** The recruitment of crystallins: new function precedes gene duplication, *Science*, 252, 1078, 1991.
216. **Pegg, A. E.,** Recent advances in the biochemistry of polyamines in eukaryotes, *Biochem. J.*, 234, 249, 1986.
217. **Tsuji, S. and Kanazawa, S.,** Chicken ornithine transcarbamylase: its unexpected expression, *Biochem. Genet.*, 25, 259, 1987.
218. **Takiguchi, M., Miura, S., Mori, M., and Tatibana, M.,** Transport of proteins into mitochondria: a high conservation of precursor uptake and procesing system, *Comp. Biochem. Physiol.*, 75B, 277, 1983.
219. **Tsuji, S.,** Chicken ornithine transcarbamylase: purification and some properties, *J. Biochem. (Tokyo)*, 94, 1307, 1983.
220. **Tsuji, S., Nakagawa, K., and Fukushima, T.,** Genetically controlled quantitative variation of ornithine transcarbamylase in the chick kidney, *Biochem. Genet.*, 21, 857, 1983.

Chapter 11

Nitrogen Excretion in Reptiles

William H. Dantzler

CONTENTS

INTRODUCTION

The major end products of nitrogen metabolism excreted in the urine of reptiles are ammonia, urea, and urate.* The percent of urinary nitrogen excreted in each form varies among orders and even among species within orders (Table 1). This variation appears to relate in part to habitat — whether aquatic or mesic or xeric terrestrial — and the requirements for conserving water (Table 1), although metabolic factors, especially acid–base regulation, may also determine the amount of each formed (see Chapter 10). The importance of each of these compounds for the excretion of nitrogen, as well as for acid–base regulation and other requirements, will be considered further in the discussion of the process by which each is excreted. However, this chapter will be concerned primarily with the renal excretory processes themselves.

As in other vertebrate classes, the excretion of amino acids in the urine of reptiles is limited because of their importance for basic metabolic processes. Therefore, their excretion probably plays a small role in eliminating excess nitrogen. In fact, the only quantitative information on this possibility in reptiles — a preliminary measurement on cloacal urine from olive sea snakes (*Aipysurus laevis*) — indicates that nitrogen excreted as amino acids only makes up about 2.4% of the total urinary nitrogen.[1] There is enough variation in the degree of excretion and the process of excretion for amino acids in reptiles compared with other vertebrate classes, however, to make this limited route of nitrogen excretion worthy of consideration in this chapter.

* The term "urate" in this chapter refers to all forms that contain the urate anion (uric acid, uric acid dihydrate, and monobasic urate salts).

0-8493-8411-7/95/$0.00+$.50

Table 1 Percent of Urinary Nitrogen as Ammonia, Urea, and Urates

Order and Species	Ammonia	Urea	Urates	Reference
Testudinea				
Wholly aquatic	20–25	20–25	5	54
Semiaquatic	6–15	40–60	5	55
Pseudemys scripta (Fresh-water turtle)	4–44	45–95	1–24	14
Wholly terrestrial				
Mesic	6	30	7	54
Xeric	5	10–20	50–60	55
Gopherus agassizii (Desert tortoise)	3–8	15–50	20–50	14
Rhynchocephalia	3–4	10–28	65–80	56
Squamata				
Sauria	insignificant –?highly significant	0–8	90	16, 20, 57, 58
Ophidia	insignificant –?highly significant	0–2	98	20, 59, 60
Crocodilia	25	0–5	70	61

Modified from Dantzler, W. H., *Structure and Function of the Kidney,* Karger, Basel, 1989, 143. With permission.

RENAL AMMONIA EXCRETION

SIGNIFICANCE

Ammonia is a principal end product of nitrogen metabolism in aquatic and semiaquatic chelonian reptiles (e.g., freshwater turtle, *Pseudemys scripta*) and crocodilians (e.g., alligators, *Alligator mississippiensis*) (Table 1). In the semiaquatic and aquatic chelonians, nitrogen excretion is about equally partitioned between ammonia and urea, although urea usually tends to predominate (Table 1). In alligators, about 75% of the excreted nitrogen is in the form of urates; about 25% is in the form of ammonia (Table 1). Moreover, the absolute amount of ammonia excreted daily by a well-hydrated alligator on a standard meat diet (about 2.4 mEq kg^{-1} d^{-1}) is greater than the amount recorded for any other vertebrate.[2,3] In any case, the primary function of ammonia excretion in these reptile groups appears to be the elimination of nitrogen.

The reason why ammonia has evolved as a major excretory end product of nitrogen metabolism in these reptiles and not in others is not clear. As a major route for removal of nitrogen, ammonia excretion has often been correlated with life in an aqueous environment where this highly toxic substance can be readily diluted in a large volume of water. Certainly, the terrestrial chelonian reptiles do not excrete significant amounts of their excess nitrogen by this route (Table 1). On the other hand, the completely aquatic ophidian reptiles also do not rid excess nitrogen as ammonia (Table 1).

In most vertebrates, the excretion of ammonia is involved in acid–base balance, the production and excretion of ammonia increasing with acidosis. There is little evidence that this happens in reptiles, but the question has not been examined closely. Indeed, in reptiles only one preliminary study on water snakes (*Nerodia* spp.) has explored the response of ammonia excretion to an acid load. It suggested that there might be some increase in ammonia excretion during an acute acid load.[4]

RENAL PRODUCTION

As in most other vertebrates, almost no ammonia is present in the systemic blood of those reptiles studied.[4] Apparently, all ammonia excreted in the urine is produced in renal tubule cells and secreted into the tubule lumen. Although the complete process of renal production of ammonia has not been studied in reptiles, it is assumed to be similar to that in mammals. In mammals (see Chapter 13), glutamine is taken up by the renal tubule cells and enters the mitochondria where it is deaminated by phosphate-dependent glutaminase I to form glutamate and NH_4^+. Glutamate is then deaminated via glutamate dehydrogenase to form α-ketoglutarate and NH_4^+. α-Ketoglutarate is then further metabolized in a process that consumes protons and produces bicarbonate.[5-7] Among reptiles, portions of these reactions have been studied only in alligators.[8] This study showed that glutaminase I is present in renal tissue with sufficient activity for ammonia production, whereas it is absent from the liver. Glutamine synthetase is found only in the liver, however. Therefore, the authors of these studies suggested that the liver might be a source of glutamine for ammonia production by the kidney. This study also showed that both glutamate dehydrogenase and alanine aminotransferase activities are high in the kidney but low in the liver and that isolated tubule fragments can produce ammonia from glutamine and alanine *in vitro*.[8] No information is available, however, as to which segments of the renal tubules generate ammonia in alligators or any other species of reptile. Nevertheless, like mammals, it seems likely that the proximal tubule is involved in reptiles.

Of particular interest with regard to ammonia production, dehydration in alligators apparently leads to decreased production of ammonia and increased production of urate.[3] In other words, the fraction of nitrogen excreted as urate increases and that excreted as ammonia decreases. The control of this process is unknown, but King and Goldstein[3] suggest that it occurs primarily through regulation of renal ammonia production. They speculate that, with low urine flow during dehydration, accumulation of ammonia in the renal tissues could drive the reversible deamination reactions in the direction of amino acid formation. They also suggest that a decrease in renal blood flow during dehydration could decrease delivery of amino acids to the renal tissue and thereby reduce the quantity of substrate available for ammoniagenesis.[3]

RENAL SECRETION

Ammonia produced in the renal tubule cells is secreted into the tubule lumen for excretion (Table 2). There is no direct information on the mechanism of secretion involved in reptiles. It could involve nonionic diffusion of free ammonia (NH_3) across the luminal membrane and trapping in the lumen as ammonium ion (NH_4^+) in those species in which the luminal fluid is significantly acidified. Although such a process occurs in mammalian nephrons, some direct secretion of NH_4^+ also appears to occur.[9] Direct secretion of NH_4^+ by the nephrons appears particularly likely in alligators because they always produce an alkaline urine containing large amounts of bicarbonate as the anion accompanying NH_4^+.[10] This secretory process may involve substitution of NH_4^+ for H^+ on a Na^+/H^+ exchanger at the brush border of the proximal tubule cells, as has been shown *in vitro* in brush-border membrane vesicles (BBMVs) from mammalian kidneys;[11] a functional Na^+/H^+ exchanger exists in BBMVs from reptilian kidneys.[12] This process may be important for ammonia secretion by the renal tubules of those chelonian species in which ammonia excretion accounts for a high proportion of the nitrogen excretion.

It is still possible that nonionic diffusion and diffusion trapping could play some role in ammonia secretion by the renal tubules in alligators despite the normally alkaline urine. Since alligators maintain a rather low blood pH (about 7.1), the initial filtrate may have a pH below that of the cells

Table 2 Apparent Normal Direction of Net Tubular Transport

	Net reabsorption	Net secretion
Ammonia		+
Urea	+	+
Urates		+
Amino Acids	+	+

Modified from Dantzler, W. H., *Structure and Function of the Kidney*, Karger, Basel, 1989, 143. With permission.

and some ammonia may enter the lumen by nonionic diffusion in the first part of the proximal tubule.[13] However, neither direct secretion of NH_4^+ nor diffusion of NH_3 with trapping in the lumen as NH_4^+ in the early proximal tubule would explain the lack of diffusion of NH_3 back into the cells further along the nephron where the tubule fluid must be highly alkaline. If ammonia secretion involves only a carrier-mediated process (e.g., on the Na^+/H^+ exchanger) in alligator nephrons, then the luminal membrane of these nephrons may simply have a much lower passive permeability for NH_3 than the luminal membrane of nephrons in other species.[10] This intriguing problem of secretion of large amounts of ammonia into an alkaline tubular urine has yet to be examined directly.

RENAL UREA EXCRETION

SIGNIFICANCE AND PROCESS OF RENAL EXCRETION

Among reptiles, urea is a major excretory end product of nitrogen metabolism only in chelonians from mesic terrestrial, semiaquatic, and aquatic habitats, and in rhynchocephalians (Table 1). It is freely filtered by the glomeruli of all those species studied. In most species (including those chelonians in which it forms a major excretory end product), urea undergoes a variable amount of passive reabsorption along the nephrons (Table 2).[4,10,14] In general, the extent of the passive reabsorption varies with hydration and urine flow, being more obvious in dehydrated animals with low urine flows.[4,14,15] Net tubular secretion of urea has also been observed during clearance studies in some lizard species (*Lacerta viridans* and *Sceloporus cyanogenys*) and in the only living representative of the rhynchocephalians (*Sphenodon punctatus*) (Table 2).[4,14,16] Such net secretion is observed only during extreme diuresis, however. Therefore, even in these species, the amount of urea excreted appears to be determined primarily by filtration and net tubular reabsorption.

RENAL URATE EXCRETION

SIGNIFICANCE

Urates are the primary form in which nitrogen is excreted in the urine of all reptiles except chelonians from aquatic, semiaquatic, and mesic terrestrial habitats (Table 1). Even in these chelonians, however, in which urea is the major excretory end product of nitrogen metabolism, urates are an important means of excreting excess nitrogen (Table 1). Although urates can be excreted with far less water than urea, there is no convincing evidence of a shift in production from urea to urate in these animals when they undergo dehydration (see Chapter 10). As noted above, an apparent shift from ammonia to urate in alligators during dehydration may largely reflect a reduction in ammonia production by the kidneys without any change in urate production.[3]

Although urate excretion is certainly a way to remove nitrogen from the body with relatively little water, it may also play a role in acid-base balance. Studies with water snakes (*Nerodia sipedon*) indicate that net secretion of urate by the renal tubules (see below for discussion of the secretory process) increases during an acute metabolic alkalosis (produced by a bicarbonate infusion).[17] Although the mechanism involved in this response is not understood, it is apparently related to an increase in the pH of the blood, not the urine, because administration of a carbonic anhydrase inhibitor, which made the urine alkaline with little or no change in the blood pH, had no effect on urate secretion by the renal tubules. Because the urine is quite alkaline during a metabolic alkalosis, almost all of the urate eliminated by the increase in tubular secretion will exist as a salt. Therefore, this increased secretion of urate will eliminate base in the same fashion as an increased excretion of bicarbonate. These data suggest that the role of urate in acid–base balance in reptiles deserves further study.

Renal excretion of urate also may play a role in the excretion of inorganic ions and water and, therefore, in the function of other aspects of the renal system. Inorganic cations, especially sodium and potassium (but sometimes calcium, magnesium, or ammonium), are found with urate precipitates in the ureteral urine of reptiles[18-20] (see below for forms of urate excretion). The predominant inorganic cations may be determined by dietary intake and, possibly, the requirements of acid–base balance in each species. In any case, inorganic cations complexed with urate precipitates in the urine are excreted without contributing to the osmotic pressure of the final urine. For this reason, the absence in reptiles of any significant ability to produce urine hyperosmotic to the plasma does not necessarily give an accurate indication of the ability of these animals to excrete inorganic ions while conserving water.

PROCESS OF RENAL EXCRETION

Urate is freely filtered by the glomeruli of all those species of reptiles studied[21,22] and undergoes net secretion by the renal tubules (Table 2).[18] Almost all the information on the net secretory process in reptiles comes from studies with isolated, perfused renal tubules of garter snakes (*Thamnophis* spp.), although some also comes from studies with BBMVs from this same species.

Transepithelial Transport Process

In these isolated, perfused nephrons, net transepithelial transport from bath to lumen (net secretion) against a concentration gradient occurs equally throughout the proximal tubule,[23] but not in the distal tubule.[24] Although there is no evidence of net urate reabsorption anywhere along these nephrons, a substantial passive unidirectional reabsorptive flux can occur in the proximal tubules.[23] The apparent permeability of the entire tubule epithelium calculated from measurements of this passive reabsorptive flux (about 2.4×10^{-5} cm s^{-1}) is about four times the value (about 0.6×10^{-5} cm s^{-1}) calculated from the independently measured peritubular and luminal membrane permeabilities shown in Figure 1A.[24] This difference between the measured and calculated epithelial permeabilities suggests that most of this reabsorptive flux moves between the cells (Figure 1A). This passive reabsorptive flux varies inversely with perfusion rate and is reflected in the corresponding observation that net secretion varies directly with perfusion rate.[23]

The net transepithelial secretory flux saturates, yielding an apparent K_t of about 150 μ*M* and an apparent J_{max} of about 140 fmol min^{-1} mm^{-1}. This apparent K_t is well below the normal plasma urate concentration of about 400 to 500 μ*M* in these animals. This finding suggests that, in these animals, the tubular secretory mechanism is normally saturated (or close to saturated) and that changes in the plasma urate concentration have little effect on net urate secretion. Because of this, changes in the rate of flow of filtrate along the lumen of the proximal tubule and, therefore, changes in the passive reabsorptive flux may be especially important in determining the rate of net secretion of urate and, in turn, the final rate of net excretion of urate.

The transepithelial secretory system for urate in snakes (and presumably other reptiles) is apparently quite separate from that for para-aminohippurate (PAH) and the other organic anions that share that system. First, the anatomical locations of the transport systems differ. As already noted, net transepithelial urate secretion against a concentration gradient occurs throughout the length of the proximal tubule, whereas net transepithelial PAH secretion against a concentration gradient occurs only in the distal portion of the proximal tubule.[23,25] Second, high concentrations of PAH in the medium bathing isolated, perfused tubules do not inhibit net urate secretion and high concentrations of urate do not inhibit net PAH secretion.[18]

Transport at Basolateral and Luminal Membranes

During the process of net transepithelial secretion in isolated, perfused snake proximal renal tubules, the concentration of urate in the tubule lumen is greater than that in the bathing medium and the concentration in the cells is greater than that in ether the lumen or the bathing medium (Figure 1A).[23] Because the inside of the cells is about 60 mV negative relative to the bath or lumen,[26] urate is an anion at physiological pH, and there is no evidence of intracellular binding of urate, these concentration measurements indicate that urate is transported into the cells against an electrochemical gradient at the basolateral membrane and moves from the cells into the tubule lumen down an electrochemical gradient (Figure 1A). However, the apparent permeability of the basolateral mem-

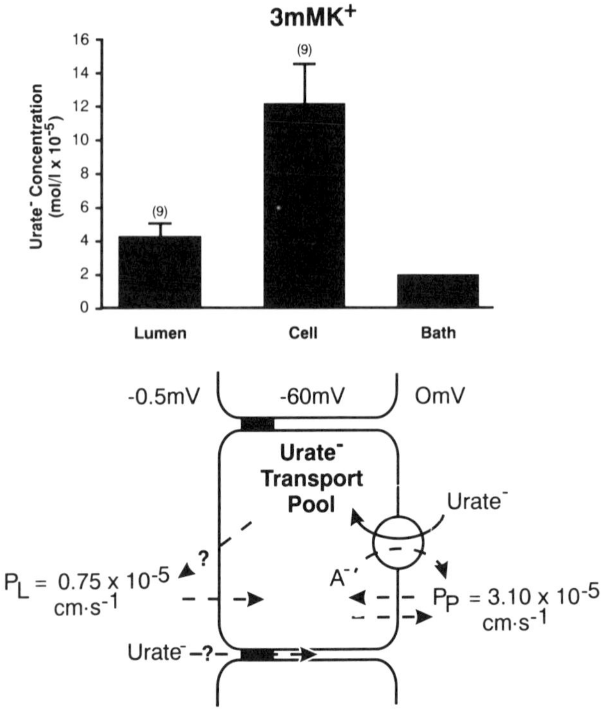

Figure 1A Model of net transepithelial urate secretion based on studies with garter snake (*Thamnophis* spp.) renal tubules and BBMV in the presence of control (3 m*M*) [K^+]. The concentrations of urate in the bath, cell water, and lumen during the net transepithelial flux are shown in the upper part of the figure. Solid circle with solid arrow indicates either primary or secondary active transport. Broken arrows indicate passive transport down an electrochemical gradient. For countertransport, the solid arrow indicates substrate moving against an electrochemical gradient and the broken arrow indicates substrate moving down an electrochemical gradient. Broken arrows with question marks indicate possible passive movements. A^- indicates anion of unspecified nature. Apparent permeabilities for luminal (P_L) and peritubular (P_P) membranes are also shown.

brane to urate (determined from efflux of radiolabeled urate from tubules with oil-filled lumens) is about four times the apparent permeability of the luminal membrane (determined from flux of labeled urate into the lumen; Figure 1A).[23,24] This appears inefficient if urate transported into the cells at the basolateral side is to move into the lumen and not back into the bathing medium. However, net secretion does occur. Moreover, the high permeability of the basolateral membrane to urate may help explain the failure of nonperfused tubules to maintain as high an intracellular concentration as perfused tubules. Perfused tubules maintain an intracellular urate concentration about three times that of nonperfused tubules with the same urate concentration in the bathing medium.[23] It appears that transport into the cells against an electrochemical gradient at the peritubular side depends, in part, on the presence of glomerular filtrate (or a similar artificial perfusate) in the lumen.[23] This apparent dependence of urate transport into the cells across the basolateral membrane on the presence of filtrate in the lumen, the relatively high permeability of the basolateral membrane to urate, and the apparent large passive reabsorptive flux of urate between the cells may be important physiologically because reptilian nephrons filter intermittently. These three features of urate transport may help prevent accumulation of insoluble urates in the lumens or cells of nephrons that are not filtering.

Basolateral Transport

All evidence indicates that the primary step in net urate secretion is transport into the cells against an electrochemical gradient at the basolateral membrane,[23,24,27] but the mechanism involved in this

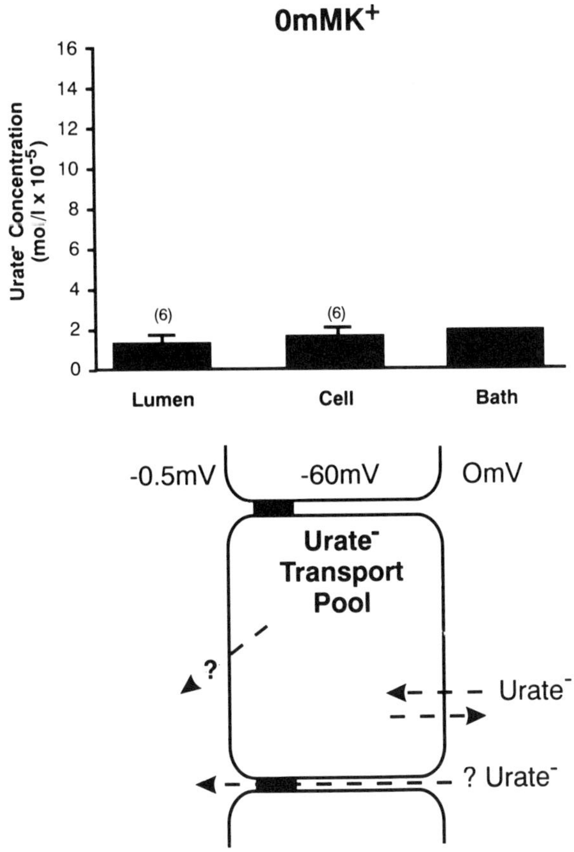

Figure 1B Model of net transepithelial urate secretion based on studies with isolated garter snake (*Thamnophis* spp.) in absence of potassium. Symbols as in Figure 1A.

process is not understood. Net urate secretion is completely unaffected by the replacement of all sodium in the bathing medium and perfusate with choline.[27] Therefore, transport into the cells across the basolateral membrane does not directly or indirectly depend on sodium. The transport step is reversibly inhibited by the disulfonic stilbene SITS, suggesting that it involves anion countertransport (Figure 1A).[28] However, the anion that might be exchanged for urate is unknown. The lack of dependency of urate transport on sodium makes it unlikely that this transport step involves countertransport for dicarboxylates, such as α-ketoglutarate, which is countertransported for PAH. Moreover, a preliminary series of studies on the effects of preloading the tubules with numerous dicarboxylates and tricarboxylates provided no evidence of significant stimulation of urate transport like the stimulation of PAH transport observed when tubules are preloaded with α-ketoglutarate.[29,30]

Net urate secretion, like net PAH secretion, however, depends on potassium. Removal of potassium from the medium bathing isolated, perfused tubules reversibly depresses net urate secretion.[27] During maximum depression of net urate secretion in the absence of potassium, the concentration of urate in the cells is lower than that in the bathing medium and the concentration in the lumen is lower than that in either the cells or the bathing medium (Figure 1B).[27] These observations suggest that, in the absence of potassium, active transport into the cells across the basolateral membrane is eliminated and that urate moves from bath to lumen by a purely passive process (Figure 1B). In the absence of potassium, the transepithelial permeability calculated from the urate flux from bath to lumen assuming purely passive movement (1.9×10^{-5} cm s^{-1}) is essentially the same as the apparent permeability determined from the passive urate flux from lumen to bath (see above). This determination indicates that removal of potassium does not influence the transepithelial permeability to urate and further supports the concept that its removal eliminates the active basolateral transport step for urate. This determination of the transepithelial permeability also suggests that much of the apparent passive movement from bath to lumen occurs between the cells (Figure 1B). The mechanism involved in this potassium effect is unknown. Although removal of potassium certainly depresses Na^+-K^+-ATPase activity, which might well be involved in urate transport across the basolateral membrane, the lack of effect of sodium removal and the inability to demonstrate a countertransport step for an anion whose transport might be sodium coupled makes it difficult to know how the energy from this enzyme might be coupled to urate transport.

Finally, net urate secretion is not influenced by the replacement of almost all calcium in the perfusate and bathing medium or by the presence of lanthanum in the bathing medium, perfusate, or both.[31] Thus, the basolateral transport step for urate, unlike that for PAH, is unaffected by calcium entry into the cells.

Luminal Transport

Although movement of urate from cells to lumen during net transepithelial secretion is clearly down an electrochemical gradient, there is no evidence that movement across the luminal membrane is carrier mediated. Neither probenecid, SITS, nor unlabeled urate in the lumen has any effect on the movement of radiolabeled urate across this membrane during the determination of net secretion in isolated, perfused renal tubules.[28,32] There is also no evidence of mediated urate transport by BBMV from these snake tubules.[33] These observations are all compatible with the movement of urate across the luminal membrane by a purely passive process (Figure 1), but they certainly do not prove it. Moreover, the normal flux across this membrane is sufficiently important to deserve additional study. Recent studies on secretion of the organic anion fluorescein by teleost renal tubules indicate that it may be sequestered in intracellular vesicles during the transepithelial transport process.[34] Perhaps, urate too is sequestered in vesicles during transepithelial transport and enters the lumen by some form of exocytosis.

Forms of Urate Excreted

Uric acid itself has a very low solubility in water (0.384 m*M*), but it is very unlikely that a significant amount of urate is excreted as uric acid in the aqueous phase of the urine in reptiles. Because the pK_a of the monobasic salt of urate is about 5.7,[35] the pK_a of the dibasic salt of urate is about 10.2,[36] and the pH of the ureteral urine of reptiles is only slightly acid (e.g., turtles, snakes, and carnivorous lizards),[14,17,20] or even alkaline (e.g., crocodilians and herbivorous lizards),[20,34] most of the urate in solution in the urine must exist as the monobasic salt.[18] The cations associated with the dissolved urate will probably be primarily sodium and potassium, the predominant one determined by whether the animal is a carnivore or herbivore. In crocodilians, in which large amounts of ammonia are excreted in an alkaline urine (see above), urate may exist in the aqueous phase of the urine primarily as the ammonium salt.

Although the aqueous solubilities of the monobasic urate salts are modest (6.76 m*M* for sodium urate; 12.06 m*M* for potassium urate),[35] they are much greater than that of uric acid itself. In some reptiles (e.g., turtles), the concentrations of sodium, potassium, ammonium, and urate in the aqueous phase of the ureteral urine are low enough that the salts of urate can exist in true solution.[14] For some lizards (*Dipsosaurus dorsalis*) and snakes (*Nerodia* spp.) studied, however, the concentrations of sodium, potassium, and urate in the ureteral urine are above those at which urate salts can exist in true solution.[17,18,38] These findings indicate that some of the urate in the aqueous phase of the ureteral urine in these animals must exist in a colloidal state.

The ability of urates to form lyophobic colloids is well known.[29,40-42] As long as the colloid is stable, the small size of the colloidal particles prevents the formation of large precipitated masses.[38]

In some reptiles, however, the concentrations of urates in the aqueous phase of the urine exceed the stability limits for lyophobic colloidal urate.[38,40] In these cases, the lyophobic urate colloidal particles may be converted to a lyophilic state by absorption to lyophilic macromolecules.[40] Such lyophilic colloids can exist in the aqueous phase of the urine at concentrations above the stability limits for lyophobic colloids. Minnich[38] obtained preliminary evidence for a urate-binding protein in the ureteral urine of a lizard (*D. dorsalis*) that may serve as an appropriate lyophilic macromolecule for the purpose of converting lyophobic urate colloids to a lyophilic form, but much more work is needed on this problem. It appears that the liquid phase of the urine in many reptiles contains, in addition to urate salts in solution, lyophilic colloids and lyophobic colloids. This permits much more urate to remain in the liquid phase of the urine than would otherwise be the case thereby minimizing some of the possible deleterious effects of excessive urate precipitates in the urinary system.

As pointed out above, however, urates are also present as precipitates in the urine of reptiles. These precipitates occur not only in the bladder or cloaca, where urate from the aqueous phase may be precipitated, but also in the ureters and collecting ducts.[43,44] In the ureteral urine, these precipitates are formed exclusively as spheres (2 to 10 μm in diameter), whereas in the cloacal urine of the same animals, these precipitates formed as uric acid crystals as well as spheres.[44] The cause of the sphere formation is unknown, but Minnich[38] suggests that the lyophilic macromolecules could be important and that the precipitates could occur by "salting out" of lyophilic colloidal urates. This process could be initiated by the presence of an increased concentration of electrolytes in the renal tubules.

Although the chemical form of these urate precipitates is not yet fully characterized, they can contain appreciable amounts of inorganic cations. Of the bound cations, potassium is the most abundant cation in the precipitates from boid snakes, most lizards (except monitor lizards, *Varanus* spp.), and the desert tortoise (*Gopherus agassizii*).[20] It is particularly abundant in herbivores. Sodium and ammonium occur to a lesser extent. In precipitates from monitor lizards and most snakes (except boids), ammonium is more prevalent than sodium or potassium.[20] In these animals, the cations occur in an appropriate ratio to suggest that monobasic urate salts are present in the urate precipitate.[20] In urate precipitates excreted by chuckwallas (*Sauromalus obesus*), however, potassium is present in concentrations above that expected from the inclusion only of monobasic urate salts.[45] Whatever the exact chemical form of these precipitates, and this certainly needs to be determined, the inclusion of inorganic cations in the precipitates permits their excretion with little water.

Reabsorption of water in the cloaca or bladder of many reptiles apparently enhances the formation of additional urate precipitates from the colloidal material present in the aqueous phase of the urine.[17,18,38] Acidification of urine in the bladder or cloaca could also initiate precipitation of urates. This process could convert urate salts to less soluble uric acid thereby allowing elimination of more acid and reabsorption of additional sodium. Even without such a conversion from urate to uric acid, acidification would also promote precipitation of urates because flocculation limits for lyophobic colloidal urates decrease with decreasing pH.[40] Acidification of urine has been demonstrated within the bladders of turtles and the cloacas of some lizards.[46-48] Thus, this process may well be significant in the formation of urate precipitates. Finally, there in some evidence to suggest that urate itself might be secreted by the cloaca of one species of lizard (*D. dorsalis*),[38] but this requires additional study.

RENAL AMINO ACID EXCRETION

SIGNIFICANCE AND PROCESS OF RENAL EXCRETION

Amino acid excretion plays a minimal role in urinary excretion of excess nitrogen (see above). However, clearance studies on a few species of reptiles (olive sea snake, *A. laevis*; garter snakes, *T. sirtalis* and *T. radix*; Burmese python, *Python molurus*; and alligator, *A. mississippiensis*) reveal some distinct differences in overall tubular transport, both between species and between reptiles and other vertebrates.[1,49,50] In all reptile species studied, amino acids are freely filtered. In alligators, as in mammals and most other nonmammalian vertebrates, reabsorption of filtered amino acids (determined with cloacal urine collections) appears to be 99% complete.[50] In snakes, however, tubular transport (determined with ureteral urine collections) appears to be quite different. In garter and sea snakes, both net reabsorption and net secretion of the sulfonated amino acid taurine, other endogenous β-amino acids (β-alanine and β-aminoisobutyric acid) and an endogenous analogue of β-amino acids (L-cysteic acid) can occur (Table 2).[1,49] Although the physiological significance of net secretion of these amino acids in snakes is unknown, the direction of net transport most clearly correlates with the filtered load of amino acid.[49] For each of these amino acids, net secretion is consistently observed when the filtered load is low and net reabsorption, when the filtered load is high.[49] The occurrence of both net reabsorption and net secretion of taurine under physiological conditions in snakes is particularly interesting because the kidneys of mammals, and most other vertebrates, exhibit only net reabsorption of taurine, and the kidneys of marine teleost fishes exhibit only net secretion of taurine.[3]

In snakes, all amino acids, other than those noted above, undergo net tubular reabsorption. In contrast to the situation in alligators, however, as well as in most other vertebrates, such reabsorption is often far from complete.[1,49] For example, in both garter and sea snakes, only about 80% of the filtered histidine, an amino acid with a high plasma concentration in all reptiles studied, is reabsorbed.[1,49] Reabsorption of filtered serine is often this low too.[1] Moreover, in garter snakes adapted to cold, only about 60% of the filtered phenylalanine and glutamic acid is reabsorbed.[1]

The apparent difference between alligators and snakes in the extent to which filtered amino acids are reabsorbed may relate to the site of urine collection because cloacal urine was collected from alligators and ureteral urine from snakes.[1,49,50] Reabsorption, or possibly even secretion, of amino acids might occur in the cloaca as well as in the renal tubules.[1] This possibility is supported by preliminary data on sea snakes in which there was less filtered amino acid in cloacal urine than in ureteral urine collected from the same animal.[1] However, confirmation of this hypothesis requires more detailed and direct studies.

Although no direct studies of the tubule sites of amino acid transport have been made in reptiles, the proximal tubule must be involved to a major extent. In snakes, net reabsorption and net secretion probably both occur in this region. However, analysis of the transport rates as a function of filtered load suggests that net secretion occurs distal to net reabsorption in this segment.[49] Confirmation of this assumption requires more direct studies.

Transport at Luminal and Basolateral Membranes

The mechanism of amino acid transport by reptilian renal tubules has been studied only with brush-border and basolateral membrane vesicles (BBMV and BLMV) from garter snake (*Thamnophis* spp.) kidneys.[51-53] Moreover, these studies have involved only taurine. The transport of this amino acid is of particular significance, however, because it can undergo both net reabsorption and net secretion in snake kidneys.

Luminal Transport

Taurine is transported into snake BBMV by an electrogenic Na^+-Cl^--cotransport system (Figure 2).[51] Although taurine can be transported into the BBMV against a concentration gradient by coupling to an inwardly directed Na^+-gradient alone, its transport also can be stimulated by a Cl^--gradient alone, and Cl^- is required as the anion for optimal stimulation.[51] The transport system is apparently shared by other β-amino acids.[51] It is not affected by a proton gradient, at least at physiological pH, and it is not inhibited by amiloride or furosemide.[51] These data indicate that transport does not involve a H^+-driven cotransport system, a Na^+/H^+ exchange system, or a Na^+-K^+-$2Cl^-$ cotransport system.

Kinetic analysis of the Na^+-Cl^--taurine cotransport system in these snake BBMV indicated that it involves two independent saturable systems: (1) a high affinity, low capacity system ($K_t \cong 70\ \mu M$; $J_{max} \cong 22$ pmol mg protein^{-1}) that functions at taurine concentrations below about 0.8 m*M*; and (2) a low affinity, high capacity system ($K_t \cong 1.1$ m*M*; $V_{max} \cong 250$ pmol mg protein^{-1}) that functions

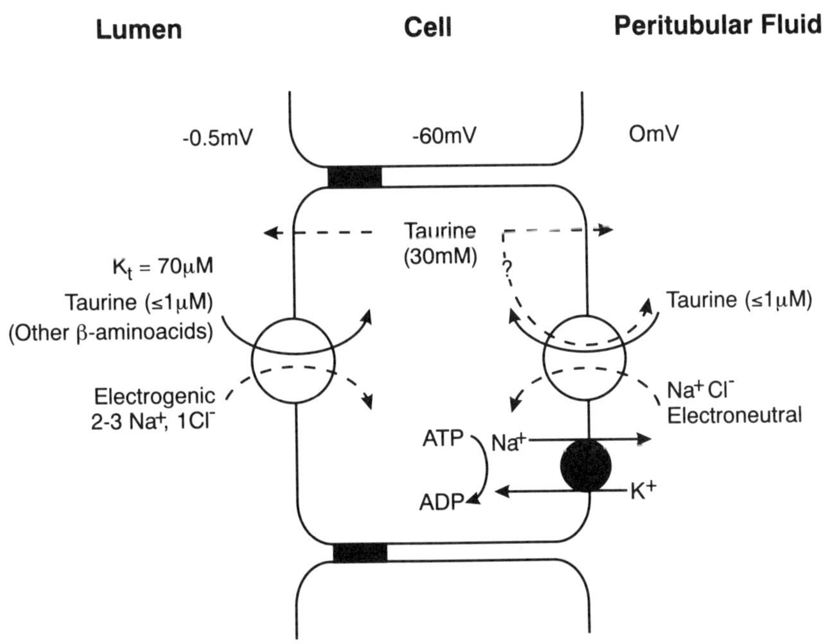

Figure 2 Model of transepithelial taurine transport in garter snake (*Thamnophis* spp.) renal tubules. Model is largely based on studies with BBMV and BLMV[51-53] with some modifications from model suggested in Benyyajati and Bay.[53] See text for further information.

at taurine concentrations above about 0.8 m*M*.[52] The initial studies with snake BBMV, showing that transport was electrogenic, suggested that more than one Na^+ was involved with one Cl^- and one taurine in the transport process.[51] Further kinetic analysis of the interaction of Na^+ with the high affinity system in these snake BBMV suggests that 2 to 3 Na^+ ions are involved in transport of each taurine molecule.[52] This coupling of more than one Na^+ ion with the transport of each taurine molecule suggests that this system would only work in the reabsorptive direction and would not play a role in the net secretory process that often predominates in these animals. In this regard, the apparent passive movement of taurine out of these snake BBMV was extremely slow, despite little evidence of intracellular binding.[51] Moreover, this efflux was not enhanced by the presence of either external taurine or NaCl.[53]

These two systems may both be involved in the reabsorptive process for taurine in the intact renal tubule. The normal plasma taurine concentration in garter snakes is about 1 μ*M*, far below the apparent K_t (≅70 μ*M*) of the high-affinity system. This suggests that under normal physiological circumstances, the high-affinity, low-capacity system functions to reclaim taurine from the filtrate.[52] During cold acclimation (e.g., winter hibernation), however, the plasma taurine concentration ranges from 0.5 to 1.8 m*M* (1), suggesting that the low-affinity, high-capacity system also may function under these circumstances to reabsorb taurine (52).[52]

Basolateral Transport

Because taurine undergoes net secretion as well as net reabsorption by snake renal tubules, the transport system at the basolateral membrane is of particular interest. Taurine is transported into snake BLMV by an electroneutral Na^+-Cl^--cotransport system of substantially smaller magnitude than that in BBMV (Figure 2).[53] Although Cl^- is the required anion for optimal taurine transport, a Cl^--gradient alone does not stimulate taurine transport, as it does in BBMV.[53] Apparently, Cl^- acts as a catalytic activator of the transporter in the basolateral membrane.[53] Moreover, analysis of the stoichiometry of Na^+-Cl^--taurine cotransport system in these snake BLMV indicated that, unlike the BBMV system, it involves only 1 Na^+ ion for each taurine transported (Figure 2).[53] Also, efflux of taurine from BLMV was much more rapid than from BBMV and was enhanced in an added fashion by both taurine and NaCl in the external medium. These last observations suggest that at least part of the exit of taurine across the basolateral membrane occurs via a mediated pathway (possibly, in part, by reversal of the uptake pathway) (Figure 2).[53] The evidence for a high-affinity electrogenic, 2-3Na^+-1Cl^--1taurine cotransporter at the luminal membrane and an electroneutral 1Na^+-1Cl^--1taurine cotransporter at the basolateral membrane combined with the evidence for more marked efflux across the basolateral than the luminal membrane suggest a process in the intact tubules that favors net reabsorption.[53] The process by which net secretion occurs has yet to be determined.

ACKNOWLEDGMENTS

Personal work reported in this chapter was supported in part by National Science Foundation Research Grant DCB-9001985.

REFERENCES

1. **Benyajati, S. and Dantzler, W. H.,** Plasma levels and renal handling of endogenous amino acids in snakes: A comparative study, *J. Exp. Zool.*, 238, 17, 1986.
2. **Coulson, R. A. and Hernandez, T.,** Nitrogen metabolism and excretion in the living reptile, in *Comparative Biochemistry of Nitrogen Metabolism. Vol. 2, The Vertebrates*, Campbell, J. W., Ed., Academic Press, New York, 1970, 640.
3. **King, P. A. and Goldstein, L.,** Renal excretion of nitrogenous compounds in vertebrates. *Renal Physiol.*, 8, 249, 1985.
4. **Dantzler, W. H.,** Renal function (with special emhasis on nitrogen excretion), in *Biology of the Reptilia, Vol. 5, Physiology A*, Gans, C. G. and Dawson, W. R., Eds., Academic Press, London, 1976, chap. 9.
5. **Goldstein, L.,** Ammonia production and excretion in the mammalian kidney, *Int. Rev. Physiol.*, 11, 284, 1976.
6. **Silbernagl, S. and Scheller, D.,** Formation and excretion of NH_3-NH_4. New aspects of an old problem, *Klin. Wochenschr.*, 64, 862, 1986.
7. **Tannen, R. L.,** Ammonia metabolism, *Am. J. Physiol.*, 235, F265, 1978.
8. **Lemieux, G., Craan, A. G., Quenneville, A., Lemieux, C., Berkovsky, J., and Lewis, V.,** Metabolic machinery of the alligator kidney, *Am. J. Physiol.*, 247, F686, 1984.
9. **Knepper, M. A., Packer, R., and Good, D. W.,** Ammonium transport in the kidney, *Physiol. Rev.*, 69, 179, 1989.
10. **Dantzler, W. H.,** *Comparative Physiology of the Vertebrate Kidney*, Springer-Verlag, Berlin, Heidelberg, New York, 1988, chap. 6.
11. **Good, D. W. and Knepper, M. A.,** Ammonia transport in the mammalian kidney, *Am. J. Physiol.*, 248, F459, 1985.
12. **Dantzler, W. H., Wright, S. H., and Brokl, O. H.,** Tetraethylammonium transport by snake renal brush-border membrane vesicles, *Pflügers Arch.*, 418, 325, 1991.
13. **Lemieux, G., Berkovsky, J., Quenneville, A., and Lemieux, G.,** Net tubular secretion of bicarbonate by the alligator kidney, antimammalian response to acetazolamide, *Kidney Int.*, 28, 760, 1985.
14. **Dantzler, W. H. and Schmidt-Nielsen, B.,** Excretion in fresh-water turtle (*Pseudemyss scripta*) and desert tortoise (*Gopherus agassizii*), *Am. J. Physiol.*, 210, 198, 1966.
15. **Schmidt-Nielsen, B. and Schmidt, D.,** Renal function of *Sphenodon punctatum*, *Comp. Biochem. Physiol.*, 44, 121, 1973.
16. **Perschmann, C.,** Uber die Bedeutung der Nierenpfortader insbesondere für die Ausschiedung von Harnstoff und Harnsäure bei *Testudo hermanni* Gml. und *Lacerta viridis* Laur. sowie über die Funktion der Harnblase bei *Lacerta viridis* Laur., *Zool. Beitr.*, 2, 447, 1956.
17. **Dantzler, W. H.,** Effect of metabolic alkalosis and acidosis on tubular urate secretion in water snakes (*Natrix sipedon*), *Am. J. Physiol.*, 215, 747, 1968.
18. **Dantzler, W. H.,** Urate excretion in nonmammalian vertebrates, in *Handbook of Experimental Pharmacology, Vol. 51, Uric Acid*, Kelley, W. N. and Weiner, I. M., Eds., Springer-Verlag, Berlin, Heidelberg, NY, 1978, chap. 8.
19. **Dantzler, W. H.,** Comparative aspects of renal function, in *The Kidney: Physiology and Pathophysiology*, Seldin, D. W. and Giebisch, G., Eds., Raven Press, NY, 1985, chap. 17.
20. **Minnich, J. E.,** Excretion of urate salts in reptiles, *Comp. Biochem. Physiol.*, 41, 535, 1972.
21. **Bordley, J., III and Richards, A. N.,** Quantitative studies of the composition of glomerular urine. VIII. The concentration of uric acid in glomerular urine of snakes and frogs, determined by an ultramicroadaptation of Folin's method, *J. Biol. Chem.*, 101, 193, 1933.
22. **Simkin, P. A.,** Uric acid binding to serum proteins: differencs among species, *Proc. Soc. Exp. Biol. (NY)*, 139, 604, 1972.

23. **Dantzler, W. H.,** Characteristics of urate transport by isolated perfused snake proximal renal tubules, *Am. J. Physiol.*, 224, 445, 1973.
24. **Dantzler, W. H.,** Comparison of uric acid and PAH transport by isolated, perfused snake renal tubules, in *Amino Acid Transport and Uric Acid Transport*, Silbernagl, S., Lang, F. and Greger, R., Eds., Theime, Stuttgart, 1976, 169.
25. **Dantzler, W. H.,** PAH transport by snake proximal renal tubules: differences from urate transport, *Am. J. Physiol.*, 226, 634, 1974.
26. **Kim, Y. K. and Dantzler, W. H.,** Relation of membrane potential to basolateral tetraethylammonium (TEA) transport in isolated snake proximal renal tubules. *FASEB J.*, 8, A836, 1994.
27. **Randle, H. W. and Dantzler, W. H.,** Effects of K^+ and Na^+ on urate transport by isolated perfused snake renal tubules, *Am. J. Physiol.*, 225, 1206, 1973.
28. **Mukherjee, S. K. and Dantzler, W. H.,** Effects of SITS on urate transport by isolated, perfused snake renal tubules, *Pflügers Arch.*, 403, 35, 1985.
29. **Chatsudthipong, V. and Dantzler, W. H.,** PAH-α-KG countertransport stimulates PAH uptake and net secretion in isolated snake renal tubules, *Am. J. Physiol.*, 261, F858, 1991.
30. **Kim, Y. K. and Dantzler, W. H.,** unpublished data, 1993.
31. **Dantzler, W. H. and Brokl, O. H.,** Lack of efect of low [Ca^{2+}], La^{3+}, and pyrazinoate on urate transport by isolated, perfused snake renal tubules, *Pflügers Arch.*, 401, 262, 1984.
32. **Dantzler, W. H. and Bentley, S. K.,** Effects of inhibitors in lumen on PAH and urate transport by isolated renal tubules, *Am. J. Physiol.*, 236, F379, 1979.
33. **Benyajati, S. and Dantzler, W. H.,** Enzymatic and transport characteristics of isolated snake renal brush-border membranes, *Am. J. Physiol.*, 255, R52, 1988.
34. **Miller, D. S., Stewart, D. E., and Pritchard, J. B.,** Intracellular compartmentation of organic anions within renal cells, *Am. J. Physiol.*, 264, R882, 1993.
35. **Gudzent, F.,** Physikalisch-chemische Untersuchungen über das Verhalten der harnsauren Salze in Lösungen, *Z. Physiol. Chem.*, 56, 150, 1908.
36. **Atsmon, A., DeVries, A., and Frank, M.,** *Uric Acid Lithiasis*, Elsevier, New York, 1963.
37. **Coulson, R. A. and Hernandez, T.,** *Biochemistry of the Alligator*, Louisiana State University Press, Baton Rouge, 1964.
38. **Minnich, J. E.,** Adaptations in the reptilian excretory system for excreting insoluble urates, *Isr. J. Med. Sci.*, 12, 854, 1976.
39. **Bechhold, H. and Ziegler, J.,** Vorstudien über Gicht. III., *Biochem. Z.*, 64, 471, 1914.
40. **Porter, P.,** Physico-chemical factors involved in urate calculus formation. II. Colloidal flocculation, *Res. Vet. Sci.*, 4, 592, 1963.
41. **Schade, H. and Boden, E.,** Uber die Anamolie der Harnsäurelöslichkeit (Kolloides Harnsäure), *Z. Physiol. Chem.*, 83, 347, 1913.
42. **Young, E. G. and Musgrave, F. F.,** The formation and decomposition of urate gels, *Biochem. J.*, 26, 941, 1932.
43. **Dantzler, W. H.,** unpublished data, 1970.
44. **Minnich, J. E. and Piehl, P. A.,** Spherical precipitates in the urine of reptiles, *Comp. Biochem. Physiol. A*, 41, 551, 1972.
45. **Nagy, K. A.,** Nitrogen requirement and its relation to dietary water and potassium content in the lizard *Sauromalus obesus*, *J. Comp. Physiol.*, 104, 49, 1975.
46. **Green, B.,** Water and electrolyte balance of the sand goanna *Varanus gouldii* (Gray), Ph.D. thesis, University of Adelaide, Australia, 1969.
47. **Minnich, J. E.,** Water and electrolyte balance of the desert iguana, *Dipsosaurus dorsalis*, in its natural habitat, *Comp. Biochem. Physiol.*, 35, 921, 1970.
48. **Steinmetz, P. R.,** Characteristics of hydrogen ion transport in urinary bladder of water turtle, *J. Clin. Invest.*, 46, 1531, 1967.
49. **Benyajati, S. and Dantzler, W. H.,** Renal secretion of amino acids in ophidian reptiles, *Am. J. Physiol.*, 250, R712, 1986.
50. **Hernandez, T. and Coulson, R. A.,** Amino acid excretion in the alligator, *Comp. Biochem. Physiol.*, 23, 775, 1967.
51. **Benyajati, S. and Johnson, J. L.,** Characteristics of renal transport of taurine in reptilian brush-border membranes, *Am. J. Physiol.*, 260, R879, 1991.
52. **Benyajati, S. and Bay, S. M.,** Sodium-taurine cotransport in reptilian brush-border membrane vesicles, *Pflügers Arch.*, 421, 168, 1992.
53. **Benyajati, S. and Bay, S. M.,** Basolateral taurine transport system in reptilian renal cells, *Am. J. Physiol.*, 266, F439, 1994.
54. **Moyle, V.,** Nitrogenous excretion in chelonian reptiles, *Biochem. J.*, 44, 581, 1949.
55. **Baze, W. B. and Horne, F. R.,** Ureogenesis in chelonia, *Comp. Biochem. Physiol.*, 34, 91, 1970.
56. **Hill, L. and Dawbin, W. H.,** Nitrogen excretion in the tuatara, *Sphenodon punctatus*, *Comp. Biochem. Physiol.*, 31, 453, 1969.
57. **Khalil, F.,** Excretion in reptiles. IV. Nitrogenous constituents in excreta of lizards, *J. Biol. Chem.*, 189, 443, 1951.

58. **Dessauer, H. C.,** Biochemical studies on the lizard, *Anolis carolinensis*, *Proc. Soc. Exp. Biol. (NY)*, 80, 742, 1952.
59. **Khalil, F.,** Excretion in reptiles. II. Nitrogen constituents of urinary concretions of the oviparous snake *Zamensis diadema*, *J. Biol. Chem.*, 172, 101, 1948.
60. **Khalil, F.,** Excretion in reptiles. III. Nitrogen constituents of urinary concretions of the viviparous snake *Eryx thebaicus*, *J. Biol. Chem.*, 172, 105, 1948.
61. **Khalil, F. and Haggag, G.,** Nitrogenous excretion in crocodiles, *J. Exp. Biol.*, 35, 552, 1958.
62. **Dantzler, W. H.,** The nephron in reptiles, in *Structure and Function of the Kidney*, Kinne, R. K. H., Ed., Karger, Basel, 1989, 143.

Chapter 12

Urea Synthesis in Mammals

Alfred J. Meijer

CONTENTS

INTRODUCTION

During the catabolism of amino acids, NH_4^+ and HCO_3^- ions are formed in almost equal amounts, in addition to CO_2. In this respect, it is of no importance whether amino acids are directly oxidized to yield ATP, or whether they are first converted into compounds such as glucose, fat, or ketone bodies in the liver, which are then oxidized elsewhere in the body. Production of these ions arises from the fact that at neutral pH, the amino groups and carboxyl groups of amino acids carry a positive and negative charge, respectively. The complete oxidation of alanine can serve to illustrate this:

$$CH_3-CHNH_3^+-COO^-+3O_2 \longrightarrow 2CO_2+HCO_3^-+NH_4^++H_2O \qquad (1)$$

Large amounts of NH_4^+ and HCO_3^- are produced in this way, in man about 1 mol/d, when the daily protein consumption is 100 g.[1] Unlike fish and other marine and aquatic organisms, which can dispose of these ions in the surrounding water, this is not possible in mammals. The pathway responsible for the removal of these two ions is the synthesis of urea in the liver (Figure 1), which results in the net consumption of equal amounts of NH_4^+ and HCO_3^-:

$$2NH_4^++HCO_3^- \xrightarrow{4ATP} \text{urea}+H^++2H_2O \qquad (2)$$

$$H^++HCO_3^- \longrightarrow CO_2+H_2O \qquad (3)$$

$$2NH_4^++2HCO_3^- \longrightarrow \text{urea}+CO_2+3H_2O \qquad (4)$$

For the synthesis of 1 mol urea, 4 mol ATP are consumed, which yields 4 protons. However, during the synthesis of ATP via oxidative phosphorylation in the mitochondria, these protons are consumed again. For this reason, they do not appear in the sum reaction. Because the rate of urea synthesis is

0-8493-8411-7/95/$0.00+$.50

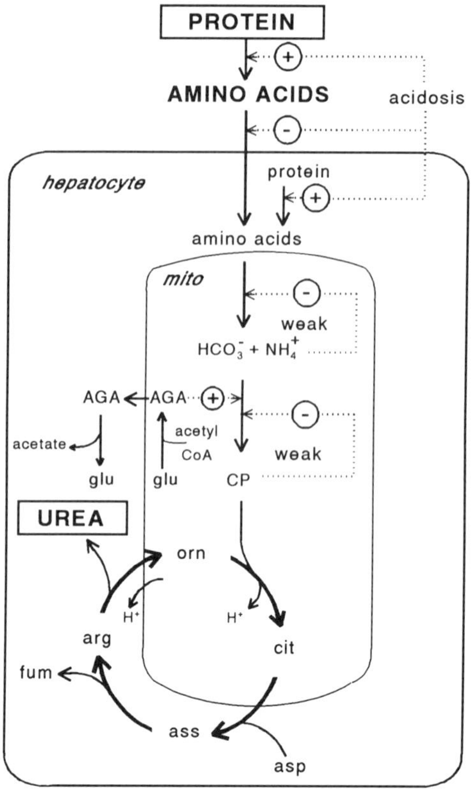

Figure 1 Urea synthesis. In the figure, NH_4^+ is depicted as a substrate for carbamoyl-phosphate synthase. In reality, NH_3 is the true substrate for the enzyme.[91] Interactions are indicated by dotted lines; + = activation, – = inhibition. Abbreviations: CP, carbamoyl phosphate; orn, ornithine; cit, citrulline; asp, aspartate; ass, argininosuccinate; arg, arginine; fum, fumarate; glu, glutamate; AGA, Nacetylglutamate; mito, mitochondrion. Entry of ornithine into the mitochondria is mediated by a specific translocator and occurs by an electroneutral 1:1 exchange of ornithine+ for citrulline and H+.[9]

high, the rate of ATP utilization is high. In the human liver, 15 to 50% of the total ATP production is used for urea synthesis.[1]

Originally it was thought that the purpose of this process was to eliminate ammonia, which is very toxic for the brain.[2,3] However, an alternative pathway for the elimination of NH_4^+ is its excretion in the urine, with glutamine as the nontoxic nitrogen carrier between the tissues and the kidneys. Atkinson and Bourke et al.[4-7] focused attention on the importance of the ornithine cycle as the *only* mechanism by which the body can get rid of the vast majority of the daily bicarbonate production. In mammals, only 2.5 to 5% of the daily HCO_3^- production can be removed in urine. If higher quantities were disposed in urine, this would inevitably lead to precipitation of $CaCO_3$.[1]

One potential example of urea synthesis as a detoxification mechanism for bicarbonate in a vertebrate is *Oreochromis alcalicus grahami*, a tilapia fish from Lake Magadi, Kenya, in which hepatic synthesis and its subsequent excretion in the surrounding water allows it to survive the very alkaline environment (pH, 9.6 to 10; total bicarbonate plus carbonate, 160 to 200 m*M*) where ammonia excretion is virtually impossible.[8] Recent evidence suggests, however, that bicarbonate may not be the overall driving force in ureogenesis in this species.[104]

The question is whether the process of urea synthesis must be carefully controlled, or whether it is simply an overflow mechanism which removes any excess of NH_4^+ and HCO_3^-. As pointed out by Krebs,[2] NH_4^+ is not only toxic for the brain but is also a metabolite in many vital metabolic processes. Thus, the intrahepatic concentration of ammonia must be kept within certain limits.

HCO_3^-, together with CO_2, constitutes the major H^+ buffer system in the body, and changes in urea synthesis can be expected to affect the pH of body fluids. In metabolic acidosis, for example, synthesis of urea slows down so that HCO_3^- is saved to neutralize the excess H^+,[1,4-7] with ammonia being secreted in the urine. It is important to realize that the metabolic pathway of urea synthesis does not start with ammonia, bicarbonate, and aspartate, even though these compounds are direct substrates for the ornithine cycle, but rather with the degradation of protein (Figure 1).[9,10] Thus, retention of bicarbonate only occurs when urea synthesis decreases relative to the rate of HCO_3^- production via protein degradation and subsequent amino acid oxidation. In metabolic acidosis, this condition is certainly fulfilled because muscle proteolysis is then increased[11] (Figure 1) and in this way further helps to correct the acidosis.[9] As discussed by Bourke and Häussinger,[7] however, partial hepatectomy does not affect acid–base homeostasis[12] since both urea synthesis and (hepatic) proteolysis decrease similarly under these conditions. In chronic liver disease, on the other hand, the plasma bicarbonate increases with progressive loss of urea synthesis,[13] as expected.

These considerations clearly demonstrate that flux through the ornithine cycle must be carefully controlled in order to buffer both the intrahepatic ammonia and bicarbonate concentration. The mechanisms by which this is achieved will now be discussed.

CONTROL OF ORNITHINE CYCLE ACTIVITY

THE RATE OF UREA SYNTHESIS *IN VIVO* IS CONTROLLED BY THE RATE OF AMINO ACID DEGRADATION

Essential in understanding control of urea synthesis is the notion that activation or inhibition of one of the ornithine cycle enzymes (e.g., by a change in its kinetic properties), will not affect the rate of urea synthesis unless the activity change feeds back to proteolysis and amino acid degradation[14] (Figure 1). Amino acid degradation is almost insensitive to inhibition by ammonia.[2,14] Degradation of glutamine is even stimulated by ammonia because ammonia is essential for mitochondrial glutaminase activity.[15,16] Thus, any change in the activity of one of the enzymes of the ornithine cycle will only cause an opposite change in the intrahepatic concentration of ammonia, with little effect on the overall rate of amino acid degradation. This is illustrated in the experiment of Table 1. Isolated hepatocytes were incubated with a physiological mixture of amino acids. Activation of carbamoyl-phosphate synthase by (permeant[17]) N-carbamoylglutamate stimulated urea production by only 14%, but caused a decrease of 70% in the steady-state concentration of ammonia. Because

Table 1 Effect of Activation of Carbamoyl-Phosphate Synthase on Urea Synthesis and Ammonia Concentration

Addition	Urea production (μmol/g dry mass)	Extracellular concentration of ammonia (μ*M*)
None	89	62
N-Carbamoylglutamate	102	20

Note: Rat hepatocytes (10 mg dry mass per ml) from a 20-h starved male Wistar rat (250 g) were incubated for 40 min in Krebs-Henseleit bicarbonate medium at 37°C and a complete mixture of amino acids in which each amino acid was present at 4× their concentration in the portal vein of a starved rat except that arginine was replaced by ornithine (see Reference 9 for composition of this mixture). N-carbamoylglutamate (Na^+-salt) was added at a final concentration of 20 m*M*. In the control, an osmotic equivalent amount of NaCl (30 m*M*) was added instead. Urea and ammonia were determined in the extracellular fluid. The extracellular concentration of ammonia presumably reflects the concentration of ammonia in the mitochondria.[9]

of the lack of significant feedback by ammonia, control of the ornithine cycle is essentially control of ammonia concentration, rather than of flux.[14,18]

In 1958, Duda and Handler[19] showed, by injection of $[^{15}N]NH_4^+$ into the tail vein of fasted rats, that urea synthesis increased almost linearly with the dose of ammonia injected up to a level where it became toxic. This demonstrated that apparently *in vivo* urea synthesis operates below its K_m for ammonia and is limited by the production of ammonia by amino acid degradation, and not by the availability of ornithine.[19] Of course, the rate of amino acid degradation, in turn, depends on the rate of degradation of protein of either endogenous or dietary origin[9,10] (Figure 1).

The concentration of ornithine in the liver is a function of the rate of supply of ornithine from intra- and extrahepatic sources, such as arginine from the diet, from intracellular proteolysis (e.g., muscle, liver), or from the kidney where citrulline synthesized in the intestine from glutamine is converted to arginine (see Reference 9 for survey). Net loss of ornithine can occur via ornithine aminotransferase. In liver, however, this enzyme is localized in the pericentral hepatocytes only.[20] The latter enzyme population also contains glutamine synthase but lacks the enzymes involved in urea synthesis (see below). Possibly, ornithine aminotransferase in pericentral hepatocytes supplies glutamate for glutamine synthase while its absence in periportal cells favors the participation of ornithine in urea synthesis.[20]

Although under normal conditions ornithine availability is not a problem, this may not be true at very high concentrations of ammonia. This is indicated by the fact that administration of arginine[21] or inhibition of ornithine degradation via ornithine aminotransferase[22] *in vivo* protects against ammonia toxicity.

A comparison of the intrahepatic concentration of metabolites involved in urea synthesis with the K_m values of their enzymes also indicates that *in vivo*, these enzymes operate below their K_m values (reviewed in Reference 9). This indicates that urea synthesis *in vivo* operates below 50% of its capacity, which is determined by the enzymes with the lowest V_{max} values (i.e., argininosuccinate synthase and argininosuccinate lyase).[9,10] Argininosuccinate synthase is always saturated with aspartate, and stoichiometric formation of carbamoyl phosphate and aspartate for urea synthesis is ensured by the fact that aspartate aminotransferase is close to equilibrium, with aspartate being automatically replaced after its utilization to form argininosuccinate.[2] Thus, *in vivo*, the rate of urea synthesis is controlled by the rate of citrulline production only, which agrees with the data of Duda and Handler.[19] It is only in experimental situations under extreme conditions, with an excess of ammonia and ornithine, that the capacity of argininosuccinate synthase and of arginosuccinate lyase is exceeded. Under these conditions, the enzymes become truly rate-limiting (i.e., flux control coefficient equal to one[18]), and citrulline and argininosuccinate will accumulate.[14,23-27] This also occurs in patients with deficiencies in one of these enzymes.[28,29]

If, as indicated above, urea synthesis *in vivo* is controlled by the production of ammonia, then the question of the source of hepatic ammonia becomes relevant. On the basis of experiments with $[^{13}N]$-labeled ammonia, glutamine, glutamate, and alanine in the adult anaesthesized fed rat, the contribution of the various sources of ammonia to urea synthesis was estimated, with the following result: portal vein ammonia, 33%; glutamine (amide-N), 6 to 13%; mitochondrial glutamate (from alanine, glutamine, histidine, proline) via glutamate dehydrogenase, 20%; other sources, including arterial ammonia and glutamine and enzymes capable of generating ammonia directly (involved in the degradation of asparagine, threonine, glycine and serine), 33 to 40%.[30] Alanine, which is a major substrate for urea synthesis,[31] is primarily metabolized via aspartate.[32,33]

UREA SYNTHESIS AT CONSTANT PHYSIOLOGICAL AMMONIA CONCENTRATION IS CONTROLLED BY CARBAMOYL-PHOSPHATE SYNTHASE

At constant, physiological, ammonia concentration (<0.5 mM), control of flux through the ornithine cycle is almost completely exerted by carbamoyl-phosphate synthase. The reason for this is that this enzyme is almost insensitive to inhibition by its product, carbamoyl phosphate. This is illustrated in Table 2. In this experiment,[34] rat-liver mitochondria were incubated under citrulline synthesizing

Table 2 Carbamoyl-Phosphate Synthase Is Not Inhibited by Its Product Carbamoyl Phosphate in Intact Mitochondria

[Norvaline] (m*M*)	Intramitochondrial carbamoyl phosphate (nmol/mg protein)	Rate of citrulline production (nmol/mg protein/min)
0	0.16	24.6
1	0.64	23.9
2	1.14	22.9
3	1.57	22.1

Note: Rat-liver mitochondria were incubated at 25°C in a medium containg 100 m*M* KCl, 50 m*M* Tris/HCl, 1 m*M* Tris/EGTA, 1 m*M* Tris/ATP, 20 m*M* Tris/succinate, 5 m*M* potassium phosphate, 5 m*M* NH_4Cl, 3 m*M* ornithine, 16.6 m*M* $KHCO_3$, 1 m*M* aminooxyacetate (to inhibit ornithine aminotransferase), rotenone (2 μg/ml; to inhibit NADH dehydrogenase), and norvaline (to inhibit ornithine carbamoyltransferase) at the concentrations indicated; pH, 7.4. Gas phase, 95% O_2/5% CO_2 (v/v). Intramitochondrial carbamoyl phosphate was determined after 5 min. (See Reference 34 for further experimental details.)

conditions, with an excess of ammonia, bicarbonate, and ornithine. Norvaline, an inhibitor of ornithine carbamoyltransferase, was added in increasing amounts. Inhibition of this enzyme was indicated by the gradual increase (up to tenfold) of the intramitochondrial concentration of carbamoyl phosphate. Yet, citrulline production, and thus flux through carbamoyl-phosphate synthase, was hardly affected. Thus, in the intact mitochondrion, activity of carbamoyl-phosphate synthase is not very sensitive to inhibition by its product carbamoyl phosphate. Channeling of carbamoyl phosphate in the carbamoyl-phosphate synthase/ornithine carbamoyltransferase complex does not affect this conclusion (see below).

Supportive evidence for the conclusion that ornithine cycle flux is, indeed, controlled by the activity of carbamoyl-phosphate synthase is the observation that activation of carbamoyl-phosphate synthase by N-carbamoylglutamate, at a constant ammonia concentration, greatly stimulated urea production in isolated hepatocytes.[14] If any of the other ornithine cycle intermediates were to limit urea synthesis, carbamoyl phosphate would accumulate.[14] Thus, in patients with inborn errors in one of the ornithine cycle intermediates distal to carbamoyl-phosphate synthase, intramitochondrial carbamoyl phosphate rises to levels which allow it to overcome the low permeability of the mitochondrial membrane to carbamoyl phosphate[35] and to escape to the cytosol, where it results in overproduction of orotic acid which is secreted in the urine.[28,29,36] This condition is also found in spf mice (sparse fur, abnormal skin and hair) with a genetic defect in ornithine carbamoyltransferase,[37,38] with arginine deficient diets,[39,40] or in experimental conditions where activity of hepatic ornithine decarboxylase is greatly stimulated.[41] Under normal conditions, *in vivo*, however, the intramitochondrial concentration of carbamoyl phosphate is low (<100 μ*M*[14]; <18 μ*M*[42]). This fact indicates that *in vivo* the production of ammonia rather than its removal via the ornithine cycle controls urea synthesis.

In man, *in vivo*, flux through carbamoyl-phosphate synthase is only a very small fraction of the actual capacity of this enzyme. Based on a diet containing 100 g protein per d, daily urea synthesis is about 500 mmol. This is equivalent to 0.3 μmol/min/g wet weight of liver (assuming a liver mass of 1 kg), which is 10 to 15% of the V_{max} of carbamoyl-phosphate synthase.[43,44] The large overcapacity of carbamoyl-phosphate synthase is also illustrated by a recent report on a patient with a 92% deficiency in carbamoyl-phosphate synthase activity which remained unnoticed for 16 years![44]

CARBAMOYL-PHOSPHATE SYNTHASE BUFFERS THE INTRAHEPATIC CONCENTRATION OF AMMONIA

Carbamoyl-phosphate synthase has an absolute requirement for N-acetylglutamate,[45] the allosteric activator of the enzyme.[46] This compound is synthesized in the mitochondria from acetylCoA and

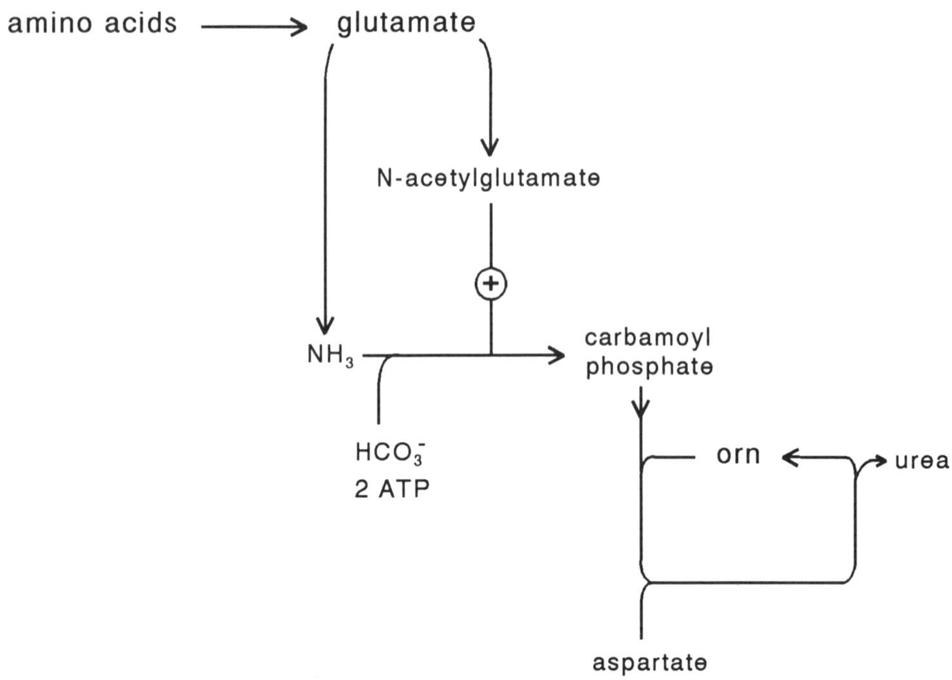

Figure 2 An increased supply of amino acids to the liver results in an increase of both the concentration of substrate (ammonia) and the number of active carbamoyl-phosphate synthase molecules.

glutamate (activated by arginine[47,48]) and degraded in the cytosol[49] after transport of this compound out of the mitochondria[50,51] (Figure 1). This transport is slow ($t_{1/2}$, 20 min[50,51]), only unidirectionally out of the mitochondria, and occurs, presumably, by diffusion.[52] There is overwhelming evidence in the literature suggesting that the activity of carbamoyl-phosphate synthase, and thus the capacity of liver mitochondria to synthesize citrulline, under various nutritional conditions is directly controlled by variations in the intramitochondrial concentration of N-acetylglutamate.[10,53-59]

Activation of carbamoyl-phosphate synthase by N-acetylglutamate makes the enzyme extremely suitable to act as a buffer for the intrahepatic ammonia concentration. Whenever the supply of amino acids to the liver increases, flux through carbamoyl-phosphate synthase increases for two reasons: (1) there is an increase in ammonia production, and (2) the number of active enzyme molecules is increased due to a rise in intramitochondrial N-acetylglutamate (Figure 2). These combined effects result in a sigmoidal relationship between the steady-state ammonia concentration and the rate of urea synthesis. Large variations in ornithine cycle flux are thus possible at a relatively constant ammonia concentration.[10,14,28] This also provides a mechanism to protect against depletion of the hepatic pool of amino acids when the supply of amino acids to the liver falls, e.g., during metabolic acidosis (see below).

The importance of N-acetylglutamate in the control of the intrahepatic concentration of ammonia is highlighted by a variety of conditions in which its formation is decreased. This may not only be due to inborn errors but also to environmental factors. In addition to a direct deficiency of N-acetylglutamate synthase,[60] problems with the synthesis of N-acetylglutamate arise with inborn errors associated with fatty acid oxidation beyond the activation to their CoA derivatives. This is due to trapping of mitochondrial CoA in fatty acylCoA, so that acetylCoA for the synthesis of N-acetylglutamate cannot be formed.[9] The same mechanism is responsible for the hyperammonemia observed after intoxication with the antiepileptic drug valproate or with hypoglycin, the short chain fatty acid responsible for Jamaican vomiting disease which is caused by eating the ackee fruit.[9] Likewise, hyperammonemia in Reye's syndrome (acute encephalopathy and fatty infiltration in many tissues[61]) may be a combination of a partial deficiency in fatty acid oxidation, viral infection, and the use of aspirin,[61,62] the latter compound being rapidly converted to its CoA derivative and which may compete with fatty acid oxidation.[63]

UREA SYNTHESIS AND HEPATIC GLUTAMINE METABOLISM

Amino acid degradation, urea synthesis and gluconeogenesis all occur in the periportal zone of the acinus.[64,65] Synthesis of glutamine, on the other hand, is exclusively localized in the cytosol of the last row of hepatocytes surrounding the terminal hepatic venules.[66] Because the affinity of carbamoyl-phosphate synthase for ammonia is rather low (K_m = 1–2 m*M*),[43] some ammonia, which can easily cross the mitochondrial and plasma membranes in its unprotonated form, will always escape the periportal cells. However, entry of ammonia into the systemic circulation is prevented by its conversion to glutamine by pericentral glutamine synthase, which not only has a high affinity for ammonia (K_m = 0.2 m*M*) but also a high capacity.[67,68] Damage to the pericentral area can thus lead to hyperammonemia without a decrease in ureogenic capacity.[69]

Even though glutamine synthase is an efficient scavenger of pericentral ammonia, it is remarkable that the localization of this enzyme is restricted only to the last row of pericentral hepatocytes. Although never considered before, pericentral glutamine synthesis may, however, serve an additional function other than as scavenger of ammonia. One possibility is that glutamine is involved in the control of nitric oxide synthesis, as recently shown in endothelial cells,[70] and thus helps to control the resistance of the terminal hepatic venules with nitric oxide being responsible for vasodilatation.[71] This would explain the presence of glutamine synthase adjacent to the terminal hepatic venules, so that a high local concentration of glutamine is ensured (see Chapter 18 for discussion of nitric oxide synthesis).

In contrast to glutamine synthase, glutaminase, the enzyme responsible for glutamine degradation, is localized in periportal hepatocytes.[67,68,72] Thus, simultaneous periportal glutamine degradation and pericentral glutamine synthesis may occur.[67,68] The advantage of such a glutamine cycle, although energetically unfavorable, is that glutamine is a very efficient donor of ammonia for carbamoyl-phosphate synthase. Both enzymes are located in the mitochondria in close association with the mitochondrial inner membrane,[73,74] and glutaminase-derived ammonia is directly channeled to carbamoyl-phosphate synthase.[75] In addition, activity of liver glutaminase absolutely depends on the presence of ammonia.[15,16] This product activation of glutaminase further enhances the efficiency of delivery of ammonia to carbamoyl-phosphate synthase.[75] Interestingly, N-acetylglutamate, the essential activator of carbamoyl-phosphate synthase, also stimulates glutaminase activity in intact, but not in lysed liver mitochondria.[75] This is possibly an indication for the close proximity of the two enzymes so that the increase in activity of carbamoyl-phosphate synthase by N-acetylglutamate can be sensed by glutaminase.

Studies with [^{15}N]glutamine, both *in vivo* in the postprandial rat[33] and *in vitro* with isolated hepatocytes[76] underscore the efficiency at which the amide-N of glutamine can be used for urea synthesis. The importance of glutamine as a donor of ammonia for urea synthesis is also illustrated by the large increase in glutaminase protein in rats fed with a high-protein diet,[77] and by the rapid activation of existing glutaminase molecules under these conditions.[78] A large increase in glutaminase activity was also found in cirrhotic human livers.[79] The latter observation has been interpreted as a compensatory mechanism for maintenance of life-compatible urea synthesis despite the decreased activities of the ornithine cycle enzymes in these livers.[79]

METABOLITE CHANNELING IN THE ORNITHINE CYCLE

An interesting development in the field of urea synthesis has been the discovery of metabolite channeling in this process. Data obtained with intact mitochondria,[34,80] permeabilized mitochondria,[81] and permeabilized hepatocytes[82] suggest that the ornithine cycle intermediates may be passed from enzyme to enzyme without being released to either the mitochondrial matrix or the cytosol. In isolated liver mitochondria, [^{14}C]citrulline production from [^{14}C]HCO_3^-, NH_4^+ and ornithine occurred without a lag, even though it took about 1 min before the entire intramitochondrial pool of carbamoyl phosphate became labeled. This situation suggests channeling of carbamoyl phosphate from carbamoyl-phosphate synthase to ornithine carbamoyltransferase, although, apparently, this channeling is not very tight.[34] Even across the mitochondrial membrane channeling is possible: extramitochondrial ornithine is directly accessible to ornithine carbamoyltransferase in the mitochondria via the ornithine-citrulline translocator without prior mixing with the ornithine pool in the mitochondrial matrix.[80] Citrulline, synthesized in the mitochondria, is channeled to argininosuccinate synthase via the same transport system and does not need to be mixed with the cytosolic pool of citrulline. However, this channeling is not as tight as the other steps in the ornithine cycle.[27,82,83]

Thus, channeling occurs throughout the entire ornithine cycle, with the enzymes being clustered around the mitochondrial membrane. The cycle can be considered as a metabolon, spanning both the mitochondrial and cytosolic compartments.[84] [^{15}N]NMR may be useful for studying metabolite channeling in the ornithine cycle in the intact liver.[27]

The occurrence of metabolite channeling does not invalidate conclusions summarized above regarding *in situ* kinetic parameters of ornithine cycle enzymes which are based on cytosolic and mitochondrial metabolite "pool" concentrations (contrast Ref. 81). If channeling is not extremely tight, the steady-state concentrations of metabolites in the bulk phases of the cytosol and the mitochondria will not be lower than those in the vicinity of the enzyme complexes.[85]

INHIBITION OF UREA SYNTHESIS IN METABOLIC ACIDOSIS

As mentioned earlier, in metabolic acidosis urea synthesis falls in order to save HCO_3^- ions for the neutralization of H^+ ions. NH_4^+, not used in urea synthesis, is then transported to the kidneys via glutamine. As will be seen below, however, glutamine does not necessarily have to be synthesized in the liver, nor is it necessary that HCO_3^- is produced there. In principle, this can occur anywhere in the body, although muscle is quantitatively the most important tissue. Not only is muscle protein an important source of amino acids in various catabolic states, but the oxidation of amino acids is also very active in this tissue.[31] As discussed previously,[9] any tissue capable of affecting the concentration of HCO_3^- and CO_2 will influence the systemic pH. These include the lungs (CO_2 elimination), liver, muscle, and other tissues. It is not at all clear why in discussions on pH homeostasis[86,87] such a role should be attributed to the kidneys only. Of course, ultimately, the amount of HCO_3^- retained by the kidney is equal to the amount of NH_4^+ excreted in the urine. However, the primary production of this HCO_3^- occurs elsewhere in the body but HCO_3^-, like NH_4^+, was subsequently packaged in the glutamine molecule. Thus, pH control is a multiorgan function. Like enzymes in a metabolic pathway,[18,88] the tissues share the control. The amount of control that the various tissues contribute to pH homeostasis varies with the conditions. A mathematical treatment of tissue control has recently been given.[89]

The key question is, what are the mechanisms responsible for the decrease in urea synthesis when the pH falls? Several mechanisms have been proposed, and these probably all contribute to some extent. They include reduced hepatic glutamine degradation at low pH due to a reduction in glutaminase activity,[68] the degree of protonation of NH_3[90] which is the true substrate for carbamoyl-phosphate synthase (not NH_4^+),[91] inhibition of carbonic anhydrase[92] (which provides HCO_3^- for carbamoyl-phosphate synthase from CO_2 produced in the citric acid cycle[93]), decreased synthesis of N-acetylglutamate,[94] and inhibition of amino acid transport across the plasma membrane.[95] According to Häussinger et al.,[68] ammonia (not used in (periportal) urea synthesis in acidosis) reaches the pericentral (glutamine synthase containing) cells where it is converted to glutamine, a conclusion based on experiments with livers perfused with physiological concentrations of glutamine and ammonia. If this were true, total hepatic amino acid consumption in acidosis would not be affected, but only the fate of the ammonia produced by amino acid degradation. Contrary to this expectation, however, total amino acid consumption by the liver *in vivo* decreases in (HCl-induced) acidosis, which is due to diminished amino acid transport across the plasma membrane.[96,97] This finding agrees with earlier data obtained with isolated hepatocytes which also showed decreased amino acid transport and consumption at low extracellular pH.[95] Thus, in acidosis, more amino acids become available for the extrahepatic tissues where they can be oxidized to generate HCO_3^-. Apparently, in acidosis, glutamine for renal utilization is not synthesized in the liver, but elsewhere in the body (e.g., muscle, lung).[9] The role of hepatic glutamine synthesis in pH homeostasis has also been questioned by Almond et al.[98] on the basis of perfusion studies with livers from chronic acidotic rats.

Although the major inhibitory effect of acidosis on hepatic amino acid degradation is at the level of transport across the plasma membrane, the other pH dependent mechanisms, as mentioned above, also contribute to the decrease in urea synthesis, and presumably serve to prevent depletion of intrahepatic glutamine and ammonia.[99]

In conclusion, pH homeostasis is achieved at various levels simultaneously. In acidosis, HCO_3^- accumulation is stimulated by increased proteolysis (see Introduction), and decreased urea synthesis is due to diminished hepatic amino acid transport; extrahepatic glutamine synthesis and renal glutamine consumption are increased; NH_4^+ ions, not used in urea synthesis, are excreted in the urine.

LONG-TERM REGULATION OF UREA SYNTHESIS

Superimposed on the short-term regulation of urea synthesis is the long-term regulation in which changes in the amount of enzyme protein, rather than in the activity of existing enzyme molecules, are involved. Coordinated increases in the ornithine cycle enzymes occur in starvation and with increasing protein content of the diet.[100,101] Literature on this topic and on the hormonal control of the changes in these enzymes has been reviewed recently.[9,102] Mechanisms involved include transcriptional control, stabilization of mRNA, and protein stabilization.[103]

REFERENCES

1. **Atkinson, D. E. and Camien, M.,** The role of urea synthesis in the removal of metabolic bicarbonate and the regulation of blood pH, *Curr. Top. Cell. Regul.*, 21, 261, 1982.
2. **Krebs, H. A., Hems, R., and Lund, P.,** Some regulatory mechanisms in the synthesis of urea in the mammalian liver, *Adv. Enzyme Regul.*, 11, 361, 1973.
3. **Ratner, S.,** Enzymes of arginine and urea synthesis, *Adv. Enzymol.*, 39, 1, 1973.
4. **Oliver, J. and Bourke, E.,** Adapatations in urea ammonium excretion in metabolic acidosis in the rat: a reinterpretation, *Clin. Sci. Mol. Med.*, 48, 515, 1975.
5. **Atkinson, D. E. and Bourke, E.,** The role of ureagenesis in pH homeostasis, *Trends Biochem. Sci.* 9, 297, 1984.
6. **Atkinson, D. E.,** Functional roles of urea synthesis in vertebrates, *Physiol. Zool.*, 65, 243, 1992.
7. **Bourke, E. and Häussinger, D.,** pH Homeostasis: the conceptual change, in *The Kidney Today. Selected Topics In Renal Science, Contrib. Nephrol.*, Vol. 100, Berlyne, G. M., Ed., Karger, Basel, 1992, 58.
8. **Randall, D. J., Wood, C. M., Perry, S. F., Bergman, H., Maloiy, G. M. O., Mommsen, T. P., and Wright, P. A.,** Urea excretion as a strategy for survival in fish living in a very alkaline environment, *Nature*, 337, 165, 1989.
9. **Meijer, A. J., Lamers, W. H., and Chamuleau, R. A. F. M.,** Nitrogen metabolism and ornithine cycle function, *Physiol. Rev.*, 70, 701, 1990.
10. **Beliveau Carey, G., Cheung, C. W., Cohen, N. S., Brusilow, S., and Raijman, L.,** Regulation of urea and citrulline synthesis under physiological conditions, *Biochem. J.*, 292, 241, 1993.
11. **May, R. C., Kelly, R. A., and Mitch, W. E.,** Metabolic acidosis stimulates protein degradation in rat muscle by a glucocorticoid dependent mechanism, *J. Clin. Invest.*, 77, 614, 1986.
12. **Almdal, T., Vilstrup, H., Bjerrum, K., and Kristensen, L. O.,** Decrease in ureagenesis by partial hepatectomy does not influence acid-base balance, *Am. J. Physiol.*, 257, F696, 1989.
13. **Häussinger, D., Steeb, R., and Gerok, W.,** Ammonium and bicarbonate homeostasis in chronic liver disease, *Klin. Wochenschr.*, 68, 175, 1990.
14. **Meijer, A. J., Lof, C., Ramos, I. C., and Verhoeven, A. J.,** Control of ureogenesis, *Eur. J. Biochem.*, 148, 189, 1985.
15. **Verhoeven, A. J., van Iwaarden, J. F., Joseph, S. K., and Meijer, A. J.,** Control of rat liver glutaminase by ammonia and pH, *Eur. J. Biochem.*, 133, 241, 1983.
16. **McGivan, J. D. and Bradford, N. M.,** Characteristics of the activation of glutaminase by ammonia in sonicated rat liver mitochondria, *Biochim. Biophys. Acta*, 759, 296, 1983.

17. **Rubio, V. and Grisolia, S.,** Human carbamoylphosphate synthetase I, *Enzyme (Basel)*, 26, 233, 1981.
18. **Westerhoff, H. V., Groen, A. K., and Wanders, R. J. A.,** Modern theories of metabolic control and their applications, *Biosci. Rep.*, 4, 1, 1984.
19. **Duda, G. D. and Handler, P.,** Kinetics of ammonia metabolism in vivo, *J. Biol. Chem.*, 232, 303, 1958.
20. **Kuo, F. C., Hwu, W. L., Valle, D., and Darnell, J. E.,** Colocalization in pericentral hepatocytes in adult mice and similarity in developmental expression pattern of ornithine aminotransferase and glutamine synthetase mRNA, *Proc. Natl. Acad. Sci. U.S.A.*, 88, 9468, 1991.
21. **Greenstein, J. P., Winitz, M., Gullino, P., and Birnbaum, S. M.,** The prevention of ammonia toxicity by L-arginine and related compounds, *Arch. Biochem. Biophys.* 59, 302, 1955.
22. **Seiler, N., Sarhan, S., Knoedgen, B., Hornsperger, J. M., and Sablone, M.,** Enhanced endogenous ornithine concentrations protect against tonic seizures and coma in acute ammonia toxication, *Pharmacol. Toxicol.*, 72, 116, 1993.
23. **Meijer, A. J., Gimpel, J. A., DeLeeuw, G. A., Tager, J. M., and Williamson, J. R.,** Role of anion translocation across the mitochondrial membrane in the regulation of urea synthesis from ammonia by isolated rat hepatocytes, *J. Biol. Chem.*, 250, 7728, 1975.
24. **Krebs, H. A., Lund, P., and Stubbs, M.,** Interrelations between gluconeogenesis and urea synthesis, in *Gluconeogenesis, Its Regulation in Mammalian Species*, Hanson, R. W. and Mehlman, M. A., Eds., Wiley, New York, 1976, 269.
25. **Triebwasser, K. C., and Freedland, R. A.,** The effect of glucagon on ureagenesis from ammonia by isolated rat hepatocytes, *Biochem. Biophys. Res. Commun.*, 76, 1159, 1977.
26. **Meijer, A. J., Gimpel, J. A., DeLeeuw, G. A., Tischler, M. E., Tager, J. M., and Williamson, J. R.,** Interrelationships between gluconeogenesis and ureogenesis in isolated hepatocytes, *J. Biol. Chem.*, 253, 2308, 1978.
27. **Geissler, A., Kanamori, K., and Ross, B. D.,** Real-time study of the urea cycle using ^{15}N n.m.r. in the isolated perfused rat liver, *Biochem. J.*, 287, 813, 1992.
28. **Walser, M.,** Urea cycle disorders and other hereditary hyperammonemic syndromes, in *The Metabolic Basis Of Inherited Disease*, Stanbury, J. B., Wyngaarden, J. B., Frederickson, D. S., Goldstein, J. L., and Brown, M. J., Eds., McGraw-Hill, New York, 1983, 402.
29. **Brusilow, S. W. and Horwich, A. L.,** Urea cycle enzymes, in *The Metabolic Basis Of Inherited Disease*, Vol. 1, Scriver, C. R., Beaudet, A. L., Sly, W. S., and Valle, D., Eds., McGraw-Hill, New York, 1989, 629.
30. **Cooper, A. J. L., Nieves, E., Filc-Dericco, S., and Gelbard, A. S.,** Short-term metabolic fate of [^{13}N]ammonia, L-[^{13}N]alanine, L-[^{13}N]glutamate and L-[amide-^{13}N]glutamine in normal rat liver in vivo, in *Advances In Ammonia Metabolism And Hepatic Encephalopathy*, Soeters, P. B., Wilson, J. H. P., Meijer, A. J., and Holm, E., Eds., Excerpta Medica, Amsterdam, 1988, 11.
31. **Felig, P.,** Amino acid metabolism in man, *Annu. Rev. Biochem.*, 44, 933, 1975.
32. **Williamson, J. R., Gimpel, J. A., Meijer, A. J., DeLeeuw, G., and Refino, C.,** Role and regulation of glutamate dehydrogenase in ureogenesis, in *Use Of Isolated Liver Cells And Kidney Tubules In Metabolic Studies*, Tager, J. M., Söling, H. D., and Williamson, J. R., Eds., Elsevier, North-Holland, Amsterdam, 1976, 339.
33. **Jahoor, F., Jackson, A. A., and Golden, M. H. N.,** In vivo metabolism of nitrogen precursors for urea synthesis in the postprandial rat, *Ann. Nutr. Metab.*, 32, 240, 1988.
34. **Wanders, R. J. A., van Roermund, C. W. T., and Meijer, A. J.,** Analysis of the control of citrulline synthesis in isolated rat-liver mitochondria, *Eur. J. Biochem.*, 142, 247, 1984.
35. **Natale, P. J. and Tremblay, G. C.,** Studies on the availability of intramitochondrial carbamoylphosphate for utilization in extramitochondrial reactions in rat liver, *Arch. Biochem. Biophys.*, 162, 357, 1974.
36. **Bachmann, C. and Colombo, J. P.,** Diagnostic value of orotic acid excretion in heritable disorders of the urea cycle and in hyperammonemia due to organic acidurias, *Eur. J. Pediatr.*, 134, 109, 1980.
37. **Qureshi, I. A., Letarte, J., and Ouellet, R.,** Ornithine transcarbamylase deficiency in mutant mice. I. Studies on the characterization of enzyme defect and suitability as animal model of human disease, *Pediatr. Res.*, 13, 807, 1979.
38. **Nelson, J., Qureshi, I. A., Vasudevan, S., and Sarma, D. S. R.,** The effects of various inhibitors on the regulation of orotic acid excretion in sparse-fur mutant mice (spf/Y) deficient in ornithine transcarbamylase, *Chem. Biol. Interact.* 89, 35, 1993.
39. **Milner, J. A. and Visek, W. J.,** Dietary protein intake and arginine requirements in the rat, *J. Nutr.*, 108, 382, 1978.
40. **Hassan, A. S. and Milner, J. A.,** Orotic acid biosynthesis in arginine-deficient rats, *Arch. Biochem. Biophys.*, 194, 24, 1979.
41. **Vasudevan, S., Laconi, E., Rajalakshmi, S., Rao, P. M., and Sarma, D. S. R.,** Interrelationship between ornithine decarboxylase and orotic acid synthesis induced by amino acids, *Fed. Proc.*, 46, 753, 1987.
42. **Cooper, A. J. L., Nieves, E., Coleman, A. E., Filc-DeRicco, S., and Gelbard, A. S.,** Short-term metabolic fate of [^{13}N]ammonia in rat liver in vivo, *J. Biol. Chem.*, 262, 1073, 1987.
43. **Meijer, A. J., and Hensgens, H. E. S. J.,** Ureogenesis, in *Metabolic Compartmentation*, Sies, H., Ed., Academic Press, New York, 1982, 259.
44. **Lo, W. D., Sloan, H. R., Sotos, J. F., and Klinger, R. J.,** Late clinical presentation of partial carbamyl phosphate synthetase I deficiency, *Am. J. Dis. Child.*, 147, 267, 1993.

45. **Hall, L. M., Metzenberg, R. L., and Cohen, P. P.,** Isolation and characterization of a naturally occurring cofactor of carbamyl phosphate biosynthesis, *J. Biol. Chem.*, 230, 1013, 1958.
46. **Lusty, C. J.,** Catalytic active monomer and dimer forms of rat liver carbamoyl-phosphate synthetase, *Biochemistry*, 20, 3665, 1981.
47. **Shigesada, K. and Tatibana, M.,** N-acetylglutamate synthetase from rat-liver mitochondria. Partial purification and catalytic properties, *Eur. J. Biochem.*, 84, 285, 1978.
48. **Kawamoto, S., Sonoda, T., Ohtake, A., and Tatibana, M.,** Stimulatory effect of arginine on acetylglutamate synthesis in isolated mitochondria of mouse and rat liver, *Biochem. J.*, 232, 329, 1985.
49. **Reglero, A., Rivas, J., Mendelson, J., Wallace, R., and Grisolia, S.,** Deacylation and transacetylation of acetylglutamate and acetylornithine in rat liver, *FEBS Lett.*, 81, 13, 1977.
50. **Morita, T., Mori, M., and Tatibana, M.,** Regulation of N-acetyl-L-glutamate degradation in mammalian liver, *J. Biochem. (Tokyo)*, 91, 563, 1982.
51. **Meijer, A. J. and van Woerkom, G. M.,** Turnover of N-acetylglutamate in isolated rat hepatocytes, *Biochim. Biophys. Acta*, 721, 240, 1982.
52. **Meijer, A. J., van Woerkom, G. M., Wanders, R. J. A., and Lof, C.,** Transport of N-acetylglutamate in rat-liver mitochondria, *Eur. J. Biochem.*, 124, 325, 1982.
53. **Shigesada, K. and Tatibana, M.,** Role of acetylglutamate in ureotelism. I. Occurrence and biosynthesis of acetylglutamate in mouse and rat tissues, *J. Biol. Chem.*, 246, 5588, 1971.
54. **Shigesada, K., Aoyagi, K., and Tatibana, M.,** Role of acetylglutamate in ureotelism. Variations in acetylglutamate level and its possible significance in control of urea synthesis in mammalian liver, *Eur. J. Biochem.*, 85, 385, 1978.
55. **Meijer, A. J. and van Woerkom, G. M.,** Control of the rate of citrulline synthesis by short-term changes in N-acetylglutamate levels in isolated rat-liver mitochondria, *FEBS Lett.*, 86, 117, 1978.
56. **Hensgens, H. E. S. J., Verhoeven, A. J., and Meijer A. J.,** The relationship between intramitochondrial N-acetylglutamate and activity of carbamoyl-phosphate synthetase (ammonia), *Eur. J. Biochem.*, 107, 197, 1980.
57. **Stewart, P. M. and Walser, M.,** Short term regulation of ureagenesis, *J. Biol. Chem.*, 255, 5270, 1980.
58. **Morimoto, B. H., Brady, J. F., and Atkinson, D. E.,** Effect of level of dietary protein on arginine-stimulated citrulline synthesis. Correlation with mitochondrial N-acetylglutamate concentrations, *Biochem. J.*, 272, 671, 1990.
59. **Felipo, V., Miñana, M. D., and Grisolia, S.,** Control of urea synthesis and ammonia utilization in protein deprivation and refeeding, *Arch. Biochem. Biophys.*, 285, 351, 1991.
60. **Colombo, J. P., Bachmann, C., and Schrämli, A.,** Inborn defects of the mitochondrial part of the urea cycle, *Ann. N.Y. Acad. Sci.*, 488, 109, 1988.
61. **Trauner, D. A.,** Reye's syndrome, *Curr. Probl. Pediatr.*, 12, 1, 1981.
62. **Fulginiti, V. A., Brunell, P. A., Cherry, J. D., Ector, W. L., Gershon, A. A., Gotoff, S. P., Hughes, W. A., Mortimore, E. A., and Peter, G.,** Aspirin and Reye syndrome (Special report), *Pediatrics*, 69, 810, 1982.
63. **Deschamps, D., Fisch, C., Fromenty, B., Berson, A., Degott, C., and Pessayre, D.,** Inhibition by salicylic acid of the activation and thus oxidation of long chain fatty acids. Possible role in the development of Reye's syndrome, *J. Pharmacol. Exp. Ther.*, 259, 894, 1991.
64. **Jungermann, K. and Katz, N.,** Functional specialization of different hepatocyte populations, *Physiol. Rev.*, 69, 708, 1989.
65. **Gebhardt, R.,** Metabolic zonation of the liver: regulation and implications for liver function, *Pharmacol. Ther.*, 53, 275, 1992.
66. **Gebhardt, R. and Mecke, D.,** Heterogeneous distribution of glutamine synthetase among rat liver parenchymal cells in situ and in primary culture, *EMBO J.*, 2, 567, 1983.
67. **Häussinger, D.,** Hepatocyte heterogeneity in glutamine and ammonia metabolism and the role of an intercellular glutamine cycle during ureogenesis in perfused rat liver, *Eur. J. Biochem.*, 133, 269, 1983.
68. **Häussinger, D., Gerok, W., and Sies, H.,** Hepatic role in pH regulation: role of the intercellular glutamine cycle, *Trends Biochem. Sci.*, 9, 300, 1984.
69. **Häussinger, D. and Gerok, W.,** Hepatocyte heterogeneity in ammonia metabolism: impairment of glutamine synthesis in CCl_4 induced liver cell necrosis with no effect on urea synthesis, *Chem. Biol. Interact.*, 48, 191, 1984.
70. **Wu, G. Y. and Meininger, C. J.,** Regulation of L-arginine synthesis from L-citrulline by L-glutamine in endothelial cells, *Am. J. Physiol.*, 265, H1965, 1993.
71. **Lowenstein, C. J., Dinerman, J. L., and Snyder, S. H.,** Nitric oxide: a physiologic messenger, *Ann. Intern. Med.*, 120, 227, 1994.
72. **Watford, M. and Smith, E. M.,** Distribution of hepatic glutaminase and messenger RNA in perivenous and periportal rat hepatocytes, *Biochem. J.*, 267, 265, 1990.
73. **McGivan, J. D., Lacey, L. H., and Joseph, S. K.,** Localization and some properties of phosphate-dependent glutaminase in disrupted liver mitochondria, *Biochem. J.*, 192, 537, 1980.
74. **Powers-Lee, S. G., Mastico, R. A., and Bendayan, M.,** The interaction of rat liver carbamoyl phosphate synthetase and ornithine transcarbamoylase with inner mitochondrial membranes, *J. Biol. Chem.*, 262, 15683, 1987.
75. **Meijer, A. J.,** Channeling of ammonia from glutaminase to carbamoyl-phosphate synthetase in liver mitochondria, *FEBS Lett.*, 191, 249, 1985.

76. **Nissim, I., Catano, C., Nissim, I., and Yudkoff, M.,** Relative role of the glutaminase, glutamate dehydrogenase, and AMP-deaminase pathways in hepatic urea synthesis: studies with ^{15}N, *Arch. Biochem. Biophys.*, 292, 393, 1992.
77. **Smith, E. M. and Watford, M.,** Rat hepatic glutaminase: purification and immunochemical characterization, *Arch. Biochem. Biophys.*, 260, 740, 1988.
78. **Ewart, H. S. and Brosnan, J. T.,** Rapid activation of hepatic glutaminase in rats fed on a single high-protein meal, *Biochem. J.*, 293, 339, 1993.
79. **Kaiser, S., Gerok, W., and Häussinger, D.,** Ammonia and glutamine metabolism in human liver slices: new aspects on the pathogenesis of hyperammonemia in chronic liver disease, *Eur. J. Clin. Invest.*, 18, 535, 1988.
80. **Cohen, N. S., Cheung, C. W., and Raijman, L.,** Channeling of extramitochondrial ornithine to matrix ornithine transcarbamylase, *J. Biol. Chem.*, 262, 203, 1987.
81. **Cohen, N. S., Cheung, C. W., Sijuwade, E., and Raijman, L.,** Kinetic properties of carbamoyl-phosphate synthase (ammonia) and ornithine carbamoyltransferase in permeabilized mitochondria, *Biochem. J.*, 282, 173, 1992.
82. **Cheung, C. W., Cohen, N. S., and Raijman, L.,** Channeling of urea cycle intermediates in situ in permeabilized hepatocytes, *J. Biol. Chem.*, 264, 4038, 1989.
83. **Meijer, A. J.,** Ureagenesis and ammoniagenesis, in *Amino Acid Metabolism In Health And Nutritional Diseases*, Cynober, L., Ed., CRC Press, Boca Raton, FL, 1994.
84. **Watford, M.,** Channeling in the urea cycle: a metabolon spanning two compartments, *Trends Biochem. Sci.*, 14, 313, 1989.
85. **Cornish-Bowden, A.,** Failure of channelling to maintain low concentrations of metabolic intermediates, *Eur. J. Biochem.*, 195, 103, 1991.
86. **Walser, M.,** Roles of urea production, ammonium excretion and amino acid oxidation in acid-base balance, *Am. J. Physiol.*, 250, F181, 1986.
87. **Halperin, M. L., Jungas, R. L., Cheema-Dhadli, S., and Brosnan, J. T.,** Disposal of the daily acid load: an integrated function of the liver, lungs and kidneys, *Trends Biochem. Sci.*, 12, 197, 1987.
88. **Kacser, H. and Burns, J. A.,** Molecular democracy: who shares the controls, *Biochem. Soc. Trans.*, 7, 1149, 1979.
89. **Brown, G. C.,** Control analysis applied to the whole body: control by body organs over plasma concentrations and organ fluxes of substances in the blood, *Biochem. J.*, 297, 115, 1994.
90. **Rémésy, C., Demigné, C., and Fafournoux, P.,** Control of ammonia distribution ratio across the liver cell membrane and of ureogenesis by extracellular pH, *Eur. J. Biochem.*, 158, 283, 1986.
91. **Cohen, N. S., Kyan, F. S., Kyan, S. S., Cheung, C. W., and Raijman, L.,** The apparent K_m of ammonia for carbamoyl phosphate synthetase (ammonia) in situ, *Biochem. J.*, 229, 205, 1985.
92. **Häussinger, D. and Gerok, W.,** Hepatic urea synthesis and pH regulation. Role of CO_2, HCO_3^-, pH and the activity of carbonic anhydrase, *Eur. J. Biochem.*, 152, 381, 1985.
93. **Dodgson, S. J., Forster, R. E., Schwed, D. A., and Storey, B. T.,** Contribution of matrix carbonic anhydrase to citrulline synthesis in isolated guinea pig mitochondria, *J. Biol. Chem.*, 258, 7696, 1983.
94. **Kamemoto, E. S. and Atkinson, D. E.,** Modulation of the activity of rat liver acetylglutamate synthase by pH and arginine concentration, *Arch. Biochem. Biophys.*, 243, 100, 1985.
95. **Boon, L. and Meijer, A. J.,** Control by pH of urea synthesis in isolated rat hepatocytes, *Eur. J. Biochem.*, 172, 465, 1988.
96. **Boon, L., Blommaart, P. J. E., Meijer, A. J., Lamers, W. H., and Schoolwerth, A. C.,** Acute acidosis inhibits amino acid uptake by the liver: implications for regulation of acid-base balance, *Contrib. Nephrol.*, 110, 133, 1994.
97. **Boon, L., Blommaart, P. J. E., Meijer, A. J., Lamers, W. H., and Schoolwerth, A. C.,** Effect of chronic acidosis on hepatic amino acid uptake and gene regulation: implications for control of acid-base balance, *Contrib. Nephrol.*, 110, 138, 1994.
98. **Almond, M. K., Smith, A., Cohen, R. D., Iles, R. A., and Flynn, G.,** Substrate and pH effects on glutamine synthesis in rat liver: consequences for acid-base regulation, *Biochem. J.*, 278, 709, 1991.
99. **Boon, L., Blommaart, P. J. E., Meijer, A. J., Lamers, W. A., and Schoolwerth, A. C.,** Decreased hepatic amino acid transport in acute acidosis: no primary role for the ornithine cycle in acid-base homeostasis, *Am. J. Physiol.*, 267, F1015, 1994.
100. **Schimke, R. T.,** Adaptive characteristics of urea cycle enzymes in the rat, *J. Biol Chem.*, 237, 459, 1962.
101. **Schimke, R. T.,** Differential effects of fasting and protein-free diets on levels of urea cycle enzymes in rat liver, *J. Biol. Chem.*, 237, 1921, 1962.
102. **Morris, S. M.,** Regulation of enzymes of urea and arginine synthesis, *Annu. Rev. Nutr.*, 12, 81, 1992.
103. **Ulbright, C. and Snodgrass, P. J.,** Coordinate induction of the urea cycle enzymes by glucagon and dexamethasone is accomplished by three different mechanisms, *Arch. Biochem. Biophys.*, 301, 237, 1993.
104. **Wood, C. M., Bergman, H. L., Laurent, P., Maina, J. N., Narahara, A., and Walsh, P. J.,** Urea production, acid-base regulation and their interactions in the lake Magadi tilapia, a unique teleost adapted to a highly alkaline environment, *J. Exp. Biol.*, 189, 13, 1994.

Chapter 13

Urea and Ammonium Transport in the Mammalian Kidney*

Mark A. Knepper and Chung-Lin Chou

CONTENTS

SYSTEMIC NITROGEN METABOLISM IN MAMMALS

Both the liver and kidney play essential roles in the regulation of nitrogen balance in mammals (Figure 1). Excess amino acids derived from dietary proteins and tissue protein catabolism are utilized in the liver for gluconeogenesis. In addition to glucose, products of this process include ammonium ions, bicarbonate ions, phosphoric acid (derived from oxidation of phosphorylated serine, threonine, and tyrosine), and sulfuric acid (derived from oxidation of the sulfur-containing amino acids methionine and cysteine). The ammonium and bicarbonate, which are produced in nearly equal molar amounts, are metabolic intermediates that are rapidly utilized in the liver for urea and glutamine production. The relative rates of glutamine and urea production are regulated (see below). Virtually all of the urea produced is excreted by the kidney. Small amounts are found, however, in sweat and some is metabolized by bacteria in the gastrointestinal tract. Glutamine, which is produced in skeletal

* This material was compiled in the authors' capacity as government employees of the Department of Health and Human Services.

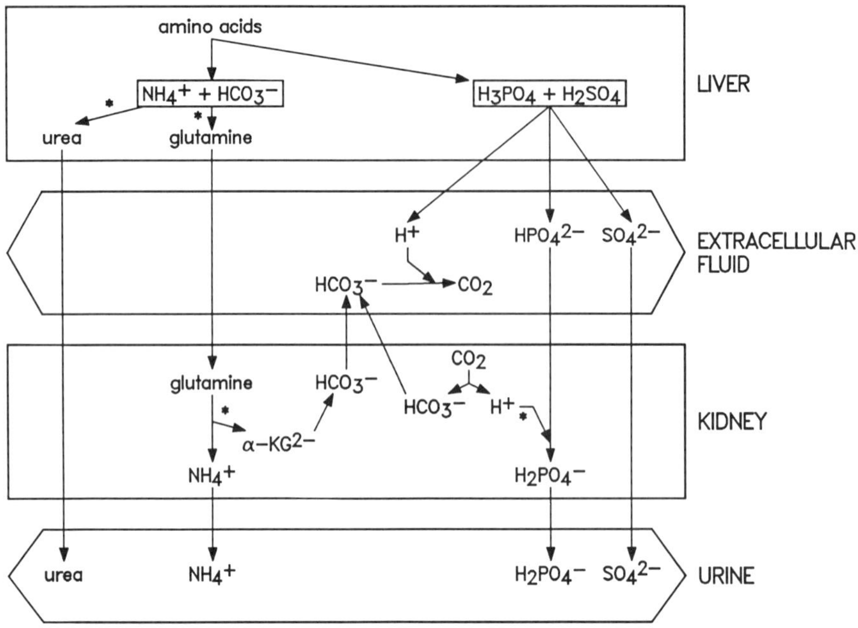

Figure 1 Roles of liver and kidney in nitrogen metabolism and systemic pH regulation. (From Knepper, M. A., Burg, M. B., Orloff, J., Berliner, R. W., and Rector, F., *Am. J. Physiol.*, 253, F199, 1987. With permission.)

muscle as well as liver is utilized in various organs including the kidney for energy metabolism. Glutamine metabolism in the kidney plays an important additional role, *viz.*, formation of ammonium, whose excretion is essential for the regulation of acid–base balance.

UREA METABOLISM

Urea is produced in the liver via the urea cycle.[1] The key components of this pathway are illustrated in Figure 2. The final step in the formation of urea is catalyzed by arginase, which hydrolyzes arginine into urea and ornithine. Arginase is present in many tissues besides liver, including kidney.[2] However, these tissues produce only limited amounts of urea from circulating arginine, because other urea cycle enzymes are not expressed. Although the urea cycle is complex (Figure 2), the overall metabolic process can be summarized by a simple metabolic equation:

$$2NH_4^+ + HCO_3^- \longrightarrow \text{urea} + H^+ + 2H_2O \tag{1}$$

The two ammonium ions are utilized in the formation of carbamoyl phosphate and aspartate. (The formation of aspartate from oxaloacetate by transamination from glutamate can be considered to consume one ammonium ion although the reaction does not directly remove a free ammonium ion.) The HCO_3^- is consumed in the formation of carbamoyl phosphate. The bicarbonate utilized in urea synthesis is dependent on intramitochondrial carbonic anhydrase which catalyzes the hydration of CO_2 to HCO_3^- and H^+. Thus, the formation of urea from ammonium can be viewed as generating two protons:

$$2NH_4^+ + CO_2 \longrightarrow \text{urea} + 2H^+ + H_2O \tag{2}$$

Because bicarbonate is the major buffer in the extracellular fluid, it may be consumed when the protons are produced. Thus, urea production is sometimes viewed as consuming two bicarbonate ions:[3,4]

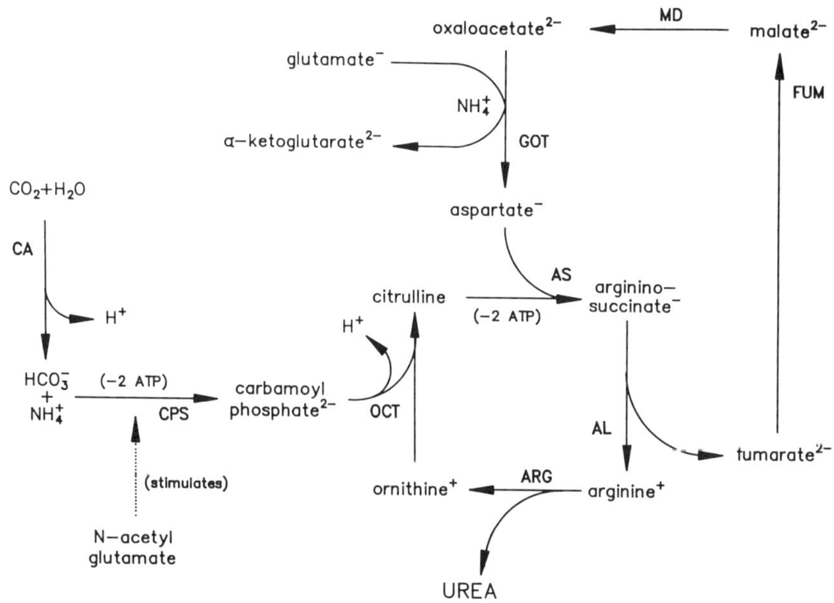

Figure 2 Urea cycle and related metabolic reactions in liver. CA, carbonic anhydrase; CPS, carbamoyl phosphate synthetase; OCT, ornithine carbamoyl transferase; AS, argininosuccinate synthetase; AL, argininosuccinate lyase; ARG, arginase; FUM, fumarase; MD, malate dehydrogenase; GOT, glutamate-oxaloacetate transaminase.

$$2NH_4^+ + 2HCO_3^- \longrightarrow urea + CO_2 + H_2O \qquad (3)$$

The metabolic cost of urea formation in mammals is substantial. Energy from four ATP molecules is required to produce one urea molecule. It has been estimated that hepatic urea production can account for up to 3% of whole body ATP utilization in humans.[4]

Under normal circumstances, more than 90% of excess amino acid nitrogen is eliminated through hepatic production followed by renal excretion of urea. Consequently, the regulation of urea production and excretion plays an essential role in the maintenance of nitrogen balance. Ammonium, the direct nitrogenous product of amino acid catabolism, is highly toxic and thus can cause serious disruption of neuromuscular function if its concentration in the peripheral blood rises substantially above 0.15 m*M*. Ammonium toxicity is seen in patients in which there is a failure of ureagenesis either due to hepatic failure or inherited deficiencies of urea cycle enzymes. Ammonium toxicity is prevented by the tight coupling of gluconeogenesis to ureagenesis, resulting in the efficient transfer of amino acid nitrogen to the relatively nontoxic nonelectrolyte urea. Thus, urea production prevents the release of ammonium into the general circulation. The liver can also extract NH_4^+ directly from the circulation, incorporating NH_4^+ into urea. Therefore, urea production in the liver serves two related functions: (1) maintenance of nitrogen balance through the production of a molecule, namely urea, which can be efficiently excreted by the kidneys; and (2) extraction and detoxification of circulating NH_4^+. When given systemically as ammonium chloride, ammonium is converted in the liver to urea.[5] When this occurs, protons are produced (Equation 2). Thus, systemic administration of ammonium chloride effectively acidifies the extracellular fluid despite the fact that ammonium, with a pKa of 9.03 at 37°C, is a poor proton donor *in vitro*. Thus, ammonium is an effective proton donor *in vivo* because of its rapid metabolic incorporation into urea (and other neutral products). The role of ammonium in net acid excretion by the kidney (see below) depends on this special role as an effective *in vivo* proton donor.

GLUTAMINE METABOLISM

Normally, nearly all nitrogen derived from excess amino acids is converted to urea in the liver. A second pathway for disposal of excess amino acid nitrogen is glutamine production which occurs not only in liver, but also in skeletal muscle. This pathway is important in the regulation of acid–base balance because it provides the substrate for ammoniagenesis in the kidney. Glutamine is formed in liver and muscle from glutamate in a reaction catalyzed by glutamine synthetase:

$$\text{Glutamate}^- + NH_4^+ \longrightarrow \text{glutamine} \tag{4}$$

The glutamate is generated in the liver largely from α-ketoglutarate via glutamate dehydrogenase:

$$\alpha\text{-ketoglutarate}^{2-} + NH_4^+ \longrightarrow \text{glutamate}^- \tag{5}$$

Production of α-ketoglutarate from glucose consumes two bicarbonate ions:

$$\text{glucose} + 2HCO_3^- \longrightarrow \alpha\text{-ketoglutarate}^{2-} + 3CO_2 \tag{6}$$

When all of the reactions leading to the formation of glutamine are summed, we obtain:

$$2NH_4^+ + 2HCO_3^- + \text{glucose} \longrightarrow \text{glutamine} + 3CO_2 \tag{7}$$

Like the production of urea, the production of each glutamine molecule consumes two ammonium ions and two bicarbonate ions.

The kidney uses glutamine for ammoniagenesis, a process that is accelerated during metabolic acidosis.[6] The increase in ammonium excretion allows net acid excretion (ammonium excretion + titratable acid excretion – bicarbonate excretion) to increase, ameliorating the acidosis. In addition to NH_4^+, the metabolism of glutamine in the kidney results in the *de novo* production of bicarbonate via the intermediate α-ketoglutarate.[7] The ability of ammonium excretion to alter systemic acid–base balance depends on the associated generation of bicarbonate. This bicarbonate, which is produced in the kidney, is returned to the general circulation, neutralizing acid formed from amino acid metabolism (Figure 1). Transport processes along the renal tubule transfer ammonium ions from their site of production in the proximal tubule to the final urine (see below). If the ammonium formed in the kidney leaked back into the general circulation, it would be incorporated into urea in the liver, thus consuming the bicarbonate originally produced from glutamine, and negating the entire regulatory process. Thus, mechanisms in the kidney that efficiently channel ammonium to the final urine are of premier importance to the overall regulation of systemic acid–base balance.

In response to systemic acid loading, urinary ammonium excretion increases at the expense of urea excretion.[3,5,8] Glutamine production rises in liver[8] and skeletal muscle[9,10] to match the increase in glutamine utilization by the kidney. Elevated hepatic glutamine production is associated with decreased urea production by the liver,[8,10] such that net utilization of amino acid nitrogen for glutamine and urea production is maintained at a level dictated by the rate of amino acid metabolism.

The argument has been made that changes in the relative rates of disposal of amino acid nitrogen through hepatic urea production vs. glutamine production could directly regulate systemic pH independent of renal function.[4,11] Because urea production consumes bicarbonate (Equation 3), it has been suggested that reciprocal changes in urea and glutamine production associated with changes in systemic pH could alter systemic bicarbonate balance so as to directly affect systemic pH. For example, this hypothesis proposes that acidosis would decrease hepatic urea production (and concomitantly increase glutamine production to maintain a constant rate of nitrogen disposal), and that an associated decrease in the rate of bicarbonate incorporation into urea would tend to correct the acidosis. A comparison of the overall processes of urea production and glutamine production (Equations 3 and 7), however, reveals that both processes consume two bicarbonate ions per neutral

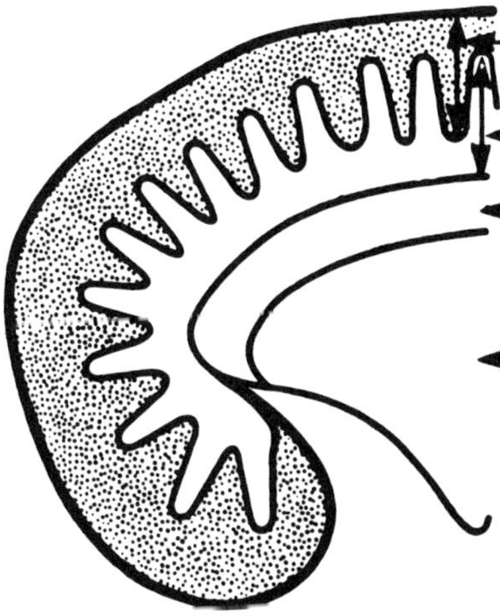

Figure 3 Regional architecture of the mammalian kidney. Diagram depicts a hemisected unipapillate kidney such as found in rat or rabbit. Kidneys of humans and many large mammals have multiple medullary units with architecture similar to that depicted here. (From Knepper, M. and Burg, M., *Am. J. Physiol.*, 244, F579, 1983.)

molecule produced. Consequently, urea production and glutamine production have the same net effects on systemic bicarbonate balance per nitrogen atom incorporated. Thus, changes in the distribution of amino acid nitrogen disposal between urea production and glutamine production have no direct effect on systemic bicarbonate balance or systemic pH. Rather, changes in glutamine production affect systemic pH only when the glutamine produced is utilized in the kidney for ammoniagenesis and the ammonium produced is excreted.

STRUCTURAL ORGANIZATION OF THE MAMMALIAN KIDNEY

The efficient excretion of urea and ammonium by the mammalian kidney depends on the complex structural organization of renal tubules and vascular elements within each of the regions of the kidney. Figure 3 is a diagram showing the distinct regional organization of the mammalian kidney. The arrangement of nephrons and collecting ducts in these regions is shown in Figure 4. The nomenclature for the renal tubule segments is summarized in Table 1.

RENAL TUBULES

Glomeruli and Proximal Convoluted Tubules

The glomerulus (Figure 4) is the site where blood plasma undergoes ultrafiltration to form the glomerular filtrate. This glomerular filtrate is progressively modified as it flows through the nephrons and collecting ducts. The initial part of the nephron is the proximal convoluted tubule (PCT), which is located totally within the cortical labyrinth (Figure 4). The PCT is comprised of two distinct subsegments called the S-1 and S-2 proximal tubules.

Loops of Henle

Two populations of nephrons join to form the collecting duct system (Figure 4). These two populations are defined by the length of their loops of Henle. One population ("long-looped" nephrons) has loops which bend in the inner medulla. In most mammalian species, more than 70% bend in the outer half of the inner medulla and relatively few loops reach the papillary tip, the deepest part of the inner medulla. The other population ("short-looped" nephrons) has loops which bend in the outer medulla. The loops of Henle receive the tubule fluid exiting the proximal convoluted tubules. They carry the tubule fluid into and out of the renal medulla establishing countercurrent flow between the two limbs of the loop. Several discrete nephron segments comprise the loop of Henle described in Figure 4 and Table 1. The descending part of the loop is made up

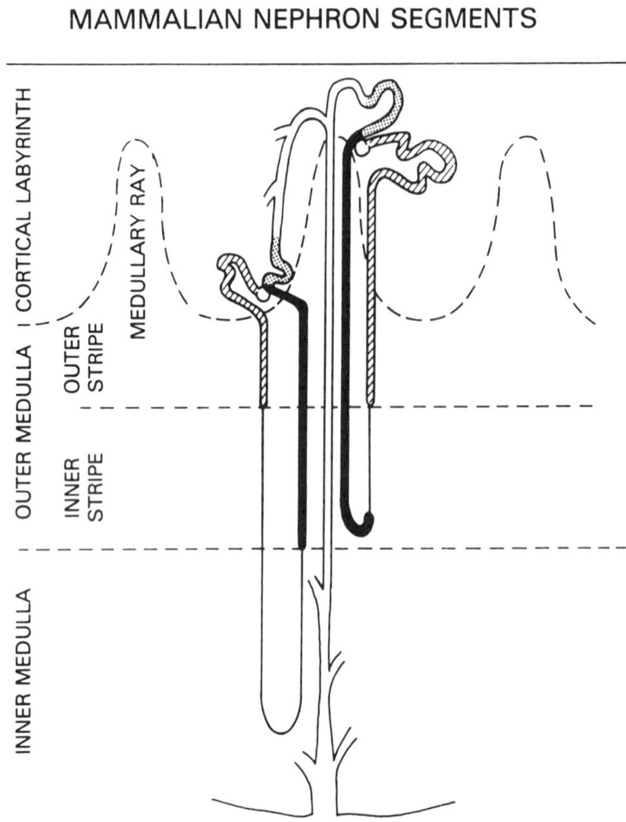

Figure 4 Mammalian renal structure. Major regions of the kidney are shown on left. Configurations of long-looped and short-looped nephrons are shown. The major portions of the nephron are: proximal tubules (hatched), thin limbs of Henle's loops (single line), thick ascending limbs of Henle's loops (solid), distal convoluted tubules (stippled), and the collecting duct system (open). (Modified from Knepper, M. A. and Stephenson, J. L., in *Physiology of Membrane Disorders*, 1986, 713. With permission.)

of the S2 proximal straight tubules in the medullary rays, the S3 proximal straight tubule in the outer stripe of the outer medulla, and the thin descending limbs in the inner stripe of the outer medulla and the inner medulla. Several distinct thin descending limb segments are recognized (Table 1). The ascending part of the loop of Henle consists of the ascending thin limbs (ATL) which are present only in long loops, the medullary thick ascending limbs (MTAL) in the outer medulla, and the cortical thick ascending limbs (CTAL) in the medullary rays of the cortex.

Distal Tubule Segments in the Cortical Labyrinth

The loop of Henle ends at the macula densa, a special portion of the thick ascending limb which is attached to the vascular pole of the glomerulus of the same nephron. After exiting the loop of Henle, tubule fluid enters the distal convoluted tubules in the cortical labyrinth. Several distal tubules merge to form a connecting tubule arcade in most mammalian species.[12] Arcades carry tubule fluid upward through the cortical labyrinth and deliver their tubule fluid to initial collecting tubules which join in the superficial cortex to form a cortical collecting duct.

Collecting Duct System

The collecting duct system crosses all the regions of the kidney from the superficial cortex to the tip of the inner medulla (Figure 4). The collecting ducts are arranged in parallel with the loop of Henle in the medulla and medullary rays. Like the loop of Henle, the collecting duct system is composed of several distinct tubule segments (Table 1), *viz.* the cortical collecting duct (CCD), the

Table 1 Nomenclature for Renal Tubule Segments

Major segment	Subsegment	Abbreviation	Region of kidney
Proximal convoluted tubule	Early proximal convolution	PCT(S1)	Cortical labyrinth
	Late proximal convolution	PCT(S2)	Cortical labyrinth
Loop of Henle	Early proximal straight	PST(S2)	Medullary ray
	Late proximal straight	PST(S3)	Outer medulla
	Descending thin limb of short loops	SDL	Outer medulla
	Descending thin limb of long loops, upper part	LDL_U	Outer medulla
	Descending thin limb of long loops, lower part	LDL_L	Inner medulla
	Ascending thin limb	ATL	Inner medulla
	Medullary thick ascending limb	MTAL	Outer medulla
	Cortical thick ascending limb	CTAL	Medullary ray
Distal	Distal convoluted tubule	DCT	Cortical labyrinth
	Connecting tubule	CNT	Cortical labyrinth
Collecting ducts	Initial collecting tubule	ICT	Cortical labyrinth
	Cortical collecting duct	CCD	Medullary ray
	Outer medullary collecting duct	OMCD	Outer medulla
	Inner medullary collecting duct (initial part)	$IMCD_i$	Inner medulla (Base)
	Inner medullary collecting duct (terminal part)	$IMCD_t$	Inner medulla (Papilla)

Note: Terminology used is that proposed by the Renal Commission of the International Union of Physiological Sciences[106] with two exceptions: (1) Terminology for descending thin limb subsegments is from Imai et al.[107] because it is more literally descriptive of the locations and topography of the segments; (2) expanded terminology for the inner medullary collecting duct is used based on recent studies demonstrating two different IMCD subsegments.[18,29]

outer medullary collecting duct (OMCD), the initial part of the inner medullary collecting duct ($IMCD_i$), and the terminal part of the inner medullary collecting duct ($IMCD_t$). The collecting ducts descend straight through the medullary rays and outer medulla without joining other collecting ducts. Repeated junctions occur in the inner medulla, however, resulting in a progressive reduction in the number of inner medullary collecting ducts toward the tip of the inner medulla. This reduction in collecting ducts, combined with the progressive decrease in the number of loops of Henle reaching successive levels of the inner medulla, is responsible for the tapered structure of the renal papilla (Figure 3). The final urine emerges from the kidney at the tip of the papilla to be delivered to the bladder via the ureter.

VASCULATURE

The vascular elements that carry blood into and out of the renal medulla are called the *vasa recta.* The descending vasa recta deliver blood to the capillary plexuses at each level of the medulla. Blood from the capillary plexus of the inner medulla feeds into ascending vasa recta. (Ascending vasa recta are never formed directly from descending vasa recta.) Ascending vasa recta from the inner medulla cross the inner stripe of the outer medulla in close physical association with the descending vasa recta in *vascular bundles.* These vascular bundles vary in size and complexity in various species.[13]

The counterflow arrangement of vasa recta in the medulla facilitates *countercurrent exchange* of solutes and water. Countercurrent exchange allows a reduction in the effective blood flow to the medulla while maintaining a high absolute perfusion rate.[14] A low effective blood flow is important to the preservation of solute concentration gradients in the medullary tissue, which are generated by *countercurrent multiplication* (see below).

In contrast to the medulla, the cortex has a high effective blood flow. The rapid blood flow in this region speeds the return of solutes and water absorbed from cortical parts of the nephron to the general circulation.

UREA HANDLING IN THE MAMMALIAN KIDNEY

DISTRIBUTION OF UREA TRANSPORT ALONG THE RENAL TUBULE

The general pattern of urea transport along the mammalian renal tubule has been determined in micropuncture experiments.[15,16] With a moderate dietary protein intake, the rate of urea excretion is generally between 30 and 50% of its rate of glomerular filtration in most mammalian species. Urea is freely filtered at the glomerulus, i.e., the urea concentration in the glomerular filtrate is the same as the concentration in the blood plasma. Forty to fifty percent of the filtered load of urea is reabsorbed along the proximal convoluted tubule. In the early distal tubules, the mass flow rate of urea is generally between 80 and 150% of filtered load indicating that there is net addition of urea along the loop of Henle. The increase in the mass flow of urea along the loop is owing to recycling of urea from the collecting ducts. Thus, urea is reabsorbed from the collecting ducts and secreted into the loops of Henle forming a recycling pathway between the two structures. Consistent with this concept, it has been demonstrated that the mass flow rate of urea near the bends of the long loops of Henle is 300 to 500% of the filtered load, suggesting that there is massive tubular secretion of urea into the descending limbs of Henle's loop. Micropuncture and microcatherization experiments have demonstrated rapid urea reabsorption along the terminal part of the inner medullary collecting duct, matching the rapid rate of urea entry into the loop of Henle.

Urea transport along the nephron occurs chiefly (if not totally) by passive permeation. The urea permeabilities of renal tubule segments are summarized in Figure 5. This figure is a diagram of the renal tubule showing the width of each segment distorted so that its width is proportional to the urea permeability coefficient in that segment. The proximal tubule is moderately permeable to urea, accounting for the proximal urea absorption seen in micropuncture studies (see above). The nature of the urea permeability pathway across the proximal tubule epithelium is not known. Among loop of Henle segments, high urea permeabilities were found in the inner medullary portion of the long-loop descending limb (IMDL), and in the ascending thin limb (tAL). Urea permeation across the thin limbs of Henle's loop is thought to occur via the paracellular pathway, i.e., via the tight junctions.[17] Among collecting duct segments, a high urea permeability has been found only in the terminal part of the inner medullary collecting duct ($IMCD_t$).[18] The urea permeability of the $IMCD_t$ increases to an extremely high value in response to vasopressin[19-22] (not shown on diagram). The $IMCD_t$ is the only tubule segment in which urea permeability is regulated. The high urea permeability of the $IMCD_t$ is due to the presence of a specialized phloretin-sensitive, vasopressin-regulated urea transporter in the apical and basolateral plasma membranes of the IMCD cells (see below).

ROLE OF UREA IN THE CONCENTRATING MECHANISM

The large amounts of urea produced in the liver must be excreted by the kidneys to keep pace with hepatic production. It might be expected that the high rate of urea excretion would obligate relatively large amounts of water excretion, owing to the osmotic effect of high luminal urea concentrations. Thus, urea would be expected to behave as an osmotic diuretic, inhibiting water absorption from the renal tubule lumen and forcing a high rate of urine flow. However, the classic studies of Gamble et al.[23] demonstrated that urea excretion obligates little water excretion above the level needed to excrete nonurea solutes. For example, rats and other rodents can increase the urinary urea concentration to greater than 1 mol while the rate of urinary flow remains low, typically less than 1/200th of the glomerular filtration rate.[24,25]

The ability of the kidneys to excrete large quantities of urea without obligating rapid water excretion is a consequence of the fact that urea accumulates to high concentrations in the renal medullary interstitium during water-conserving states. Chemical analysis of tissue from the medul-

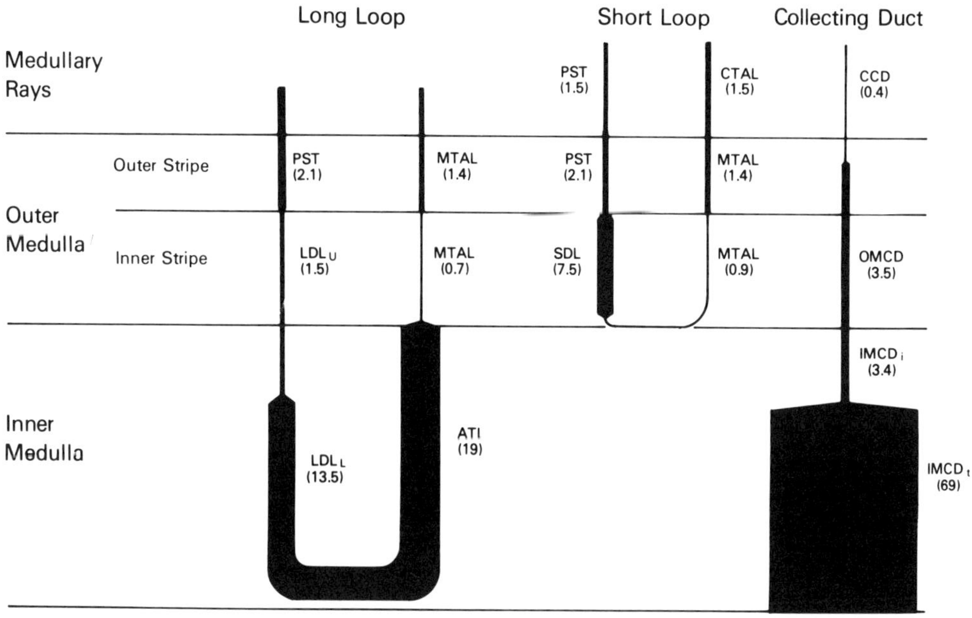

Figure 5 Urea permeability along the renal tubule(values in cm/s × 10^5). The width of each renal tubule segment is distorted to represent the relative urea permeability coefficient in that segment. (From Knepper, M. A. and Rector, F. C., Jr., in *The Kidney*, 1991, 445. With permission.)

las of rats has demonstrated that tissue urea concentration in the inner medulla can be 200 times higher than the urea concentration in the circulating blood (Figure 6). The high urea concentration in the inner medullary interstitium balances the osmotic effect of urea in the collecting duct lumen, permitting water to be reabsorbed osmotically from the collecting ducts driven by the medullary sodium chloride gradient.[14] Thus, the high concentration of urea in the medullary interstitium "cancels out" the osmotic effect of urea in the collecting duct lumen allowing water transport to be regulated independent of the rate of urea excretion.[26]

The maintenance of a high concentration of urea in the renal inner medulla requires that urea be delivered rapidly and continuously to the medullary interstitium from one of the structures that traverses the inner medulla. The structure that is responsible for most of the urea delivery is the inner medullary collecting duct.[27] The mechanism of accumulation is diagrammed in Figure 7. The tubule fluid entering the the collecting duct system from the distal convoluted tubule has a relatively low urea concentration. During antidiuresis, however, water is reabsorbed from the urea-impermeable parts of the collecting duct in the renal cortex and outer medulla. This water absorption is driven by the osmotic gradients across the collecting duct epithelium which result mainly from active NaCl transport in the thick ascending limb of Henle's loop. The absorption of water causes a progressive increase in the luminal urea concentration along the cortical and outer medullary collecting ducts. Subsequently, when the collecting duct fluid reaches the terminal part of the inner medullary collecting duct, which is highly permeable to urea (Figure 5), urea can exit rapidly. The urea is trapped in the interstitium of the inner medulla because the effective blood flow is low as a consequence of countercurrent exchange by the main blood vessels of the medulla, the *vasa recta*.[14] Because the urea permeability of the terminal IMCD is extremely high, particularly in the presence of vasopressin, urea nearly equilibrates across the IMCD epithelium when a steady state is established. This allows urea in the interstitium to almost completely balance osmotically the high urea concentration in the collecting duct lumen, and prevents the osmotic diuresis that would otherwise occur as discussed above. Thus, rapid vasopressin-regulated urea transport out of the inner medullary collecting ducts is vital to the process that concentrates urea in the renal medulla.

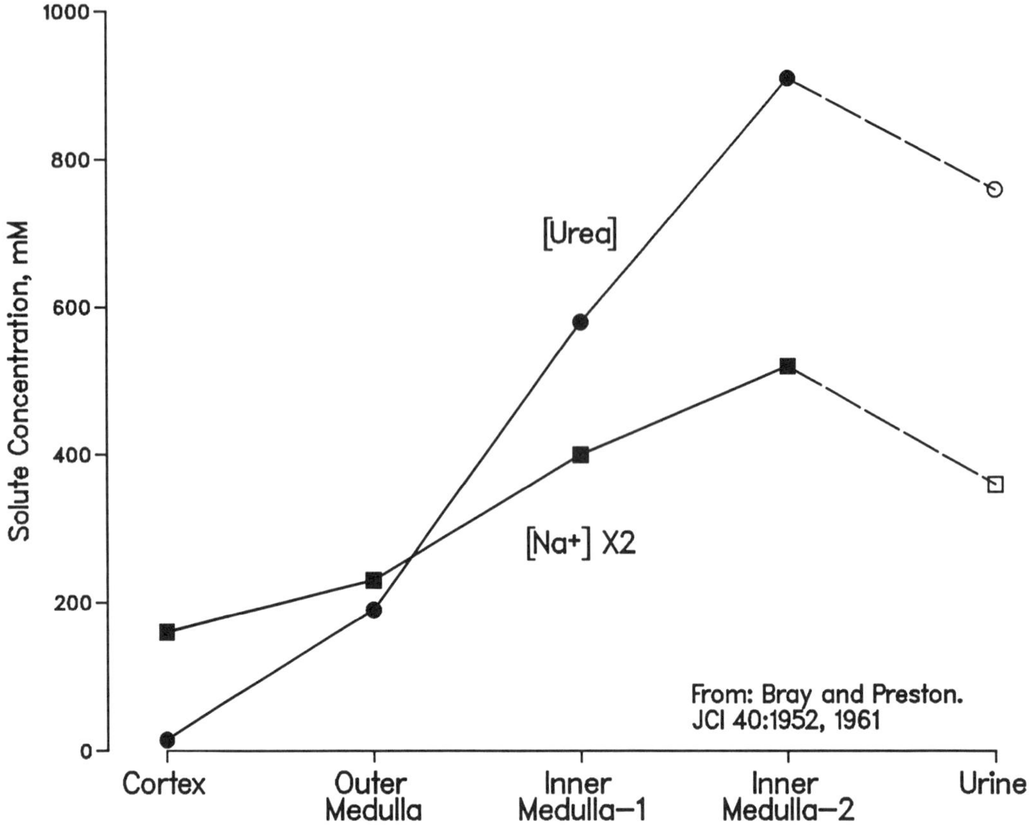

Figure 6 Urea and sodium ion gradients in the mammalian kidney. Sodium concentrations are multiplied by 2 to represent the effective osmolality of sodium plus its associated anion.

THE VASOPRESSIN-REGULATED UREA TRANSPORTER

The ability of vasopressin to increase the urea permeability of the IMCD was originally demonstrated by Morgan et al. in 1968.[28] It was not until the late 1980s, however, that investigators began to study the mechanism of rapid vasopressin-regulated urea transport in the IMCD. Sands and Knepper[18] showed that the high urea permeability was limited to the distal two-thirds of the IMCD in both rat and rabbit, i.e., the region possessing IMCD cells rather than principal cells.[29] In 1987, Sands et al.[30] reported that, in response to vasopressin, the urea permeability of the terminal IMCD rises to a level at least 200-fold greater than expected from lipid-phase diffusion and at least 25-fold greater than expected from paracellular transport alone. Thus, it was proposed[30] that an intrinsic membrane protein was present in the plasma membranes of IMCD cells that functions as a urea carrier, and that the transport properties of this carrier were modified by vasopressin. This view was supported by Kondo and Imai,[31] who showed that phloretin, an inhibitor of facilitated urea transport in erythrocytes,[32] decreases the urea permeability in the hamster IMCD without affecting diffusional water permeability. Similarly, Chou and Knepper[33] showed in isolated perfused rat terminal IMCD segments that phloretin blocks basal and vasopressin-stimulated urea transport without affecting osmotic water permeability, supporting the notion that the urea pathway is not the vasopressin-regulated water channel. Imai et al.[34] reported that the reflection coefficient for urea in the terminal IMCD ranged from 0.68 to 0.71, however, a value which would suggest that urea transport may occur via the water permeation pathway. In contrast, Knepper et al.[35,36] carried out measurements of reflection coefficients for urea (from solvent drag and water flux studies) which gave values that were not significantly different from theoretical value (0.96) for complete independence of water

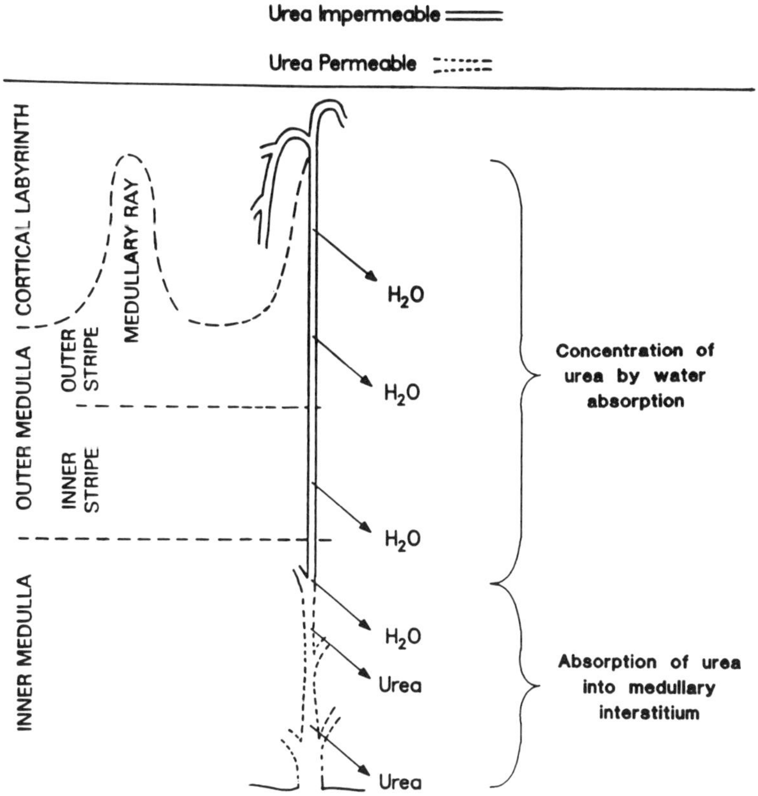

Figure 7 Mechanism of urea accumulation in mammalian renal medulla. Water is absorbed throughout the length of the collecting duct system. Because the cortical and outer medullary collecting duct are essentially impermeable to urea, water absorption concentrates the urea in the lumen. When the tubule fluid reaches the inner medullary collecting duct, which is highly permeable to urea, urea rapidly exits. This urea is trapped in the inner medulla as a result of the countercurrent exchange function of the vasa recta. (From Knepper, M. A. and Star, R. A., *Am. J. Physiol.*, 259, F393, 1990. With permission.)

and urea transport. Subsequently, Hamada et al.[37] have postulated the view that the low reflection coefficient found in their studies was due to restriction of urea diffusion through the cytosol of the collecting duct cells, rather than due to the penetration of urea through water channels.

The fact that phloretin inhibits transepithelial urea transport when added either to the luminal perfusate or peritubular bath suggests that phloretin-sensitive urea transporters may be present in both apical and basolateral membranes of the IMCD cells. Star[38] solidified this conclusion by direct measurements of urea permeation across apical and basolateral membranes using differential interference contrast microscopy in isolated perfused IMCD segments. He also demonstrated in the same study that the apical plasma membrane is normally rate limiting for transcellular urea transport and therefore is presumably the site where vasopressin-induced urea permeability changes occur.

Chou and Knepper[33] demonstrated that various structural analogues of urea which had been shown to be competitive inhibitors of urea transport in erythrocytes (*viz.* thiourea, acetamide, and methylurea) inhibit basal and vasopressin-stimulated urea transport in the terminal IMCD. As was the case for phloretin, these agents inhibit urea transport from either the apical or basolateral side of the IMCD epithelium. The half-maximal concentration for urea transport inhibition by thiourea (ID_{50}) was the same in the presence or absence of vasopressin, suggesting that vasopressin works by increasing the number of functional transporters, rather than by altering the urea transporter's affinity for urea.[33] Many of these findings were confirmed by Zhang and Verkman[39] who studied the effects of phloretin and structural analogs of urea on urea efflux in rat IMCD cells in suspension.

In addition, they demonstrated that the apparent activation energy for urea transport in these cells is very low (4.1 kcal/mol), strongly supporting the concept that urea moves across cell membranes via a specialized transporter rather than via the lipid phase.

Imai et al.[34] and Chou et al.[40] showed that under normal conditions, even high concentrations of urea (>300 m*M*) do not saturate the urea transport pathway. Under "zero-trans" conditions, however, in which the urea concentration was varied simultaneously in the luminal perfusate and peritubular bath, urea transport did show saturation with a K_m of 218 m*M*.

The evidence cited above had established with considerable certainty by 1990 that vasopressin-regulated urea transport in the IMCD is due to a specialized urea transporter which is likely to be an intrinsic membrane protein.[41] In 1993, You et al.[42] reported the cloning from rabbit medulla of a cDNA for a urea transporter which they call "UT-2". When the UT-2 cRNA was expressed in *Xenopus* oocytes, rapid urea transport was observed. The properties of this transport activity corresponded closely to the urea transport observed in the IMCD including phloretin-sensitivity, inhibition by structural analogues of urea, and lack of saturability of transport at submolar urea concentrations. UT-2 is a glycoprotein of 397 amino acids. The primary structure includes two long hydrophobic regions of approximately 120 amino acids each which presumably constitute membrane-spanning regions. This primary structure is quite unusual and is apparently unique among known intrinsic membrane proteins. *In situ* hybridization studies in rabbits showed labeling restricted to the terminal part of IMCD consistent with the known localization of vasopressin-regulated urea transport, but showed no labeling of earlier parts of the collecting duct. Northern blots revealed hybridization to mRNAs present in kidney and colon. The overall evidence gives strong support to the view that UT-2 is the vasopressin-regulated urea transporter. Interestingly, the northern blot analysis revealed no hybridization to mRNA from anemic spleen which was expected to be a site where red cell urea transporter mRNA would be present. This finding suggests that the red cell urea transporter and the renal urea transporter may be distinct proteins.

REGULATION OF RENAL UREA EXCRETION

The rate of renal excretion of urea is determined largely by its rate of production in the liver. When dietary protein intake is high, the urea concentration of the blood plasma rises,[43] increasing the amount of urea filtered by the glomerulus and as a consequence elevating urea excretion. Theoretically, the control of urea excretion could be entirely passive, resulting only from changes in filtered load of urea due to changes in concentration of urea in circulating blood plasma. The mammalian kidney has intrinsic control mechanisms, however, that contribute to the regulation of urea excretion.[25] These control mechanisms maintain the plasma urea concentration within narrower bounds than would otherwise be required to excrete urea at a rate equal to the urea "appearance rate" (which is equal to production rate minus the urea degradation rate by intestinal tract microorganisms). Maintenance of plasma urea concentration within a relatively narrow range is critical for renal water conservation. A plasma urea concentration that is too low is associated with a defect in renal water conservation owing to the central role that urea plays in the urinary concentrating mechanism.[23,44,45] A high plasma urea concentration causes osmotic diuresis,[46] which impairs renal water conservation. Control of urea excretion by the kidney involves both the control of glomerular filtration rate and the regulation of renal tubular transport of urea, as discussed below.

Regulation of Urea Excretion Via Changes in Glomerular Filtration Rate

An elevated dietary protein intake accelerates glomerular filtration rate.[47,48] Such elevations in glomerular filtration increase urea excretion by increasing the filtered load of urea.[49] Hormones associated with increased utilization of amino acids for gluconeogenesis and increased ureagenesis, e.g., glucagon and glucocorticoids, also have been shown to increase glomerular filtration rate,[50,51] and have been postulated to be in part responsible for variations in glomerular filtration rate with protein intake.

Regulation of Urea Excretion by Changes in Transport

Urea excretion has long been recognized to be controlled in part through regulated transport in the renal tubules.[25] Restriction of dietary protein results in a marked reduction in the fraction of filtered urea excreted compared to the fractional excretion with a normal protein intake.[24,43,52,53] This decline in fractional urea excretion is believed to be primarily due to increased renal tubular reabsorption of urea. Free-flow micropuncture experiments and microinjection studies in rats[54,55] have demonstrated that increases in tubular reabsorption of urea associated with dietary protein restriction occur predominantly in the renal collecting ducts. Microperfusion studies *in vivo* (standing droplet method) have demonstrated an increase in the apparent urea permeability of inner medullary collecting ducts of low-protein-fed rats, which suggests that an increase in passive or active reabsorption of urea may occur in this part of the collecting duct system.[56] Studies using isolated perfused inner medullary collecting ducts have demonstrated accelerated urea transport from the lumen of tubules dissected from rats given a low protein diet, possibly because of the induction of active urea transport.[57] The increased transport was seen only in collecting ducts from the outer part of the inner medulla, suggesting a mechanism for reducing medullary urea accumulation with dietary protein restriction by short-circuiting urea from the collecting ducts back to the general circulation.

Glucagon and glucocorticoids, hormones involved in the regulation of hepatic urea production, have been shown to increase the fraction of filtered urea excreted by rats maintained on a low protein diet.[24,58] These findings are consistent with possible roles for glucagon and glucocorticoids in the renal tubular regulation of urea excretion.

AMMONIUM HANDLING IN THE MAMMALIAN KIDNEY

AMMONIUM EXCRETION AND ACID–BASE BALANCE

The renal role in systemic acid–base balance is generally quantified in terms of net acid excretion (NH_4^+ excretion + titratable acid excretion – bicarbonate excretion). Balance studies done in humans by Relman et al.[59,60] revealed the essential role that ammonium excretion plays in the regulation of net acid excretion (Figure 8). Systemic acid loading resulted in an increase in the rate of net acid excretion adequate to excrete the acid load. Almost all of the increase in net acid excretion was due to an increase in ammonium excretion. In contrast, when an alkaline load was administered, rates of net acid excretion fell, primarily because of an increase in bicarbonate excretion. However, reduced ammonium excretion accounted for a significant fraction (about 25%) of the reduction in net acid excretion following alkali loading (Figure 8). Thus, it is clear that regulation of ammonium excretion is necessary for a physiologically appropriate response to both acid and alkali loading.

RENAL AMMONIUM PRODUCTION: ROLE OF THE PROXIMAL TUBULE

Ammonium concentrations are very low in circulating blood plasma (<0.1 m*M*). Consequently, very little ammonium is filtered at the renal glomerulus. Instead, the ammonium excreted in the urine arises from net production by renal tubule cells.[6,61] Ammonium is derived from metabolism of amino acids (predominantly glutamine) which are extracted from the blood by renal cells.[6,61]

The chief site of renal ammonium production in the kidney is the proximal tubule. Indeed, the proximal tubule appears to be the only site where ammonium production is regulated.[62-64] The metabolic processes associated with ammonium production in proximal tubule cells are shown in Figure 9. The first step, deamidation of glutamine by glutaminase, forms an ammonium ion and a glutamate ion. Next, glutamate is deaminated by glutamic dehydrogenase resulting in the formation of a second ammonium ion and the divalent anion, α-ketoglutarate. Alternatively, the glutamate can be consumed in a transamination reaction which also produces α-ketoglutarate but does not generate a free ammonium ion. The α-ketoglutarate is either oxidized via the citric acid cycle or it is utilized for gluconeogenesis depending on the animal species and the *in vivo* conditions. Both processes form bicarbonate ions from the carboxyl groups of the α-ketoglutarate.[7] This newly produced

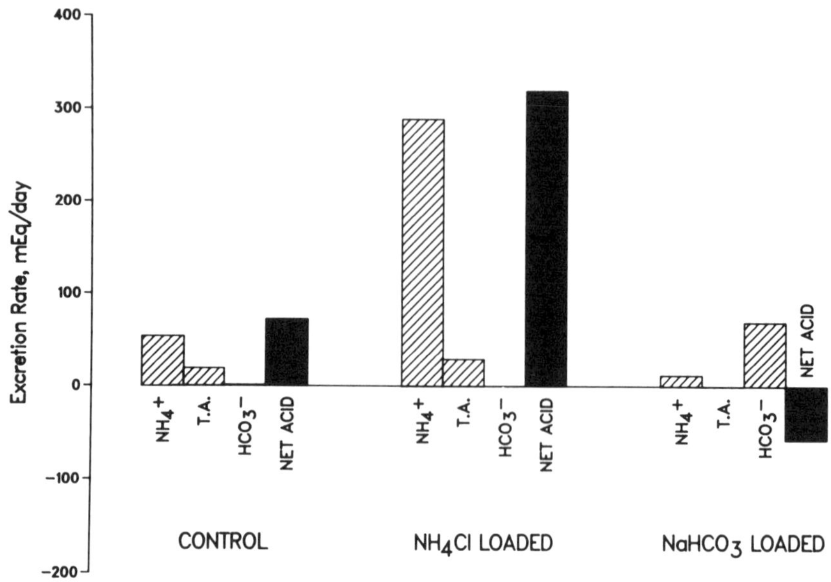

Figure 8 Components of net acid excretion in normal human subjects subjected to an acid load (NH_4Cl loaded) and an alkali load ($NaHCO_3$ loaded). T.A., titratable acidity. (From Knepper, M. A., Packer, R., and Good, D. W., *Physiol. Rev.*, 69, 179, 1989. With permission.)

bicarbonate exits the proximal tubule cells via the basolateral membrane, probably by sodium/bicarbonate cotransport,[65] to enter the general circulation. There it neutralizes acids formed elsewhere in the body or those ingested. Thus, through the regulation of glutamine metabolism and the associated production of bicarbonate, the kidney regulates systemic acid–base balance. (Note that some of the α-ketoglutarate produced from glutamate is transported across the basolateral plasma membrane via a specialized organic anion transporter and is either oxidized or utilized for gluconeogenesis elsewhere in the body.)

For the process of glutamine metabolism and the resulting production of new bicarbonate to effectively alter systemic pH, however, the ammonium ions generated in the proximal tubule cells must be excreted in the urine. If instead the ammonium formed in the proximal tubule was returned to the general circulation, it would be utilized for the production of urea and other neutral products such as glutamine in the liver and other organs. Since these processes release protons or utilize bicarbonate (see above), they would essentially cancel the buffering effect of the bicarbonate formed during ammoniagenesis in the kidney.[7,66] Thus, the transport processes that transfer ammonium from its site of production in the proximal tubule to the final urine are critical to the overall process allowing the kidney to eliminate systemic acid loads.

Figure 9 also emphasizes the fact that when protons are excreted as ammonium, the protons and the buffer molecules with which they are excreted (NH_3) are formed in the same metabolic pathway within the proximal tubule cells. (It is not clear whether NH_4^+ is formed directly or whether H^+ and NH_3 are formed independently and combine in the proximal tubule cell to form NH_4^+. This distinction, however, is inconsequential physiologically.) Transport processes must account for the transfer of NH_4^+ *in toto*, not simply the movement of NH_3, in order to fulfill the role of proximal ammonium production in systemic pH regulation. Thus, to eliminate the protons formed in the proximal tubule as ammonium, renal transport processes must either move NH_4^+ directly or must transfer equal quantities of NH_3 and H^+ so that the overall transport is equivalent to that of an ammonium ion. This is our reason for using the term "ammonium transport" rather than "ammonia transport" as is common in the mammalian acid–base transport literature.

If ammonium produced in the proximal tubule cells exited the cells solely by diffusion of its NH_3 component, most of the ammonium would be expected to exit the basolateral membrane *in vivo*

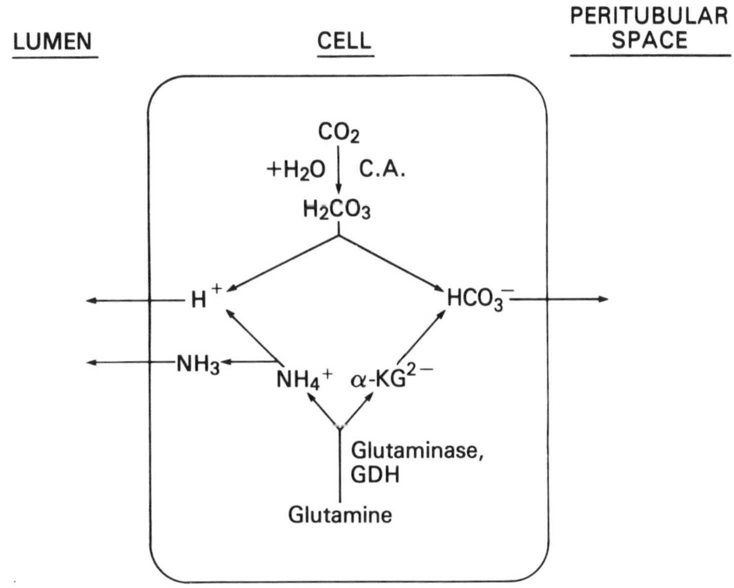

Figure 9 Cellular chemical reactions in proximal tubule cells involved in acid-base transport. C.A., carbonic anhydrase; α-KG^{2-}, α-ketoglutarate; GDH, glutamic dehydrogenase. (From Knepper, M. A. and Burg, M. B., in *Diseases of the Kidney*, 1988, 211. With permission.)

owing to the fact that the rapid blood flow in the cortical labyrinth would provide a better sink than the relatively slow-flowing tubule fluid.[66] Most of the ammonium exiting across the basolateral membrane would be lost to the general circulation. Yet, a large fraction of ammonium produced in the proximal tubule is excreted in the urine. This is explained in part by the fact that the ammonium produced in the proximal tubule is preferentially secreted across the apical plasma membrane into the tubule fluid as demonstrated in studies by Nagami and Kurokawa.[67,68] They perfused mouse proximal tubules *in vitro* in the presence of glutamine, but in the absence of exogenous ammonium, and observed that most of the ammonium produced by the cells entered the lumen. Much more ammonium entered the lumen than could be explained by diffusion trapping. In fact, the luminal ammonium concentration exceeded the ammonium concentration in the peritubular bath by approximately 300-fold. Because the proximal tubule can lower the luminal pH by no more than 1 pH unit, a lumen-to-bath NH_4^+ concentration ratio no greater than 10 could result from simple diffusion trapping (i.e., with equal NH_3 concentrations in lumen and peritubular fluid). The results of Nagami and Kurokawa[67] imply that the luminal NH_3 concentration rises to well above that of the peritubular fluid as a result of preferential secretion of ammonium across the apical membrane.

One possible explanation for the preferential transport of ammonium across the apical plasma membrane of the proximal tubule is that the transport may occur by NH_3 diffusion and that the apical NH_3 permeability may greatly exceed that of the basolateral membrane.[66] Preisig and Alpern[69] showed, however, that the NH_3 permeabilities of the apical and basolateral membranes of the rat proximal convoluted tubule are nearly equal (apical, 6×10^{-2} cm/s; basolateral, 7×10^{-2} cm/s).

Nagami[68] suggested that the preferential secretion of ammonium in the proximal tubule results from direct NH_4^+ transport across the apical plasma membrane. This hypothesis was based on the observation of Kinsella and Aronson[70] in rabbit renal brush-border membrane vesicles that NH_4^+ can substitute for H^+ on the apical Na^+-H^+ exchanger. Nagami[68] examined this possibility in isolated mouse proximal tubules by perfusing the lumen with a solution containing 0.1 m*M* amiloride and only 10 m*M* Na^+ (steps that should effectively block the apical Na-H exchanger). Under these conditions, preferential secretion of ammonium produced by the tubules did not occur. He concluded that this effect was not simply due to inhibition of luminal acidification by Na^+-H^+ exchange because reducing the perfusate pH to 6.2 did not eliminate the inhibitory effect of the combination of amiloride and low luminal sodium. Consequently, he drew the conclusion that the preferential secretion of ammonium in mouse proximal tubules occurs in part by direct apical NH_4^+-Na^+ exchange.

It is possible that other specialized NH_4^+ transport processes in the apical membrane could be partly responsible for the preferential secretion of ammonium in the proximal tubule. For example, it is conceivable that ammonium secretion may occur via an organic base secretion pathway. Another possibility is that a primary NH_4^+ pump may be present in the apical membrane similar to the NH_4^+ transporter present in many bacteria and fungi.[66] Finally, inwardly directed NH_4^+ uptake mechanisms in the basolateral plasma membrane could help bias net ammonium transport toward the lumen. One such transport process that has been described is NH_4^+ active transport into cells by substitution of NH_4^+ for K^+ on the Na^+-K^+-ATPase,[71] which normally transports Na^+ out of the cell and K^+ into the cell.

ROLE OF AMMONIUM TRANSPORT IN REGULATION OF AMMONIUM EXCRETION: DIFFUSION TRAPPING VS. DIRECT AMMONIUM ION TRANSPORT

Studies by Tizianello et al.[72] quantified ammonium production and excretion in humans. These studies demonstrated that under control conditions, the kidneys excreted 55% of the ammonium produced. After acid loading for 1 d, ammonium production increased by 63% and the efficiency of excretion increased so that 76% of the ammonium produced was excreted in the urine. Thus, transport processes in the kidney provide an efficient mechanism by which a very large proportion of ammonium produced in the kidney is excreted. These studies also demonstrate that regulation of ammonium excretion by the kidney depends not only on regulation of ammonium production, but also on regulation of ammonium transport.

For many years, the model utilized to explain the transfer of ammonium to the urine has been the diffusion trapping model.[73] Briefly, this model assumes that NH_4^+ cannot move across renal tubule epithelia and that NH_3 diffuses so rapidly that NH_3 concentration gradients are obliterated. Consequently, this model predicts that NH_4^+ will accumulate in various renal compartments in proportion to the free H^+ concentration, i.e., in regions of low pH. Thus, acidic urine would be expected to accumulate NH_4^+ to a relatively high concentration, as observed. Although the diffusion trapping model successfully predicts the qualitative behavior of the kidney (i.e., an increase in ammonium excretion when the urine is more acidic), the model fails to provide a quantitative explanation for the levels of ammonium excretion actually observed. For example, in the study of Tizianello et al.[72] cited above, the diffusion trapping model predicts that in the control condition, 10% of the ammonium formed in the kidney should have appeared in the urine rather than the 55% actually observed.[74] Thus, some other mechanism must be present to explain how such a large fraction of the ammonium produced in the kidney can be excreted. Extensive studies carried out over the last 10 years have revealed that the excretion of ammonium by the kidney involves a process that concentrates ammonium in the interstitium of the renal medulla, viz. countercurrent multiplication of ammonium.

COUNTERCURRENT MULTIPLICATION OF AMMONIUM IN THE RENAL MEDULLA

Micropuncture experiments in the rat kidney[75,76] have demonstrated that the NH_3 concentration in the urine and renal medulla markedly exceeds that in the renal venous blood or in the tubule segments of the renal cortex. The high NH_3 concentration in the medulla is because ammonium accumulates in the renal medulla, resulting in the elevation of the concentrations of both NH_3 and NH_4^+. Studies by Robinson and Owen[77] and Sullivan[78] in 1965 were the first to demonstrate total ammonia accumulation in the renal medulla. These authors proposed independently that the medullary ammonium accumulation occurs by countercurrent multiplication. At that point in time, however, an active "single effect" that would provide the energy for such a countercurrent multiplier for ammonium had not been identified. Consequently, the possibility of countercurrent multiplication of ammonium was not widely accepted until 1984 when Good et al.[79] showed that NH_4^+ is absorbed directly across the epithelium of the thick ascending limb of Henle's loop. The experiments demonstrated that direct NH_4^+ transport by cells of the thick ascending limb epithelium decreased the luminal ammonium concentration to well below the concentration surrounding the tubule. Based on these observations, it was proposed that active NH_4^+ absorption from the thick ascending limb drives the countercurrent multiplication of NH_4^+ by a process similar to the countercurrent multiplication of NaCl.[79,80]

The proposed model for countercurrent multiplication of ammonium is illustrated in Figure 10. The ammonium absorbed by the thick ascending limb must be secreted into the descending limb of Henle's loop, creating a recycling pathway between the two limbs of the loop. Flessner et al.[81] have demonstrated relatively high NH_4^+ and NH_3 permeabilities in the outer medullary descending limbs of long loops of Henle, compatible with the requirement for recycling. In addition, the outer stripe portion of the descending limb (the S-3 proximal tubule) in the rabbit has been studied directly by *in vitro* microperfusion methodology.[82] These studies have demonstrated that net ammonium secretion in this segment occurs by parallel H^+ and NH_3 secretion (nonionic diffusion), a process which is enhanced by the presence of an acidic luminal disequilibrium pH. Thus, the properties of the loop of Henle segments in the outer medulla are consistent with requirements for the countercurrent multiplication of ammonium.

To account for the ammonium accumulation in the inner medullary interstitium and the transfer of ammonium from the loops to the collecting ducts in the inner medulla, ammonium must exit either the thin descending limb or the thin ascending limb (Figure 4). Recently, studies by Flessner et al.[83] have shown that the NH_4^+ permeability of the thin ascending limb of the rat is very high (>10^{-4} cm/s), as is true for other monovalent cations. Direct studies of NH_4^+ and NH_3 transport in the inner medullary part of the thin descending limb suggest that there is little or no transport of either entity in this portion of the loop of Henle.

MECHANISM OF ACTIVE NH_4^+ TRANSPORT IN THE THICK ASCENDING LIMB

NH_4^+ is directly absorbed across the thick ascending limb epithelium.[79] The NH_4^+ absorption occurs largely by direct active transport of NH_4^+.[84,85] The active transport is a secondary active process that is a consequence of coupled Na^+-NH_4^+-$2Cl^-$ entry across the apical plasma membrane resulting from substitution of NH_4^+ for K^+ on the apical Na^+-K^+-$2Cl^-$ cotransporter (Figure 10). The transport function of this transporter is blocked by diuretics such as furosemide and bumetanide.

Several studies support the role of the apical Na^+-K^+-$2Cl^-$ transporter in active ammonium transport in the thick ascending limb. First, studies in isolated apical membrane vesicles from rabbit thick ascending limbs demonstrated that an NH_4^+ gradient can drive bumetanide-sensitive ^{22}Na uptake.[86] Second, isolated perfused tubule studies in rat thick ascending limbs[85,87] showed that increasing K^+ concentrations in the lumen competitively inhibited NH_4^+ absorption. Third, isolated perfused tubule studies in rabbit thick ascending limbs[84] showed that furosemide inhibits active absorption of ammonium independent of its effect on transepithelial voltage and that active chloride absorption (which depends on apical Cl^- entry via the Na^+-K^+-$2Cl^-$ cotransporter) persists when NH_4^+ replaces K^+ in solutions perfusing and bathing isolated perfused thick ascending limbs. Finally, when NH_4^+ was added to the solution perfusing mouse thick ascending limb cells, the intracellular pH underwent a marked decrease that was attenuated by adding furosemide.[88] The acidification occurred because NH_4^+ enters across the apical membrane via NH_4^+-Na^+-$2Cl^-$ cotransport and dissociates in the cell to release protons when NH_3 exits across the basolateral membrane.

The paracellular pathway in the thick ascending limb is cation selective and has a high permeability to several cations.[89] Thus, it can be expected that NH_4^+ can penetrate the tight junctions and that passive NH_4^+ fluxes can contribute to the net transport of ammonium across the thick ascending limb epithelium. Direct measurements have demonstrated that the transepithelial NH_4^+ permeability is indeed relatively high (15×10^{-5} cm/s in the medullary thick ascending limb of the rabbit[84]; 6×10^{-5} cm/s in the medullary thick ascending limb of the rat[85]). Therefore, it has been proposed that the lumen positive transepithelial potential normally present in the thick ascending limb can drive passive NH_4^+ absorption which would augment the active flux. Passive paracellular transport was responsible for about 1/3 of net ammonium absorption in isolated perfused thick ascending limbs from both rat and rabbit.

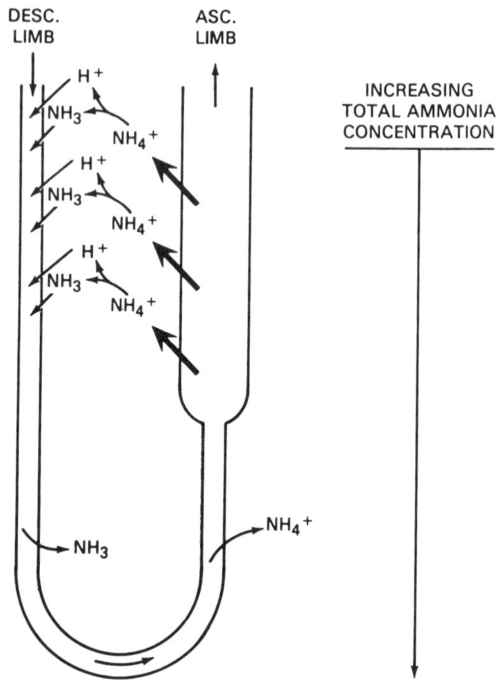

Figure 10 Countercurrent multiplication of ammonium in the mammalian renal medulla. (Adapted from Good, D. W. and Knepper, M. A., *Am. J. Physiol.*, 248, F459, 1985. With permission.)

Measurements of the transepithelial NH_3 permeability of isolated perfused thick ascending limbs[84] showed an extraordinarily low value — 1/20th the value found in rabbit proximal convoluted tubules. The low NH_3 permeability of the thick ascending limb is important to the active transport process in this segment because it prevents NH_3 backflux into the lumen and therefore allows direct NH_4^+ absorption to result in the net absorption of total ammonia which lowers both the NH_3 and NH_4^+ concentrations in the lumen. Studies by Kikeri et al.[88] suggested that the low transepithelial NH_3 permeability is attributable to an extremely low apical membrane NH_3 permeability. It seems possible that the low NH_3 permeability of the apical membrane is the result of a specialized lipid composition that limits NH_3 diffusion.

MECHANISM OF AMMONIUM TRANSPORT ACROSS THE COLLECTING DUCT EPITHELIUM

Collecting ducts have little capacity for ammonium production.[62] Nevertheless, extensive ammonium secretion has been observed in rat collecting ducts *in vivo*,[90-93] which is due mainly to transepithelial transport. The mechanism of transepithelial ammonium transport in collecting ducts has been studied extensively in isolated perfused tubules.[94-100] The main conclusion from these studies is that transepithelial ammonium secretion in all collecting duct segments occurs primarily (if not exclusively) by parallel H^+ and NH_3 secretion. This process somewhat differs from the classical diffusion trapping model due to the fact that the observed NH_3 secretion depends on the presence of an interstitium-to-lumen NH_3 concentration gradient resulting from countercurrent multiplication of total ammonia (NH_4^+ + NH_3) in the renal medullary interstitium.[76] Thus, the transepithelial NH_3 concentration gradient that drives NH_3 entry into the collecting ducts is present mainly because there is a high NH_3 concentration in the medullary interstitium. The transepithelial NH_3 gradient is enhanced by a low luminal pH which shifts the NH_3-NH_4^+ buffer reaction toward NH_4^+ reducing the luminal NH_3 concentration. Consequently, luminal acidification influences the rate of NH_3 secretion although it is not the sole determinant of the rate.

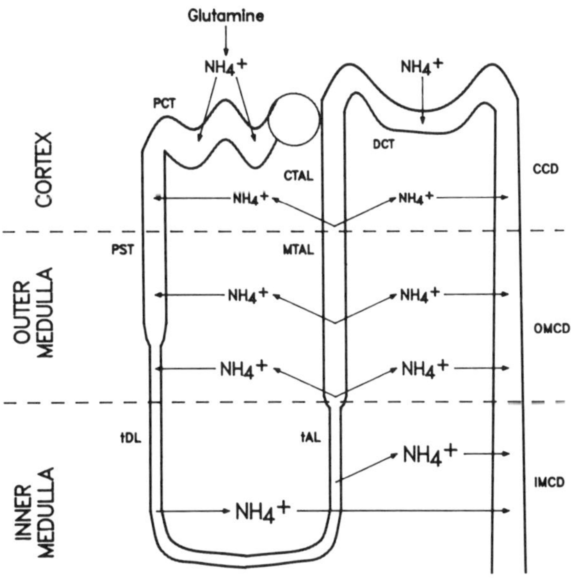

Figure 11 Ammonium pathways in the mammalian kidney. This figure is not intended to represent ammonia/ammonium transport mechanisms; where NH_4^+ movement is indicated, NH_4^+ could be transported as ionic ammonium directly, or by a combination of NH_3 and H^+ transport as in the collecting ducts. Transport out of the loops of Henle generates a cortico-medullary NH_4^+ concentration gradient in the renal interstitium. See text for full description. PCT, proximal convoluted tubule; PST, proximal straight tubule; tDL, thin descending limb; tAL, thin ascending limb; MTAL, medullary thick ascending limb; CTAL, cortical thick ascending limb; DCT, distal convoluted tubule; CCD, cortical collecting duct; OMCD, outer medullary collecting duct; IMCD, inner medullary collecting duct. (From Knepper, M. A., Packer, R., and Good, D. W., *Physiol. Rev.*, 69, 179, 1989. With permission.)

NH_4^+ and NH_3 permeability measurements in collecting duct segments have supported the view that net NH_4^+/NH_3 transport into the collecting ducts occurs primarily as a result of diffusion of NH_3.[94-100] The NH_4^+ permeability of isolated perfused collecting ducts from rats was immeasurably low (<10^{-5} cm/s). In contrast, the NH_3 permeability coefficients were quite large in rat collecting ducts:cortical, 23×10^{-3} cm/s; outer medullary, 13×10^{-3} cm/s; inner medullary, 2×10^{-3} cm/s. Protons, secreted into the collecting duct lumina, react with the secreted NH_3 to form NH_4^+.

Proton secretion in collecting ducts is mediated largely by direct active H^+ across the apical membrane catalyzed by a specialized H^+-ATPase. In addition, the H^+ secretion in collecting ducts is believed to occur in part via an H^+-K^+-ATPase similar to that present in the gastric mucosa. These pumps[101] can move H^+ against large concentration gradients and can lower the luminal pH in collecting ducts 2 pH units or more below the peritubular fluid.

Most of the collecting duct lacks luminal carbonic anhydrase, an enzyme that catalyzes the dehydration of carbonic acid in other renal tubule segments such as the proximal tubule and thick ascending limb. Carbonic acid is formed when actively secreted proton titrate bicarbonate. Because of the accumulation of carbonic acid in the lumen of the collecting duct, a so-called disequilibrium pH develops which lowers the luminal pH more than would be expected from a given rate of proton secretion in the presence of carbonic anhydrase. Several studies have demonstrated that the presence of an acidic disequilibrium pH can markedly accelerate the rate of NH_3 secretion in collecting ducts.[95,99,102-105] The enhancement occurs because the disequilibrium pH lowers the luminal NH_3 concentration and consequently increases the NH_3 gradient that drives NH_3 entry.

SUMMARY: DIRECT NH_4^+ TRANSPORT PROCESSES IN THE KIDNEY

The overall process that accounts for the efficient transfer of ammonium from its production site in the proximal tubule cells to the final urine is complex (Figure 11). As described above, it involves preferential secretion of NH_4^+, countercurrent multiplication of ammonium by the loops of Henle (which concentrates both NH_4^+ and NH_3 in the renal medulla), and parallel H^+ and NH_3 secretion into the collecting ducts from the medullary interstitium.

We have described several examples of physiologically important direct NH_4^+ transport in the kidney: (1) Na^+-NH_4^+ exchange across the apical plasma membrane of the proximal tubule is responsible for the preferential secretion of ammonium in the proximal tubule; (2) coupled Na^+-NH_4^+-$2Cl^-$ transport across the apical membrane of the thick ascending limb of Henle's loop mediates secondary active transport of ammonium, which drives countercurrent multiplication of ammonium in the renal medulla; (3) part of the NH_4^+ uptake across the apical membrane of the thick ascending limb may occur as a result of penetration by NH_4^+ through apical K^+ channels; (4) NH_4^+ can be actively transported into cells by substitution of NH_4^+ for K^+ on the Na^+-K^+-ATPase; and (5) NH_4^+ penetrates the paracellular pathway in some nephron segments such as the proximal tubule and thick ascending limb. Simple passive diffusion of NH_4^+ via the paracellular pathway is thought to be physiologically important in the thick ascending limb where it contributes to net NH_4^+ absorption.

We emphasize that the classical diffusion trapping model does not provide an accurate overall description of renal ammonium handling except in the collecting duct. Rather, a combination of countercurrent multiplication of ammonium in the renal medulla driven by active NH_4^+ transport in the thick ascending limb, and diffusion trapping of ammonium in the collecting ducts driven by active proton secretion appear to be the major processes responsible for the concentration of ammonium in the urine.

REFERENCES

1. **Krebs, H. A. and Henseleit, K.,** Untersuchungen ueber die Harnstoffbildung im Tierkoerper, *Hoppe-Seylers Z. Physiol. Chem.*, 210, 33, 1932.
2. **Robinson, R. R. and Schmidt-Nielsen, B.,** Distribution of arginase within the kidneys of several vertebrate species, *J. Cell. Comp. Physiol.*, 62, 147, 1963.
3. **Oliver, J., Koeltz, A. M., Costello, J., and Bourke, E.,** Acid-base induced alterations in glutamine metabolism and ureagenesis in perfused muscle and liver of the rat, *Eur. J. Clin. Invest.*, 7, 445, 1977.
4. **Atkinson, D. E. and Camien, M. N.,** The role of urea synthesis in the removal of metabolic carbonate and the regulation of blood pH, *Curr. Top. Cell. Regul.*, 21, 261, 1982.
5. **Adolph, E. F.,** The metabolism of ammonium salts and of urea in man, *Am. J. Physiol.*, 71, 355, 1924.
6. **Pitts, R. F.,** Control of renal production of ammonia, *Kidney Int.*, 1, 297, 1972.
7. **Halperin, M. L. and Jungas, R. L.,** Metabolic production and renal disposal of hydrogen ions, *Kidney Int.*, 24, 709, 1983.
8. **Phromphetcharat, V., Jackson, A., Dass, P. D., and Welbourne, T. C.,** Ammonia partitioning between glutamine and urea: interorgan participation in metabolic acidosis, *Kidney Int.*, 20, 598, 1981.
9. **Schroeck, H. and Goldstein, L.,** Interorgan relationships for glutamine metabolism in normal and acidotic rats, *Am. J. Physiol.*, 240, E519, 1981.
10. **Welbourne, T. C.,** Effect of metabolic acidosis on hindquarter glutamine and alanine release, *Metabolism*, 35, 614, 1986.
11. **Haussinger, D. and Gerok, W.,** Hepatic urea synthesis and pH regulation, Role of CO_2, HCO_3^-, pH and the activity of carbonic anhydrase, *Eur. J. Physiol.*, 152, 381, 1985.
12. **Sperber, I.,** Studies on the mammalian kidney, *Zool. Bidrag. Uppsala*, 22, 249, 1944.
13. **Bankir, L. and DeRouffignac, C.,** Urinary concentrating ability: insights from comparative anatomy, *Am. J. Physiol.*, 249, R643, 1985.
14. **Berliner, R. W., Levinsky, N. G., Davidson, D. G., and Eden, M.,** Dilution and concentration of the urine and the action of antidiuretic hormone, *Am. J. Med.*, 27, 730, 1958.
15. **Marsh, D. J. and Knepper, M. A.,** Renal handling of urea, in *Handbook of Physiology. Section 8: Renal Physiology*, Oxford University Press, New York, 1992, 1317.
16. **Knepper, M. A. and Rector, F. C., Jr.,** Urinary concentration and dilution, in *The Kidney*, Saunders, Philadelphia, PA, 1991, 445.
17. **Chou, C.-L. and Knepper, M. A.,** *In vitro* perfusion of chinchilla thin limb segments: urea and NaCl permeabilities, *Am. J. Physiol.*, 264, F337, 1993.
18. **Sands, J. M. and Knepper, M. A.,** Urea permeability of mammalian inner medullary collecting duct system and papillary surface epithelium, *J. Clin. Invest.*, 79, 138, 1987.

19. **Morgan, T. and Berliner, R. W.,** Permeability of the loop of Henle, vasa recta, and collecting duct to water, urea, and sodium, *Am. J. Physiol.*, 215, 108, 1968.
20. **Rocha, A. S. and Kudo, L. H.,** Water, urea, sodium, chloride, and potassium transport in the *in vitro* isolated perfused papillary collecting duct, *Kidney Int.*, 22, 485, 1982.
21. **Sands, J. M., Nonoguchi, H., and Knepper, M. A.,** Vasopressin effects on urea and H_2O transport in inner medullary collecting duct subsegments, *Am. J. Physiol.*, 253, F823, 1987.
22. **Star, R. A., Nonoguchi, H., Balaban, R., and Knepper, M. A.,** Calcium and cyclic adenosine monophosphate as second messengers for vasopressin in the rat inner medullary collecting duct, *J. Clin. Invest.*, 81, 1879, 1988.
23. **Gamble, J. L., McKhann, C. F., Butler, A. M., and Tuthill, E,** An economy of water in renal function referable to urea, *Am. J. Physiol.*, 109, 139, 1934.
24. **Knepper, M. A., Danielson, R. A., Saidel, G. M., and Johnston, K. H.,** Effects of dietary protein restriction and glucocorticoid administration on urea excretion in rats, *Kidney Int.*, 8, 303, 1975.
25. **Schmidt-Nielsen, B.,** Urea excretion in mammals, *Physiol. Rev.*, 38, 139, 1958.
26. **Knepper, M. and Burg, M.,** Organization of nephron function, *Am. J. Physiol.*, 244, F579, 1983.
27. **Knepper, M. A. and Roch-Ramel, F.,** Pathways of urea transport in the mammalian kidney, *Kidney Int.*, 31, 629, 1987.
28. **Morgan, T., Sakai, F., and Berliner, R. W.,** *In vitro* permeability of medullary collecting ducts to water and urea, *Am. J. Physiol.*, 214, 574, 1968.
29. **Clapp, W. L., Madsen, K. M., Verlander, J. W., and Tisher, C. C.,** Morphologic heterogeneity along the rat inner medullary collecting duct, *Lab. Invest.*, 60, 219, 1989.
30. **Sands, J. M., Nonoguchi, H., and Knepper, M. A.,** Vasopressin effects on urea and H_2O transport in inner medullary collecting duct subsegments, *Am. J. Physiol.*, 253, F823, 1987.
31. **Kondo, Y. and Imai, M.,** Effects of glutaraldehyde fixation on renal tubular function. I. Preservation of vasopressin-stimulated water and urea pathways in rat papillary collecting duct, *Pflügers Arch.*, 408, 479, 1987.
32. **Macey, R. I. and Farmer, R. E. L.,** Inhibition of water and solute permeability in human red cells, *Biochim. Biophys. Acta*, 211, 104, 1970.
33. **Chou, C.-L. and Knepper, M. A.,** Inhibition of urea transport in inner medullary collecting duct by phloretin and urea analogues, *Am. J. Physiol.*, 257, F359, 1989.
34. **Imai, M., Taniguchi, J., and Yoshitomi, K.,** Osmotic work across inner medullary collecting duct accomplished by difference in reflection coefficients for urea and NaCl, *Pflügers Arch.*, 412, 557, 1988.
35. **Knepper, M. A., Sands, J. M., and Chou, C.-L.,** Independence of urea and water transport in rat inner medullary collecting duct, *Am. J. Physiol.*, 256, F610, 1989.
36. **Chou, C.-L., Sands, J. M., Nonoguchi, H., and Knepper, M. A.,** Urea gradient-associated fluid absorption with σ_{urea}=1 in rat terminal collecting duct, *Am. J. Physiol.*, 258, F1173, 1990.
37. **Hamada, Y., Taniguchi, J., and Imai, M.,** The urinary concentrating mechanism: the countercurrent multiplication between thin ascending limb and inner medullary collecting duct, in *Vasopressin*, John Libbey Eurotext, Paris, 1993, 407.
38. **Star, R. A.,** Apical membrane limits urea permeation across the rat inner medullary collecting duct, *J. Clin. Invest.*, 86, 1172, 1990.
39. **Zhang, R. and Verkman, A. S.,** Urea transport in freshly isolated and cultured cells from rat inner medullary collecting duct, *J. Membr. Biol.*, 117, 253, 1990.
40. **Chou, C.-L., Sands, J. M., Nonoguchi, H., and Knepper, M. A.,** Concentration dependence of urea and thiourea transport in rat inner medullary collecting duct, *Am. J. Physiol.*, 258, F486, 1990.
41. **Knepper, M. A. and Star, R. A.,** The vasopressin-regulated urea transporter in renal inner medullary collecting duct, *Am. J. Physiol.*, 259, F393, 1990.
42. **You, G., Smith, C. P., Kanai, Y., Lee, W.-S., Stelzner, M., and Hediger, M. A.,** Cloning and characterization of the vasopressin-regulated urea transporter, *Nature*, 365, 844, 1993.
43. **Schmidt-Nielsen, B. and Osaki, H.,** Renal response to changes in nitrogen metabolism in sheep, *Am. J. Physiol.*, 193, 657, 1958.
44. **Levinsky, N. G. and Berliner, R. W.,** The role of urea in the urine concentrating mechanism, *J. Clin. Invest.*, 38, 741, 1959.
45. **Bray, G. A. and Preston, A. S.,** Effect of urea on urine concentration in the rat, *J. Clin. Invest.*, 40, 1952, 1961.
46. **Kauker, M. L., Lassiter, W. E., and Gottschalk, C. W.,** Micropuncture study of effects of urea infusion on tubular reabsorption in the rat, *Am. J. Physiol.*, 219, 45, 1970.
47. **Shannon, J. A., Jollife, N., and Smith, H. W.,** The excretion of urine in the dog. IV. The effect of maintenance diet, feeding, etc. upon the quantity of glomerular filtrate, *Am. J. Physiol.*, 101, 625, 1932.
48. **Pitts, R. F.,** The effects of infusing glycin and of varying the dietary protein intake on renal hemodynamics in the dog, *Am. J. Physiol.*, 142, 355, 1944.
49. **O'Connor, W. J. and Summerhill, R. A.,** The excretion of urea by dogs following a meat meal, *J. Physiol.*, 256, 81, 1976.
50. **Garrod, O., Davies, S. A., and Cahill, G., Jr.,** The action of cortisone and deoxycorticosterone acetate on glomerular filtration rate and sodium and water exchange in the adrenalectomized dog, *J. Clin. Invest.*, 34, 761, 1955.

51. **Pullman, T. N., Lavender, A. R., and Aho, I.,** Direct effects of glucagon on renal hemodynamics and excretion of inorganic ions, *Metabolism*, 16, 358, 1967.
52. **Schmidt-Nielsen, B., Osaki, H., Murdaugh, H. V., and O'Dell, R.,** Renal regulation of urea excretion in sheep, *Am. J. Physiol.*, 194, 221, 1958.
53. **Truniger, B. and Schmidt-Nielsen, B.,** Intrarenal distribution of urea and related compounds: effects of nitrogen intake, *Am. J. Physiol.*, 207, 971, 1964.
54. **Danielson, R. A., Schmidt-Nielsen, B., and Hohberger, C.,** Micropuncture study of the regulation of urea excretion by the collecting ducts in rats on high and low protein diets, in *Urea and the Kidney*, Excerpta Medica, Amsterdam, 1968, 375.
55. **Danielson, R. A. and Schmidt-Nielsen, B.,** Recirculation of urea analogs from renal collecting ducts of high- and low-protein-fed rats, *Am. J. Physiol.*, 223, 130, 1972.
56. **Ullrich, K. J., Rumrich, G., and Schmidt-Nielsen, B.,** Urea transport in the collecting duct of rats on normal and low protein diet, *Pflügers Arch.*, 295, 147, 1967.
57. **Isosaki, T., Verlander, J. N., and Sands, J. M.,** Low protein diet alters urea transport and cell structure in rat initial inner medullary collecting duct, *J. Clin. Invest.*, 92, 2448, 1993.
58. **Knepper, M. A., Gunter, C. V., and Danielson, R. A.,** Effects of glucagon on renal function in protein-deprived rats, *Surg. Forum*, 27, 29, 1976.
59. **Relman, A. S., Lennon, E. J., and Lemann, J., Jr.,** Endogenous production of fixed acid and the measurement of the net balance of acid in normal subjects, *J. Clin. Invest.*, 40, 1621, 1961.
60. **Lemann, J., Jr., Lennon, E. J., Goodman, A. D., Litzow, J. R., and Relman, A. S.,** The net balance of acid in subjects given large loads of acid or alkali, *J. Clin. Invest.*, 44, 507, 1965.
61. **Pitts, R. F.,** Renal production and excretion of ammonia, *Am. J. Med.*, 36, 720, 1964.
62. **Good, D. W. and Burg, M. B.,** Ammonia production by individual segments of the rat nephron, *J. Clin. Invest.*, 73, 602, 1984.
63. **Nonoguchi, H., Uchida, S., Shiigai, T., and Endou, H.,** Effect of chronic metabolic acidosis on ammonia production from L-glutamine in microdissected rat nephron segments, *Pflügers Arch.*, 403, 229, 1985.
64. **Good, D. W. and DuBose, T. D.,** Ammonia transport by early and late proximal convoluted tubule of the rat, *J. Clin. Invest.*, 79, 684, 1987.
65. **Alpern, R. J.,** Mechanism of basolateral membrane $H^+/OH^-/HCO_3^-$ transport in the rat proximal convoluted tubule: a sodium-coupled electrogenic process, *J. Gen. Physiol.*, 86, 6133, 1985.
66. **Knepper, M. A., Packer, R., and Good, D. W.,** Ammonium transport in the kidney, *Physiol. Rev.*, 69, 179, 1989.
67. **Nagami, G. T. and Kurokawa, K.,** Regulation of ammonia production by mouse proximal tubules perfused *in vitro*, Effect of luminal perfusion, *J. Clin. Invest.*, 75, 844, 1985.
68. **Nagami, G. T.,** Luminal secretion ammonia in the mouse proximal tubule perfused *in vitro*, *J. Clin. Invest.*, 81, 159, 1988.
69. **Preisig, P. A. and Alpern, R. J.,** Apical and basolateral membrane NH_3 and NH_4^+ pathways in the rat proximal convoluted tubule, *Kidney Int.*, 37 (Abstr.), 543, 1990.
70. **Kinsella, J. L. and Aronson, P. S.,** Interaction of NH_4^+ and Li^+ with the renal microvillus membrane Na^+-H^+ exchanger, *Am. J. Physiol.*, 241, C220, 1981.
71. **Garvin, J. L., Burg, M. B., and Knepper, M. A.,** Ammonium replaces potassium in supporting sodium transport by the Na-K-ATPase of renal proximal straight tubules, *Am. J. Physiol.*, 249, F785, 1985.
72. **Tizianello, A., Deferrari, G., Garibotto, G., Robaudo, C., Acquarone, N., and Ghiggeri, G. M.,** Renal ammoniagenesis in an early stage of metabolic acidosis in man, *J. Clin. Invest.*, 69, 240, 1982.
73. **Pitts, R. F.,** The renal excretion of acid, *Fed. Proc.*, 7, 418, 1948.
74. **Knepper, M. A.,** NH_4^+ transport in the kidney, *Kidney Int.*, 40, S95, 1991.
75. **Stern, L., Backman, K. A., and Hayslett, J. P.,** Effect of cortical-medullary gradient for ammonia on urinary excretion of ammonia, *Kidney Int.*, 27, 652, 1985.
76. **Good, D. W., Caflisch, C. R., and DuBose, T. D., Jr.,** Transepithelial ammonia concentration gradients in inner medulla of the rat, *Am. J. Physiol.*, 252, F491, 1987.
77. **Robinson, R. R. and Owen, E. E.,** Intrarenal distribution of ammonia during diuresis and antidiuresis, *Am. J. Physiol.*, 208, 1129, 1965.
78. **Sullivan, L. P.,** Ammonia secretion during stopped flow: a hypothetical ammonium countercurrent system, *Am. J. Physiol.*, 209, 273, 1965.
79. **Good, D. W., Knepper, M. A., and Burg, M. B.,** Ammonia and bicarbonate transport by thick ascending limb of rat kidney, *Am. J. Physiol.*, 247, F35, 1984.
80. **Good, D. W. and Knepper, M. A.,** Ammonia transport in the mammalian kidney, *Am. J. Physiol.*, 248, F459, 1985.
81. **Flessner, M. F., Mejia, R., and Knepper, M. A.,** Ammonium and bicarbonate transport in isolated perfused rodent long-loop thin descending limbs, *Am. J. Physiol.*, 264, F388, 1993.
82. **Kurtz, I., Star, R., Balaban, R. S., Garvin, J. L., and Knepper, M. A.,** Spontaneous luminal disequilibrium pH in S3 proximal tubules: role in ammonia and bicarbonate transport, *J. Clin. Invest.*, 78, 989, 1986.

83. **Flessner, M. F. and Knepper, M. A.,** Ammonium and bicarbonate transport in isolated perfused rodent ascending limbs of the loop of Henle, *Am. J. Physiol.*, 264, F837, 1993.
84. **Garvin, J. L., Burg, M. B., and Knepper, M. A.,** Active NH_4^+ absorption by the thick ascending limb, *Am. J. Physiol.*, 255, F57, 1988.
85. **Good, D. W.,** Active absorption of NH_4^+ by rat medullary thick ascending limb: inhibition by potassium, *Am. J. Physiol.*, 255, F78, 1988.
86. **Kinne, R., Kinne-Saffran, E., Schuetz, H., and Schloelermann, B.,** Ammonium transport in medullary thick ascending limb of rabbit kidney: Involvement of the Na^+, K^+, Cl^--cotransporter, *J. Membr. Biol.*, 94, 279, 1986.
87. **Good, D. W.,** Effects of potassium on ammonia transport by medullary thick ascending limb of the rat, *J. Clin. Invest.*, 80, 1358, 1987.
88. **Kikeri, D., Sun, A., Zeidel, M. L., and Hebert, S. C.,** Cell membranes impermeable to NH_3, *Nature,* 339, 478, 1989.
89. **Burg, M. B.,** Thick ascending limb of Henle's loop, *Kidney Int.*, 22, 454, 1982.
90. **Graber, M. L., Bengele, H. H., Mroz, E., Lechene, C., and Alexander, E. A.,** Acute metabolic acidosis augments collecting duct acidification rate in the rat, *Am. J. Physiol.*, 241, F669, 1981.
91. **Sajo, I. M., Goldstein, M. B., Sonnenberg, H., Stinebaugh, B. J., Wilson, D. R., and Halperin, M. L.,** Sites of ammonia addition to tubular fluid in rats with chronic metabolic acidosis, *Kidney Int.*, 20, 353, 1981.
92. **Sonnenberg, H., Cheema-Dhadli, S., Goldstein, M. B., Stinebaugh, B. J., Wilson, D. R., and Halperin, M. L.,** Ammonia addition into the medullary collecting duct of the rat, *Kidney Int.*, 19, 281, 1981.
93. **Buerkert, J., Martin, D., and Trigg, D.,** Ammonium handling by superficial and juxtamedullary nephrons in the rat: evidence for an ammonia shunt between the loop of Henle and the collecting duct, *J. Clin. Invest.,* 70, 1, 1982.
94. **Flessner, M. F., Wall, S. M., and Knepper, M. A.,** Permeabilities of rat collecting duct segments to NH_3 and NH_4^+, *Am. J. Physiol.*, 260, F254, 1991.
95. **Knepper, M. A., Good, D. W., and Burg, M. B.,** Mechanism of ammonia secretion by cortical collecting ducts of rabbits, *Am. J. Physiol.*, 247, F729, 1984.
96. **Knepper, M. A., Good, D. W., and Burg, M. B.,** Mechanisms of ammonia secretion by rabbit cortical collecting ducts: quantitative considerations, *Contrib. Nephrol.*, 47, 116, 1985.
97. **Knepper, M. A., Good, D. W., and Burg, M. B.,** Ammonia and bicarbonate transport by rat cortical collecting ducts perfused *in vitro, Am. J. Physiol.*, 249, F870, 1985.
98. **Hamm, L., Trigg, D., Martin, D., Gillespie, C., and Buerkert, J.,** Transport of ammonia in the rabbit cortical collecting tubule, *J. Clin. Invest.*, 75, 478, 1985.
99. **Star, R. A., Burg, M. B., and Knepper, M. A.,** Luminal disequilibrium pH and ammonia transport in outer medullary collecting duct, *Am. J. Physiol.*, 252, F1148, 1987.
100. **Wall, S. M., Sands, J. M., Flessner, M. F., Nonoguchi, H., Spring, K. R., and Knepper, M. A.,** Net acid transport by isolated perfused inner medullary collecting ducts, *Am. J. Physiol.*, 258, F75, 1990.
101. **Knepper, M. A. and Flessner, M. F.,** Renal acid-base transport, in *Diseases of the Kidney*, Little Brown, Boston, 1993, 207.
102. **Knepper, M. A., Good, D. W., and Burg, M. B.,** Ammonia and bicarbonate transport by rat cortical collecting ducts perfused *in vitro, Am. J. Physiol.*, 249, F870, 1985.
103. **Star, R. A., Kurtz, I., Mejia, R., Burg, M. B., and Knepper, M. A.,** Disequilibrium pH and ammonia transport in isolated perfused cortical collecting ducts, *Am. J. Physiol.*, 253, F1232, 1987.
104. **DuBose, T. D., Jr.,** Application of the disequilibrium pH method to investigate the mechanism of urinary acidification, *Am. J. Physiol.*, 245, F535, 1983.
105. **Graber, M. L., Bengele, H. H., Schwartz, J. H., and Alexander, E. A.,** pH and PCO_2 profiles of the rat inner medullary collecting duct, *Am. J. Physiol.*, 241, F659, 1981.
106. **Kriz, W. and Bankir, L.,** A standard nomenclature for structures of the kidney, *Am. J. Physiol.*, 254, F1, 1988.
107. **Imai, M., Taniguchi, J., and Tabei, K.,** Function of thin loops of Henle, *Kidney Int.*, 31, 565, 1987.
108. **Knepper, M. A., Burg, M. B., Orloff, J., Berliner, R. W., and Rector, F. C.,** Ammonium, urea, and systemic pH regulation [letter], *Am. J. Physiol.*, 253, F199, 1987.
109. **Knepper, M. A. and Stephenson, J. L.,** Urinary Concentrating and Diluting Processes, in *Physiology of Membrane Disorders* (2nd ed.), Plenum, New York, 1986, 713.
110. **Knepper, M. A. and Burg, M. B.,** Renal Acid-Base Transport, in *Diseases of the Kidney*, Little Brown, Boston, 1988, 211.

Chapter 14

Gene Expression and Development of Hepatic Nitrogen Metabolic Pathways

Maria A. Dingemanse and Wouter H. Lamers

CONTENTS

INTRODUCTION

Screening the Medline database reveals that the developmental appearance of ornithine cycle enzymes was not an intensely studied field in the last 10 years. This is not a surprising conclusion since many of the studies that were done up to then,[1-12] and since then,[13-23] seem to show that the basic regulatory mechanisms do not differ in the pre- and postnatal period. In this respect the fetus therefore merely represents an interesting hormonal or metabolic condition that can help to identify regulatory factors for specific genes. Rather than summarizing the data obtained in these studies, we want to focus on the peculiarities of the early stages of gene expression. We will start with defining a couple of concepts, and then will use two enzymes to illustrate the topics that will be addressed in this chapter. The more metabolic aspects of prenatal urea synthesis were recently reviewed.[24]

CYTODIFFERENTIATION, MATURATION, AND MODULATION: CONCEPTS AND DEFINITIONS USED TO DESCRIBE DEVELOPMENTAL ASPECTS OF THE REGULATION OF GENE EXPRESSION

Cytodifferentiation is defined as the process by which cells acquire the capacity to express a characteristic set of genes. This multistage process causes a qualitative change in the pattern of gene expression. The multistep properties of the differentiation process probably explain why the process is usually considered to be irreversible and why the presence of inducing factors appears to be only temporarily required to maintain the terminally differentiated phenotype. The protracted development towards terminal differentiation is also revealed by the observation that the maximally attainable rate of expression of tissue-specific genes is usually lower during the period immediately following cytodifferentiation (protodifferentiated stage[25]) than later, in the fully grown, mature stage. The gradual acquisition of the capacity to express a gene at its maximal (mature) rate is called *maturation*. Changes in the rate of expression of characteristic genes in differentiated cells represent the adaptive response called *modulation*. In this reversible metabolic process, an increase in the rate of gene expression depends on the concentration, and hence, on the permanent presence of inducing

0-8493-8411-7/95/$0.00+$.50

factors. Thus, whereas cytodifferentiation addresses qualitative aspects of development, both maturation and modulation address quantitative aspects. Yet, they are distinct processes: maturation increases the upper limit of the rate of tissue-specific gene expression (i.e., the capacity; it is therefore part of the differentiation process), whereas modulation changes the actual rate of tissue-specific gene expression without affecting the capacity. It is clear that these processes operate simultaneously during development.

MONITORING THE DEVELOPMENTAL PROCESSES: PARAMETERS

Although the progress of development can be monitored with both simple and complex structural and/or functional parameters, our definition of differentiation and maturation already imply that a particular protein, or its corresponding mRNA, can serve as an adequate parameter to follow the spatio-temporal development of these processes in an embryo. It has to be kept in mind, however, that the developing liver is a complex tissue in which, in addition to parenchymal cells, stromal and hematopoietic cells are abundant, so that data obtained from homogenates have to be recalculated to hepatocytes. As an example, the developmental appearance of carbamoylphosphate synthase (CPS, Figure 1) and of glutamate dehydrogenase (GDH, Figure 2) are shown. These genes were chosen because within the liver they are only expressed in hepatocytes,[26-30] and because data on the tissue concentration of their protein and mRNA in the developing and adult liver are available.[7,11,22,31-33] The data were recalculated to concentrations per hepatocyte, using available morphometric data.[34-39] The averaged parameter values from these morphometric studies that were used for the preparation of Figures 1 and 2 are shown in Figure 3.

Figure 3 reveals that the volume fraction of hepatocytes in the developing liver increases from approximately 40% shortly after the appearance of the liver primordium to approximately 90% shortly after birth. This increase in the volume fraction of hepatocytes is largely due to the disappearance of hematopoietic tissue in the last few days before birth. Concomitant with the disappearance of the hematopoietic cells from the liver in the late fetal period, the cell volume of the hepatocyte increases approximately fourfold. Immediately after birth the hepatocyte volume shrinks to about 50% of its prenatal size and after weaning it increases again approximately fourfold. The latter increase in cell size is accompanied by the appearance of binucleation[40] and subsequently of polyploidization[41,42] in the hepatocyte population. Due to the developmental changes in the hepatocellular volume and the volume fraction of the hepatocytes in the liver, the number of hepatocytes per gram of tissue decreases twofold in the late fetal period, increases neonatally to virtually the same number as before birth (as a result of compensatory changes in the volume fraction and the size of the hepatocytes), and decreases fourfold after weaning.

Figures 1 and 2 both show that the cellular CPS and GDH mRNA and protein concentrations increase in the prenatal period and again in the *pre*weaning period. The intervening dip is most pronounced in the first postnatal week and more evident for the mRNAs than for the proteins, reflecting the longer half-life of the latter macromolecules. It is not known why this apparent interruption of the maturation of the liver exists, and what the benefits for the suckling animal are, but a comparison with the spiny mouse, a murinoid species that differs from the rat and mouse mainly in being a precocial rather than an altricial species, seems to be of interest. The developmental profiles of cell-specific enzymes in spiny mice show a more gradual prenatal increase but do not show the postnatal interruption in the developmental rise that is seen in the rat.[18,43,44] In agreement with the biochemical data, the developmental stage of newborn spiny mice is comparable to that of 8 to 10-d-old neonatal rats. These data suggest that the steep prenatal rise in CPS and GDH expression followed by a dip after birth, which is shared with virtually all other hepatocyte-specific genes,[45] is typical for altricial species and reflects an accelerated rate of maturation in the late fetal period of those organs that are vital for postnatal survival. Interestingly, such organs (i.e., lungs, gastrointestinal tract, liver) are all of endodermal origin. Accordingly, the developmental appearance of enzymes in the central nervous system does not appear to differ between altricial and precocial species.[46]

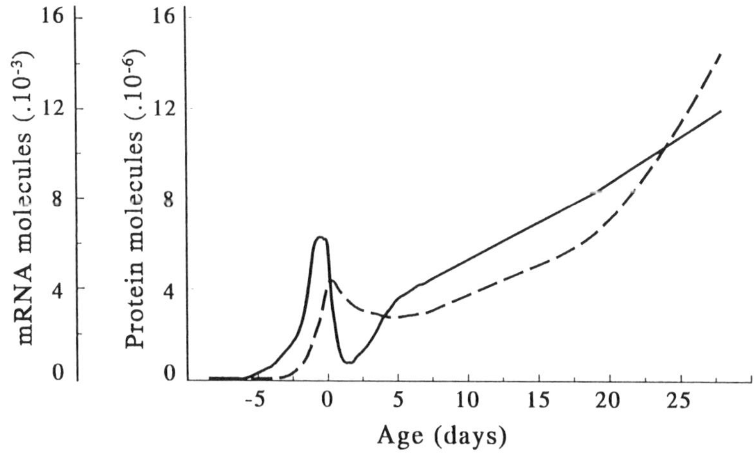

Figure 1 Developmental appearance of CPS in rat liver. The number of CPS mRNA (solid line) and protein molecules (dashed line) per hepatocyte are related to gestational age.

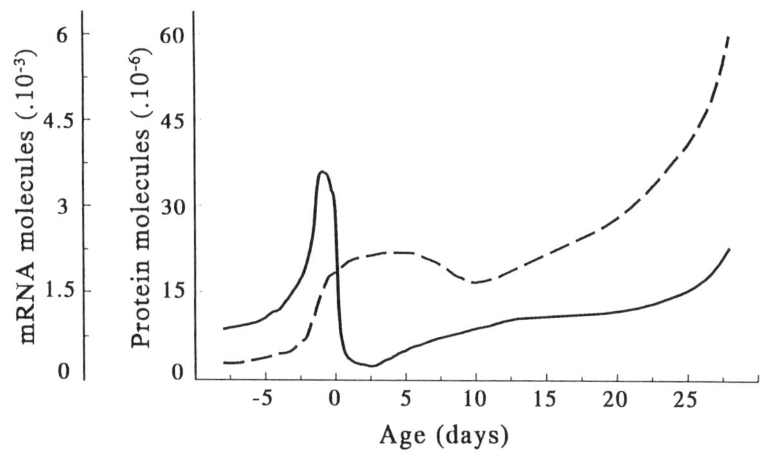

Figure 2 Developmental appearance of GDH in rat liver. The number of GDH mRNA (solid line) and protein molecules (dashed line) per hepatocyte are related to gestational age.

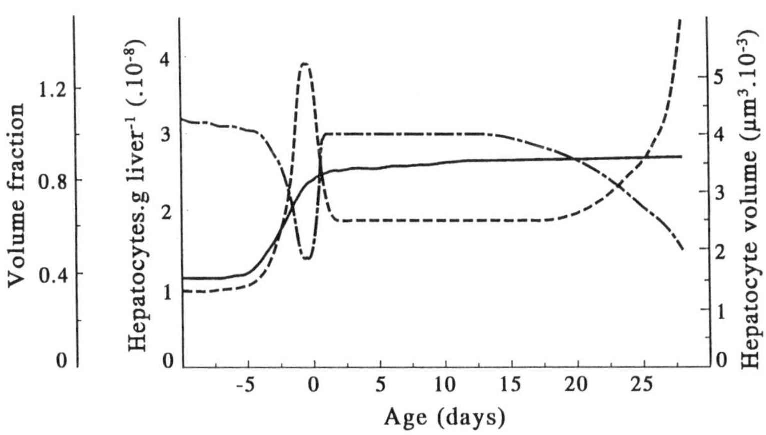

Figure 3 Developmental changes in morphometric parameters of the liver. The volume fraction (solid line), the average cell volume (dashed line), and the number of hepatocytes per gram of tissue (dashed-dotted line) are shown.

The cellular concentrations of CPS and GDH proteins are 10^3- to 10^4-fold higher than the cellular concentrations of the respective mRNAs (Figures 1 and 2). Taking into account the longer half-life of the CPS protein (approximately 4 to 5 d[20,47]) relative to the CPS mRNA (approximately 4 to 6 h[48]), these data indicate that translation is responsible for a more than 200-fold amplification of CPS gene expression. Since no data are available for the half-life of GDH mRNA (the half-life of the GDH protein is approximately 1 d[47]), the effect of translation on gene expression can only be calculated indirectly. Using the molar ratio of GDH and CPS protein in the liver (at 28 d approximately 3.5), and the ratio of the CPS and GDH half-lives (approximately 4.5), the rate of synthesis of GDH can be calculated to be approximately 16-fold higher than that of CPS (under steady-state conditions $dE/dt = 0$; $[E] = k_s/k_d$). Since the molar ratio of GDH and CPS mRNAs at 28 d is approximately 0.2, it follows that the efficiency of translation of GDH mRNA is approximately 80-fold higher than that of CPS mRNA (i.e., translation amplifies the expression of GDH approximately 16,000-fold). This simple computation reveals the quantitative importance of the process of translation for gene expression.

In view of the above conclusion it is perhaps not surprising that an increase in translational efficiency appears to be one of the mechanisms underlying the process of maturation. Figures 1 and 2 also show the developmental changes in the *ratio* of the cellular mRNA and protein concentrations for GDH and CPS. Assuming that no developmental change occurs in the half-lives of either protein (which is the case for the CPS protein[20,47]) and that the developmental decrease in the mRNA/protein ratio is not caused by the shorter half-life of the mRNAs compared to the proteins,[33,49] the developmental change in this ratio reflects a substantial and significant increase in translational efficiency in the prenatal period. Such a relatively low translational efficiency before birth may also explain the more than 10-year-old observation that the hepatic concentration of several hepatocyte-specific proteins that can be attained upon hormonal stimulation, is appreciably lower in term fetuses than in neonates, despite direct injection of the hormones into the fetuses.[8-10,12,18] Many studies have shown that the hormonal regulation of the expression of genes is the same before and after birth. As a result, the phenomenon was ascribed to the intrauterine environment.[1,6,15,17,18,50] Interestingly, a similarly low translational efficiency was also observed for arginase mRNA in premetamorphic *Xenopus* liver.[51]

CPS enzyme activity is first detectable at 4 d before birth.[7,9] It can be calculated that at this time an average hepatocyte already contains approximately 50 to 100 × 10^3 enzyme molecules. At 8 d before birth an average hepatocyte contains 2 × 10^6 GDH monomers or 250 × 10^3 octameric enzyme molecules. These data raised the question regarding at what stage of liver development CPS and GDH expression arises. Since earlier biochemical measurements were not feasible for technical reasons, the question was addressed immunohistochemically.[29] This study showed that GDH is expressed in all hepatocytes as soon as the liver has differentiated from the embryonic foregut, whereas CPS expression is not seen until embryonic Day 15 (ED 15). Furthermore, CPS is expressed initially only in a small fraction of hepatocytes in the median lobe of the liver. Gradually, more hepatocytes including those from the other lobes of the liver are recruited for CPS biosynthesis so that all hepatocytes have accumulated CPS protein at 1 to 2 d before birth. This recruitment of hepatocytes for CPS biosynthesis was already predicted from the exponentially increasing content of CPS enzyme activity.[9]

These observations raised a number of questions, aspects of which are addressed in the remainder of this chapter.

- What are the signals that induce the expression of hepatocyte-specific gene expression during cytodifferentiation?
- What is the molecular mechanism that is responsible for the delayed onset of CPS expression in the liver when compared to GDH, for example?
- What is the molecular mechanism that underlies the observed recruitment of hepatocytes that participate in CPS biosynthesis during the fetal period?
- Is a late onset of CPS expression in the liver and, subsequently, a recruitment of hepatocytes for CPS biosynthesis specific for rodents or a more general phenomenon?

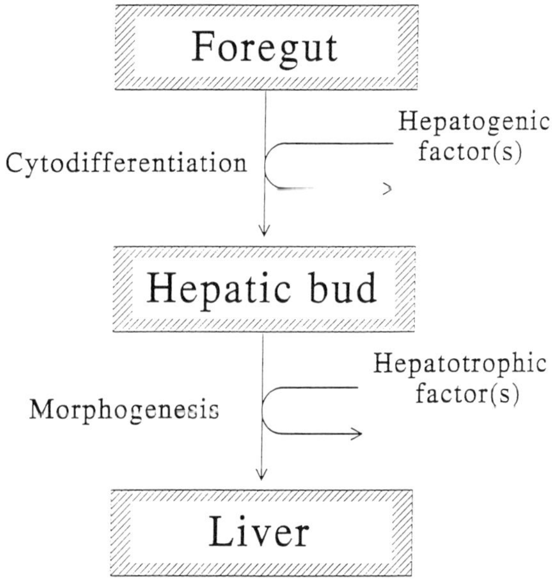

Figure 4 Sequential steps in liver organogenesis. Hepatogenic factor(s), produced by the hepatocardiac mesoderm, induce hepatocyte differentiation in the endoderm of the embryonic foregut. Subsequently, hepatotrophic factor(s), produced by lateral plate mesoderm, induce proliferation of the newly differentiated hepatocytes and morphogenesis of the liver.

EXPRESSION OF HEPATOCYTE-SPECIFIC GENES AND CYTODIFFERENTIATION

The development of the liver has been the subject of numerous descriptive studies. The endodermal hepatic bud becomes visible in the floor of the caudal part of the foregut in both mammalian and avian embryos, when they have developed 15 to 20 somites (ED 10.5; Carnegie stage 10).[52-55] The proliferating primordial hepatocytes then invade the surrounding mesenchyme of the transverse septum, leading to the formation of one of the typical features of the architecture of the liver — the three-dimensional network of hepatocytes arranged into a muralium. An extensive body of experimental results indicates that these steps in hepatic organogenesis result from two consecutive inductive processes (Figure 4). The first inductive process involves an interaction between the early hepatocardiac mesoderm and the *pre*hepatic endoderm. As a result of gastrulation, these tissues evolved as two apposed layers in the periphery of the anterior part of the embryonic disc (Carnegie stage 9).[56-58] The factors responsible for this inductive process are solely produced by the mesoderm, while only the peripheral part of the anterior endoderm of the embryonic disc appears to be responsive to these hepatogenic factors.[58,59] The induction of hepatocyte differentiation resulting from interaction between these tissues is completed at the 5 to 10 somite stage.[59-62] The ability of the "cardiac" mesoderm to induce hepatic endoderm *in vitro* is highest during the early stages of cardiogenesis (Carnegie stages 9 to 13) and gradually declines to background levels thereafter (Figure 5).[61] The hepatogenic factors are well conserved in evolution; irrespective of whether the "cardiac" mesoderm is taken from chick, quail, or mouse embryos, it can induce hepatocyte differentiation in the prehepatic endoderm of chick *and* mouse embryos.[61]

Although the hepatic endodermal epithelium, when cultured alone, shows a rapid cytological[63] and enzymological[64] maturation, it fails to grow. Proliferation of the hepatocytes and the development of the characteristic histological architecture of the liver (Figure 4) require the interaction of the newly differentiated hepatocytes with the mesenchymal cells of the transverse septum.[60,62-64] The stimulatory activity of the mesodermal liver component is also present in other mesenchymes derived from the lateral plate, but not from paraxial (somitic) mesoderm.[62] The stimulatory capacity of these mesenchymes persists during the entire embryonic period[62] and is not species specific,

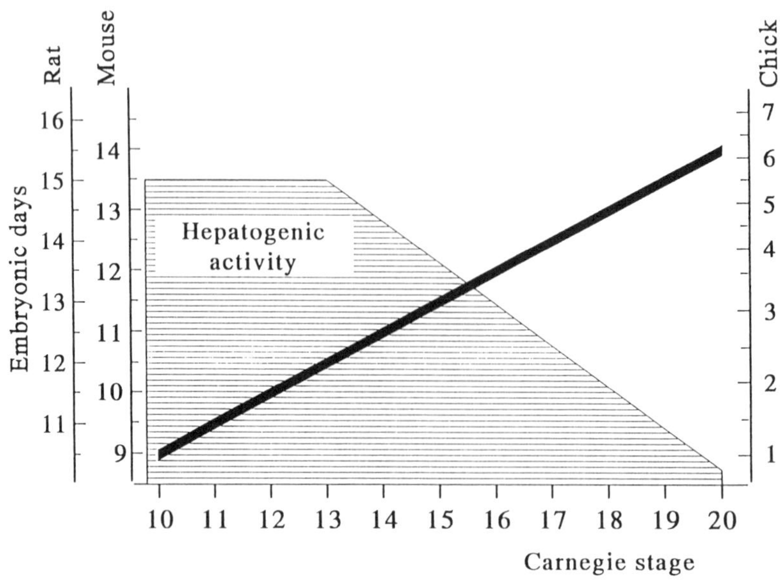

Figure 5 Relation of the activity of hepatogenic factors to Carnegie stages of development and gestational age of chicken, rat, and mouse. The experimentally determined developmental changes in hepatogenic activity of the cardiac mesoderm are represented by the horizontally hatched area. The black ascending line relates Carnegie stage to postcopulatory age (rat, mouse) or incubation age (chicken) of embryos.

allowing the formation of chimeric livers with hepatocytes from mouse embryos and mesenchyme from chick embryos.[60]

These observations show that the induction of cytodifferentiation in the embryonic foregut requires exposure of the endoderm to an *instructive* interaction with the mesoderm, whereas proliferation and early morphogenesis of the liver depend on a mesodermally derived growth factor that serves a *permissive* rather than an instructive role.[63-66] In other words, cytodifferentiation of hepatic endodermal cells results in their cytological maturation, but the development of a histiotypic architecture is, in all likelihood, not necessary.[64,65] It therefore seems possible to study the effects of the signals that induce the expression of hepatocyte-specific genes during cytodifferentiation in cell cultures. For this reason we are in the process of developing an *in vitro* model system by immortalization of the cells of the (pre-)hepatic endoderm at successive stages in the differentiation process. So far, we have only partially succeeded because we were unable to obtain immortalized cell lines from the prehepatic endoderm (i.e., cells in the process of cytodifferentiation). To immortalize cells from this stage, their requirements for growth *in vitro* will have to be defined. An alternative approach to study liver differentiation, however, is to isolate the hepatogenic factors by molecular methods and to analyze their effects on competent cells much as one studies the differentiative effects of the *myoD* family of transcription factors.[67]

THE TEMPORAL REGULATION OF THE DELAYED ONSET OF CPS EXPRESSION IN THE LIVER

As mentioned earlier, GDH is expressed in hepatocytes as soon as they differentiate from the embryonic foregut.[29,68] In addition to GDH, hepatocyte-specific proteins such as arginase,[29] albumin,[69-71] and α-fetoprotein[69,72] are expressed in the hepatic rudiment *in vivo* as soon as, or shortly after, it becomes recognizable at the 20-somite stage (ED 10.5). In contrast, other hepatocyte-specific enzymes are not expressed until much later in development; e.g., CPS at ED 15 or phospho*enol*pyruvate carboxykinase (PEPCK) at birth, respectively. Interestingly, both enzymes can be induced within 12 to 24 h in monolayers of hepatocytes from ED 11 embryos[19,73] if the cultures are treated with glucocorticosteroids and cyclic AMP. To establish whether this difference is due to a lack of hormonal factors *in vivo* or to a loss of a normal cell environment in cell cultures, intact rat embryos of ED 10 were cultured and exposed to inducing hormones.[74] Despite an undisturbed morphological development of the embryos in culture and a normal appearance of GDH and α-fetoprotein in their livers, no induction of CPS could be demonstrated immunohistochemically. Induction of CPS could be demonstrated, however, in monolayers of hepatocytes that were isolated from embryos that had previously been cultured for 48 h. In addition, an adequate uptake of hormones by the embryo during culture could be demonstrated. Therefore it has to be postulated that the presence of factors within the embryo or the embryonic liver, or the stable modification of the CPS gene in the embryonic liver is suggestive of a condition that inhibits CPS expression.

The initiation (hormone-induced) of transcription requires the association of the promoter and enhancer sequences of a gene with regulatory proteins. DNA methylation influences the DNA conformation and, hence, its affinity and/or accessibility for such transcription factors[75] and the ability of RNA polymerase to continue mRNA synthesis.[76] Demethylation will facilitate transcription by decreasing the lower limit of the concentration of transcription factors that are required for gene expression or by removing the obstacles for RNA polymerase. We have investigated the methylation state of the CPS gene during development of the rat liver *in vivo* with respect to the developmental appearance of the capacity of hepatocytes to express CPS.[77] The use of the isoschizomeric restriction enzymes MspI and HpaII (only the latter of which is sensitive to methylation), allowed us to identify a site within the CPS enhancer that becomes demethylated in the liver between ED 13 and the adult stage, but remains methylated in spleen (Figure 6). Figure 6 shows that this site becomes demethylated only after CPS expression in hepatocytes has reached adult levels (see Figure 1). Similar observations were made for PEPCK.[78,79] Contrary to general wisdom, these developmental changes in methylation pattern suggest that demethylation of CG nucleotide pairs follows, rather than causes, increases in the rate of gene expression.[80]

THE MOLECULAR MECHANISM UNDERLYING RECRUITMENT OF HEPATOCYTES FOR BIOSYNTHESIS OF CPS IN THE FETAL PERIOD

During the initial phases of gene expression no intercellular heterogeneity in GDH concentration is observed in rat liver *in vivo* at either the mRNA or protein level.[29,30] In contrast, CPS is initially expressed in a small fraction of hepatocytes and only gradually more hepatocytes are recruited for

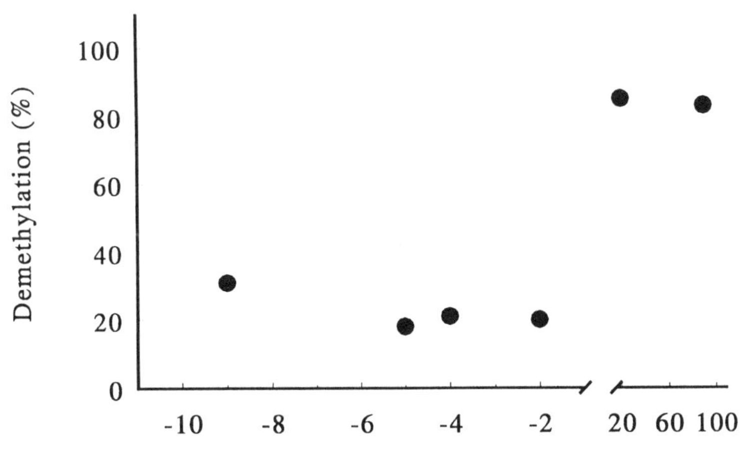

Figure 6 Developmental changes in the degree of methylation of CG dinucleotide pairs in the upstream enhancer of the rat CPS gene. The major change in methylation at this site occurs after birth.

CPS biosynthesis. This observation is made *in vivo*[29] and *in vitro*,[19,21,81] while a similar observation was made for CPS[23,82] and ornithine transcarbamoylase[83] in developing human liver. To gain insight into the mechanisms that govern the first steps of hormone-induced gene expression in hepatocytes, the *in vitro* observations were explored further. When embryonic hepatocytes are cultured in the presence of glucocorticosteroids, cyclic AMP, and thyroid hormones, they start to accumulate hepatocyte-specific enzymes like CPS and PEPCK within 8 h.[21,73,81] Unexpectedly, both microscopic and flow-cytometric analysis showed that initially four groups of hepatocytes can be distinguished: (1) negative for both CPS and PEPCK protein, (2) only positive for CPS protein, (3) only positive for PEPCK protein, and (4) positive for both CPS and PEPCK protein. The percentage of cells expressing both CPS and PEPCK protein increases during prolonged culture in the presence of hormones. These observations have also been made for other sets of enzymes like CPS and arginase, or PEPCK and arginase. These results show that the initial expression of these genes is a stochastic event, i.e., lacks the coordination *within* cells that can be expected from the common hormonal regulation of these genes.[81] The coordination is gradually established during continued cultivation of the hepatocytes. We have hypothesized that this initial intercellular heterogeneity in the rate of accumulation of a particular enzyme results from competition of different genes for an initially limiting amount of (a) common regulatory factor(s), leading to random differences in the rate of accumulation of the respective genes.[81] Likely candidates for such regulatory factors are intermediates of the signal transduction pathway of the inducing hormones.

To further elucidate the molecular events during the initial phases of hepatocyte-specific gene expression, we have addressed the question at what regulatory level the observed interhepatocyte heterogeneity in protein synthesis and accumulation is generated.[48] In this regard, the cellular accumulation of CPS and PEPCK mRNAs upon exposure of cultured embryonic rat hepatocytes to hormones has been visualized hybridocytochemically. This analysis revealed that the observations at the protein level reflected those at the mRNA level. We have also visualized the pattern of accumulation of the nuclear CPS mRNA precursors, this being the closest approximation to evaluate transcription rates of specific genes in individual cells (Figure 7). The data indicate that the intercellular heterogeneity in CPS mRNA content upon exposure to hormones is generated by intercellular differences in the timing of the onset of transcription. This conclusion provides a good basis for the time-dependent increase in the number of CPS-positive cells seen under these conditions.[21] However, this latter study[21] had also shown that the number of hepatocytes that express CPS, increases up to 80 to 85% during the first 20 h of exposure to hormones, but rapidly levels off thereafter, revealing that accumulation of CPS in the remaining 10 to 15% of cells is controlled by a different process. After reviewing these and previous data on the accumulation of proteins, it became clear that the intercellular heterogeneity in (pre-)mRNA and protein content that persisted for more than 24 h was inversely related to the half-life of the gene product (i.e., higher for PEPCK than for CPS and higher for mRNAs than for the respective proteins) and to the concentration of inducing hormones, subsaturating doses prolonging the presence of intercellular phenotypic heterogeneity. Since we observed no intercellular differences in the overall rate of degradation of the mRNAs, we have hypothesized that, although all hepatocytes can and do express the genes, transcription of a gene in a particular cell is initially a discontinuous process.[48]

THE (FUNCTIONAL) SIGNIFICANCE OF THE DEVELOPMENTAL TIMING OF THE ONSET OF CPS EXPRESSION IN THE LIVER

A still poorly understood problem is why ornithine cycle enzymes like CPS already accumulate in human liver shortly after the organ has formed,[23,82,83] whereas in the rat and mouse it takes until the onset of the fetal period before a majority of hepatocytes express the enzymes. The difference does not appear to depend on a different hormonal regulation of the expression of the gene in both species.[19,23] Furthermore, both human and rat hepatocytes acquire the capacity to express CPS soon

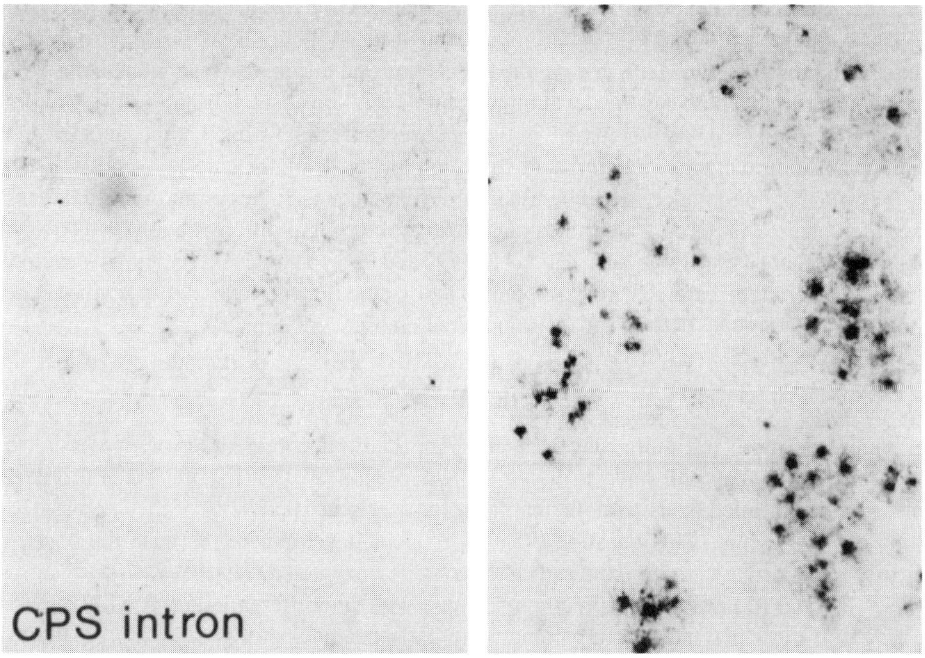

Figure 7 Heterogeneity in hormone-induced transcription of the CPS gene in cultured embryonic Day 14 hepatocytes. Cultures were induced with dexamethasone, cyclic AMP, and triiodothyronine and fixed after 24 h. Nuclear CPS mRNA precursors were visualized hybridocytochemically with intron-specific cDNA probes. Left panel, noninduced control; right panel, induced with hormones.

after they differentiate from the embryonic foregut.[19,23] In the rat, however, the expression of CPS appears to be temporarily suppressed *in vivo*, with embryonic rat hepatocytes having to be cultured in monolayer to expose this capacity.[19,74] It is thought that towards the end of the embryonic period, substantial levels of ornithine cycle enzymes have accumulated in human liver and urea biosynthesis takes place (see Reference 24 for a review on the indirect evidence). A relatively early appearance and high concentrations of ornithine cycle enzymes are also observed in pigs[84] and sheep,[85,86] but not in guinea pigs,[87] mice,[88] and spiny mice.[18] Although the developmentally early or late appearance of ornithine cycle enzymes has been linked to such developmental characteristics as placental permeability, renal development, and maturity at parturition,[84,86] we hypothesized that these interspecies differences in the developmental timing of the appearence of ornithine cycle enzymes depend on whether organisms utilize the available amino acids mainly for growth.[24] Rapid prenatal growth as seen in rodents would be associated with a late appearance of ornithine cycle enzymes, whereas a more protracted prenatal growth as seen in human and ungulate species, would allow the oxidation of amino acids for energy production and would require an early presence of the capacity to detoxify ammonia and, hence, to synthesize urea.

To elaborate on this hypothesis we have recently studied the developmental appearance and distribution of several ammonia-metabolizing enzymes in the liver, the mesonephros, and the gut of human embryos.[82] The mesonephros was included because we had observed the expression of CPS in this temporary embryonic organ, while the gut was included because of its ability to produce citrulline in the adult.[89] The analysis revealed an intercellular heterogeneity in enzyme content for both CPS and GDH in the liver (the latter observation contrasting with that in the rat[29,30]), but not for arginase and phosphate-dependent glutaminase. As in the rat, this intercellular heterogeneity disappears with ongoing development. The expression of CPS in the gut differs in two respects from that in the liver: (1) both in rodents and in humans the gene is first expressed at Carnegie stage 22-

23, i.e., later in development than in the liver; and (2) all enterocytes accumulate the mRNA[30] (and unpublished observations) and the protein[29] simultaneously. Both observations support the hypothesis that the expression of the CPS gene in the gut is regulated differently than in the liver.[90] Finally, a correlation was found between the developmental timing of the onset of ornithine cycle expression in the liver and the degree of development of the mesonephros, suggesting that the latter organ might function as an important site of amino acid oxidation and possibly for gluconeogenesis. This study shows that when animal models are to be used for developmental studies of amino acid metabolism, only results obtained in species with rates of growth and enzymic differentiation of liver, gut, and mesonephros that are comparable to the human embryo and fetus, are relevant for human development. One should therefore be careful with extrapolations of data on the regulation of growth, amino acid metabolism, and ammonia detoxification from rodent embryos to humans.

FUTURE DIRECTIONS

In our view, the more intriguing developmental aspects of the regulation of expression of the ornithine cycle enzymes will have to focus on two general areas. The first general area where progress is much awaited deals with the regulation of the tissue specificity of the expression of the ornithine cycle enzymes. This question directly involves development, because the onset of the capacity of endodermal cells to express ornithine cycle enzymes differs between liver and small intestine,[29,30,82] and is, in the liver, directly associated with cytodifferentiation. It will be extremely interesting to learn which factors are paramount to the appearance of this capacity. To address this question it will be necessary to identify and characterize the hepatogenic and hepatotrophic factors that were mentioned earlier so as to be able to manipulate cells *in vitro*. Another direct consequence of this approach is the possibility to manipulate the pattern of expression of the genes and, potentially, of entire metabolic pathways ("metabolic reprogramming"). Such studies are not only interesting from the standpoint of gene regulation, but they also represent the only way to assess the functional importance of the zonal distribution of (ornithine cycle) enzymes in the liver.[24]

The second topic deals with the presumably very different regulation of the expression of ornithine cycle enzymes in the gut and the liver. The different developmental timing of the onset of expression and the apparent absence of recruitment of cells for expression in the gut have already been discussed. After birth, however, developmental changes in the expression of ornithine cycle enzymes in the gut are very different from those in liver. Prior to weaning, endogenous arginine biosynthesis appears to take place in the gut, rather than in the kidney as in the adult. This conclusion is mainly based on a comparison of the developmental profiles of relevant enzymes in the gut and the kidney. The intestinal concentration of carbamoyl phosphate synthase, N-acetylglutamate synthase, ornithine transcarbamoylase, argininosuccinate synthase, and argininosuccinate lyase (i.e., all ornithine cycle enzymes *except* arginase) in rodents and pigs all peak shortly after birth and gradually decline until adult levels are reached at weaning[90-94] (Moorman et al., in preparation). Similarly, the enzymes that convert glutamine to ornithine (phosphate-dependent glutaminase, pyrroline-5-carboxylate synthase and ornithine aminotransferase) are elevated well above adult levels in the suckling period,[92,94-96] whereas the activity of pyrroline-5-carboxylate reductase, the enzyme that can divert pyrroline-5-carboxylate towards proline, shows a reciprocal pattern.[96,97] Furthermore, ornithine decarboxylase activity, which diverts ornithine towards polyamine biosynthesis, is very low during the suckling period, but shows a sharp, transient burst in activity at the onset of weaning.[95,98] Finally, intestinal arginase activity is virtually absent in enterocytes until it appears abruptly at weaning.[91,97,99] In accordance with these data, findings have been presented that indicate that the neonatal small intestine possesses the *capacity* for *de novo* arginine biosynthesis,[91,99] whereas it is now well established that in the intestine of the adult, the synthetic pathway stops at citrulline.[89,100,101] Meanwhile, the developmental appearance in the kidney of the enzymes that convert citrulline to arginine resembles that in the liver (i.e., it gradually appears in the perinatal period to reach adult levels at weaning).[91,102]

These data have been interpreted to mean that in the suckling period, arginine biosynthesis mainly occurs in the gut and that this biosynthesis is gradually transferred to the kidney at weaning. One reason for this complex developmental regulation of arginine metabolism may be that the flux through the hepatic ornithine cycle *in vivo* is extremely low during the suckling period,[103] whereas the adult liver actively degrades portal arginine to urea and ornithine. The complementary question that also needs to be addressed, however, is why *de novo* arginine biosynthesis needs to be so high in the gut during the suckling period and what factors cause the gut to be different from the liver vis à vis the genes of the ornithine cycle enzymes.

REFERENCES

1. **Husson, A., Bouazza, M., Buquet, C., and Vaillant, R.,** Role of dexamethasone and insulin on the development of the five urea-cycle enzymes in cultured rat foetal hepatocytes, *Biochem. J.*, 225, 271, 1985.
2. **Husson, A., Bouazza, M., Buquet, C., and Vaillant, R.,** Precocious induction of arginase in primary cultures of fetal rat hepatocytes, *In Vitro*, 20, 314, 1984.
3. **Husson, A. and Vaillant, R.,** Effects of glucocorticosteroids and glucagon on argininosuccinate synthetase, argininosuccinase, and arginase in fetal rat liver, *Endocrinology,* 110, 227, 1982.
4. **Husson, A., Bouazza, M., Buquet, C., and Vaillant, R.,** Hormonal regulation of two urea-cycle enzymes in cultured foetal hepatocytes, *Biochem. J.*, 216, 281, 1983.
5. **Husson, A. and Vaillant, R.,** Hormonal regulation of three urea cycle enzymes in rat fetal liver, *Biol. Neonate*, 35, 74, 1979.
6. **Gautier, C. and Vaillant, R.,** Role of insulin on the control of postnatal increase in ornithine transcarbamylase activity in rat liver, *Biochem. Biophys. Res. Commun.*, 123, 317, 1984.
7. **Charles, R., de Graaf, A., Lamers, W. H., and Moorman, A. F. M.,** Control of the changes in rat liver carbamoylphosphate synthetase (ammonia) protein levels during ontogenesis. Evidence for changes in immunoreactivity of the protein, *Mech. Ageing Dev.*, 22, 193, 1983.
8. **Lamers, W. H. and Mooren, P. G.,** Role of thyroid hormones in the normal and glucocorticosteroid hormone-induced evolution of carbamoylphosphate synthetase (ammonia) and arginase activity in perinatal rat liver, *Biol. Neonate*, 37, 264, 1980.
9. **Lamers W. H. and Mooren, P. G.,** Role of glucocorticosteroid hormones on the levels of rat liver carbamoylphosphate synthetase (ammonia) and arginase activity during ontogenesis, *Biol. Neonate*, 37, 113, 1980.
10. **Lamers W. H. and Mooren, P. G.,** Multihormonal control of enzyme clusters in rat liver ontogenesis. II. Role of glucocorticosteroid and thyroid hormone and of glucagon and insulin, *Mech. Ageing Dev.*, 15, 93, 1981.
11. **Lamers W. H. and Mooren, P. G.,** Multihormonal control of enzyme clusters in rat liver ontogenesis. I. Effects of adrenalectomy and gonadectomy, *Mech. Ageing Dev.*, 15, 77, 1981.
12. **Lamers W. H. and Mooren, P. G.,** Role of sex steroid hormones in the normal and glucocorticosteroid hormone-induced evolution of carbamoylphosphate synthetase (ammonia) and arginase activity in rat liver ontogenesis, *Biol. Neonate*, 40, 78, 1981.
13. **Renouf, S., Buquet, C., Fairand, A., Benamar, M., and Husson, A.,** Changes in levels of argininosuccinate lyase mRNA during induction by glucagon and cyclic AMP in cultured foetal-rat hepatocytes, *Biochem. J.*, 291, 609, 1993.
14. **Benamar, M., Gautier, C., Renouf, S., Fairand, A., and Husson, A.,** Changes in argininosuccinate lyase gene expression in the rat liver during development, *Biol. Neonate*, 61, 381, 1992.
15. **Husson, A., Renouf, S., Fairand, A., Buquet, C., Benamar, M., and Vaillant, R.,** Expression of argininosuccinate lyase mRNA in foetal hepatocytes. Regulation by glucocorticoids and insulin, *Eur. J. Biochem.*, 192, 677, 1990.
16. **Husson, A., Buquet, C., and Vaillant, R.,** Induction of the five urea-cycle enzymes by glucagon in cultured foetal rat hepatocytes, *Differentiation*, 35, 212, 1987.
17. **Husson, A., Guechairi, M., Fairand, A., Bouazza, M., Ktorza, A., and Vaillant, R.,** Effects of pancreatic hormones and glucocorticosteroids on argininosuccinate synthetase and argininosuccinase activities of rat liver during the perinatal period: in vivo and *in vitro* studies, *Endocrinology*, 119, 1171, 1986.
18. **Lamers, W. H., Mooren, P. G., de Graaf, A., and Charles, R.,** Perinatal development of the liver in rat and spiny mouse: its relation to altricial and precocial timing of birth, *Eur. J. Biochem.*, 146, 475, 1985.
19. **Lamers, W. H., Zonneveld, D., and Charles, R.,** Inducibility of carbamoylphosphate synthetase (ammonia) in cultures of embryonic hepatocytes: Ontogenesis of the responsiveness to hormones, *Dev. Biol.*, 105, 500, 1984.
20. **van Roon, M. A., Charles, R., and Lamers, W. H.,** Synthesis, accumulation and turnover of carbamoylphosphate synthetase and phosphoenolpyruvate carboxykinase in cultures of embryonic rat hepatocytes, *Eur. J. Biochem.*, 165, 229, 1987.
21. **van Roon, M. A., Charles, R., and Lamers, W. H.,** The initial accumulation of carbamoylphosphate synthetase in embryonic rat hepatocytes and the cell cycle, *Differentiation*, 41, 139, 1989.

22. **van Roon, M. A., Zonneveld, D., Charles, R., and Lamers, W. H.,** Accumulation of CPS and PEPCK mRNA in cultures of embryonic rat hepatocytes, *Eur. J. Biochem.*, 178, 191, 1988.
23. **van Roon, M. A., Zonneveld, D., de Boer, P. A. J., and Moorman, A. F. M., Charles, R., and Lamers, W. H.,** Regulation of hepatocyte-specific gene expression in cultures of human embryonic hepatocytes, *Biol. Neonate*, 58, 152, 1990.
24. **Meijer, A. J., Lamers, W. H., and Chamuleau, R. A. F. M.,** Nitrogen metabolism and ornithine cycle function, *Physiol. Rev.*, 70, 701, 1990.
25. **Rutter, W. J., Kemp, J. D., Bradshaw, W. S., Clark, W. R., Ronzio, R. A., and Sanders, T. G.,** Regulation of specific protein synthesis in cytodifferentiation, *J. Cell Physiol.*, 72(Suppl.), 1, 1968.
26. **Gaasbeek Janzen, J. W., Lamers, W. H., Moorman, A. F. M., de Graaf, A., Los, J. A., and Charles, R.,** Immunohistochemical localization of carbamoylphosphate synthetase (ammonia) in adult rat liver, *J. Histochem. Cytochem.*, 32, 557, 1984.
27. **Lamers, W. H., Gaasbeek Janzen, J. W., Moorman, A. F. M., Charles, R., Knecht, E., Martinez-Ramon, A., Hernandez-Yago, J., et al.,** Immunohistochemical localization of glutamate dehydrogenase in rat liver; plasticity of the distribution during development and with hormone treatment, *J. Histochem. Cytochem.*, 36, 41, 1988.
28. **Moorman, A. F. M., de Boer, P. A. J., Geerts, W. J. C., van de Zande, L. P. W. G., Charles, R., and Lamers, W. H.,** Complementary distribution of CPS (ammonia) and GS mRNA in rat liver acinus is regulated at a pretranslational level, *J. Histochem. Cytochem.*, 36, 751, 1988.
29. **Gaasbeek Janzen, J. W., Westenend, P. J., Charles, R., Lamers, W. H., and Moorman, A. F. M.,** Gene expression in derivatives of the embryonic foregut during prenatal development of the rat, *J. Histochem. Cytochem.*, 36, 1223, 1988.
30. **Moorman, A. F. M., de Boer, P. A. J., Das, A. T., Labruyère, W. T., Charles, R., and Lamers, W. H.,** Expression patterns of mRNAs for ammonia-metabolizing enzymes in the developing rat: the ontogenesis of hepatocyte heterogeneity, *Histochem. J.*, 22, 457, 1990.
31. **de Groot, C. J., ten Voorde, C. H. J., van Andel, R. E., te Kortschot, A., Gaasbeek Janzen, J. W., Wilson, R. H., Moorman, A. F. M., et al.,** Reciprocal regulation of glutamine synthetase and carbamoylphosphate synthetase levels in rat liver, *Biochim. Biophys. Acta*, 908, 231, 1987.
32. **Das, A. T., Arnberg, A. C., Malingré, H., Moerer, P., Charles, R., Moorman, A. F. M., and Lamers, W. H.,** Isolation and characterization of the rat gene encoding glutamate dehydrogenase, *Eur. J. Biochem.*, 211, 795, 1993.
33. **Das, A. T., Salvadó, J., Boon, L., Biharie, G., Moorman, A. F. M., and Lamers, W. H.,** Regulation of glutamate dehydrogenase expression in developing rat liver: control at different levels in the prenatal period, *Eur. J. Biochem.*, in press, 1993.
34. **Rohr, H. P., Wirtz, A., Henning, L. C., Riede, U. N., and Bianchi, L.,** Morphometric analysis of the rat liver cell in the prenatal period, *Lab. Invest.*, 24, 128, 1971.
35. **Vassy, J., Kraemer, M., Chamuleau, M. T., and Foucrier, J.,** Development of the fetal rat liver: ultrastructural and stereological study of hepatocytes, *Cell Differ.*, 24, 9, 1988.
36. **Kanamura, S., Kanai, K., and Watanabe, J.,** Fine structure and function of hepatocytes during development, *J. Electron Microsc. Tech.*, 14, 92, 1990.
37. **Greengard, O., Federman, M., and Knox, W. E.,** Cytomorphometry of developing rat liver and its application to enzymic differentiation, *J. Cell Biol.*, 52, 261, 1972.
38. **Daimon, T., David, H., Zglinicki, T., and Marx, I.,** Morphometric study on the absolute volume of differentiating hepatocytes in rats, *Exp. Pathol.*, 26, 141, 1984.
39. **Daimon, T., David, H., Zglinicki, T., and Marx, I.,** Correlated ultrastructural and morphometric studies on the liver during prenatal development of rats, *Exp. Pathol.*, 21, 237, 1982.
40. **Wheatley, D. N.,** Binucleation in mammalian liver, *Exp. Cell Res.*, 74, 455, 1972.
41. **Carriere, R.,** The growth of liver parenchymal nuclei and its endocrine regulation, *Int. Rev. Cytol.*, 25, 201, 1964.
42. **Tsanev, R., and Sendov, B.,** Possible molecular mechanism for cell differentiation in multicellular organisms, *J. Theor. Biol.*, 30, 337, 1971.
43. **Oosterhuis, W. P., Mooren, P. G., Charles, R., and Lamers, W. H.,** Perinatal development of the lung in rat and spiny mouse: its relation to altricial and precocial timing of birth, *Biol. Neonate*, 45, 236, 1984.
44. **Lamers, W. H., Mooren, P. G., de Graaf, A., and Charles, R.,** Perinatal development of the small intestine and pancreas in rat and spiny mouse: its relation to altricial and precocial timing of birth, *Biol. Neonate*, 47, 153, 1985.
45. **Greengard, O.,** The developmental formation of enzymes in rat liver, in *Biochemical Actions of Hormones*, Litwack G., Ed., Academic Press, New York, 1970, 53.
46. **Lamers, W. H., Mooren, P. G., and de Graaf, A., et al.,** Perinatal organ development in rat and spiny mouse: its relation to altricial and precocial timing of birth, in *Physiological Development of Fetus and Newborn*, Jones, C. T., Ed., Academic Press, London, 1985, 41.
47. **Nicoletti, M., Gueri, C., and Grisolía, S.,** Turnover of carbamoylphosphate synthetase, of other mitochondrial enzymes and of rat tissues, *Eur. J. Biochem.*, 75, 583, 1977.

48. **Dingemanse, M. A., de Boer, P. A. J., Moorman, A. F. M., Charles, R., and Lamers, W. H.,** The expression of liver-specific genes within rat embryonic hepatocytes is a discontinuous process, *Differentiation*, 56, 153, 1994.
49. **de Groot, C. J., Zonneveld, D., de Laaf, R. T. M., Dingemanse, M. A., Mooren, P. G., Moorman, A. F. M., Lamers, W. H., et al.,** Developmental and hormonal regulation of carbamoylphosphate synthetase gene expression in rat liver: evidence for control mechanisms at different levels in the perinatal period, *Biochim. Biophys. Acta*, 866, 61, 1986.
50. **Gautier, C., Habechi, Z., Belbekouche, M., and Vaillant, R.,** Effects of corticosteroids and pancreatic hormones on carbamyl phosphate synthetase-I and ornithinc transcarbamylase activities in fetal rat liver, *Biol. Neonate*, 48, 285, 1985.
51. **Xu, Q., Baker, B. S., and Tata, J. R.,** Developmental and hormonal regulation of the Xenopus liver-type arginase gene, *Eur. J. Biochem.*, 211, 891, 1993.
52. **Kingsbury, J. W., Alexanderson, M., and Kornstein, E. S.,** The development of the liver in the chick, *Anat. Rec.*, 124, 165, 1956.
53. **Theiler, K.,** *The House Mouse, Development and Normal Stages from Fertilization to 4 Weeks of Age*, Berlin, Springer-Verlag, 1972.
54. **Severn, C. B.,** A morphological study of the development of the human liver. I. Development of the hepatic diverticulum, *Am. J. Anat.*, 131, 133, 1978.
55. **Butler, H. and Juurlink, B. M. J.,** *An Atlas for Staging Mammalian and Chick Embryos*, CRC Press, Boca Raton, FL, 1987.
56. **Fukuda-Taire, S.,** Location of pre-hepatic cells in the early developmental stages of quail embryos, *J. Embryol. Exp. Morph.*, 64, 73, 1981.
57. **Rosenquist, G. C.,** Location and movements of cardiogenic cells in the chick embryo: the heart forming portion of the primitive streak, *Dev. Biol.*, 22, 461, 1970.
58. **Rosenquist, G. C.,** The origin and movements of hepatogenic cells in the chick embryo as determined by radioautographic mapping, *J. Embryol. Exp. Morphol.*, 24, 497, 1970.
59. **Fukuda, S.,** The development of hepatogenic potency in the endoderm of quail embryos, *J. Embryol. Exp. Morphol.*, 52, 49, 1979.
60. **Houssaint, E.,** Differentiation of the mouse hepatic diverticulum. I. An analysis of tissue interactions in hepatocyte differentiation, *Cell Differ.*, 9, 269, 1980.
61. **Fukuda-Taire, S.,** Hepatic induction in the avian embryo: specificity of reactive endoderm and inductive mesoderm, *J. Embryol. Exp. Morphol.*, 63, 111, 1981.
62. **Le Douarin, N. M.,** An experimental analysis of liver development, *Med. Biol.*, 53, 427, 1975.
63. **Sherer, G. K.,** Tissue interaction in chick liver development: a reevaluation, I. Epithelial morphogenesis: the role of vascularity in mesenchymal specificity, *Dev. Biol.*, 46, 281, 1975.
64. **Sherer, G. K.,** Tissue interaction in chick liver development: a reevaluation. II. Parenchymal differentiation: mesenchymal influence and morphogenetic independence, *Dev. Biol.*, 46, 296, 1975.
65. **Spooner, B. S., Cohen, H. I., and Faubion, J.,** Development of the embryonic mammalian pancreas: the relationship between morphogenesis and cytodifferentiation, *Dev. Biol.*, 61, 119, 1977.
66. **Ronzio, R. A. and Rutter, W. J.,** Effects of a partially purified factor from chick embryos on macromolecular synthesis of embryonic pancreatic epithelia, *Dev. Biol.*, 30, 307, 1973.
67. **Weintraub, H., Davis, R., Tapscott, S., Thayer, M., Krause, M., Benezra, R., Blackwell, T. K., et al.,** The myoD gene family: nodal point during specification of the muscle cell lineage, *Science*, 251, 761, 1991.
68. **Vassall-Adams, P. R.,** The development of the atrioventricular bundle and its branches in the avian heart, *J. Anat.*, 134, 169, 1982.
69. **Monaco, A. P., Bertolson, C. J., Liechti-Gallati, S., Moser, H., and Kunkel, L. M.,** An explanation for the phenotypic difference between patients bearing partial deletions of the DMD locus, *Genomics*, 2, 90, 1988.
70. **Shiorjiri, N.,** Analysis of differentiation of hepatocytes and bile duct cells in developing mouse liver by albumin immunofluorescence, *Dev. Growth Differ.*, 26, 555, 1984.
71. **Cascio, S. and Zaret, K. S.,** Hepatocyte differentiation initiates during endodermal-mesenchymal interactions prior to liver formation, *Development*, 113, 217, 1991.
72. **Shiojiri, N.,** Analysis of differentiation of hepatocytes and bile duct cells in developing mouse liver by albumin immunofluorescence, *Dev. Growth Differ.*, 26, 555, 1984.
73. **Lamers, W. H., van Roon, M. A., Mooren, P. G., de Graaf, A., and Charles, R.,** Amino acid environment determines expression of carbamoylphosphate synthetase and phosphoenolpyruvate carboxykinase in embryonic rat hepatocytes, *In Vitro Cell Dev. Biol.*, 21, 606, 1985.
74. **Westenend, P. J., Dahmen, R., Charles, R., and Lamers, W. H.,** Hormonal inducibility of liver-specific enzymes in cultured rat embryos, *Acta Morphol. Neerl.-Scand.*, 24, 165, 1986.
75. **Wright, W. E.,** BUdR, probability and cell variants: towards a molecular understanding of the decision to differentiate, *Bioessays*, 3, 245, 1987.
76. **Kassavetis, G. A. and Geiduschek, E. P.,** RNA polymerase marching backward, *Science*, 259, 944, 1993.

77. **van den Hoff, M. J. B., van de Zande, L. P. W. G., Dingemanse, M. A., Das, A. T., Moorman, A. F. M., Charles, R., and Lamers, W. H.,** Isolation and characterization of the rat carbamoylphosphate synthetase gene, *Eur. J. Biochem.*, in press, 1993.
78. **Benvenisty, N., Mencher, D., Meyuhas, O., Razin, A., and Reshef, L.,** Sequential changes in DNA methylation patterns of the rat phospohenolpyruvate carboxykinase gene during developments, *Proc. Natl. Acad. Sci. U.S.A.*, 82, 267, 1985.
79. **Benvenisty, N. and Reshef, L.,** Developmental acquisition of DNAse I sensitivity of the phosphoenolpyruvate carboxykinase (GTP) gene in rat liver, *Proc. Natl. Acad. Sci. U.S.A.*, 84, 1132, 1987.
80. **Benvenisty, N., Szyf, M., Mencher, D., Razin, A., and Reshef, L.,** Tissue-specific hypomethylation and expression of rat phosphoenolpyruvate carboxykinase gene induced by in vivo treatment of fetuses and neonates with 5-azacytidine, *Biochemistry*, 24, 5015, 1985.
81. **van Roon, M. A., Aten, J. A., Charles, R., and Lamers, W. H.,** The initiation of hepatocyte-specific gene expression within embryonic hepatocytes is a stochastic event, *Dev. Biol.*, 136, 508, 1989.
82. **Dingemanse, M. A. and Lamers, W. H.,** Expression patterns of ammonia-metabolizing enzymes in the liver, mesonephros and gut of human embryos and their possible implications, *Anat. Rec.*, 238, 480, 1994.
83. **Holzgreve, W., Fisk, A. S., Goldsmith, P. C., and Golbus, M. S.,** Histochemical demonstration of ornithine transcarbamylase activity in fetal liver, *Biol. Neonate*, 47, 339, 1985.
84. **Kennan, A. C. and Cohen, P. P.,** Biochemical studies of the developing mammalian fetus. I. Urea cycle enzymes, *Dev. Biol.*, 1, 511, 1959.
85. **Rubio, V. V., Tan, C. B., Andrews, W. H. H., Nixon, D. A., Alexander, D. P., and Britton, H. G.,** Nitrogen metabolism in the sheep fetus, *Biol. Neonate*, 43, 86, 1983.
86. **Rattenbury, J. M., Jeacock, M. J., and Shepherd, D. A. L.,** Urea synthesis in the liver and kidney of developing sheep, *Biochim. Biophys. Acta*, 630, 210, 1980.
87. **Raiha, N. C. R., and Schwartz, A. L.,** Development of urea biosynthesis and factors influencing the activity of the arginine synthetase system in perinatal mammalian liver, in *Inborn Errors of Metabolism*, Hommes, F. A. and van den Berg C. J., Eds., Academic Press, New York, 1973, 221.
88. **Morris, S. M., Kepka, D. M., Sweeney, W. E., and Avner, E. D.,** Abundance of mRNAs encoding urea cycle enzymes in fetal and neonatal mouse liver, *Arch. Biochem. Biophys.*, 269, 175, 1989.
89. **Windmüller, H. G.,** Glutamine utilization by the small intestine, *Adv. Enzymol.*, 53, 210, 1982.
90. **Ryall, J. C., Quantz, M. A., and Shore, G. C.,** Rat liver and intestinal mucosa differ in the developmental pattern and hormonal regulation of carbamoylphosphate synthetase I and ornithine carbamoyl transferase gene expression, *Eur. J. Biochem.*, 156, 453, 1986.
91. **Hurwitz, R. and Kretchmer, N.,** Development of arginine-synthesizing enzymes in mouse intestine, *Am. J. Physiol.*, 251, G103, 1986.
92. **Yamada, E. and Wakabayashi, Y.,** Development of pyrroline-5-carboxylate synthase and N-acetylglutamate synthase and their changes in lactation and aging, *Arch. Biochem. Biophys.*, 291, 15, 1991.
93. **Dubois, N., Cavard, C., Chasse, J.-F., Kamoun, P., and Briand, P.,** Compared expression levels of ornithine transcarbamylase and carbamylphosphate synthetase in liver and small intestine of normal and mutant mice, *Biochim. Biophys. Acta*, 950, 321, 1988.
94. **Wu, G., Knabe, D. A., and Flynn, N. E.,** Synthesis of citrulline from glutamine in pig enterocytes, *Biochem. J.*, 299, 115, 1994.
95. **Riby, J. E., Hurwitz, R. E., and Kretchmer, N.,** Development of ornithine metabolism in the mouse intestine, *Pediatr. Res.*, 28, 261, 1990.
96. **Schiller, C. M., Southern, J. T., and Walden, R.,** Glutamine and glutamate utilization in the hamster small intestine, *J. Appl. Biochem.*, 3, 147, 1981.
97. **Herzfeld, A. and Raper, S. M.,** Enzymes of ornithine metabolism in adult and developing rat intestine, *Biochim. Biophys. Acta*, 428, 600, 1976.
98. **Luk, G. D., Marton, L. J., and Baylin, S. B.,** Ornithine decarboxylase is important in intestinal mucosal maturation and recovery from injury in rats, *Science*, 210, 195, 1980.
99. **Blachier, F., M'Rabet-Touil, H., Posho, L., Darcy-Vrillon, B., and Duée, P.-H.,** Intestinal arginine metabolism during development. Evidence for de novo synthesis of L-arginine in newborn pig enterocytes, *Eur. J. Biochem.*, 216, 109, 1993.
100. **Windmüller, H. G. and Spaeth, A. E.,** Uptake and metabolism of plasma glutamine by the small intestine, *J. Biol. Chem.*, 249, 5070, 1974.
101. **Windmüller, H. G., and Spaeth, A. E.,** Source and fate of circulating citrulline, *Am. J. Physiol.*, 241, E473, 1981.
102. **Morris, S. M., Sweeney, W. E., Kepka, D. M., O'Brien, W. E., and Avner, E. D.,** Localization of arginine biosynthetic enzymes in renal proximal tubules and abundance of mRNA during development, *Pediatr. Res.*, 29, 151, 1991.
103. **Illnerova, H.,** Urea formation in rats during postnatal development, *Biol. Neonate*, 9, 197, 1966.

Chapter 15

Maternal-Fetal Transfer of Nitrogen in Chondrichthyans

Gregg A. Kormanik

CONTENTS

INTRODUCTION

The literature on nitrogen metabolism, nitrogenous waste synthesis, and excretion in fishes is voluminous[1-4] with some reviews dealing specifically with elasmobranchs and their close relatives.[5,6] The chondrichthyans, or cartilaginous fishes, including the sharks, skates, rays, and chimeras are of special interest for several reasons. In the marine environment, these primitive fishes retain high concentrations of urea and trimethylamine N-oxide (TMAO) as osmolytes to help offset the water balance problems associated with hypoionic regulation.[7] The origin of our vertebrate ancestors, whether freshwater or seawater, is not clear; physiological explanations, in addition to the paleontological, are used to support various hypotheses regarding vertebrate origins.[8,9] The detoxification of ammonia via urea synthesis is relevant to understanding the aquatic to terrestrial transition of the vertebrates. In addition, the high concentration of urea (300 to 400 m*M*) typically maintained by marine elasmobranchs could be expected to disrupt protein structure and therefore enzyme function if it were not offset by the stabilizing effects of a high (50 to 100 m*M*) TMAO concentration.[10,11] The adaptive strategies and the biochemical means by which the chondrichthyans accumulate, regulate, retain, and tolerate these high concentrations of urea and TMAO are therefore of considerable interest.

0-8493-8411-7/95/$0.00+$.50

Accumulation and retention of urea have also played a significant role in the discussions regarding the evolution of viviparity in this group.[12-17] The literature on maternal-fetal relations in the elasmobranchs has also been well reviewed.[18-21] Strategies of reproduction within this group range from species that lay eggs to those that develop a placental-like connection to the mother. Among the chondrichthyans it has been estimated that viviparity evolved from oviparity on as many as 18 different occasions.[22] Thus, development processes in elasmobranchs provide a wealth of "experiments" to help understand the evolution of nitrogen metabolism and excretion. One difficulty that complicates attempts to understand the evolution of viviparity is that the fossil record contains very little information regarding the developmental biology of the viviparous vertebrates.[23,24] Studies regarding evolutionary scenarios of development and related physiology may be undertaken on extant species, where a comparative approach might prove especially fruitful.[23]

This chapter specifically considers nitrogen metabolism and transfer in embryonic elasmobranchs, and the special conditions that dictate the way they handle nitrogenous metabolites. In addition to showing gaps in our knowledge and avenues for future research with respect to maternal-fetal aspects of nitrogen transfer, I hope to point out some physiological and evolutionary implications of nitrogen transfer strategies.

NITROGEN ACQUISITION BY EMBRYOS

CHONDRICHTHYAN MODES OF REPRODUCTION

Chondrichthyan reproductive strategies range from egg laying to live birth, with myriad variations in between. Wourms et al.[20,25] distinguish oviparity, or egg laying, from viviparity in that, in the latter case, eggs are fertilized and retained internally, developing to an advanced state in the maternal reproductive system. These categories have been subdivided.[20,25] Compagno[21] distinguishes four modes of viviparity, or live bearing. Yolksac viviparity (=ovoviviparity) is widely used to denote a viviparous fish, with the primary source of nutrients being the egg, which is retained until term in the oviduct or uterus of the mother.[21,26] More advanced forms of viviparity (uterine, placental, and cannibal viviparity) denote various specializations for providing nutrients to embryos. The definition of viviparity has been the subject of some debate[20,21,23,26,27] with some basis being the time during development, and the amount and manner by which nutrients are acquired by, or provided for, the developing embryo. All chondrichthyans initially depend on yolk reserves. While there is a fair amount of data on the extent to which many elasmobranch embryos may or may not acquire elemental constituents from the mother (as measured directly by changes in mass or by ash weights of the embryos compared to the egg[20]), few data exist regarding the amounts, nature, or manner by which specific metabolites are transferred from mother to developing embryo.

The diversity and complexity of how elasmobranch mothers transfer nutrients to developing embryos and our limited knowledge of the processes preclude making too many generalizations. While the general mechanism of oviparity may be similar among chondrichthyans, viviparity is varied, and little is known regarding its physiology. Only general aspects and a few examples of maternal-fetal transfer of nitrogen in these fishes are described below.

OVIPARITY

About 42% of living chondrichthyans (including possibly all the living chimeroids) deposit on a substrate, large, cased demersal eggs which hatch into a miniature adult.[21] In oviparous species, embryos gain water and salts from the environment, and possibly some organics, such as amino acids, although this latter aspect has not been examined. These kinds of eggs would not be considered truly "cleidoic" by Needham[28] (see below) because the aquatic environment provides water as well as many inorganic constituents.

Egg Contents

While many examples of egg content exist in the literature for teleosts, few exist for the chondrichthyans.[29,30] Lechenault et al.[31] report that the yolk in egg cells of *Scyliorhinus cannicula* is made up of elliptical yolk platelets, with glycogen particles and smaller oil droplets; however the composition of platelets and elasmobranch vitellogenins has not been adequately analyzed. *Raja oxyrhyncus* eggs contain 28% protein and 6.7% fat, while *Heptranchias deani* eggs contain 25% protein and 23% fat.[28] Lipovitellin-like proteins have been identified from the eggs of a dogfish, *Scyliorhinus settaris*.[32] Given this variation and the fact that composition of eggs is species specific in the teleosts, there is every reason to expect the same for the chondrichthyans. Generally, in teleosts, protein, which is composed of lipovitellin and phosvitin, is the most abundant constituent by dry weight (30 to 80%), followed by lipids (20%) and carbohydrates (5%).[30,33,34] Protein mainly provides amino acids for growth and also as an energy source. Thus, catabolism of protein and other amino acids represents a potential source of excess ammonia and bicarbonate in a 1:1 ratio that must be eliminated.[35] The low permeability of teleost eggs to nitrogenous metabolites has made it difficult to measure protein catabolism.[30] Although there are no direct measurements, the same probably can be said for elasmobranchs, since the oocyte (yolk) membranes (plasma + vitelline) have a very low permeability to osmolytes, at least as evidenced by the ability of the egg cell to retain molecules.[36,37]

Cleidoic Egg and Nitrogen Retention

The "cleidoic" egg has been considered to be primarily important in the evolution of urea retention and has figured prominently in discussions by several authors regarding the evolution of viviparity in the elasmobranchs and other primitive fishes.[14-17] An enclosed egg requires that the embryonic system deal with nitrogenous end products of metabolism. The more the system is enclosed, the greater is the need to detoxify these endproducts for storage and/or to recycle them for growth. For example, animals using less cleiodic eggs (e.g., aquatic) like the frog (*Rana temporaria*), the salmon (*Salmo salar*), and the squid (*Loligo vulgaris*) can combust relatively large portions of their protein stores (29.6, 36.3, and 15%, respectively) as a source of energy, compared to the chick (*Gallus domesticus*) which uses less than 5%. The chick, like other birds and insects, derives most of its energy (80%) from lipid stores compared to eggs of fish and amphibians (30%).[28]

While the chondrichthyan egg capsule itself does not present a substantial barrier to the movement of small osmolytes (see below), the egg or oocyte membranes do by restricting the loss of urea, TMAO, and other metabolites initially endowed by the mother (Figure 1). Thus the egg cell membranes, and the nature and origin of their relative impermeance to urea and TMAO must have played an important role in the evolution of urea retention.

Needham and Needham[12] postulated that for *Scyllium* (=*Scyliorhinus*) *canicula* to have evolved an enclosure approaching that of the cleidoic egg the elasmobranchs evolved a tolerance to severe uremia, enclosed the egg, but then later opened the capsule for respiratory or other purposes. The Needhams[12] noted that extant elasmobranchs do not now possess a true cleidoic egg. Having adapted

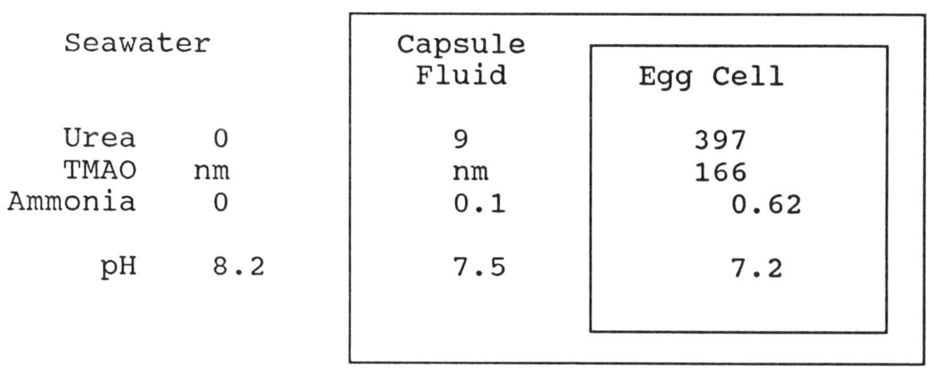

Figure 1 This diagram shows an encapsulated egg of *Squalus acanthias* a few days after deposition. Concentrations of nitrogenous osmolytes are in m*M* (nm = not measured). The plasma and vitelline membranes are the barriers between the egg cell and capsule fluid/jelly. The capsule wall is the boundary between the capsular fluid and seawater. (From Kormanik, G. A., *Bull. Mt. Desert Isl. Biol. Lab.*, 28, 12, 1989. With permission.)

Table 1 Representative Permeabilities (cm sec^{-1}) of Ammonia and Urea in Tissues of Several Elasmobranch Species

		Ammonia		
	Urea	**NH_3**	**NH_4^+**	**Reference**
Egg capsule		NH_3	NH_4^+	
Scyliorhinus cannicula				
	5×10^{-5}			43, 44
	6×10^{-4}			
Raja erinacea	2×10^{-5}	2×10^{-5}	2×10^{-5}	37, 52
Mustelis canis	4×10^{-4}			46
Adult branchial				
Squalus acanthias	8×10^{-8}			103
Raja erinacea	1×10^{-7}			61
Endometrial				
Squalus acanthias	6×10^{-7}			75
Embryo branchial				
Squalus acanthias		4×10^{-4}	0.4×10^{-6}	73

to urea retention, the elasmobranchs simply retained it. Smith[13] suggested that the elasmobranchs possess a cleidoic egg as an adaptation secondary to urea retention. With respect to urea retention, Smith argued that viviparity and intrauterine development protected the embryo against loss of urea better than oviparity could. The notion that this elasmobranch egg capsule acts as a barrier to the movement of urea (and other molecules) was supported by many authors (see References 36 and 37 for discussion) and in fact has been purported to be either a prerequisite for, or a feature associated with, urea retention by developing elasmobranch embryos.[15,16,38,39] Forster[40] and Forster and Goldstein[41] suggested that detoxification of nitrogenous wastes via urea synthesis was necessary in the encapsulated environment of the embryo, with the retention of urea (and TMAO) later becoming important for osmoregulatory purposes. These hypotheses were promulgated in spite of the original data of Needham and Needham[12] that suggested that the membranes of the embryo and not the egg capsule wall were the barriers to urea loss. It has been shown many times that the egg capsule is quite permeable to urea, ammonia, and other small molecules for several species of oviparous elasmobranchs (Table 1).[37,42-46] The hypothesis that urea is synthesized for ammonia detoxification is nonetheless a viable one[40,41] if we consider that the egg cell or embryonic system membranes, rather than the capsule, pose the barrier to its elimination.

High concentrations of urea and TMAO are provided in the egg by the mother and retained by the egg cell membrane due to its low permeability.[12,42,47,48] Needham and Needham[12] demonstrated that embryos of *Scyllium* and *Pristiurus melanostoma* could synthesize urea, since urea content increased severalfold as embryos grew in size before the time of hatching. They also reported that urea increased in *Acanthias vulgaris* (=*S. acanthias*) through at least the early portions of development, but not to as great an extent. Read[49] showed that early embryos of *Raja binoculata* and *Squalus sucklii* possess enzymes of the ornithine-urea cycle and can synthesize urea. Read[42] also showed that TMAO content increased during development of *R. binoculata*, and concluded that this species could synthesize TMAO as well (Table 2). Previously, Baker et al.[50] found no trimethylamine oxidase activity in the livers of the ray (*Torpedo californica*), the dogfish (*S. acanthias*), or the skate (*R. binoculata*). Kormanik et al.[48,51,52] showed that, unlike urea, TMAO did not increase in either *R. erinacea* (oviparous) or *S. acanthias* (yolksac viviparous) embryos during development. The only TMAO available to these embryos is that placed into the egg cell by the mother. These latter observations support those of Goldstein et al.[53] who were unable to show TMAO synthesis in adults of these two species. Some warmer water species, however, show some synthetic activity for TMAO.[6,54] In the absence of synthesis, TMAO is acquired in the diet, since it is abundant in many marine organisms.[55,56]

At moderate levels, tissue tolerance to urea and TMAO does not appear to be a major obstacle to retention of these molecules for osmoregulatory or other purposes. Neither urea nor TMAO,

Table 2 Comparison of the Major Nitrogenous Osmolytes and Total Nitrogen in Hatchlings and Egg Cells of Three Oviparous Elasmobranchs

	Wet wt. (g)	Urea (μmol)	TMAO (μmol)	Total N (mg)	Reference
Squalidae					
Scyliorhinus canicula					
Egg cell	2.3	400	nm	196	45
Hatchling	3.5	1000	nm	136	45
Rajidae					
Raja erinacea					
Egg cell	4.3	990	410	284	48
Hatchling	6.0	1690	380	171	52
Raja binoculata					
Egg cell	19	4700	1200	nm	42
Hatchling	45	12000	1700	nm	42

Note: Expressed per individual; nm, not measured.

which may reach concentrations of 70 μmol g^{-1} in Alaska pollock (*Theragra chalcogramma*) or Atlantic cod (*Gadus morhua*) is very toxic to teleosts.[15,57] Griffith[58] has shown that TMAO may reach levels of 97 m*M* in some deep-sea teleosts, while urea remains at only a few m*M*. Thus, ancestors to the chondrichthyans might have easily tolerated initial accumulation of urea or TMAO.

Nitrogen Loss During Development

Since the embryonic system is open to the environment, developing embryos of oviparous chondrichthyans are characterized by a net loss of nitrogen to their surroundings. Mellinger et al.[45] report that *S. canicula* loses 31% of the nitrogen initially stored in the egg during the course of development. Kormanik et al.[48,52] reported similar losses for *R. erinacea* (40%; see Table 2). Nitrogen probably is lost mainly as urea, ammonia, and TMAO, although there are no data showing the excretion rates of TMAO in the earliest stages of embryos. Both early- and late-term embryos of *S. acanthias* lose urea and ammonia.[51,59] Urea and TMAO represent 10 and 2% respectively, of the total nitrogen initially contributed by the mother to the egg cell, and increase to 25 and 3% respectively at the time of hatching.[48] The ratio of urea to TMAO in the egg cell at oviposition is about 2.4:1, typical of elasmobranchs in general.[2,7] This ratio approaches 4.4:1 at the time of hatching, still below the ratios seen in adult red blood cells (11.5:1) or muscle cells (6.2:1) for this species.[60,61]

No data for amino acid composition of chondrichthyan egg cells have been published. The egg cell of a chondrichthyan is probably like a typical muscle or other elasmobranch cell with respect to osmolytes.[6] Some amino acids found in the tissues of elasmobranchs must be obtained from the diet since (at least as adults) they are unable to synthesize them from the usual precursors.[55] Many of these free amino acids as well as other nitrogenous compounds are not incorporated into proteins and represent important cellular osmolytes.[6,55] For example, taurine, a sulfur-containing amino acid, appears to play an osmoregulatory role in elasmobranchs, and is accumulated by heart and other tissues.[62,63] Embryos of *Raja eglanteria* cannot synthesize taurine, which must be obtained from yolk stores. The egg cell provides sufficient taurine to the embryo during the course of development. Goldstein et al.[63] found that the volume-activated taurine release mechanism is present in the embryonic skate heart before hatching. The latter seems a reasonable observation, since *R. eglanteria*, an oviparous species, is likely exposed to seawater salt concentrations early in its development, due to the high permeability of the egg capsule.

Chondrichthyan embryos might obtain energy from the catabolism of free amino acids, as do marine teleosts. Free amino acid content in developing eggs and larvae of some marine teleosts (*G. morhua* and *H. hippoglossus*) declines precipitously during development.[64,65] No comparable data exist for the chondrichthyans.

YOLKSAC VIVIPARITY

S. acanthias provides one of the most complete data sets we have for the physiology of a viviparous elasmobranch species. Classically, *S. acanthias* is considered to exhibit yolksac viviparity (=ovoviviparous) since eggs are produced, fertilized internally, and develop to term in expanded terminal portions of the oviducts, referred to as paired uteri.[18,20,25] Previous authors assumed that the embryos receive no organic nutrients from the mother other than those stored in the egg cell prior to ovulation, based on the observation that organic content of the embryo decreases throughout development.[18,20,25]

Uterine Environment

The conditions under which these embryos develop are of special interest with respect to nitrogen budgets. The developing embryos are incubated initially in a solution that resembles a plasma exudate with high concentrations of the osmolyte urea, and ammonia concentrations typical of elasmobranch blood.[36,66,67] By four to six months incubation, salt concentrations are somewhat elevated over those of maternal plasma, and urea concentration is reduced, presumably due to at least some seawater leakage into the uterine horns.[36]

By 14 to 16 months into the 22-month period of incubation, the embryos are bathed by a solution that resembles seawater with respect to the major ions, but has a pH of about 6, a urea (and TMAO) concentration of only a few millimolar, but a high ammonia concentration, averaging about 10 m*M*, but at times reaching up to 22 m*M*.[67-69] Files and Lombardi[70] have shown that the only free amino acid in this solution is taurine, which is found at very low (15 μ*M*) concentrations. Embryos, still with pendant yolk sacs can be removed from the mother and will survive to normal term in seawater provided that care is taken to prevent abrasion and infection.[66] While its presence has not been demonstrated in uterine tissues or seawater, squalamine, the potent steroid antibiotic recently found in *Squalus* may play an important role.[71]

Handling Uterine Ammonia

Kormanik et al.[48,51,52,67,69] have examined several aspects of acid–base balance, urea and ammonia excretion, and nitrogen budgets in this species during gestation. This level of ammonia in the ambient medium is usually considered toxic, but the toxicity is ameliorated by the low pH of the uterine seawater that reduces the concentration of NH_3 to levels below those of both the maternal and embryonic blood (Figure 2).[67] Thus the low pH of the uterine seawater, established by the mother, serves as a "sink" for ammonia, which is generated by both the mother and the embryo. The partial pressure gradients for NH_3, the predominant excretory form of ammonia, are directed from both maternal and fetal blood to the uterine seawater.[67,72] Permeability of elasmobranch epithelia to

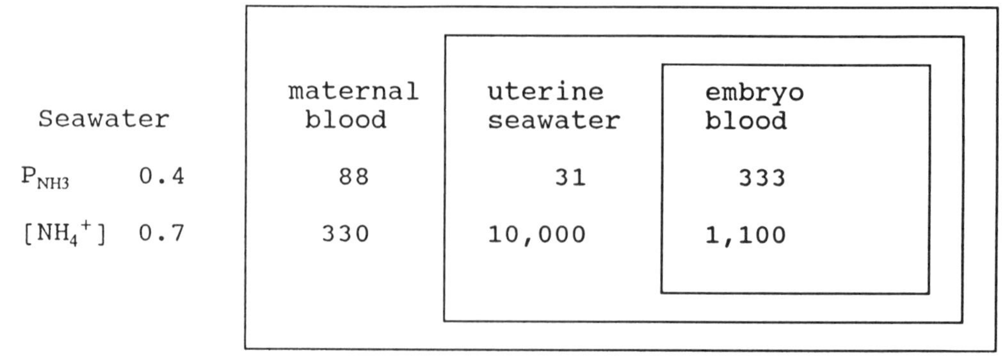

Figure 2 Concentration and partial pressure gradients of ammonia in its two diffusive forms developed between the various blood and seawater compartments of late-term *Squalus acanthias* embryos incubated *in utero*. The partial pressure (P_{NH3}) is expressed in μTorr and the ionic form, NH_4^+, in μ*M*. (From Kormanik, G. A. and Evans, D. H., *J. Exp. Biol.*, 125, 173, 1986. With permission.)

NH_3 is about 1100 times that of the charged form, NH_4^+ (see also Table 1).[73] Since blood ammonia is elevated (approximately threefold) in embryos compared to adults, these levels would not appear to be toxic to the fetal brain. It is not known whether the fetal brain has a greater tolerance to ammonia compared to the adult, or whether these levels are within the normal range of tolerance. Ovine fetal plasma has about twice the ammonia level of the adult, which appears to be readily tolerated.[74] This high uterine ammonia might arise for several reasons. First, since the mother apparently does not frequently flush the uterine horns with seawater,[69] embryonic excretory ammonia must be detoxified by conversion to the less permeable and less toxic form, NH_4^+. Most of this ammonia comes from maternal excretion into the uterine seawater because these levels of ammonia can be established in the absence of embryos.[69] It may be that the mother cannot impermeabilize the uterine endometrium to NH_3. Extensive surface area and vascularization must be maintained for oxygen and carbon dioxide transfer. She therefore must detoxify ammonia by acidification. It is possible that some of this ammonia may arise from urea of maternal origin. While the permeability of embryonic epithelia[59,66] and maternal endometrium to urea is low[75] (see Table 1), the gradients and surface area are substantial. Although some urea might leak into the uterine environment, conversion to ammonia by bacterial degradation is probably minimal since high bacterial content would infect the embryos.

Nevertheless, the presence of high uterine ammonia concentrations increases blood ammonia of these late-term embryos and the potential exists for the use of this maternal ammonia in the synthesis of urea or other nitrogenous metabolites by the embryos. While late-term *S. acanthias* embryos lose some nitrogen during development, they do not seem to lose as much as do oviparous species (see Table 2).[48,51,69] Normal blood ammonia of elasmobranchs is typically 200 to 400 μ*M*.[2] Embryonic blood ammonia is about 1200 μ*M*,[67] or several fold higher, and might be expected to stimulate both glutamine and urea production via glutamine synthetase.[76] The ornithine-urea cycle of elasmobranchs uses carbamoyl-phosphate synthetase (CPS III), which accepts nitrogen from glutamine.[77] Shankar and Anderson[78] have shown that the Km of glutamine synthetase for ammonia is 0.015 m*M* (although the species of ammonia was not determined), and might be expected already to be saturated at normal (200 to 400 μ*M*) levels of blood ammonia.[1] CPS III can accept ammonia directly as a substrate for urea synthesis with a Km of 2 m*M*, within the range of the elevated embryonic blood ammonia (1.1 m*M*). Synthesis using ammonia as a substrate for CPS III would proceed more slowly, at about 10% (0.05 μmol g^{-1} h^{-1}) of the rate estimated for glutamine as a substrate (0.5 μmol g^{-1} h^{-1}).[77,79] The gestation period of *S. acanthias* is long (ca. 22 mo.), and it would take only a small increase in urea synthesis rates to be of benefit to the developing embryo; an increase in whole body urea synthesis rate by only 0.01 μmol g^{-1} h^{-1} could increase total body urea by 10% in six months.[52]

Mommsen and Walsh[3] suggest that ammonia detoxification might be the prime objective of the late-term pup urea cycle, since urea excretion accounts for 98% of nitrogen loss from early embryos, and 93% from late-term embryos.[51] In adult elasmobranchs urea usually accounts for about 80% or less of the nitrogen excreted as ammonia or urea.[6] An alternate interpretation is that ammonia is retained by embryos for metabolite synthesis. Theoretically, showing whether external ambient ammonia is used by embryos for net accumulation of nitrogen could be done by incubating them in high and low ammonia concentrations to modify blood P_{NH3}, and monitoring embryo growth and total nitrogen through time. One problem certainly would be maintaining embryos *ex utero* long enough for definitive results. Lack of a suitable tracer for monitoring the movement of ammonia nitrogen and its incorporation into urea or other nitrogen pools of embryonic tissues makes experimentation difficult for short-term studies. The stable isotope ^{15}N may provide a tool to track the movement of nitrogen in this and other systems.[80] The studies of Read,[42,49] Mellinger et al.,[45] and Kormanik[48,51,52] and others on oviparous species provide a baseline to examine the changes in total urea, TMAO, and nitrogen during development in oviparous and viviparous elasmobranchs to determine the role uterine incubation may play in nitrogen acquisition and retention by developing embryos.

The Coelacanth

The coelacanth (*Latimeria chalumnae*; Smith, 1939), while not a chondrichthyan, deserves mention because of its ureosmotic regulation,[15,81] and because it exhibits the reproductive mode of yolksac viviparity (=ovoviviparity).[26,82] Weight loss through development may be –16 to –60% of dry mass,[26] thus there is no evidence for supplemental nutrient provision by the mother. No data exist regarding the conditions under which the eggs and embryos are incubated *in utero*, so it is not known if embryos must osmoregulate in uterine seawater or face ammonia accumulation during development like *Squalus*, or if uterine fluids resemble maternal plasma throughout gestation.

ADVANCED FORMS OF VIVIPARITY

Wourms and others have written extensively regarding the maternal specializations in structure involved in the transfer of nutrients to the developing embryos.[18,20,22,25,83,84] While the contents and amounts of egg nutrients initially are established by the ovaries, the walls of the uterus are the boundaries of the embryonic environment and help to establish uterine conditions. The uterus has evolved many specialized properties for secretory activity, enhanced surface area, compartmentalization, and placental attachment in various species.[20,83,84] The ultrastructural changes that occur in the epithelia of the embryonic system and the mother during placentation in sharks has been reviewed.[83,84]

Uterine Fluid Composition

In some species of viviparous elasmobranchs, uterine fluids resemble maternal plasma (*M. canis*[38]; *Carcharinus leucas*[85]), at least with respect to the major ionic constituents. Parker[72] observed that urea was present in the pseudoamniotic fluid of the viviparous *Mustelus antarcticus*. Wourms[20] summarized some published and unpublished studies on the content of the complex histotrophe or "uterine milk" (1 to 2% organic content) of the butterfly ray, *G. micrura*. With respect to the major nitrogen-containing components, Wourms et al. found a total protein content of 2.4 mg ml^{-1}; likely representing various albumins and globulins. Free amino acid (FAA) analysis found phosphoethanolamine, urea, taurine, gamma-aminobutyric acid (GABA), citrulline, alanine, glutamine, and valine. The FAA composition was reportedly similar to that of mammalian amniotic fluid as well as some viviparous teleost histotrophe.

Files and Lombardi[70] examined the occurrence of amino acids in the uterine fluids of several species of elasmobranchs representing various degrees of maternal-fetal trophic relationships. In viviparous *Carcharhinus plumbus*, the most abundant amino acid was taurine (788 μM), with others at lower levels (0 to 60 μ*M*). In contrast, uterine fluids of *Rhizoprionodon terraenovae*, whose embryos possess elaborate absorptive structures (appendiculae), had FAA concentrations often several times that of the blood. The highest levels of FAAs were norleucine (8 m*M*), taurine (5.6 m*M*), glutamate (2.7 m*M*), valine (1.5 m*M*), and phenylalanine (1 m*M*). These high levels (about 6 to 21 times blood concentrations) are likely the result of some mechanism for active concentration or secretory activity. Files and Lombardi[70] suggest that some of these ketogenic amino acids (e.g., norleucine, phenylalanine, isoleucine, and leucine) may serve as metabolic fuel for the developing elasmobranch embryos. Alanine (a good gluconeogenic substrate) and leucine are both good oxidative substrates for *R. erinacea* liver.[86,87] Both are found in uterine fluids of several viviparous species albeit at somewhat lower levels (0.5 m*M* in *R. terranovae*).[70] Viviparous teleosts like the blenny (*Zoarces viviparous*) extract alanine from the ovarian fluids.[88] While one might presume that the mother is secreting these uterine constituents for use by developing embryos, the source of these metabolites (i.e., whether from the mother or from the embryo) is not always clear.

Metabolite Transfer

While much has been done to correlate physical attributes of uterine tissue with the nature and amount of nutrient transfer to developing embryos,[25,89,90] little has been published regarding either the physiological mechanisms of transfer, or the biochemical nature of the constituents. Several studies have examined the movement of large molecules such as horseradish peroxidase by endocytosis that could account for the movement of large molecules like proteins.[83] I am unaware of any studies tracing specifically the movement of small metabolites like free amino acids, urea, or TMAO from mother to embryo, comparable to the many investigations on mammalian placentae.[91] In studies examining nutrient transfer between mother and embryo of several species exhibiting varying degrees of maternal support, Lombardi et al.[92] found that the protein content (also lipid and reducing sugars) was correlated with the degree of accessory absorptive structures.

In placental mammals, urea produced by the developing fetus passes to the mother for excretion. In species of elasmobranchs that establish placental or paraplacental transport between the mother and fetus, the origins for urea and TMAO found in the neonate have not been shown. It is likely that urea and TMAO can pass through placental routes to the developing embryo. One might assume that elasmobranch embryos soon develop the capability to synthesize urea, but those embryos that derive substantial nutritional support from the mother likely receive a large portion of urea and TMAO from the mother as well, although evidence is lacking.

Taurine transfer from mother to embryo is of special interest since this molecule is apparently not synthesized by the elasmobranchs nor by any fish.[55] In addition to its use as an osmolyte[6,63] taurine plays a role in cell growth and movement[93] and is transported across the human placenta to meet the needs of the developing embryo.[94] The trimethylamine, sarcosine, another major cellular osmolyte, is synthesized by skates[95] and may play a role with TMAO in the stabilization of protein in the presence of urea.[10,11]

EVOLUTIONARY AND DEVELOPMENTAL CONSIDERATIONS

ORIGINS OF UREA RETENTION

Goldstein et al.[96] suggested, and Pang and others agree, that production and retention of urea are primitive characteristics of the vertebrates.[3,97,98] But why and how did urea production and retention arise? Urea synthesis may provide a nontoxic means to eliminate nitrogenous waste or an osmoregulatory advantage when associated with urea retention. It may be useful in acid–base balance and regulation of HCO_3^-.[2,3,76] While urea production and retention may have arisen for any of these reasons, present use of the system may of course be different. In addition, there is a danger in extrapolation of data regarding nitrogen metabolism from adult to embryonic systems. Eggs, embryos, and larvae often undergo a very different set of environmental constraints than do adults. They therefore may use different physiological mechanisms from those of adults. We may obtain glimpses of those mechanisms in the activities of genes that adults possess but ordinarily remain silent, but direct study of embryonic systems may yield more important clues.

Many teleosts retain the genes for urea synthesis via the ornithine-urea cycle. Indeed, the biochemical characteristics of the ornithine-urea cycle appear to be identical in all fishes, with the except of the lungfish.[3] While teleosts have not lost the genes responsible for synthesis of urea via the ornithine-urea cycle, with the exception of a few species, notably those in alkaline lakes and air-breathers/intertidal species,[3,98-100] urea synthesis plays little if any physiological role in adult teleosts. In adult teleosts, ureagenesis is activated only under special circumstances.[76] Embryonic development may represent another special set of circumstances (see Chapter 16). Griffith[39] has already suggested several scenarios regarding the evolution of urea retention in fishes, and called for an extension of the study by Depeche et al.[101] on urea retention by teleost embryos to a broad variety of both teleost and nonteleost fishes, in order to examine the phenomenon of urea retention by embryos and its role in nitrogen detoxification.

DIFFUSIONAL EXCHANGE DURING DEVELOPMENT

Some authors have hypothesized that ureosmotic regulation and retention have been related to large size, sluggishness, internal fertilization, constancy of the environmental salinity, and the ability to synthesize and tolerate high concentrations of urea.[15,16,58] Griffith[58] also reviewed characteristics that he considered were prerequisites for the evolution of urea retention. The large size requirement for urea retention is also extended to embryos.[15,58] Of key importance is the nature of the epithelia conferring a low permeability to urea. While surface area to mass ratios of fishes may vary by an order of magnitude, urea permeabilities of tissues may vary by several orders of magnitude.[7] In some cases (e.g., the elasmobranch kidney), urea retention may be attributable to active resorption.[102] In the gill of elasmobranchs, and probably other external epithelia, the low permeability to urea is more likely due to passive or physical properties of the membrane.[7,103] Unfortunately the nature of this low permeability to urea remains unknown.

Egg Cell Permeability and Size

The egg capsule or envelope has received undue attention as the barrier to the diffusion of ions and osmolytes. Because the capsule is permeable to respiratory gases,[104] salts, urea, and other small molecules (see above), clearly the egg membranes and the epithelia of the early embryonic system are the limiting barriers to diffusion of ions and osmolytes. While many investigations have examined changes in the permeability of teleost eggs to salt and especially to water,[105] hardly any have been done with the eggs or early embryonic systems of elasmobranchs. Elasmobranch external epithelia are characterized by high-water permeability and low permeability to Na^+, Cl^-, urea, and TMAO.[7] While values have not been measured directly, egg membranes also exhibit low permeability to ions and molecules the size of TMAO and urea, since large gradients for these osmolytes are maintained between the egg cell and the capsular/seawater fluids.[36] We might direct our attention to the roles these barriers could have played in the evolution of urea retention. Some authors have related the large size of elasmobranch eggs and those of the coelacanth to urea retention.[15] A spherical egg provides a minimum surface to volume ratio, and all things being equal, larger eggs should better retain endowed osmolytes and other molecules than smaller ones. But what do relative differences in size and permeability make? Measurements of egg cell permeability and their relation to volume have been considered by several authors (see Reference 105 for a discussion). Teleost eggs tend to be spherical and somewhat uniform in size, falling within a range of 1 to 10 mm and averaging about 1 to 2 mm in size.[22] Elasmobranch eggs (oocytes) range in size from 1 mm diameter for a viviparous species (*Scoliodon laticaudus*, which receives nearly all its nutrients from the mother[106]) to 2 to 3 cm in the oviparous species *R. erinacea*, and about 4 to 5 cm for the yolksac viviparous *S. acanthias*.[68] The largest oocytes of the chondrichthyans are probably those of ovoviviparous *Ginglymostoma* and *Chlamydoselachus*, at 10 cm.[18] The ovoviviparous coelacanth *Latimeria* also possesses large eggs of up to 8 to 9 cm in diameter.[26,107]

Figure 3 shows a simple, hypothetical analysis of the relationship of egg size, permeability, and the time it takes to lose one-half of the egg contents due to diffusion down a concentration gradient. A large egg typical for *S. acanthias* (4 cm), could retain half the urea (or any other molecule) with which it was endowed, for nearly 100 d with a permeability typical of elasmobranch epithelia (see Table 1). If permeability increases by two orders of magnitude, comparable to that of an egg capsule (Table 1), retention drops to about 1 d. These simple calculations assume maintenance of the theoretical gradients and ignore the effects of unstirred boundary layers as well as internal cytoplasmic diffusion, which could affect the gradients on either side of the membrane.[105,108] The figure is nevertheless useful in pointing out the importance of both size and membrane permeability in urea retention.

Embryonic Permeability

Urea synthesis has been proposed as a means of eliminating excess CO_2 or HCO_3^- (see below). Respiratory gas exchange has been best modeled, but as Rombaugh[108] points out, in spite of the many papers published, there has been a lack of a systematic approach to the study of molecular movement through the developing embryonic system. Pelster and Bemis[109] examined the role of respiratory filaments in the embryos of the oviparous *R. erinacea* during development. Apparently body/yolk sac surface is insufficient for oxygen exchange, because the embryos develop gill filaments as outgrowths from the gill arches. These filaments are later absorbed, presumably when

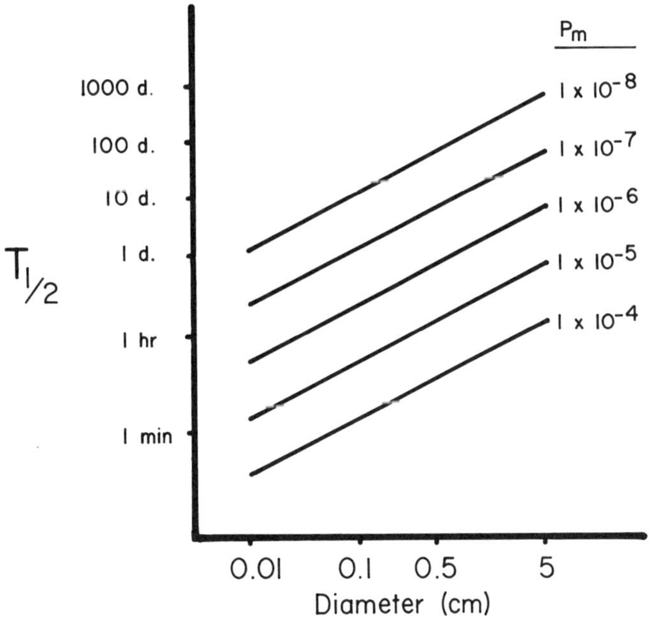

Figure 3 The rate of uncharged solute diffusing across a membrane bounding a sphere (an egg cell) is described by the Fick equation. This net flux is divided by the amount of solute in a sphere, yielding units of 1/sec. This can be calculated as a "half-time" ($T_{1/2}$), or the time required to lose one-half of the sphere's content of solute. "Half-time" ($T_{1/2}$) is presented in the figure versus the diameter for several permeabilities (Pm, in cm sec^{-1}). This calculation is theoretical because it ignores the very important components of internal and external convective mixing, as well as unstirred and boundary layers of fluids that may affect the gradients across the membrane.

regular gill respiratory surface area is sufficient to meet the needs of the embryo. Pelster and Bemis[109] estimate that these gill filaments, when fully developed, may account for 50% of the oxygen uptake by the embryo. These data suggest that respiratory surface for oxygen uptake (but not necessarily carbon dioxide excretion) may be limited during at least some of the early developmental period. It appears reasonable to hypothesize that low membrane permeability and a relatively low surface to volume ratio are embryonic characteristics which have been retained by adults.[39]

ENERGETICS OF UREA RETENTION

A few authors have considered the metabolic cost of ureosmotic regulation vs. hypoionic regulation in marine vertebrates.[15,110] Comparisons are difficult due to incomplete data. Kirschner[110] recently concluded, based on the data he could obtain, that both modes are about equally efficient. The question has been applied as well to the advantages and disadvantages of oviparity and viviparity with respect to urea retention. Wourms[18] suggests that osmoregulation can be accomplished with more efficiency in the uterine environment, with less expenditure of embryonic energy, and Smith[13] stated that viviparity protects against urea loss better than oviparity. This is likely true due to potentially reduced osmotic and chemical gradients.[36,66] The question is, however, how much energy is saved? For an oviparous species like *R. erinacea*, we will consider only the cost of urea retention, since TMAO, while retained, does not appear to be synthesized by this species.[48,52,53] *Raja* egg cells are initially endowed with 157 mg nitrogen/g dry weight (see Table 2).[48,52] By the time of hatching, these embryos have lost 40% of the total nitrogen. If all this nitrogen is lost as urea (ignoring the urea that was originally endowed in the egg, about 10% of total nitrogen), it represents 2.2 mmol urea/g. Since it takes 5 ATP molecules to synthesize a molecule of urea[110] and under standard conditions, it takes 7.3 kcal/mol to synthesize ATP,[111] this nitrogen lost before hatching represents 82 cal/g dry weight. Teleost eggs[30] and likely elasmobranch eggs have about 5.7 kcal/g.[20] Thus, the synthetic cost of nitrogen lost as urea represents less than 2% of the energy content of an egg.

Incubation of an egg in a high urea uterine environment would appear to provide little energetic benefit over seawater incubation of oviparous eggs, at least with respect to urea retention. This very low loss rate for urea and other nitrogenous osmolytes is afforded by the low permeability of the egg membranes and later the external epithelia of the embryonic system, not the egg envelope.

ACID–BASE BALANCE

One aspect of recent discussion is the role of urea synthesis in vertebrate acid–base balance and pH homeostasis.[112-114] Meijer et al.[35] (see also Chapter 12) support a role for urea synthesis in the stabilization of HCO_3^- concentrations, and Walsh and Henry[76] speculate that urea synthesis, at least in the short term, in some fish, may be regulated by the availability of HCO_3^- and not ammonia. Walsh and Henry[76] argue that urea synthesis likely does not play a role in acid–base balance in most teleost fish. Embryos catabolize considerable amounts of protein stored in the egg cell and NH_4^+ and HCO_3^- are produced in equimolar ratios.[35] Since NH_4^+ and HCO_3^- are both charged ions, one might expect egg cell membranes to have a low permeability to them, which would hinder their elimination by diffusion. However, elasmobranch epithelia are far more permeable to NH_3 than NH_4^+ (1100:1).[73] Diffusive loss of NH_3 would leave behind a proton to titrate the HCO_3^- to CO_2 and H_2O. Both CO_2 and NH_3 could leave the egg cell in gaseous form, dissolved in water, with no net pH change consequences. It is possible that urea synthesis could have evolved as a means (albeit metabolically expensive) to regulate HCO_3^- in early embryonic systems. On the one hand, surface area for gas exchange appears limiting for metabolically active elasmobranch embryos, since developing embryos must elaborate gill filaments to obtain sufficient oxygen.[109] On the other hand, however, carbon dioxide excretion should be much less of a problem in aquatic organisms.[115,116] Because carbon dioxide is about 30 times more soluble in water than is oxygen, fish are continually hyperventilating with respect to carbon dioxide excretion.[116] The partial pressure of carbon dioxide in fish blood is typically 5 to 10% of that found in air-breathing vertebrates, and HCO_3^- is lower.[76,117,116] Late-term embryonic elasmobranchs would appear to be no exception.[67] We have no data for ammonia and HCO_3^- levels in embryonic systems early in their development. It would be interesting to examine acid–base balance in eggs and embryos of both teleosts and elasmobranchs under steady-state and varying conditions to determine if, under any sets of conditions, either ammonia or CO_2 excretion is limited, or has an adverse effect on pH regulation.

CONCLUDING REMARKS

Little is known regarding the physiology of teleost or chondrichthyan embryonic systems. Low membrane permeability as well as synthesis and retention of urea may be embryonic characteristics that proved physiologically advantageous to ancestors of the fishes. These characteristics were retained in adult chondrichthyans and the coelacanth *Latimeria*. In the teleosts, some embryos may use urea synthesis, but genetic machinery for urea synthesis has been silenced in adults, unless required in some species because of environmental extremes. The study of fish embryos with respect to acid–base balance, and the role of ammonia and bicarbonate retention and pH regulation in urea synthesis could prove beneficial. Knowledge of the biochemical nature of the membranes' low permeability to urea would prove useful as well. Modern molecular techniques should be used on the controls of urea synthesis as well as the production and distribution of enzymes used in the ornithine-urea cycle in these fascinating and physiologically diverse groups of fishes.

ACKNOWLEDGMENTS

I thank the editors for inviting me to contribute to this volume. I would also like to thank my many undergraduate students who have worked with me through the years on various aspects of elasmobranch physiology. Much of my work has been supported by grants from the National Science Foundation (DCB-8502251, DCB-8904429).

REFERENCES

1. **Campbell, J. W. and Anderson, P. W.,** Evolution of mitochondrial enzyme systems in fish: the mitochondrial synthesis of glutamine and citrulline, in *Biochemistry and Molecular Biology of Fishes*, Vol. I, Hochachka, P. W. and Mommsen, T. P., Eds., Elsevier, Amsterdam, New York, 1991, 43.
2. **Wood, C. M.,** Ammonia and urea metabolism and excretion, in *The Physiology of Fishes*, Evans, D. H., Ed., CRC Press, Boca Raton, FL, 1993, 379.
3. **Mommsen, T. P. and Walsh, P. J.,** Urea synthesis in fishes: evolutionary and biochemical perspectives, in *Biochemistry and Molecular Biology of Fishes*, Vol. I., Hochachka, P. W. and Mommsen, T. P., Eds., Elsevier, Amsterdam, New York, 1991, 137.
4. **Mommsen, T. P. and Walsh, P. J.,** Biochemical and environmental perspectives on nitrogen metabolism in fishes, *Experientia*, 48, 583, 1992.
5. **Brown, G. W., Jr. and Brown., S. G.,** On urea formation in primitive fishes, in *Evolutionary Biology of Primitive Fishes*, Foreman, R. E., Gorbman, A., Dodd, J. M., and Olsen, R., Eds., Plenum Press, New York, 1985, 321.
6. **Pearlman, D. F. and Goldstein, L.,** Nitrogen metabolism, in *Physiology of Elasmobranch Fishes*, Shuttleworth, T. J., Ed., Springer-Verlag, New York, 1988, 253.
7. **Shuttleworth, T. J.,** Salt and water balance — Extrarenal mechanisms, in *Physiology of Elasmobranch Fishes*, Shuttleworth, T. J., Ed., Springer-Verlag, New York, 1988, 171.
8. **Bray, A. A.,** The evolution of the terrestrial vertebrates: environmental and physiological considerations, *Philos. Trans. R. Soc. London Ser. B*, 309, 289, 1985.
9. **Griffith, R. W.,** Freshwater or marine origin of the vertebrates?, *Comp. Biochem. Physiol.*, 87A, 3, 523, 1987.
10. **Yancy, P. H. and Somero, G. N.,** Counteraction of urea destabilization of protein structure by methylamine osmoregulatory compounds of elasmobranch fishes, *Biochem. J.*, 183, 317, 1979.
11. **Yancy, P. H. and Somero, G. N.,** Methylamine osmoregulatory solutes of elasmobranch fishes counteract urea inhibition of enzymes, *J. Exp. Zool.*, 212, 205, 1980.
12. **Needham, J. N. and Needham, D. M.,** Nitrogen excretion in selachian ontogeny, *J. Exp. Biol.*, 7, 7, 1930.
13. **Smith, H. W.,** The retention and physiological role of urea in the elasmobranchii, *Biol. Rev.*, 11, 49, 1936.
14. **Griffith, R. W. and Thomson, K. S.,** *Latimeria chalumnae*: reproduction and conservation, *Nature,* 242, 617, 1973.
15. **Griffith, R. W. and Pang, P. K. T.,** Mechanisms of osmoregulation in the coelacanth: evolutionary implications, in *The Biology and Physiology of the Living Coelacanth*, McCosker, J. E. and Lagios, M. D., Eds., Occ. Pap. Calif. Acad. Sci. 134, San Francisco, 1979, 79.
16. **Griffith, R. W.,** Habitat, phylogeny and the evolution of osmoregulatory strategies in primitive fishes, in *Evolutionary Biology of Primitive Fishes*, Foreman, R. E., Gorbman, A., Dodd, J. M., and Olsson, R., Eds., Plenum Press, New York, 1985, 69.
17. **Fange, R.,** Regulation of blood and body fluids in primitive fish groups, in *Evolutionary Biology of Primitive Fishes*, Foreman, R. E., Gorbman, A., Dodd, J. M., and Olsen, R., Eds., Plenum Press, New York, 1985, 253.
18. **Wourms, J. P.,** Reproduction and development in chondrichthyan fishes, *Am. Zool.*, 17, 379, 1977.
19. **Hamlett, W. C.,** Prenatal absorptive structures in selachians, in *Indo-Pacific Fish Biology: Proceedings of the Second International Conference on Indo-Pacific Fishes*, Uyeno, T., Arai, R., Taniuchi, T., and Matsuura, K., Eds., Ichthyological Soc. Japan, Tokyo, 1986, 333.
20. **Wourms, J. P., Grove, B. D., and Lombardi, J.,** The maternal-embryonic relationship in viviparous fishes, in *Fish Physiology*, Vol. 11B, Hoar, W. S. and Randall, D. J., Eds., Academic Press, San Diego, 1988, 1.
21. **Compagno, L. J. V.,** Alternative life histories of cartilaginous fishes in time and space, *Environ. Biol. Fish.*, 28, 33, 1990.
22. **Wourms, J. P. and Lombardi, J.,** Reflections on the evolution of piscine viviparity, *Am. Zool.*, 32, 276, 1992.
23. **Wake, M. H.,** Evolutionary scenarios, homologies and convergence of structural specializations for vertebrate viviparity, *Am. Zool.*, 32, 256, 1992.
24. **Wourms, J. P. and Callard, I. P.,** A retrospective to the symposium on evolution of viviparity in vertebrates, *Am. Zool.*, 32, 251, 1992.
25. **Wourms, J. P.,** Viviparity: the maternal fetal relationship in fishes, *Am. Zool.*, 21, 473, 1981.
26. **Heemstra, P. C. and Greenwood, P. H.,** New observations on the visceral anatomy of the late-term fetuses of the living coelacanth fish and the oophagy controversy, *Proc. R. Soc. Lond. B.*, 249, 49, 1992.
27. **Balon, E. K.,** Probable evolution of the coelacanth's reproductive style: lecithotrophy and orally feeding embryos in cichlid fishes and in *Latimeria chalumnae*, *Environ. Biol. Fish.*, 32, 249, 1991.
28. **Needham, J.,** *Biochemistry and Morphogenesis*, Cambridge University Press, Cambridge, 1950, chap. 3.
29. **Mommsen, T. P. and Walsh, P. J.,** Vitellogenesis and oocyte assembly, in *Fish Physiology*, Vol. 11B, Hoar, W. S. and Randall, D. J., Eds., Academic Press, San Diego, 1988, 347.
30. **Heming, T. A. and Buddington, R. K.,** Yolk absorption in embryonic and larval fishes, in *Fish Physiology*, Vol. 11A, Hoar, W. S. and Randall, D. J., Eds., Academic Press, New York, 1988, 407.
31. **Lechenault, H., Wrisez, F., and Mellinger, J.,** Yolk utilization in *Scyliorhinus cannicula*, an oviparous dogfish, *Environ. Biol. Fish.*, 38, 241, 1993.

32. **Fuji, T.,** Comparative biochemical studies on the egg-yolk proteins of various animal species, *Acta Embryol. Morphol. Exp.*, 3, 260, 1960.
33. **Jared, D. W. and Wallace, R. A.,** Comparative chromatography of the yolk proteins of teleosts, *Comp. Biochem. Physiol.*, 24, 437, 1968.
34. **Market, J. R. and Vanstone, W. E.,** Egg proteins of coho salmon (*Oncorhynchus kisutch*): chromatographic separation and molecular weights of the major proteins in the high density fraction and their presence in salmon plasma, *J. Fish. Res. Bd. Can.*, 28, 1853, 1971.
35. **Meijer, A. J., Lamers, W. H., and Chamuleau, R. A. F. M.,** Nitrogen metabolism and ornithine urea cycle function, *Physiol. Rev.*, 70, 701, 1990.
36. **Kormanik, G. A.,** Ion and osmoregulation in prenatal elasmobranchs: evolutionary implications, *Am. Zool.*, 32, 294, 1992.
37. **Kormanik, G. A.,** Ionic and osmotic environment of developing elasmobranch embryos, *Environ. Biol. Fish.*, 38, 233, 1993.
38. **Price, K. S., Jr. and Daiber, F. C.,** Osmotic environments during fetal development of dogfish, *Mustelus canis* (Mitchell) and *Squalus acanthias* Linnaeus, and some comparisons with skates and rays, *Physiol. Zool.*, 40, 248, 1967.
39. **Griffith, R. W.,** Guppies, toadfish, lungfish, coelacanths and frogs: a scenario for the evolution of urea retention in fishes, *Environ. Biol. Fish.*, 32, 199, 1991.
40. **Forster, R. P.,** Osmoregulatory role of the kidneys in cartilagenous fishes (Chondrichthyes), in *Sharks, Skates and Rays*, Gilbert, P. W., Mathewson, R. F., and Rall, D. P., Eds., The Johns Hopkins Press, Baltimore, 1967, 187.
41. **Forster, R. P. and Goldstein, L.,** Formation of excretory products, in *Fish Physiology*, Vol. 1, Hoar, W. S. and Randall, D. J., Eds., Academic Press, New York, 1969, 313.
42. **Read, L. J.,** Urea and trimethylamine oxide levels in elasmobranch embryos, *Biol. Bull. Mar. Biol. Lab. Woods Hole*, 135, 537, 1968.
43. **Foulley, M. and Mellinger, J.,** La diffusion de l'eau tritiée, de l'urée-^{14}C et d'autres substances à travers la coque de l'oeuf de Roussette, *Scyliorhinus canicula. C.R. Acad. Sci. Paris*, 290, 427, 1980.
44. **Hornsey, D. J.,** Permeability coefficients of the egg-case membrane of *Scyliorhinus cannicula* L., *Experientia*, 34, 1596, 1978.
45. **Mellinger, J., Wrisez, F., and Alluchon-Gerard, M.-J.,** Developmental biology of an oviparous shark, *Scyliorhinus cannicula*, in *Proceedings, Second International Conference on Indo-Pacific Fishes*, Uyeno, T., Arai, R., Taniuchi, T., and Matsuura, K., Eds., Ichthyological Soc. Japan, Tokyo, 1986, 310.
46. **Lombardi, J. and Files, T.,** Egg capsule structure and permeability in the viviparous shark, *Mustelus canis*, *J. Exp. Zool.*, 267, 76.
47. **Mellinger, J. and Wrisez, F.,** Biologie et physiologie compareés du développement de deux sélaciens ovipares les roussettes *Scyliorhinus canicula* et *Scyliorhinus stellaris*. Evolution de la matière sèche, de l'eau et des ions (Cl^-, Na^+, K^+) dans le vitellus de *S. cannicula* au cours du développement, *Bull. Soc. Zool. France*, 114, 51, 1988.
48. **Kormanik, G. A. and Totten, R. K.,** Total nitrogen, trimethylamine N-oxide and urea in developing embryos of the little skate, *Raja erinacea*, *Bull. Mt. Desert Isl. Biol. Lab.*, 32, 9, 1993.
49. **Read, L. J.,** Ornithine-urea cycle enzymes in early embryos of the dogfish *Squalus suckleyi* and the skate *Raja binoculata*, *Comp. Biochem. Physiol.*, 24, 669, 1968.
50. **Baker, J. R., Struempler, A., and Chaykin, S.,** A comparative study of trimethylamine N-oxide biosynthesis, *Biochim. Biophys. Acta*, 71, 58, 1963.
51. **Kormanik, G. A.,** Nitrogen budget in developing embryos of the spiny dogfish *Squalus acanthias*, *J. Exp. Biol.*, 144, 583, 1989.
52. **Kormanik, G. A., Lofton, A., and O'Leary-Liu, N.,** Nitrogen budget in developing elasmobranch embryos, *Bull. Mt. Desert Isl. Biol. Lab.*, 31, 44, 1992.
53. **Goldstein, L., Hartman, S. C., and Forster, R. P.,** On the origin of trimethylamine oxide in the spiny dogfish, *Squalus acanthias*, *Comp. Biochem. Physiol.*, 21, 719, 1967.
54. **Goldstein, L. and Funkhouser, D.,** Biosynthesis of trimethylamine oxide in the nurse shark, *Ginglymostoma cirratum*, *Comp. Biochem. Physiol.*, 42A, 51, 1972.
55. **Van Waarde, A.,** Biochemistry of non-protein nitrogenous compounds in fish including the use of amino acids for anaerobic energy production, *Comp. Biochem. Physiol.*, 91B, 207, 1988.
56. **Benoit, G. J. and Norris, E. R.,** Studies on trimethylamine oxide II. The origin of TMAO in young salmon, *J. Biol. Chem.*, 158, 438, 1945.
57. **Wekell, J. C. and Barnett, H.,** New method for analysis of trimethylene oxide using ferrous sulfate and EDTA, *J. Food Sci.*, 56, 132.
58. **Griffith, R. W.,** Composition of the blood serum of deep-sea fishes, *Biol. Bull.*, 160, 250, 1981.
59. **Evans, D. H. and Kormanik, G. A.,** Urea efflux from the *Squalus acanthias* pup: the effect of stress, *J. Exp. Biol.*, 119, 375, 1985.
60. **Forster, R. P. and Goldstein, L.,** Intracellular osmoregulatory role of amino acids and urea in marine elasmobranchs, *Am. J. Physiol.*, 230, 925, 1976.

61. **Payan, P., Goldstein, L., and Forster, R. P.,** Gills and kidneys in ureosmotic regulation in euryhaline skates, *Am. J. Physiol.*, 224, 367, 1973.
62. **Goldstein, L., Luer, C. A., and Blum, P. C.,** Taurine accumulation by the heart of embryonic skates, *Raja eglanteria, J. Exp. Biol.*, 150, 449, 1990.
63. **Goldstein, L., Luer, C. A., and Blum, P. C.,** Taurine transport characteristics of the embryonic skate (*Raja eglanteria*) heart, *J. Exp. Biol.*, 182, 291, 1993.
64. **Fyhn, H. J.,** Energy production in marine fish larvae with emphasis on free amino acids as a potential fuel, in *Nutrition in Wild and Domestic Animals*, Mellinger, J., Ed., Karger, Basil, 1990, 176.
65. **Fyhn, H. J. and Serigstad, B.,** Free amino acids as energy substrate in developing eggs and larvae of the cod *Gadus morhua, Mar. Biol.*, 96, 335, 1987.
66. **Evans, D. H., Oikari, A., Kormanik, G. A., and Mansberger, L.,** Osmoregulation by the prenatal spiny dogfish, *Squalus acanthias, J. Exp. Biol.*, 101, 295, 1989.
67. **Kormanik, G. A. and Evans, D. H.,** The acid-base status of prenatal pups of the dogfish, *Squalus acanthias*, in the uterine environment, *J. Exp. Biol.*, 125, 173, 1986.
68. **Hisaw, F. L. and Albert, A.,** Observations on the reproduction of the spiny dogfish, *Squalus acanthias, Biol. Bull.*, 92, 187, 1947.
69. **Kormanik, G. A.,** Time course of the establishment of uterine seawater conditions in late-term pregnant spiny dogfish (*Squalus acanthias*), *J. Exp. Biol.*, 137, 443, 1988.
70. **Files, T. and Lombardi, J.,** Free amino acids in the uterine fluids of four species of viviparous sharks, *Comp. Biochem. Physiol.*, 104B, 583, 1993.
71. **Moore, K. S., Wehrli, S., Roder, H., Rogers, M. J., and Forrest, J. N.,** Squalamine: An aminosterol antibiotic from the shark, *Proc. Natl. Acad. Sci. U.S.A.*, 90, 1354, 1993.
72. **Parker, T. J.,** Note on the foetal membranes of *Mustelus antarcticus, Trans. N.Z. Inst.*, 22, 331, 1889.
73. **Evans, D. H. and More, K.,** Modes of ammonia transport across the gill epithelium of the dogfish pup (*Squalus acanthias*), *J. Exp. Biol.*, 138, 375, 1988.
74. **Holzman, I. R., Lemons, J. A., Meschia, G., and Battaglia, F. C.,** Ammonia production by the pregnant sheep uterus, *Pediatr. Res.*, 11, 407, 1977.
75. **Bissonnette, J. M., Montgomery, L. B., and Miller, D. S.,** Urea permeability in pregnant endometrium of spiny dogfish (*Squalus acanthias*), *Bull. Mt. Desert Isl. Biol. Lab.*, 19, 66, 1979.
76. **Walsh, P. J. and Henry, R. P.,** Carbon dioxide and ammonia metabolism and exchange, in *Biochemistry and Molecular Biology of Fishes*, Vol. I, Hochachka, P. W. and Mommsen, T. P., Eds., Elsevier, New York, 1991, 181.
77. **Anderson, P. M.,** Glutamine and N-acetylglutamate-dependent carbamoyl phosphate synthetase in elasmobranchs, *Science*, 208, 291, 1980.
78. **Shankar, R. A. and Anderson, P. M.,** Purification and properties of glutamine synthetase from liver of *Squalus acanthias, Arch. Biochem. Biophys.*, 239, 248, 1985.
79. **Anderson, P. M.,** Purification and properties of the glutamine- and N-acetyl-L-glutamate-dependent carbamoyl phosphate synthetase from liver of *Squalus acanthias, J. Biol. Chem.*, 256, 23, 12228, 1981.
80. **Carter, C. G., Owen, S. F., He, Z.-Y., Watt, P. W., Scrimgeour, C., Houlihan, D. F., and Rennie, J.,** Determination of protein synthesis in rainbow trout, *Oncorhynchus mykiss*, using stable isotope, *J. Exp. Biol.*, 189, 279, 1994.
81. **Pickford, G. E. and Grant, F. B.,** Serum osmolarity in the coelacanth, *Latimeria chalumnae*: urea retention and ion regulation, *Science*, 155, 568, 1967.
82. **Smith, C. L., Rand, C. S., Schaeffer, B., and Atz, J. W.,** *Latimeria*, the living coelacanth, is ovoviviparous, *Science*, 190, 1105, 1975.
83. **Hamlett, W. C.,** Evolution and morphogenesis of the placenta in sharks, *J. Exp. Zool.*, 2 (Suppl.), 35, 1989.
84. **Hamlett, W. C., Eulitt, A. M., Jarrell, R. L., and Kelly, M. A.,** Uterogestation and placentation in elasmobranchs, *J. Exp. Zool.*, 266, 347, 1993.
85. **Thorsen, T. B. and Gerst, J. W.,** Comparison of some parameters of serum and uterine fluid of pregnant, viviparous sharks (*Carcharhinus leucas*) and serum of their near-term young, *Comp. Biochem. Physiol.*, 42A, 33, 1972.
86. **Mommsen, T. P. and Moon, T. W.,** The metabolic potential of hepatocytes and kidney tissue in the little skate, *Raja erinacea, J. Exp. Zool.*, 244, 1, 1987.
87. **Moon, T. W. and Mommsen, T. P.,** Enzymes of intermediary metabolism in tissues of the little skate, *Raja erinacea, J. Exp. Zool.*, 244, 9, 1987.
88. **Korsgaard, B.,** Amino acid uptake and metabolism by embryos of the blenny *Zoarces viviparous, J. Exp. Biol.*, 171, 315, 1992.
89. **Needham, J.,** *Biochemistry and Morphogenesis*, Cambridge University Press, New York, 1942, chap. 3.
90. **Amoroso, E. C.,** Viviparity in fishes, *Symp. Zool. Soc. (London)*, 1, 153, 1960.
91. **Jones, C. T. and Rolph, T. P.,** Metabolism during fetal life: a functional assessment of metabolic development, *Physiol. Rev.*, 65, 357, 1985.
92. **Lombardi, J., Jones, K., Garrity, C., and Files, T.,** Chemical composition of uterine fluid in four species of viviparous sharks, *Comp. Biochem. Physiol.*, 105A, 91, 1993.

93. **Sturman, J. A., Gargano, A., Messing, J., and Imaki, H.,** Feline maternal taurine deficiency: effect on mother and offspring, *J. Nutr.*, 116, 655, 1985.
94. **Phillips, A. F., Holtzman, I. R., Teng, C., and Battaglia, F. C.,** Tissue concentrations of free amino acids in term human placentas, *Am. J. Obstet. Gynecol.*, 131, 881, 1988.
95. **King, P. A., Cha, C. J., and Goldstein, L.,** Amino acid metabolism and cell volume regulation in the little skate, *Raja erinacea* II. Synthesis, *J. Exp. Zool.*, 212, 79, 1980.
96. **Goldstein, L., Harley-DeWitt, S., and Forster, R. P.,** Activities of urea cycle enzymes and of trimethylamine oxidase in the coelacanth, *Latimeria chalumnae*, *Comp. Biochem. Physiol.*, 44B, 357, 1973.
97. **Pang, P. K. T., Griffith, R. W., and Atz, J. W.,** Osmoregulation in elasmobranchs, *Am. Zool.*, 17, 365, 1977.
98. **Mommsen, T. P. and Walsh, P. J.,** Evolution of urea synthesis in vertebrates: the piscine connection, *Science*, 243, 72, 1989.
99. **Wood, C. M., Perry, S. F., Wright, P. A., Bergmann, H. L., and Randall, D. J.,** Ammonia and urea dynamics in the lake Magadi tilapia, a ureotelic teleost fish adapted to an extremely alkaline environment, *Resp. Physiol.*, 77, 1, 1989.
100. **Wood, C. M., Bergman, H. L., Laurent, P., Maina, J. N., Narahara, A., and Walsh, P. J.,** Urea production, acid-base regulation and their interactions in the lake Magadi tilapia, a unique teleost adapted to a highly alkaline environment, *J. Exp. Biol.*, 189, 13, 1994.
101. **Depeche, J., Gilles, R., Daufresne, S., and Chaipello, H.,** Urea content and urea production via the ornithine-urea cycle during the ontogenic development of two teleost fishes, *Comp. Biochem. Physiol.*, 63A, 51, 1979.
102. **Henderson, I. W., O'Toole, L. B., and Hazon, N.,** Kidney function, in *Physiology of Elasmobranch Fishes*, Shuttleworth, T. J., Ed., Springer-Verlag, New York, 1988, 201.
103. **Boylan, J. W.,** Gill permeability in *Squalus acanthias*, in *Sharks, Skates and Rays*, Gilbert, P. W., Mathewson, R. F., and Rall, D. P., Eds., Johns Hopkins, Baltimore, 1967, 197.
104. **Diez, J. M. and Davenport, J.,** Embryonic respiration in the dogfish (*Scyliorhinus canicula* L.), *J. Mar. Biol. Assoc. U.K.*, 67, 249, 1987.
105. **Alderice, D. F.,** Osmotic and ionic regulation in teleost eggs and larvae, in *Fish Physiology*, Vol. 11A, Hoar, W. S. and Randall, D. J., Eds., Academic Press, New York, 1988, 163.
106. **Wourms, J. P.,** Maximization of evolutionary trends for placental viviparity in the spadenose shark, *Scoliodon laticaudus*, *Environ. Biol. Fish.*, 38, 269, 1993.
107. **Balon, E. K., Bruton, M. N., and Fricke, H.,** A fiftieth anniversary reflection on the living coelacanth, *Latimeria chalumnae*: some new interpretations of its natural history and conservation status, *Environ. Biol. Fish.*, 23, 4, 241, 1988.
108. **Rombaugh, P. J.,** Respiratory gas exchange, aerobic metabolism, and effects of hypoxia during early life, in *Fish Physiology*, Vol. 11A, Hoar, W. S. and Randall, D. J., Eds., Academic Press, New York, 1988, 59.
109. **Pelster, B. and Bemis, W. E.,** Structure and function of the external gill filaments of embryonic skates (*Raja erinacea*), *Respir. Physiol.*, 89, 1, 1992.
110. **Kirschner, L. B.,** The energetics of osmotic regulation in ureotelic and hypoosmotic fishes, *J. Exp. Zool.*, 267, 19, 1993.
111. **Lehninger, A. L., Nelson, D. L., and Cox, M. M.,** Principles of bioenergetics in *Principles of Biochemistry*, 2nd ed., Worth, New York, 1993, 364.
112. **Portner, H.-O.,** The importance of metabolism in acid base regulation and acid-base methodology, *Can. J. Zool.*, 67, 3005, 1989.
113. **Walser, M.,** Roles of urea production, ammonia excretion, and amino acid oxidation in acid-base balance, *Am. J. Physiol.*, 250, F181, 1986.
114. **Atkinson, D. E.,** Functional roles of urea synthesis in vertebrates, *Physiol. Zool.*, 65, 243, 1992.
115. **Rahn, H.,** Aquatic gas exchange theory, *Respir. Physiol.*, 1, 1, 1966.
116. **Perry, S. F. and McDonald, G.,** Gas exchange, in *The Physiology of Fishes*, Evans, D. H., Ed., CRC Press, Boca Raton, FL, 1993, 251.
117. **Heisler, N.,** Acid-base regulation, in *Physiology of Elasmobranchs*, Shuttleworth, T. J., Ed., Springer-Verlag, New York, 1988, 215.
118. **Kormanik, G. A.,** The egg case of *Raja erinacea* plays only a minimal role as an ionic/osmotic barrier, *Bull. Mt. Desert Isl. Biol. Lab.*, 28, 12, 1989.

Chapter 16

Nitrogen Excretion in Teleostean Fish: Adaptive Relationships to Environment, Ontogenesis, and Viviparity

Bodil Korsgaard, Thomas P. Mommsen, and Patricia A. Wright

CONTENTS

INTRODUCTION

Teleostean fishes turn over large amounts of nitrogenous compounds and thus have to excrete nitrogenous end products. Independent of their environmental conditions, fish must dispose of the various end products derived from amino acid turnover, purine metabolism, methylamine metabolism, and creatine turnover. Many of the compounds in question (urate, allantoin, allantoate, creatine, creatinine, etc.) play only an ancillary function in nitrogen excretion and their overall physiological roles are under debate.[1] As a consequence, focus is most often placed on production and excretion of ammonia and urea only — on ammonia because of its dominating role in excretion, potential toxicity, interesting physical chemistry, and widespread occurrence as an environmental pollutant — and on urea, because of its pivotal position in the catabolism of purines and arginine and its fascinating route of *de novo* synthesis and energy dependence.

A discussion on urea excretion in teleostean fishes is not complete without deliberations on potential driving forces behind the choice of urea, as opposed to or supplemented by, other

0-8493-8411-7/95/$0.00+$.50

nitrogenous excretory products. By necessity, this discussion focuses on ammonia, not only because of its perceived toxicity and overwhelming role as excretory product in teleosts, but also because ammonia and urea are biochemically interconvertible. While urea breakdown by urease may be of physiological relevance in unique cases[2] and lead to interesting,[3] if misguided,[4] working hypotheses, ammonia ultimately provides the building blocks for *de novo* synthesis of urea. Urea is by no means a trivial excretory product and, certainly, not in all cases is ammonia voided from the fish quickly or without difficulty. Therefore, initially and throughout this chapter, we feel obliged to deal with ammonia, its transport, toxicity, relationship to urea, and other compounds, and its multifaceted physiology. To make our point, we present three complementary approaches to the problem of nitrogenous substance excretion in fishes: (1) environmental constraints to ammonia production and/or excretion, (2) ontogenetic changes in expression of urea cycle enzymes, and (3) limitations imposed by viviparity.

From an evolutionary point of view it would be interesting to know whether production and excretion of urea is a more widespread phenomenon for fish than previously suggested. In teleostean fishes, biochemical adaptations to extreme environments have been reported several times in relation to exceptional excretory patterns involving a shift from ammonia to urea production. However, only a few investigations have dealt with an evolutionary evaluation of the beneficial function of urea production in aquatic organisms.[3,5] The amphibians are interesting in this respect. The pattern of nitrogen excretion by embryos and larvae of amphibians depends on the mode of reproduction. Thus, larvae of amphibians with terrestrial reproductive adaptations are generally ureotelic, whereas aquatic larvae excrete mostly ammonia with a subsequent change to ureotelism at metamorphosis.[6] Similar adaptive excretory patterns may be observed in teleosts, which are frequently subjected to periods of emersion. A shift in nitrogenous excretion from ammoniotelism to ureotelism has thus been reported in a variety of teleosts by prolonged exposure to air. In these air-breathing fish, the synthesis of urea may serve primarily to generate protons that contribute to pH homeostasis.[5] The classical ornithine-urea cycle (OUC) in which ammonia and the aspartate amino group (also derived from ammonia) are incorporated into urea is present in elasmobranch fish.[7] Teleostean fish, however, have normally very low levels or nondetectable activities of OUC enzymes. Only a few ureotelic teleosts have significant activities of enzymes in this pathway, and it is unknown to what extent teleostean fish may have the genetic capacity for a functional OUC.[3,8]

For the purpose of this chapter, we distinguish between NH_3 (the gas), NH_4^+, and ammonia (the total of NH_3 and NH_4^+). Because of its chemistry, (pK_a around 9.5), ammonia assumes at least three dependent functions: (1) in aqueous solutions, it can behave as a weak base, and thus as a limited buffer; (2) it can be considered a respiratory gas; and (3) NH_4^+ is an ionized species with the inherent repercussions on membrane potential and transport phenomena. Since the equilibrium strongly depends on numerous other parameters, including temperature, salinity, gas tensions, water hardness, and not least proton concentration, all of these factors must be considered when dealing with aspects of ammonia excretion or toxicity.

NITROGEN METABOLISM AND EXCRETION-ENVIRONMENTAL CONSTRAINTS

AMMONIA EXCRETION AND TOXICITY

One of the long-standing principles of teleost physiology is that these animals produce and excrete nitrogenous waste largely as ammonia, clearly as a consequence of their aquatic existence. In water breathers, the bulk of ammonia is excreted across the gills, with minor contribution from urine, skin, and feces. Although toxic even in low concentrations (see below), ammonia is supposed to be the waste of choice for several reasons: (1) it represents the most reduced form of nitrogen; (2) it does not normally occur in appreciable amounts in bulk bodies of water, establishing a large, natural concentration gradient between the fish and the *milieu extérieur*; (3) ammonia is highly soluble either as gas or as the ammonium ion; (4) excretion, while partially relying on specific transporters and potentially impacting the membrane potential, is not considered to be a major energy expense;

and (5) ammonia might be the preferred excretory vehicle because its catabolic release does not require energetic inputs. In fact, production of ammonia from amide groups as found in adenylates, asparagine, or glutamine is by simple hydrolysis, without energy input.[9] Alternatively, the production of ammonia via transdeamination, which as the final step involves mitochondrial glutamate dehydrogenase, is sensitive to the redox status of the mitochondrion and does not demand input of energy in the form of high energy phosphates either, but *produces* NADH instead.

The clear drawback of this strategy is the toxic nature of the molecule. Consistently, the unionized NH_3 has been implicated in the toxic effects of ammonia, with NH_4^+ being largely dismissed in this respect. However, some studies have noted increased toxicity of total ammonia with decreasing pH. Since low pH changes the ammonia equilibrium in favor of the cation, and abundant evidence indicates that NH_4^+ can replace K^+ in nerve K^+ channels,[10] NH_4^+ is clearly not as innocuous as often assumed. Regardless of this discussion, it is well established that ammonia is toxic to fishes, even at relatively small concentrations. Rainbow trout (*Oncorhynchus mykiss*), for instance will show stress symptoms[11] at water ammonia concentrations around 300 μmol/l. These (reversible) stress symptoms are followed by convulsions and death (irreversible!) at slightly higher concentrations. Ammonia is acutely toxic to marine fish in the range of 0.1 to 3.5 mg/l NH_3-N (6 to 200 μmol/l). Continual, high levels of environmental ammonia will eventually lead to reduced excretion of ammonia and may even result in accumulation and net uptake of ammonia in the fish.[8] As detailed below, the degree of ammonia toxicity, or ammonia tolerance, depends not only on the fish species, but also on the developmental stage, exposure time, and other environmental parameters. For instance, high salinity and other factors (see below), may protect fish from ammonia toxicity, or at least ameliorate or delay the onset of obvious toxic effects.

Although a fair number of publications have dealt with ammonia effects on teleostean fishes, the exact mechanisms by which ammonia exerts its toxic actions are still being debated. Primary responses to ammonia even at sublethal concentrations, noticeable within hours, include significant increases in plasma stress hormones such as cortisol,[12,13] changes in ventilation rate,[14] acid–base imbalance, and increases in dorsal aortic pressure.[15] Interestingly, a study on killifish, *Fundulus heteroclitus,* plasma cortisol showed that total ammonia and NH_3 could not be distinguished in their ability to induce stress,[13] thus underscoring the contention that NH_3 is indeed the "active ingredient." In addition, ammonia exposure significantly increases the renin activity in renal tissue.[16] At this point, the correlation of this increase with the observed ammonia-dependent diuresis in freshwater fish[17] is not yet clear. Secondary responses, which generally take longer to manifest themselves, encompass damage to gill architecture[18] as well as changes in mucus production, growth, and stamina.[14] Further, in fishes with urinary catheters, ambient ammonia increases the rate of urine flow in a dose-dependent matter,[17] possibly indicative of generally increased water permeability. However, Wood and Patrick[19] explain that such catheters render the urinary sphincters incompetent, resulting in the sampling of "ureteral" urine and artifactually preventing urine storage by the bladder. Therefore, bladder transport processes modifying urine volume and composition are thwarted.

Unfortunately, to date no data are available on ammonia's effects on selected biochemical pathways, although a number of indirect routes, mediated through hormones and secondary effects through changes in respiration and gill structure, are likely. Even if reliable numbers were available for metabolites including glucose, lactate, free fatty acids, or ketone bodies, for which changes have been reported in mammalian systems,[20] the question remains whether these are primary or responses mediated by other metabolites. In fish, for instance, environmental ammonia results in a mild, dose-dependent hypercortisolemia,[12,13] likely imposing stress-oriented metabolic changes. Similarly, some of ammonia's hyperglycemic actions noted in mammals could be traced back to increased epinephrine titers, another hormone (not entirely unconnected to cortisol for that matter) which profoundly affects metabolic performance. Some intriguing relationships to other metabolic pathways may exist. For instance, carnitine and trimethylamine *N*-oxide counteract the toxic effects of ammonia,[21] a fact that may explain the generally higher tolerance of marine fishes to ammonia.

Often, the exact sites of actions for these compounds are unknown but are certainly worth pursuing experimentally and in a broad comparative framework.

Superimposed on these effects are neurological effects, possible exacerbated by partial cerebral hypoxia, leading to equilibrium problems, hyperventilation, and finally death. For mammals, an excellent brief summary of possible mechanisms of ammonia toxicity has been presented by Meijer et al.,[22] and there is no reason to believe that fish brain is substantially different. Nevertheless, there are a few dissimilarities. The mean plasma concentration for ammonia in some 15 species of bony fishes listed by Wood[23] is about 0.55 mmol/l, although careful sampling procedures may result in somewhat lower values.[24,25] Fish brains maintain high ammonia concentrations: for instance, in the lemon sole (*Parophrys vetulus*) brain ammonia fluctuates between 2.8 and 5.0 μmol/l, while values for rainbow trout, mudskipper, and air-breathing catfish are 1.2 to 1.4, 5.6, and 6.7 μmol/g tissue, respectively.[26-28] By comparison, in mammals normal brain ammonia concentrations are well below 1 mmol/l and plasma concentrations exceeding 0.20 mmol/l are considered as "hyperammonemia," with widespread clinical repercussions.[20]

It is possible that an ammonia-induced block of oxidative metabolism leads to local hypoxic conditions. We can also postulate that these artificially created hypoxic conditions are ameliorated by the presence of cerebral glycogen stores. Such stores are not uncommon in fishes,[29,30] and it can be envisaged that species with substantial glycogen deposits in their cerebral tissue such as elasmobranchs and many cyprinids, are less prone to succumb to transient ammonia toxicity than species lacking such glycogen stores. Salmonid fishes, for instance, store little or no glycogen in brain and are known to be exceptionally susceptible to ammonia toxicity.

As can be seen in Figure 1, in some circumstances, ammonia exposure may lead to the depletion of cerebral glutamate, and thus remove an important neurotransmitter. Further, the increased ATP demand for the glutamine synthetase reaction may account for the decrease in cerebral ATP noted in ammonia-exposed rainbow trout.[26] Since in all cases, the cerebral pool of glutamine plus glutamate increases substantially, it is clear that anaplerotic reactions must be activated to refill the pool of C5-metabolites, including the reductive synthesis of glutamate, thus levying additional stress on the oxidative demands of the brain (see below).

In mammals, nonneural tissues such as the liver are also sensitive to ammonia toxicity. Interestingly, the same is not hold true for the fishes where the toxic effects of ammonia appear to vary widely between tissues. Although we hesitate to generalize, fish liver appears to be highly resistant to ammonia toxicity. Hepatocytes of the Japanese eel (*Anguilla japonica*), for example, can be cultured without ill effects in NH_4Cl concentrations that are highly detrimental to mammalian hepatocytes. In fact, visually and metabolically, eel cells seem to perform better in the presence of high ammonia concentrations (23 mmol/l) than in its absence.[31] Although under normal conditions, liver is considered to be the main site of production of ammonia destined for excretion, the ammonia-exposed eel hepatocytes synthesize protein more rapidly than control cells and actively remove ammonia from the medium. Furthermore, the apparent concomitant induction of glutamate dehydrogenase (GDH) implies that under high ammonia conditions, the intracellular conditions are drastically changed (pH?, ADP, NAD[P]H?) to favor the glutamate synthesizing direction.

Under normal (low) ammonia conditions, as shown for channel catfish (*Ictalurus punctatus*) hepatocytes, this regulatory enzyme is pictured to catalyze the — terminal — deamination step in amino acid catabolism. Further, in most cases, mass action is likely to prevent ammonia removal by GDH. On the one hand, in the ureogenic toadfish, hepatic concentrations of glutamate in the 25 μmol/g range[32] are likely to saturate the enzyme with an inhibitory product. On the other hand, a high millimolar Km of GDH for ammonia[33] precludes the efficient utilization of ammonia for consumption of ammonia, unless the intramitochondrial NAD/NADH ratio is unusually skewed in favor of reduced conditions. Thus, GDH functions in the reverse direction, resulting in mitochondrial ammoniagenesis.[34] It has to be remembered that this situation is in unmitigated contrast to the mammalian set-up, where nitrogen is delivered to the liver as glutamine and alanine, subsequently to be deaminated by glutaminase to present ammonia intramitochondrially to carbamoyl-phosphate

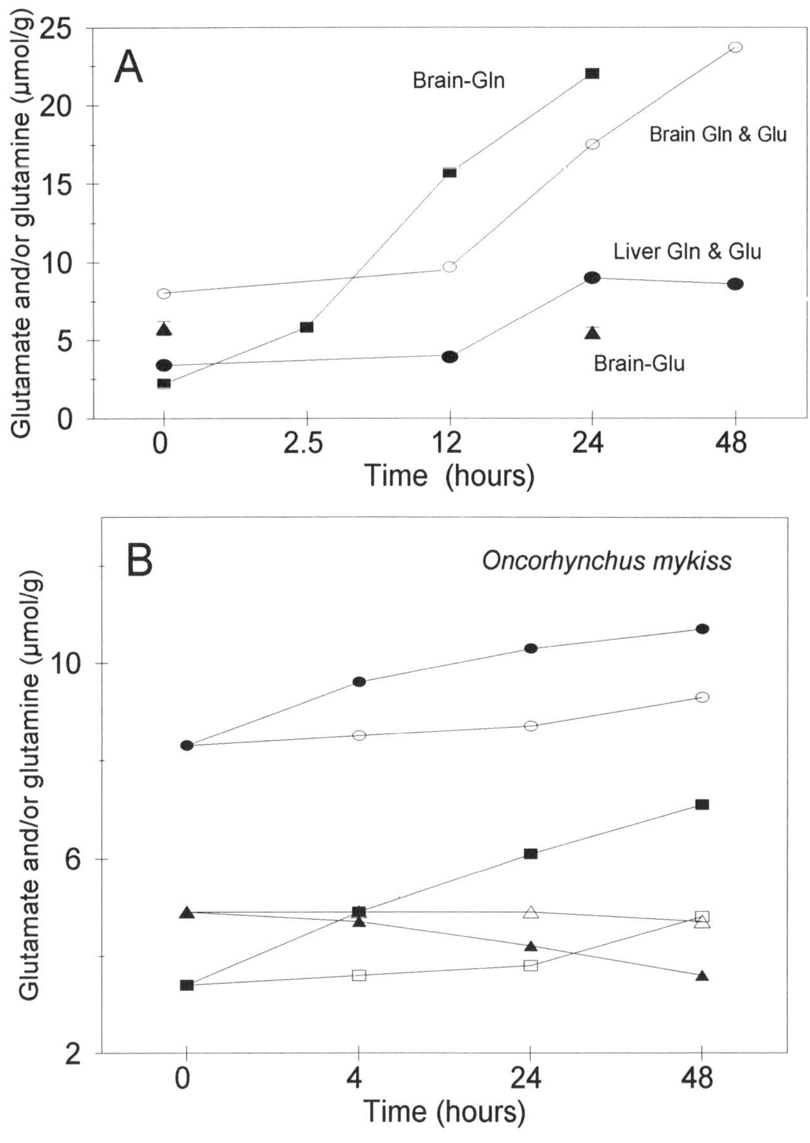

Figure 1 Build-up of the glutamine/glutamate pool in fish brain and liver. Mudskippers were exposed to air while goldfish and rainbow trout were exposed to different concentrations of ambient ammonia. (A) Mudskippers (*Periophthalmus cantonensis*) exposed to air (sum of glutamate plus glutamine, ○ brain; ● liver); goldfish (*Carassius auratus*) exposed to 0.75 m*M* ammonia (▲ brain glutamate; ■ brain glutamine). (B) rainbow trout exposed to ammonia. Brain glutamate (△), glutamine (□) and sum of glutamate and glutamine (○) at 0.14 m*M*/l (open symbols) and 0.28 m*M*/l (filled symbols), respectively at pH 7.5. (From References 26, 27, and 141.)

synthetase I (CPS I). Glutamine synthetase, in turn, functions predominantly as a scavenging enzyme, preventing release of excess ammonia from the liver into the systemic circulation.

While initially, treatment of hepatocytes with such high concentrations of ammonia may appear excessively harsh and nonphysiological, a critical reading of the literature convinced us otherwise. Even without external ammonia loading or airexposure, the hepatic concentration of ammonia in a mudskipper (*Periophthalmus cantonensis*) is around 11 µmol/g of tissue.[27] Assuming a water content of the liver of 50% or below, the intracellular concentration can be calculated to be at least 22 m*M*! Although this particular species is known for its high ammonia tolerance[27,35,36] and survives exposure to 15 mmol/l ammonia in the ambient water for 7 d without any ill effects,[27] similar high concentrations have been described for an (ammonia-tolerant) air-breathing catfish (*Heteropneustes fossilis*; 15 μmol/g)[28] and an ammonia-intolerant, alkaline lake cyprinid (*Chalcaburnus tarichi*; 13 μmol/g).[37] Even the ammonia-sensitive rainbow trout contains liver ammonia levels in the 10 m*M* range.[26] By comparison, hepatic ammonia concentration in mammalian liver[22] is well below 1 mmol/l, leading some researchers to question the saturation of CPS-I (apparent Km for NH_3 in isolated mitochondria is about 13 μmol/l).[38]

MECHANISMS OF AMMONIA EXCRETION

The exact mechanism by which ammonia is excreted across the gill epithelium is still being debated. Several different routes have been described in fish systems and these routes are not necessarily exclusive of each other. Some of these routes (e.g., diffusion of NH_3, Na^+/NH_4^+ exchange and H^+/NH_4^+ exchange) are electroneutral, while at least one other (NH_4^+ extrusion), which is important in marine species, relies on an electrochemical gradient across the apical branchial membrane.

With the assumption that aerobic metabolism of nitrogenous compounds produces largely NH_3, NH_3 diffusion out of the fish will not affect the fish's acid–base balance; neither will electroneutral H^+/NH_4^+ exchange. Under this condition, NH_4^+ extrusion will result in net excretion of H^+. If, however, NH_4^+ ion is the primary product of amino acid metabolism, neither NH_3 diffusion nor H^+/NH_4^+ exchange will be pH neutral, and both might potentially lead to acid loading of the fish. In saltwater-adapted toadfish, *Opsanus beta*, for instance, NH_3 diffusion accounts for about 57% of the ammonia nitrogen released with ionic diffusion of NH_4^+ and basolateral Na^+/NH_4^+ contributing 21 and 22%, respectively.[39] Data on 5%-saltwater adapted toadfish indicate that, as expected, the cation leakiness of the gill decreased, which is reflected in a decreased contribution of NH_4^+-diffusion to overall ammonia excretion.[39] Of course, the entire discussion is compounded where and by which mechanism CO_2 and HCO_3^- are shuttled out of the fish. CO_2-dependent acidification of the boundary layer, for instance, will trap NH_3 that has diffused out of the gill as NH_4^+, thus maintaining the NH_3-gradient across the gill.[40] The reader is referred to recent, unresolved discussions on points of ammonia excretion[23,41] and metabolic NH_3 vs. NH_4^+-production.[9]

PATTERNS OF AMMONIA EXCRETION

Often, the long-spaced sampling intervals have led to the contention that ammonia is released at a relatively constant rate, but is merely disturbed by feeding-dependent peaks.[8,42] Using continuous sampling of ammonia (and urea) excretion in unfed blennies (*Blennius pholis*), however, Sayer[43] detected distinct, well-defined peaks of ammonia release (Figure 2) superimposed on a relatively constant baseline. Notably, these peak release periods for ammonia coincide with bursts of urea discharge into the ambient water. While these data clearly impart the importance of time-scale considerations for measuring nitrogenous output by fishes, they may also hint at some common principles in the release of these two most important nitrogenous waste products, and the ability of the fish to regulate the release of both substances.

AMMONIA UPTAKE

For several reasons, the opposite direction of ammonia movement, (i.e., from the environment into the fish), has to be considered in this discussion. First, ammonia is increasingly relevant as an environmental pollutant, largely because of industrial and agricultural sources in addition to conditions of crowding of fish and other water breathers. Second, experimental analysis of ammonia uptake will give valuable insights into the mechanisms of substance uptake. Third, as the following example shows, the toxic effects of ammonia can be severely compounded by exposing the fish to additional stressors. Atrazine, an S-triazine herbicide which is known to affect kidney morphology,[44] fundamentally alters amino acid metabolism in tilapia (*Oreochromis mossambicus*) fishes already at sublethal concentrations.[45] In addition to the accumulation of urea and ammonia, especially in the tilapia liver (urea from 3 to 4.3 μmol/g; ammonia from 12.6 to 17.5 μmol/g), a number of metabolic enzymes incur substantial changes in activity in fish exposed to atrazine. In the current context, most

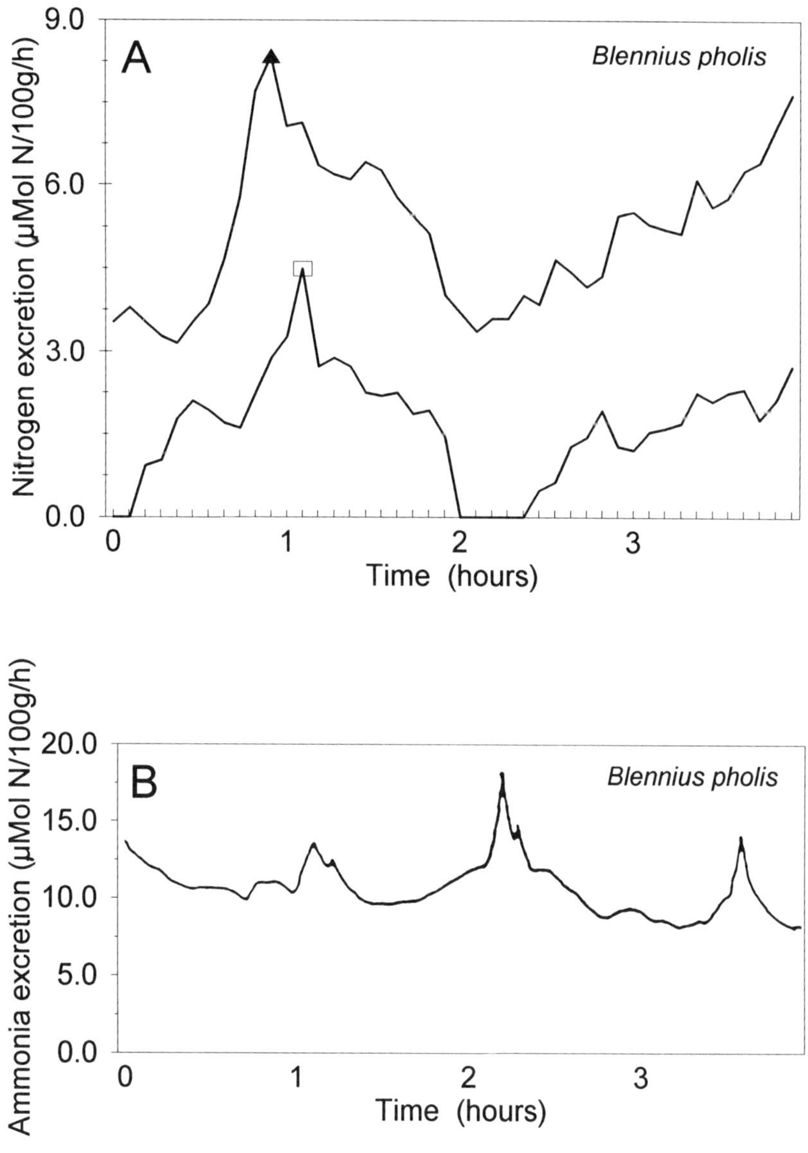

Figure 2 Burst-type excretion of ammonia and urea in fasting blennies, *Blennius pholis.* (A) Outflow from a 180 ml chamber with a single fish was collected for 5 min (flow rate 5 ml/min) and analyzed for ammonia (▲) and urea (□); (B) outflow from the tank was monitored continuously with an ammonia sensitive electrode (flow rate 2.5 ml/min). Fish had been fasted for 6 d and acclimated to the chamber for 6 h prior to sampling or monitoring. (From Sayer, M. D. J., *Comp. Biochem. Physiol. [A]*, 89, 359, 1988. With permission.)

relevant are the decrease in hepatic and cranial glutamate dehydrogenase, pivotal in the detoxication of ammonia in brain and liver (see below), and glutaminase as opposed to increases in alanine and aspartate transferases and liver arginase.[45] Since these changes are diametrically opposite to those induced by ammonia exposure, it is obvious that this added stressor will seriously affect the ability of fish to cope with even minor ammonia contamination of their environment.

Exposure of sculpins (*Myoxocephalus octodecimspinosis*) to artificially high ambient ammonia (about 1.2 mmol/l total) exerts no influence on plasma pH or total CO_2, but leads to rapid, drastic increases in plasma ammonia.[46] Total plasma ammonia does not equilibrate with the external concentrations, since, plasma concentrations seem to have stabilized at around 71% of medium concentration even after 6 h of exposure. Within the first hour of ammonia exposure, however, the diffusion gradient for NH_3 between plasma and seawater is indistinguishable from zero.[46] Three phases of ammonia flux can be distinguished. In the pre-exposure (control) period, fish lose about 0.27 mmol of ammonia/kg/h. In the first hour of exposure, fish gained almost 1.2 mmol/kg. During the last 2 h of ammonia exposure, fish returned to an efflux rate of 0.44 mmol/kg/h. The changing diffusion gradients and careful analysis of acid–base parameters during these different phases suggest that ammonia enters the sculpin predominantly as the ammonium ion.[46] These results support the idea that gills of saltwater fishes reveal a fairly "leaky" for cations, which contrasts with freshwater species with their low permeability to ions. Independent of the mechanism of uptake, the fact remains that fish will take up ammonia from a contaminated marine or freshwater[28] environment and have to cope, at least transiently, with increased intracellular concentrations of the toxicant.

AMMONIA DETOXICATION

Brain Glutamine Formation

In addition to excretion, a number of different strategies are available to fishes to deal with increasing concentration of internal ammonia, namely the direct removal by synthesis of glutamine and glutamate, and the more indirect removal through the synthesis of urea.

The first strategy, which usually seems to be associated with high ammonia tolerance in noncerebral tissues, is best exemplified by the (ammonia-intolerant) brain. In brain, ammonia is transiently "fixed" into glutamine by the glutamine synthetase reaction. In fact, in most fishes, brain clearly has the highest tissue concentration of this enzyme. The brain enzyme is also inducible, and as a result, fish under ambient ammonia stress tend to accumulate glutamine in brain (see Figure 1). The shortcoming of this strategy is an increased energy demand by the brain and potential depletion of the tissue glutamate pool. Some fish species take this scheme one step further and back glutamine synthetase up with increased activity of glutamate dehydrogenase[27] functioning in the glutamate synthesizing direction. While meeting the primary objective of removing additional ammonia from the tissue, this strategy also serves to replenish and buffer the glutamate pool resulting in substantial increases in the glutamine/glutamate pool. This route to detoxify brain ammonia has been shown to occur in a number of fish species, well exemplified in rainbow trout, goldfish (*Carassius auratus*), and mudskipper (*P. cantonensis*) (Figure 1).

Two negative side effects of this two-step route to detoxify ammonia are indisputable: (1) depletion of 5-carbon intermediates has merely been relocated to the level of α-ketoglutarate and thus expanded to the entire Krebs cycle, and (2) the redox-balance of the brain will be affected, because additional energy for the reductive synthesis [as NAD(P)H] of glutamate is required. This view is supported by data on succinate depletion in brain of *O. mykiss* exposed to very high concentrations of ambient ammonia. Since overall the pool of brain succinate (at 0.2 μmol/g) is minute compared with the increases in the glutamine/glutamate pool size in the brain,[26] from this limited data set we can speculate that the brain succinate pool turns over rather rapidly and must be sensitive to redox imbalance. It can also be envisaged that the increased energy demand results in partial or localized hypoxia, likely impacting normal brain functions.

In some species, the same phenomenon is also observed in noncerebral tissues. Under conditions of ammonia loading, air-breathing mudskippers (*P. cantonensis*) experience significant increases in liver and muscle glutamate dehydrogenase activities.[33] Nevertheless, large concentrations of ammonia accumulate in most tissues, except the plasma and the brain. The animals display an unusual tolerance for total ammonia, coinciding, as pointed out above, with an outstanding ability to synthesize and accumulate glutamine in the brain.[27]

From a strictly biochemical point of view, two interesting aspects should be mentioned. First, although brain glutamine synthetase is essential in this context and also to hepatic and renal urea synthesis delineated below, the kinetic parameters of the fish enzyme remain to be elucidated. Unfortunately, the enzyme is generally assayed as a γ-glutamyltransferase (i.e., in the direction of glutamine use), and thus the kinetic parameters for the physiologically relevant direction of the enzyme have been largely ignored. Extrapolating from the Km for ammonia other glutamine synthetases (e.g., mammalian brain)[47]: 0.18 mmol/l; liver[48]: 0.2 mmol/l, elasmobranch liver[7]: 0.015 mmol/l, the teleostean enzyme should be saturated with this substrate. Binding constants of mammalian, bird, and elasmobranch enzymes for glutamate range from 5 to 11 mmol/l,[7] and thus are also in the physiological range for both fish tissues. These considerations have to remain speculative, however, until binding constants for teleostean glutamine synthetases have been determined.

Hepatic Urea Synthesis

In the second strategy, ammonia is removed by *de novo* synthesis of urea. Urea is a nontoxic, highly soluble, uncharged compound that is chemically inert. The strategy is best illustrated by a number of species, including a soda lake tilapia, toadfishes of the genus *Opsanus*, and some air-breathing fishes (see below). These species use the OUC, a monophyletic trait in the vertebrate line,[3,49] comprising four enzymes of the cycle proper (ornithine carbamoyltransferase, OCT, argininosuccinate synthetase and lyase, and arginase). In addition, two fish-specific ancillary enzymes are essential, namely mitochondrial carbamoyl-phosphate synthetase III (CPS III, glutamine as nitrogen donor),[7] and hepatic glutamine synthetase (GNS),[7] functioning as a feeder enzyme for CPS III.

Again, two possible drawbacks of the urea-synthesis route are apparent. The first is that the *de novo* synthesis occurs at an appreciable metabolic cost, requiring an input of at least 2.5 ATP per nitrogen excreted. The second is that urea synthesis may play a role in acid–base balance, primarily by removing bicarbonate derived from protein catabolism. Since its inception for terrestrial vertebrates by Atkinson et al.,[5] this hypothesis has remained controversial in mammals (see discussion in Reference 22 and Chapters 12 and 13), but received no supporting evidence in studies on two ureogenic fishes, the Lake Magadi tilapia[50] and the Gulf toadfish.[51]

The metabolic conversion of ammonia to urea may provide a more permanent adaptive advantage for teleostean fish, which are subjected to environmental stress or changes, such as being exposed to crowding, air, or alkaline waters. In teleosts with predominant ammoniotelism, periodically critical elevations in ambient ammonia and alkalinity may also lead to increased excretion of urea such as observed in goldfish,[11] tilapia,[8] and rainbow trout.[25] In these fish, however, ammonia production will usually still predominate and urea excretion rates will amount to an average of only about 10% of values reported in ureotelic teleosts.[8]

UREA

Production and Excretion

Before we can deal with possible driving forces behind the occurrence of the OUC in some teleostean fishes, we have to point out that these animals normally excrete substantial amounts of urea, amounting to about 10 and 30% of the total nitrogenous waste.[23] In most fishes, this urea is not procured through *de novo* synthesis, but largely derived by catabolic routes from the degradation of purines via urate and arginine. The hydrolytic enzymes actually liberating urea are the ubiquitous arginase[52] and allantoicase. Allantoicase is located primarily in the hepatic peroxisomes.[53]

In the predominantly ammoniotelic tidepool sculpin (*Oligocottus maculosus*), 74% of excretory urea is disposed of by the gill. It appears that branchial urea excretion relies on simple diffusion[54] and not on carrier-mediated phenomena, an observation supporting earlier evidence supplied for the urea-loaded, freshwater eel.[55] Therefore, not surprisingly, tidepool sculpins have low or below millimolar levels of tissue urea.[54] On the one hand, this mechanism of voiding urea is straightforward and does not incur additional metabolic expenditure. However, it also leads to the conclusion that, at least in the gill tissue and compartments in equilibrium with plasma (see below for a differing plot), the fish do not have much control over urea concentrations. This type of scenario precludes the branchial "excretory bursts" of urea noticed in some species (see below), unless, of course, some specific intracellular transport mechanisms are involved. As Wright et al.[54] have shown, tidepool sculpins do not change their nitrogen excretory pattern (i.e., the partitioning between ammonia and urea) under stress conditions. On the other hand, passive diffusion of urea has an intrinsic drawback, especially in these days of anthropogenic contamination of natural environments. Without a carrier-mediated mechanism for small uncharged molecules such as urea, it is possible that fishes with this strategy are easy targets for environmental toxins which behave equal or similar to urea. Once taken up adventitiously by this passive route (in most teleostean fishes gill tissue exposes a large surface area to the ambient medium), it is not far-fetched to postulate that, secondarily, such toxins are trapped inside the fish by metabolic conversion into charged species or similar mechanisms. Active secretion of urea via the kidney tubules is likely partially for regulation of water movement in the kidney or urinary bladder, although recent divided chamber experiments clearly show that even the ureogenic *Opsanus beta* does not use urine as a vehicle for urea excretion.[56] Recently, Walsh et al. noticed that although urea was passively distributed between plasma, red blood cells, and hepatocytes of several elasmobranch and teleostean fishes, two toadfish species displayed marked-phloretin-sensitivity of urea movement, suggestive of the presence of a specific urea carrier system.[57]

A number of observations cannot be reconciled with the idea of passive diffusion of urea out of or into fish. This suggests a reevaluation of the existing data and possibly a new working hypothesis for urea distribution and transport which includes an active component. These observations include: (1) urea is not equally distributed among tissues[58] (see Reference 59 for discussion of amphibians); (2) during starvation, urea accumulates in tissues at unequal rates; (3) plasma levels are normally in the 1 to 10 mmol/l range, somewhat unexpected considering the slow rates of urea production, especially contrasted to ammonia production rates (about 5 times higher) and plasma ammonia concentration (about 5 to 10-times lower); (4) labeled urea was found to accumulate in eel chloride cells[60] and putative urea-transporting chloride cells have been described for *Oreochromis alcalicus*[61]: (5) the observation of phloretin-sensitive urea efflux from toadfish hepatocytes[57]; and (6) fish are observed to release urea across the gill not at a constant rate, but instead in excretory bursts.[43,56]

As shown by Grubinko et al.,[62] starvation in the carp (*Cyprinus carpio*) leads to four- to fivefold increases of tissue (liver, muscle, intestine, gills, plasma) urea levels from below 0.3 mmol kg^{-1} (or L^{-1}) to between 1.2 and 2 mmol urea kg^{-1} (or L^{-1}). This increase is accompanied by a 20-fold increase in liver arginase (reaching about 0.7 units g^{-1} h^{-1}) and by induction of intestinal arginase (from below the limit of detection to 0.3 units g^{-1} h^{-1}),[62] leading the authors to conclude that proteolysis and thus endogenous arginine are the primary precursors of tissue urea. In starvation, however, all tissue compounds are mobilized, even if at differential rates, which likely results in a nontrivial contribution of the breakdown of nucleic acids to the different urea pools. The question remains: Why and how these apparently trifling amounts of urea are retained in some species, considering an overall tissue osmolarity of about 360 mOsm/l. However, urea cannot entirely be dismissed as an insignificant contributor to overall osmolarity. While numbers for fish tissues are few and, depending on sample preparation and urea assay, potentially flawed, sufficient reliable numbers can be found for teleostean plasma to make this point. In this compartment, urea concentrations range from 2 to 10 m*M* for rainbow trout,[25] mudskipper,[35] the ureogenic soda lake tilapia,[24] and marine toadfish,[63] and can reach as high as 18 to 25 m*M* for an ammoniogenic cyprinid, *Chalcaburnus tarichi*.[37]

With respect to the excretory bursts, Sayer[43] documented that urea is released in intermittent bursts between intervals where its release could not be detected. These bursts coincide with periods of increased ammonia release (see Figure 2). Similarly, the largely ammoniotelic *Heteropneustes fossilis* excretes 30% more urea at night than during the day, while ammonia excretion appears fairly constant throughout a 24-h interval.[64] As in many other studies, however, the sampling periods for monitoring nitrogenous waste in the latter study (12-h periods) preclude the detection of better defined excretory bursts for both compounds.

Regardless of the mechanisms of excretion, the rate of urea release depends on the ambient conditions. In freshwater-adapted eels, for instance, urea was excreted across the gills (and incidentally also the kidney) at three times the rate of saltwater-adapted specimens.[55] Similarly, in the tidepool sculpin, net urea efflux is increased 2.6-fold during an acute 50% drop of external salinity.[54] Further, epinephrine exposure results in an increased urea flux across the trout gill, together with increased water flux.[65]

At this point, the role and excretion of urea in teleostean fishes remains enigmatic until thorough studies of these phenomena have been conducted. For instance, excretory bursts of urea should be monitored in cannulated fish (which unfortunately excludes the tiny alkaline lake tilapia) while simultaneously measuring plasma urea. Similar studies and turnover analyses should be designed to provide a better picture of urea distribution and dynamics within the fish. They may also give insights into whether plasma serves as a transient storage site for urea. Otherwise, plasma may merely represent a compartment that tissue urea has to traverse on its way out of the fish with entry and exit being controlled by gill and other tissues.

Environmental Urea Loading

Similarly conflicting results are obtained when analyzing the uptake of environmental urea. On the one hand, tidepool sculpins immersed in urea containing water will take up ambient urea in a simple concentration-dependent manner.[54] Griffith et al., on the other hand, have shown that although external urea is taken up rapidly by saltwater-adapted killifish (*F. heteroclitus*), it fails to equilibrate fully. Even when exposed to almost half-molar concentrations of urea, the fish maintain a plasma concentration of about 76% of the ambient concentration.[66] In any event, even at several hundred millimolar concentration, environmental urea fails to exert obvious toxic or osmotic stress on euryhaline, saltwater-adapted killifish,[66] although a definite diuretic action of urea was noted in the aglomerular toadfish.[67]

LIMITS TO AMMONIA EXCRETION

Since teleostean fish excrete mainly ammonia, with urea (derived mostly from catabolic routes), a distant second in importance, it can be appreciated that conditions limiting or complicating the removal of ammonia might have pronounced effects on fish nitrogen metabolism. Four different strategies come to mind: (1) increased ammonia tolerance, likely connected with glutamine synthesis in the brain and perhaps glutamine formation in noncerebral tissues, (2) transient detoxification through glutamine synthesis in the brain, (3) decrease in nitrogen turnover, and (4) ammonia removal in liver and possibly kidney[28] through the OUC.

Naturally Alkaline Conditions

Exposure of fishes to high environmental hydroxyl-ion concentrations can occur in marine environments, in intertidal pools, and in brackish environments. To date, most attention has been devoted to alkaline conditions in freshwater lakes and streams, mainly under conditions of reduced water flow and increased evaporation. Further, unique rock chemistry in a particular drainage system may result in the formation of alkaline lakes. Finally, fishes may be exposed accidentally or intermittently to high water pH, for instance, as a result of excessive addition of lime to counteract water acidity or exceptional blooms of phytoplankton in eutrophic aquatic systems.[68,69] Often, these alkaline conditions occur in combination with extremes in temperature, gas tensions, and salinity. Theoretically, fish species living under such harsh environmental conditions should incur similar problems with respect to ammonia toxicity (i.e., the equilibrium is shifted towards NH_3 and difficulties to release ammonia into the surrounding water), due to alkalinization of the boundary layer and, in some cases, high external bicarbonate and carbonate concentrations.[70,71] The best studied cases to date are fishes living in the highly alkaline salt lakes, such as Lake Magadi in Kenya (pH 9.98, temperature from 30 to 43°C, salinity 525 mOsm/l; Na^+ = 342 mmol/l),[24] Lake Van in Eastern Turkey (pH 9.81, temperature up to 22°C, salinity 551 mOsm/l; Na^+ = 337 mmol/l),[37] or Pyramid Lake in Nevada, USA (pH 9.39, temperature up to 23°C, salinity 135 mOsm/l; Na^+ = 58 mmol/l).[72]

It was initially thought that fish would respond to such alkaline conditions by minimizing or abolishing ammonia excretion and favoring detoxification through the *de novo* synthesis of urea. This view found full support in several publications dealing with nitrogen excretion in the alkaline lake tilapia *Oreochromis alcalicus grahami*. The overall excretory rate for nitrogen by this exclusively ureotelic species[24,73] seems extraordinarily high at 7.6 μmol N g^{-1} h^{-1}. Urea is derived largely by synthesis through the OUC.[50,73] By comparison, most "normal" water teleosts excrete nitrogen at a rate of 0.2 to 1.0 μmol g^{-1} h^{-1} with urea accounting for less than 30% of the total, and, as mentioned above, the bulk of urea being contributed through catabolic routes.

In contrast, Lahontan cutthroat trout, *O. clarki henshawii*, living in Pyramid lake and *C. tarichi*, a cyprinid endemic to Lake Van, follow a surprising route. Both species are largely ammoniotelic with urea contributing less than 40% of the total nitrogen.[37,72,74] In the apparent absence of OUC activity, urea is derived through catabolic routes only.[58] In the cutthroat trout,[75] OCT activity is more than three orders of magnitude below that of the ureogenic toadfish,[3] and this enzyme is below the limit of detection in *C. tarichi*.[37] Surprisingly, *C. tarichi* actually accumulate higher concentrations of urea in plasma (18 mmol/l)[37] and tissues than the exclusively ureotelic tilapia *O. grahami alcalicus* (plasma: 5.5 mmol/l).[24] Further, *C. tarichi* display brain GNS activity commensurate with ammonia-intolerant teleosts.[37] In fact, the fish die if exposed to 500 μmol/l NH_3 (at pH 9.8), a concentration which is not harmful to the tilapia. These observations are supported by studies of high-pH exposure in ammonia-intolerant species, including the rainbow trout. In the initial phases of alkaline water exposure, plasma ammonia levels increase and ammonia excretion is blocked. Within 2 d of exposure, in the surviving specimens, excretion rates have returned to normal and plasma ammonia is at a stable, albeit sixfold increased, concentration.[25] The transient shortfall in nitrogen excretion is partially compensated by an increase in urea excretion from 12 to 45% of nitrogen excretion. Considering the short time-scale of the increase, however, urea's relatively small contribution to overall nitrogen excretion,[25] and slow rate of urea synthesis in adult rainbow trout,[76] we can in essence exclude the OUC as a source of urea in this species.

Although fish exposed to alkaline conditions seem to maintain plasma pH at normal levels, it should be pointed out that metabolic alkalosis, induced by sodium-lactate infusion, drastically (up to 50%) decreases ammonia efflux.[77] An analysis whether the increased retention of ammonia, if maintained, will induce brain GNS, tissue glutamate dehydrogenase, or perhaps hepatic OUC might provide insights into the mechanisms of pathway induction.

External Ammonia Loading

As pointed out by Haywood,[78] ammonia exists as a nontrivial environmental pollutant contributed by anthropogenic activity. As in the case of naturally alkaline water, two different responses of teleostean fishes to imposed ammonia conditions have been described. One strategy is extreme ammonia tolerance coincident with induction of OUC enzymes (as exemplified by an air-breathing catfish, *Heteropneustes fossilis*). To date, this freshwater catfish appears to be the champion of ammonia tolerance in that it can withstand weeks of exposure to 75 mmol/l external ammonium chloride. While initially irritable by what must be an osmotic as well as ammonia (and chloride ?) insult, the fish settle down and are ultimately able to maintain plasma ammonia concentrations below 5 mmol/l. At the same time, the fish compensate for the unfavorable conditions to excrete ammonia by increasing the titers of key enzymes of the OUC in liver and to a lesser degree in kidney.[28] *H. fossilis* increase their rate of urea excretion and plasma urea from a pretreatment level of 0.7 to about 3 mmol/l within one week. They are able to maintain plasma urea below 4 m*M* throughout further exposure.[28] Main targets of ammonia-dependent induction are carbamoyl-phosphate synthetase (the authors used ammonia as a substrate and thus cannot distinguish between CPS I and CPS III), OCT, and the enzymes responsible for synthesis and degradation of argininosuccinate. While in the case of the catfish, the final enzyme of the cycle, arginase, remains unaffected[28]; in other ammoniotelic fish species, this enzyme is known to be regulated by additional physiological conditions, including substantial induction during starvation.[8,52,62]

A contrasting approach is seen in the *P. cantonensis*. As outlined above, these fish accumulate large amounts of ammonia during air exposure and are exceptionally ammonia tolerant (at least a week at 15 mmol/l NH_4Cl) due to efficient, multitiered mechanisms to detoxify ammonia into glutamine in brain and possibly glutamate in other tissues.[27,79] In the face of only minor increases in plasma and stable body urea load, the authors conclude that the mudskippers do not switch to urea synthesis. This conclusion agrees with one report on air exposure in the same species,[35] but does not agree with another.[36]

Crowding/Confinement

Environmental ammonia levels sufficiently high to cause harm may also occur under circumstances of high crowding of ammoniotelic species, including teleostean fishes. Such conditions are associated with demersal fish eggs and viviparity involving a prolonged maternal-fetal trophic relationship. The production and excretion of urea, which is a common phenomenon in viviparous elasmobranchs (see also Chapter 15), may serve to protect prehatch larvae and the embryos from toxic levels of ammonia.[80] Further, urea production may in itself be functionally useful, since urea retention also contributes to maintaining osmotic balance with the hypertonic environment.

In the facultatively ureogenic toadfish, crowding/confinement results in activated urea synthesis and an apparent switch from predominant ammoniotely to partial ureotely. The same change, which has also been described for ammonia-exposed catfish,[28] can be induced by air exposure and immersion in high ambient ammonia[63] in the toadfish.[32] While the data on the catfish, with their associated increases in OUC enzyme activities,[28] support data on mammalian liver and the idea of intra-cycle control sites,[22] the more subtle treatments in the toadfish lead to different conclusions and consistently identify control sites located upstream of the OUC. First, in all these situations, toadfish liver glutamine synthetase — the enzyme delivering nitrogen to CPS-III — undergoes large (up to fivefold) increases in activity. Second, aspartate aminotransferase, an additional enzyme involved in nitrogen delivery, was significantly upregulated, while, of the OUC enzymes only OCT incurred minor increases. Third, hormonal treatment with glucagon and/or dexamethasone (which alters OUC activity in mammalian liver) is without effect in the toadfish, but doubles the activity of hepatic glutamine synthetase, while also leading to minor increases in the activities of hepatic aspartate aminotransferase and glutamate dehydrogenase.[81] These data, together with specific changes in amino acid concentrations in plasma and liver,[32] coalesce into the surprising, but consistent, picture that the control of urea production is at the level of nitrogen delivery to the cycle.

Furthermore, several independent lines of evidence support the conjecture that stress-related changes, possibly involving glucocorticoids, redirect endogenous nitrogen flux into ureogenesis. First, crowded toadfish quickly elevated the activity of liver GNS over uncrowded control fish, even if maintained in flowing water (i.e., under low ammonia conditions),[32] thus excluding ammonia as the primary stressor. Second, similar increases in GNS can be induced by exposure to dexamethasone, a synthetic glucocorticoid.[81] Finally, cortisol is considered to be a primary stress indicator in fishes.[82] Incidentally, fish exposed to sublethal levels of ammonia significantly increase plasma cortisol titers.[13]

NITROGEN METABOLISM AND EXCRETION DURING EARLY DEVELOPMENT IN OVIPAROUS TELEOSTS

Oviparous teleosts deposit eggs to the external environment where the eggs are fertilized, develop, and eventually hatch. During the embryonic stage (the period between fertilization and hatching), the developing organism must be protected from environmental perturbations.

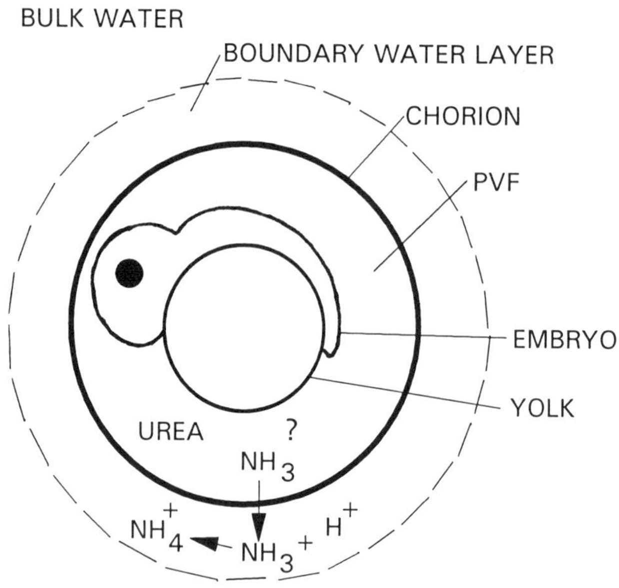

Figure 3 Macro-architecture of a fish embryo. Ammonia is released by the embryo into the perivitelline fluid (PVF) and may cross the chorion as NH_3. It may subsequently pick up a proton in the boundary layer surrounding the chorion and form NH_4^+. See text for further discussion.

The outermost acellular coat, egg capsule or chorion (Figure 3), composed of protein and glycoproteins,[83] undergoes a calcium-dependent hardening after fertilization which increases the mechanical strength of the eggshell by tenfold.[84,85] Thus, the chorion functions to isolate and protect the embryo from adverse environmental conditions.[83] The perivitelline fluid (PVF) lies between the chorion and the embryonic tissue with its attached yolk (Figure 3). In early embryos, the PVF is the major resistance to gas exchange, but once the embryos start to move inside the egg, this resistance decreases.[86] The yolk, mostly composed of protein (82%) with some fat (17%), is the main source of fuel during the endogenous feeding period.[87-89] The three major constituents of rainbow trout egg yolk protein are the lipoprotein, lipovitellin (a highly phosphorylated protein containing large amounts of serine), phosvitin, and a simple serum protein, β′-component.[88]

Excretion in both the embryonic and larval stages is mostly cutaneous. A significant boundary water layer or unstirred layer exists next to the chorion of the embryo where excretory wastes are probably more concentrated than in the bulk water (Figure 3; see Reference 86). Hence, embryo-to-environment diffusion gradients and, therefore, nitrogenous waste excretion will be influenced by water velocity and proximity of other embryos.

At hatching, cutaneous excretion may still predominate in the larvae since the respiratory pump is not fully developed[90] and respiratory movements are infrequent and irregular.[91,92] The larval stage can be divided into two stages: (1) sac-fry larvae (alevin) are hatched and feed endogenously, and (2) fry larvae, where the yolk sac has been absorbed and feeding is external. Following the disappearance of the yolk sac and skin thickening (about 20 d after hatching), the gills take over as the primary site of gas exchange.[90]

SYNTHESIS AND EXCRETION OF NITROGENOUS WASTE PRODUCTS IN EMBRYOS AND LARVAE

One of the first studies of nitrogen balance in developing rainbow trout was performed by Smith.[87] He reported that nitrogen content of the yolk decreased with development as the dry weight of the embryo and total nitrogen excretion increased. Rice and Stokes[93] found that ammonia, urea, and arginine concentrations increased several fold during rainbow trout embryonic development, while uric acid levels were relatively low and unchanging. After hatching, there was a further increase in ammonia and arginine levels, but in the fry stage, these levels declined.[93] In recent studies, Wright et al.[94] confirmed the time course of changes in ammonia and urea levels during rainbow trout development reported by Rice and Stokes.[93] Changes in urea concentrations in the embryo, however, closely paralleled changes in ammonia levels, with ammonia and urea levels peaking at 1.7 and 2.5 mmol-N/l of tissue water, respectively.[94] Synthesis of urea during early development may function to prevent toxic levels of ammonia from accumulating.[95] In resting adult trout, plasma ammonia levels vary between 0.05 and 0.50 mmol/l (depending on feeding history, stress, exercise, among other factors), whereas, intracellular concentrations are approximately 2 mmol/l tissue water.[96] It is likely that ammonia is also compartmentalized in embryonic and larval tissues, although this has not been determined. Nevertheless, the ability to convert ammonia to the less toxic urea may be an essential "safeguard" to protect the animal from the deleterious effects of elevated ammonia on survival,[97,98] growth,[99-101] and to the developing nervous system.[20] In support of this hypothesis, when sac fry were exposed for 4 h to conditions which inhibited ammonia excretion (elevated water ammonia levels or alkaline water), urea excretion was increased by sixfold (Land, M. and Wright, P. A., unpublished data).

Fertilized eggs of *F. heteroclitus* contain normal amounts of glucose (5 mmol/l), but extremely high concentrations of both lactate (40 to 50 mmol/l) and alanine (60 mmol/l).[102] Similarly, hatched eggs of Atlantic halibut (*Hippoglossus hippoglossus*) contain a total pool of free amino acids of around 150 mmol/l.[103] Although lactate utilization and turnover appear to predominate during the first stages after fertilization,[102] it is now clear that amino acids are mobilized to a great extent during the remaining stages of embryonic development, leading to substantial production of nitrogenous waste.[104] Arginine, likely the most important direct precursor of urea, accounts for less than 5% of the total amino acid pool at hatching in halibut.[105] Since it is normally considered to be an essential amino acid in teleostean fishes (in part because of the alleged absence of the ornithine urea cycle), it can be ruled out that arginine would account for an unproportionate fraction of the 40% of the amino acid pool funneled into oxidative routes.[105]

It is clear that both ammonia and urea accumulate in the developing embryo and significant diffusion gradients across the chorion must exist. Smith[87] reported that ammonia, but not urea, was excreted by embryonic trout. Rice and Stokes[93] measured ammonia excretion in both embryonic and larval trout, but could not detect urea excretion to the water in any stage of development. Wright et al.[94] also found that trout embryos excrete ammonia, but in contrast to Rice and Stokes,[93] they showed that urea excretion was initiated just after hatching in sac-fry larvae. The percentage of nitrogenous wastes excreted as urea increased during development, from 0% in embryos [0 to 40 d postfertilization (DPF)] to 7% in sac-fry (45 to 63 DPF) to 14% in fry (71 to 93 DPF). By comparison, adult rainbow trout excrete between 10 to 15% of their nitrogen wastes as urea.[25]

It is possible that a large component of urea excretion is branchial, as excretion rates were extremely low in sac fry at a time when tissue concentrations were relatively high (2.5 mmol-N/l),[94] but the gills were not fully functional (see above). After the yolk sac was absorbed, urea excretion rates increased markedly, despite a decrease in the gradient for diffusion (i.e., tissue levels <1.0 mmol-N/l).[94] The increase in gill function in more mature fry may facilitate urea clearance from the blood. Unfortunately, this hypothesis would be difficult to test on such small fish. Divided chamber studies are typically performed on individual fish surgically fitted with bladder catheters and rubber masks to separate branchial, cutaneous, and renal excretory pathways.[72]

Trout embryos alkalinize the external water if left in a small volume of water over a few hours (Land, M. and Wright, P. A., unpublished data). In adult trout, carbon dioxide excretion as molecular CO_2 hydrates to form HCO_3^- and H^+ in the gill water boundary layer, thereby increasing water hydrogen ion concentration.[70,106] As well, direct H^+ excretion via the H^+-ATPase also contributes to water acidification.[107,108] Excretion of NH_3 is facilitated by an acid boundary water layer next to the gill surface.[40,109] Furthermore, NH_3 excretion has an alkalinizing effect, in that H^+ are consumed when excreted NH_3 is converted to NH_4^+ in the external water (Figure 3). Whether H^+, CO_2 and NH_3 interactions in the water boundary layer adjacent to the embryo are similar to those in the gill boundary layer of adult fish are unknown. Investigations are warranted, however, as they would provide valuable insights into how external water chemistry influences nitrogen balance and excretion in the developing embryo.

METABOLIC PATHWAYS FOR UREA SYNTHESIS DURING EARLY LIFE STAGES

It is curious that the OUC is functional in only a few teleostean species that survive periods of air exposure or alkaline environments (see above). Why would the genes for the OUC enzymes be conserved during teleost evolution if they no longer function in the majority of adult fish? One possibility is that the OUC is expressed at higher levels during early life stages in all teleosts, but is repressed in most mature fish. Two OUC enzymes, ornithine carbamoyl transferase (OCT) and arginase (but not CPS and argininosuccinate synthetase) were detected in larval rainbow trout tissue.[93] In contrast, Wright et al.[94] found that four enzymes of the OUC or associated with it, (i.e., glutamine synthetase, OCT, CPS III, and arginase) were induced just after hatching in sac fry and the levels of activity were relatively high compared to adult trout liver tissue (with the exception of arginase). CPS III and OCT activities declined in fry larvae (90 DPF) and were not detected or were very low in adult liver tissue.[94] It is likely that Rice and Stokes[93] failed to detect CPS activity because the assay conditions used in their study were designed to measure the "mammalian-type" CPS I enzyme rather than the "fish-type" CPS III enzyme (for reviews, see References 7 and 49). Chiu et al.[76] reported that the full complement of OUC enzymes was present in rainbow trout fingerlings (25 to 40 g), and although the fingerlings were efficient at incorporating ureido-labeled citrulline into arginine, they did not produce labeled arginine when offered ^{14}C-radiolabeled bicarbonate. Once again, the assay conditions were not appropriate for determination of mitochondrial CPS III activity and it is more likely that the ubiquitous CPS II enzyme, involved with pyrimidine synthesis in the cytosol, was measured,[49] although the positive correlation between CPS activity and protein content of the diet may hint at the involvement of CPS III.[76]

To demonstrate a functional OUC in trout during early life stages, researchers have incubated tissue with the radiolabeled OUC substrate, ^{14}C-$NaHCO_3$ and attempted to recover the label in urea. Rice and Stokes[93] did not find incorporation of ^{14}C-$NaHCO_3$ into urea in embryos and sac fry. In contrast, Dépêche et al.[110] reported ^{14}C-$NaHCO_3$ incorporation in trout embryos (8 DPF), but very little incorporation in sac-fry larvae (50 DPF). No clear explanation exists for the discrepancy between these two studies, although such coarse experiments are clearly hampered by the potential dilution of label along the synthetic pathway, especially at the level of ubiquitous bicarbonate. Obviously, the use of molecular probes to assess the expression of OUC specific enzymes is likely to provide a more accurate picture about presence and timing of the OUC during early development.

In summary, there is evidence for a functional OUC in trout during early life stages as (1) OUC enzymes are induced and activities are relatively high compared to adult liver tissue,[94] and (2) radiolabeled substrates are incorporated into urea.[110] The presence of the OUC in immature teleosts may explain, in part, why fish retain the genes for the OUC enzymes despite the lack of a functional OUC in most adult teleosts.

NITROGEN METABOLISM AND EXCRETION IN VIVIPAROUS FISH: MATERNAL-FETAL INTERACTIONS

MATERNAL-FETAL TRANSFER IN VIVIPAROUS TELEOSTS

In matrotrophic relationships in viviparous teleosts the yolk supply is inadequate for the nutritional requirements of the embryos. This implies transfer of nitrogenous compounds (protein and amino acids), from the maternal organism to the embryo. The subsequent retention and assimilation of nitrogen is crucial for the embryos, and maternal and/or embryonic structural specializations may develop to facilitate physiological transfer in extensive maternal-fetal matrotrophic relationships. Nitrogenous compounds may derive from the maternal circulatory system or may be synthesized by the follicular and ovarian epithelium to be assimilated by the embryos incubated in the ovarian embryotrophe. In viviparous teleostean fish, embryogenesis always occurs in the ovary, intrafollicularly or directly in the ovarian lumen, and the embryotrophe (i.e., ovarian fluid) is normally the medium for maternal-fetal metabolic exchange.

The embryos obtain nutrients from the surrounding medium in two ways: (1) by dermatrophic uptake across the embryonic integument including skin and pericardial modifications, or (2) by enterotrophic uptake, occurring by various modifications of the embryonic gut.[111,112] These modifications may provide a well-adapted absorptive surface for macromolecules such as cellular material and proteins and for absorption and assimilation of low molecular organic substances such as amino acids, as will be illustrated in the following few examples.

In viviparous *goodeid* fish, protein tracers administered into the ovarian cavity of the pregnant fish are swallowed by the embryos and transported in the embryonic digestive tract. In some goodeids, trophotaenial cells (i.e., extraembryonal anal appendages) appear to be specialized for endocytosis of exogenous proteins dissolved in the ovarian fluid and for their subsequent enzymatic degradation. These cells have the ability to take up protein tracers, horseradish peroxidase (HRP), a medium-sized protein of 44 Kda, at a rate of approximately 13 ng per mg trophotaenial protein per min.[113] Binding of protein to the trophotaenial absorptive cells has been shown to be competitively inhibited by a large number of unlabeled polypeptides or proteins.[114]

Nitrogenous compounds for embryonic growth may also be provided in the form of free amino acids. Embryonic uptake of amino acids has been demonstrated by the use of autoradiographic and tracer techniques. In viviparous fish of the family *Clinidae* a large variety of amino acids are present in the follicular fluid at very high concentrations, whereas the protein concentration is considerably lower than in the maternal plasma, indicating that nitrogenous compounds for embryonic growth are supplied mainly in the form of low molecular compounds.[115,116] In viviparous *Embiotocidae*, polypeptides occurring in the ovarian fluid may be synthesized by the inner ovarian epithelium. The amino acid concentration in the ovarian fluid is lower than that of maternal plasma, but relative composition is similar in both fluids.[117] In the poecilid *Heterandia formosa* exhibiting intrafollicular gestation, proteins from the follicular fluid have been shown to comigrate with the maternal serum proteins, and the embryonic body have accordingly developed anatomical features indicative of macromolecular transport via absorptive epithelia of body skin.[118,119]

In viviparous teleosts, maternal-fetal trophic relationships have often been identified merely by the often dramatic increase in embryonic biomass, which in itself may not account for anabolic and/or catabolic processing of nitrogenous compounds, and thus may seriously underestimate the embryonic mass converted into energy as well as the mass of the maternal nitrogen transfer.

An analysis of energetic costs has shown that a high percentage of catabolic energy is contributed by the maternal organism from fertilization to birth in some of the viviparous *Sebastes* species.[120-122] Studies of protein and nitrogen budgets of the viviparous *Sebastes schlegeli* have indicated that the primary substances taken up during embryonic development are nitrogenous, and by using the protein tracer HRP it could be demonstrated that the protein tracer was ingested by the embryos and subsequently endocytosed in the epithelial absorptive cells in the embryonic hindgut.[123] In *Sebastes melanops* the embryos possess the ability to take up and assimilate exogenous substrates in the form of amino acids.

In the viviparous blenny (eel-pout) *Zoarces viviparus* ovarian fluid is produced abundantly after hatching of the embryos in the ovarian lumen. Nitrogenous compounds are found in the form of amino acids, which occur at a significantly lower level than in the maternal plasma. The fluid frequently contains red blood cells and other cellular components,[124] indicating that cellular material may contribute an additional nutritional source as particulate feed for the embryonic growth. Dissolved protein may be found at very low levels, and is probably derived from minor bleedings in the inner ovarian wall. At present, it is not possible to conclude whether protein represents an important source of nitrogen for the embryos. However, since amounts of nitrogen per embryo show linear increases in parallel with the increase in embryonic dry weight,[125] it is indicated that nitrogenous energy is involved in maternal contribution to embryonic growth and thus rates of flux of nitrogenous substrates through the ovarian fluid must be appreciable.

NITROGEN METABOLISM AND EXCRETION IN MATERNAL-FETAL TROPHIC RELATIONSHIPS IN VIVIPAROUS TELEOSTS

Surprisingly few investigations have been carried out to elucidate protein and amino acid metabolism and nitrogen excretory patterns in relation to matrotrophic growth of embryos in viviparous teleosts. Nitrogen metabolism during development is crucial for activities related to growth, differentiation, and maintenance. The physiological significance of maternal-fetal nitrogen transfer in relation to embryonic energetics, anabolic and catabolic processing of nitrogenous compounds, as well as an eventual role of nitrogen in osmoregulation, has been investigated in a few different families of viviparous teleosts, most extensively, in the embiotocids, the clinids, the poecilids, the scorpaenids, and the zoarcids.

In *Z. viviparus*, nitrogenous compounds have been shown to be assimilated and metabolized by embryos in the ovary. After loading amino acids into the ovarian fluid, initial high levels of amino acids were recorded, followed by a gradual turnover to the maternal serum. Subsequently, exponential disappearance was observed with half-life values of 5 h for excess serum amino acids and 3.8 h for excess ovarian fluid amino acids.[126] Thus, the ovarian fluid appears to be a dynamic pool with rapid circulation and rapid exchange of low molecular weight nitrogenous compounds. Tracer experiments performed *in vivo* or *in vitro* with individual radiolabeled amino acids confirmed these preliminary observations. During late embryonic development, little of the injected tracer amino acid ever appeared in the maternal serum. One day after the intraovarian injection, distribution of label showed a significant accumulation in the embryos.[127] Experiments performed *in vitro* revealed that net flux of labeled amino acids as measured by HPLC closely followed uptake of the tracer. Catabolism of the assimilated amino acids was indicated by production of labeled CO_2 by the embryos after preincubation with the tracer *in situ* as well as after preincubation of embryos *in vitro*. Anabolic metabolism of the tracer was observed by a sizable incorporation of radioactivity into molecules insoluble in trichloroacetic acid (most likely protein). By autoradiographic experiments, applying dark-field illumination, it was indicated that embryos were able to assimilate labeled amino acid already during their yolk-sac stages.[128]

In the viviparous scorpaenid *S. melanops*, it was also demonstrated that the embryos at various stages of development had the capability to take up labeled amino acids. Uptake capacity was shown to increase during development of the embryos, and older embryos were able to oxidize the labeled amino acid into carbon dioxide.[129] The question of whether this increase was due to higher numbers of amino acids transporters or enlarged surface area for absorbance was not addressed.

In agreement with the ability of embryos in matrotrophic relationships to assimilate and metabolize amino acids from naturally occurring concentrations in the ovarian fluid *in vivo* as well as *in vitro*, the embryos are likely to also excrete nitrogen during their development *in ovario*. The ovarian fluid, which is produced abundantly during post-yolk-sac embryonic development, is the medium into which nitrogenous end products are excreted. Since the embryos of viviparous teleosts are often extremely crowded during their development in the ovary (comprising sometimes more than 1/3 of the maternal body weight), conversion of ammonia to another nitrogenous end product such as urea may provide an adaptive advantage not only for the embryos, but also for the maternal organism. The maternal organism may benefit from embryonic urea production provided that the concentrations of embryonic excreted products are high enough and result in transovarian passage. Thus, potentially, the excretory products may constitute a physiological stress phenomenon for the maternal organism. A significant fetal-maternal transfer of excretory nitrogen may further imply that the subsequent excretory mechanisms utilized by the maternal organism may themselves be affected by the product imposed by the excretory pattern of the embryos during various phases of development.

A preliminary analysis of the ovarian fluid of *Z. viviparous* for nitrogenous compounds showed that nitrogen constituted only approximately 0.05% of the dry mass with an average concentration of 466 μg N/ml of ovarian fluid. The nitrogen fraction of the uncentrifuged fluid represented nitrogen derived from various cellular components with a predominance of red blood cells, amino acids, protein, urea, and ammonia. In the centrifuged supernatant of the fluid, nitrogen derived from amino acid nitrogen (NPS) was measured at approximately 40 μg/ml, protein-nitrogen at 5 μg/ml, and urea N and ammonia N levels at 120 and 3 μg/ml, respectively. During post-yolk-sac development of the embryos the level of urea in the ovarian fluid increased significantly from 3.7 mmol/l during early development to 6.1 mmol/l urea N in late development. The level of urea N was on average a factor 15 times higher than the level of ammonia N, increasing from about 0.25 to 0.45 mmol/l during post-yolk-sac development.[130] deVlaming et al.[117] also found large quantities of urea in the ovarian fluid of the pregnant embiotocid *Cymatogaster aggregata* and concluded that the embiotocid embryos may have the capacity to synthesize and retain urea. In this fish, the maternal serum level of urea was generally 30 to 50% lower than the level observed in the ovarian fluid during embryonic development. The physiological implications of this observation have not been further elucidated, but the production of urea in addition to ammonia may serve to protect embryos as well as the maternal organism from the severe stress of ammonia accumulation during embryonic development. The loss of urea from the ovarian fluid to the maternal serum in *Cymatogaster* seems to be very small, however, considering that most membranes are highly permeable to urea. The authors explain this by a possible involvement of an active reabsorptive mechanism such as observed in the kidneys of elasmobranchs.

Fetal-maternal transfer of urea in *Z. viviparus* was first indicated by the fact that the level of urea N in the maternal plasma showed an increase corresponding to increase of urea in the ovarian fluid during embryonic post–yolk-sac development. No significant differences were noted, however, between levels of urea or ammonia N in the maternal plasma when compared to levels of the ovarian fluid. Thus, high correlation coefficients ($r = 0.99$) were found by comparing the two fluids.[130] Kristoffersson et al.[124] observed a striking difference in the concentration of free amino acids in the maternal plasma and the ovarian fluid and explained the lower concentration in the ovarian fluid by a very active uptake of amino acids by the embryos. The authors also detected no significant differences between the mean ammonia level in the maternal plasma and the ovarian fluid. Based on their results, Kristoffersson et al.[124] concluded that catabolic reactions apparently do not dominate during embryonic development. These authors, however, did not consider urea production.

In order to further elucidate aspects of fetal-maternal transfer implicating a role of the ovarian sac in transfer of nitrogenous waste products away from the embryonic compartment, exogenous urea was injected into the ovary of *Z. viviparus*. The urea load was quickly eliminated from the ovarian fluid during a time course of 6 h and a concomitant significant increase could be observed in the urea concentration of the maternal plasma. Interestingly, no additional excretion of urea by the maternal part could be detected. When pregnant fish were placed in aquaria with a limited amount of water (5 l/kg fish body wt) it was only possible to detect output of ammonia-N, not urea-N, into the ambient seawater during a 24-h time course (B. Korsgaard, unpublished results).

The increasing concentration of urea-N in the ovarian fluid during embryonic development may be a consequence of an increase in embryonic nitrogen excretion into the fluid. Urea may be transferred to the maternal circulatory system via the ovarian epithelium and its numerous supplying blood vessels. The large quantities of urea N in the ovarian fluid indicate that the embryos of *Z. viviparus* have the capacity to produce urea in addition to ammonia as a nitrogenous waste with urea predominating over ammonia production. This was further elucidated by investigating the excretory pattern of embryos incubated *in vitro*. For this purpose the embryos were excised from the ovary during late post–yolk-sac development, rinsed, and incubated in a nitrogen-free isoosmotic

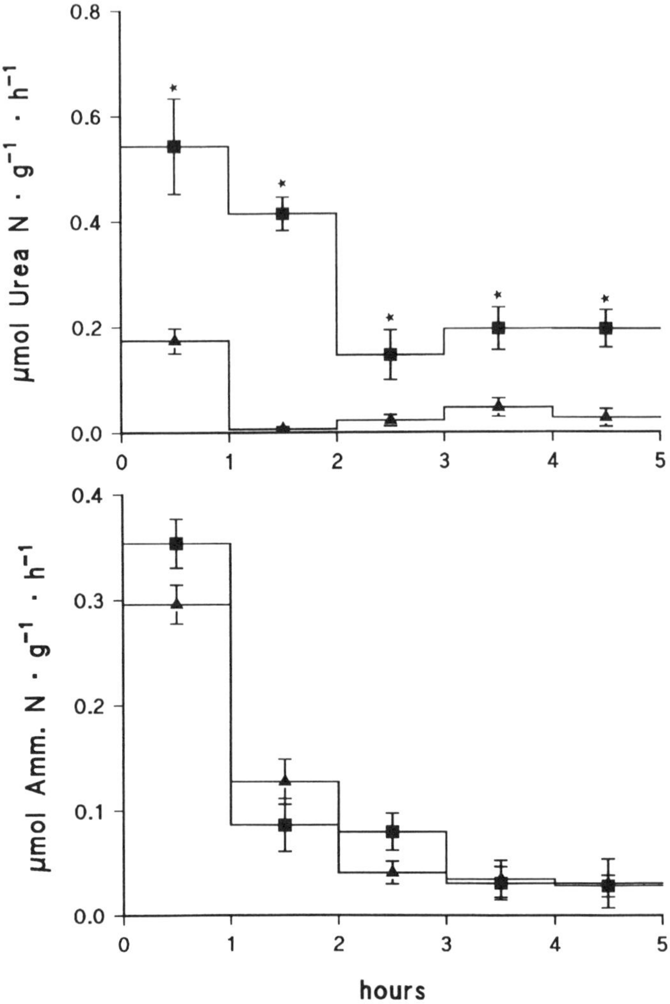

FIGURE 4 *In vitro* nitrogen excretion in larvae and post-yolk-sac embryos of the viviparous blenny *Zoarces viviparus*. Excretion rates for urea (A) and ammonia (B) were followed for a 5-h interval. Larvae (▲); embryos (■). Values represent means ± SEM; * significantly different ($p < 0.05$) from value measured for larvae. From Korsgaard, B., *J. Comp. Physiol. [B]*, 164B, 42, 1994. With permission.)

Ringer solution in acid-cleaned glass vials. Under these experimental conditions urea-N constituted an average of 65% of total N (ammonia-N plus urea-N) excretion during a 5-h time course (Figure 4). This high percentage basically rules out that this urea is derived from uricolysis and/or argininolysis. Urea nitrogen was excreted at maximum rates during the first 2 h of the experiment. The excretion rates of both urea and ammonia declined to lower levels during the subsequent time course (Figure 4). Interestingly, embryos, which had been acclimated to seawater for 96 h before being transferred to the experimental vials, only excreted urea at an average of 35% of the total nitrogen excretion.

The high initial excretion of urea, which is observed in the *in vitro* experiments may be explained by an accentuated blood-water urea gradient. As the blood-water gradient for urea excretion decreases, a subsequent decline in urea excretion may eventually be expected. The pattern of nitrogen excretion which is observed *in vitro* may then be explained by the hypothesis that, during their residence in the ovarian fluid, the embryos presumably have accumulated large stores of urea with a resultant lower blood-water urea gradient. Thus, nitrogen excretion *in vitro* may be a result of gradients favoring nitrogen loss by diffusion. In contrast, the net loss of nitrogen from the embryos *in vivo* may be reduced as a consequence of the high concentrations of urea in the ovarian fluid. Likewise, the lower urea excretion rates by the embryos, which had been acclimated to seawater, may be due to a wash-out effect of the urea load during their residence in the seawater (Figure 4).

These aspects of nitrogen excretion have been further elucidated by loading the ambient medium of the *in vitro* system with either ammonia or urea at concentrations which are normally found in the ovarian fluid. These experiments showed that embryos incubated in Ringer solution with urea excreted ammonia-N at rates similar to the control group. On the other hand when exposed *in vitro* to ammonia-N concentrations similar to normal values in the ovarian fluid, the rates of urea excretion by the embryos were not affected as compared to the control group. Thus, urea excretion rates are not affected by initial high levels of ammonia in the ambient medium and likewise, ammonia excretion rates are not affected by the initial presence of urea in the medium, provided that the respective ambient concentrations are similar to the concentrations found in the ovarian fluid under natural conditions (Korsgaard, B., unpublished data).

In *Z. viviparus* the concentration of urea and ammonia in the ovarian fluid increased during the progressive development of the embryos, and preliminary data indicate accordingly that the rate of excretion of nitrogen *in vitro* increases in the course of embryonic development. Dépêche and Schoffeniels[131] found relatively high amounts of amino acids and urea in eggs and early embryos of the guppy *Poecilia reticulata.* They suggested that the embryonic ability to produce urea may be lost in the adult fish. Dépêche et al.[110] further investigated these aspects of urea production during the ontogenic development of the ovoviviparous guppy, and observed that the urea content in the oocytes increased markedly towards the end of vitellogenesis. The level of urea remained high in the young embryos but decreased at the end of the embryonic development toward the low values characteristic of the adult fish. Kormanik[132] observed that late-term embryos from the dogfish *S. acanthias* exhibited excretion rates of urea and ammonia which were several fold higher than those observed in early term embryos. Interestingly, Kaushik et al.[133] found that mean urea N excretion rate during ontogenesis of the common carp (*Cyprinus carpio*) amounted to approximately 22% of total nitrogen excretion and that there was no significant difference in excretion rates between embryonic and postembryonic stages. In *Z. viviparus* embryos, acclimation to seawater reduced the percentage of urea excreted relatively to ammonia excretion in the *in vitro* systems. In the salamander *Salamandra salamandra*, ureotelism also predominated during their intrauterine development.[134] During their extrauterine larval life ammonia excretion was found to be slightly higher than urea excretion.

The only information regarding the possible function of urea for osmoregulatory purposes in viviparous teleosts with the benefit of experimental evidence is derived from studies on pregnant poecilids. The presence of urea and ammonia in embryos of *P. reticulata* during early development may indicate a role of urea as an important osmotic effector during very early stages of embryonic development. This hypothesis was based on the observation that urea concentrations increased in embryos by acclimation of the pregnant female to hyperosmotic media.[135] An increase in the ambient salinity did not have any effect on the urea content of the maternal organism or of the oocytes in early development. In oocytes at the end of vitellogenesis as well as in embryos, however, acclimation to a concentrated medium significantly increased their respective urea content. The possible osmotic role of urea has also been the subject of investigations in viviparous elasmobranchs. Mechanisms for ionic and urea regulation are present during early development of embryos of *S. acanthias*, and accordingly late-term embryos of this species were able to osmoregulate *in utero*.[136] It has also been indicated that the ability of elasmobranch embryos to osmoregulate increases with embryonic age. Veith[116] and Cornish and Veith[137] observed urea to be present in the follicular fluid of the pregnant clinids (*Clinus superciliosus* and *Clinus dorsalis*). Small embryos were capable of some degree of osmoregulation and full-term embryos were able to survive concentrations of seawater in the range from 40 to 100% of full strength.[116] The importance of the observed presence of urea in the ambient medium during embryonic development remains to be elucidated in these fish, however.

Further experiments are needed to elucidate by which pathways urea is produced by the maternal organism and/or by her embryos in viviparous fish. Urea cycle enzymes have been demonstrated to be present at all stages of development in elasmobranch embryos,[138] but, of course, these enzymes are retained by the adults to produce urea as an important osmolyte.[139] Preliminary experiments with *Z. viviparus* have indicated that one of the key enzymes (OCT) in the OUC is present in the liver of the pregnant fish, albeit at a low level compared with toadfish. This may indicate that urea may be produced primarily by the uricolytic pathway or directly from arginine; however, further experiments are needed to elucidate these regulatory aspects. Since there are no indications of a maternal excretion of urea into ambient seawater, it would also be interesting to investigate whether urea is retained by the maternal organism and used as an additional source of nitrogen during development of the embryos *in ovario*. The pathway for an eventual urea production by the embryos of *Z. viviparus* has not yet been investigated. Preliminary data from experiments performed on the guppy *P. reticulata*[110] indicated that developing embryos detoxify ammonia into urea and display relatively high rates of bicarbonate incorporation into urea, an ability which is apparently decreased or lost in the older embryos and the adult organism. Dépêche et al.[110] observed that urea production from ^{14}C-labeled carbon dioxide could not be detected in embryos at an early staging, but that urea production increased during later embryonic development then decreased again as development proceeded and finally reached very low levels in the larva. Comparison with results from similar experiments with the rainbow trout leads us to suggest that the urea cycle has been kept in the viviparous species in order to assist the intensive amino-group metabolism taking place in the pregnant female.

In conclusion there appears to be very little general information on the pattern of nitrogen excretion in viviparous teleosts. Embryos of different families of viviparous teleosts may produce and accumulate urea with osmotically functional implications. The maternal regulation of fetal metabolic wastes during progressive embryonic development *in ovario* also remains elusive. In *Z. viviparus* embryonic ureagenesis seems to predominate during development as indicated by urea excretion rates, which are relatively high as compared to ammonia, by embryos *in vitro*, and by the high levels of urea in the ovarian fluid — the ambient medium of the embryos. In a few viviparous teleosts, it has been indicated that urea may be removed from the ovarian fluid by transovarian migration into the maternal vascularized system. However, no evidence for further elimination of urea from the maternal organism has been presented. Few data exist to indicate metabolic pathways for urea production in viviparous fish. In *P. reticulata* developing embryos may detoxify ammonia into urea as observed by relatively high rates of bicarbonate incorporation into urea.[110]

CONCLUSIONS AND FUTURE DIRECTIONS

It is clear that fish can take two, not necessarily exclusive, routes to deal with conditions limiting ammonia disposal. The common strategy is to go with the drift, so to speak, and let ammonia build up internally. Eventually, a new equilibrium will be attained which, once again, lets the fish dispose of ammonia. This wait and see attitude is made possible by efficient means to remove ammonia from sites where it is potentially most detrimental, primarily the brain. Excellent examples for this strategy are two species living in alkaline lakes (pH >9.5), the Lahontan cutthroat trout and the cyprinid *Chalcaburnus tarichi*, and other specialized fishes such as mudskippers, goldfish, and rainbow trout. The only difference noticed between fish species appears to be the initial tolerance to increasing internal concentrations of ammonia. Irrespective of their life style, fish possess high titers of brain glutamine synthetase to deal with spillover into the brain as plasma ammonia increases. In contrast to mammals, fish have a generally large tolerance for plasma and tissue ammonia.

The less common strategy is found in an alkaline lake tilapia, facultatively in toadfish of the genus *Opsanus*, some air-breathing fishes, possibly in fish eggs before hatching, and in embryos of viviparous species. It involves detoxication of ammonia through the synthesis and subsequent excretion of urea. This route is metabolically expensive, and except in the case of the tilapia, not exclusive of ammonia excretion. Since control of urea synthesis appears to lie with nitrogen delivery to the OUC and not within the cycle itself (as it does in mammals), the enzymes involved deserve increased attention. Hepatic glutamine synthetase, for example, serves as a feeder enzyme to the cycle and not as an enzyme "cleaning-up" after the OUC and is enhanced under conditions of air exposure, ammonia exposure, or crowding/confinement. Its regulation by corticosteroids such as dexamethasone suggests transcriptional control of the enzyme, although at this point more direct control and perhaps permissive effects on nonsteroid hormones should also be considered. Brain glutamine synthetase occupies a similarly critical position to ammonia detoxication as in mammalian brain, but seems to operate at much higher concentrations of ammonia and possibly in tandem with glutamate dehydrogenase. From a biochemical point of view, the latter enzyme has been largely neglected, yet three challenging scenarios involving GDH and GNS can be envisaged. In brain, GDH functions as a helper, delivering glutamate for GNS. In liver, GDH is supposed to hold the key to amino acid transdeamination, enabling fish to dispose of ammonia. Yet in toadfish liver, the two enzymes seem to work independently of each other, while in the eel, GDH is induced and serves to take ammonia up. Since the enzyme is unlikely to beat the odds of mass action, it is clear that our understanding of enzyme and substrate compartmentation and intracellular conditions both in liver and brain is rudimentary at best. The same is true for the kinetics and regulation of both enzymes.

The question remains: What exactly are the determining factors that make fishes adapt acutely, or on an evolutionary time scale to conditions limiting the disposal of ammonia? As shown above, different strategies are employed — all of which have their selective advantages. Yet, theoretically not all avenues have been exploited experimentally. When rendering systems more alkaline, the experimenter can simply add a strong base to the medium, or add larger amounts of a weaker base. While a given internal pH alters the relative proportions of NH_3 and NH_4^+, other weak bases (amino acids) and weak acids (carbonic acid) also change, requiring a multifactorial experimental design to sort out primary and secondary effects. Unfortunately, to date, few studies have attempted such an integrated approach and usually have relied on manipulating merely one or two parameters. Further, the choice of base is likely to influence the experimental outcome. Obviously, alkalinization with ammonium hydroxide will elicit different responses from fish than alkalinization with bicarbonate, carbonate, or hydroxide. Especially in a freshwater situation, the counter-ion is likely to exert its own influence on ion-exchange, NH_4^+ efflux, and metabolic expenditure. Similarly, ambient ammonia loading has been shown to have different toxic effects, depending on the counter-ion used.[140]

When considering the physiological role of urea in teleostean viviparity, a number of factors have to be considered: (1) it is important to know how much urea the embryos excrete during their progressive development in the ovary; (2) how much urea can be accumulated in the ambient medium, the ovarian fluid; (3) to which extent the excretory nitrogenous products are exchanged with the maternal system; and, (4) how the fetal-maternal transfer may affect the maternal excretory pattern during pregnancy. All these aspects need to be elucidated experimentally to derive an understanding of the physiological and adaptive significance of urea production in viviparous teleosts.

When analyzing metabolic changes during ontogenesis or embryo development, expression of glutamine synthetase and aspartate aminotransferase may give additional insights, especially if performed in conjunction with expression of OUC enzymes and possibly an analysis of hormonal control during early development. The massive induction of the OUC in the Indian catfish can serve as an ideally suited model of enzyme induction. Just as interesting, however, would be the de-induction of OUC enzymes (and glutamine synthetase?) after a bout of ammonia exposure in this fish. The half-lives of enzymes, half-lives of messenger RNAs for key enzymes, and associated transporters could be measured. One aspect that is often overlooked when analyzing the OUC in fishes is the relative amount of space occupied by CPS. In mammals, CPS alone may account for 15 to 26% of the total mitochondrial matrix protein.[22] If the enzyme is indeed induced heavily during ontogenesis or ammonia exposure, ultimately achieving urea production rates close to those of mammals, is this reflected in mitochondrion morphology and does this go hand-in-hand with changes in urea transport mechanisms? Are similar changes in urea transport mechanisms observed in nonureogenic species during periods of fasting or starvation, which often lead to increased urea retention?

Many questions come mind regarding the relationship between brain glutamine synthetase, ammonia tolerance, and development. For instance, it can be postulated that the development of brain tissue functions as as a key timing device switching on detoxication mechanisms for ammonia and/or release of ammonia. Is breakage of the chorion possibly timed with cranial development? Do fish become susceptible to ammonia toxicity at brain development? What is the relative timing of ontogenic induction of GNS vs. ammonia toxicity vs. induction of the urea cycle? Perhaps even bicarbonate deserves renewed attention. In toadfish, bicarbonate possibly limits the activity of the OUC. Of course, ammonia and bicarbonate are not only intricately linked to membrane transport as counterions and important in acid–base regulation, they are also directly (bicarbonate) and indirectly (ammonia via glutamine and aspartate) substrates for urea synthesis.

Compared with mammals, fishes seem to operate well under conditions of sustained hyperammonemia or high tissue ammonia loads. Therefore, secondary responses to internal ammonia deserve further attention. As already pointed out, ammonia exposure induces hypercortisolemia, but other hormones may be affected as well. In mammals, ammonia directly affects the release of insulin from the beta cells of the endocrine pancreas, superimposed on indirect effects, with ammonia decreasing the glucose-dependent insulin release in pancreatic islets. Obviously, a little ammonia may go a long way in redirecting metabolism through changes in the endocrine system.

REFERENCES

1. **Van Waarde, A.,** Biochemistry of non-protein nitrogenous compounds in fish including the use of amino acids for anaerobic energy production, *Comp. Biochem. Physiol. [B]*, 91, 207, 1988.
2. **Knight, I. T., Grimes, D. J., and Colwell, R. R.,** Bacterial hydrolysis of urea in tissues of carcharinid sharks, *Can. J. Fish. Aquat. Sci.*, 45, 357, 1988.
3. **Mommsen, T. P. and Walsh, P. J.,** Evolution of urea synthesis in vertebrates: the piscine connection, *Science*, 243, 72, 1989.
4. **Walsh, P. J., Blackwelder, P., Gill, K. A., Danulat, E., and Mommsen, T. P.,** Carbonate deposits in marine fish intestines: a new source of biomineralization, *Limnol. Oceanogr.*, 36, 1227, 1991.
5. **Atkinson, D. E.,** Functional roles of urea synthesis in vertebrates, *Physiol. Zool.*, 65, 243, 1992.
6. **Alcocer, I., Santacruz, X., Steinbeisser, H., Thierauch, K.-H., and del Pino, E. M.,** Ureotelism as the prevailing mode of nitrogen excretion in larvae of the marsupial frog *Gastrotheca riobambae* (Fowler) (Anura, Hylidae), *Comp. Biochem. Physiol. [A]*, 101, 229, 1992.
7. **Campbell, J. W. and Anderson, P. M.,** Evolution of mitochondrial enzyme systems in fish: the mitochondrial synthesis of glutamine and citrulline, in *Biochemistry and Molecular Biology of Fishes*, Hochachka, P. W. and Mommsen, T. P., Eds., Elsevier, Amsterdam, New York, 1991, 43.
8. **Wright, P. A.,** Nitrogen excretion and enzyme pathways for ureagenesis in freshwater tilapia (*Oreochromis niloticus*), *Physiol. Zool.*, 66, 881, 1993.
9. **Walsh, P. J. and Henry, R. P.,** Carbon dioxide and ammonia metabolism and exchange, in *Biochemistry and Molecular Biology of Fishes*, Hochachka, P. W. and Mommsen, T. P., Eds., Elsevier, New York, 1991.
10. **Hille B.,** Potassium channels in myelinated nerve, *J. Gen. Physiol.*, 61, 669, 1973.
11. **Olson, K. R. and Fromm, P. O.,** Excretion of urea by two teleosts exposed to different concentrations of ambient ammonia, *Comp. Biochem. Physiol. [A]*, 40, 999, 1971.
12. **Tomasso, J. R., Davis, K. B., and Simco, B. A.,** Plasma corticosteroid dynamics in channel catfish (*Ictalurus punctatus*) exposed to ammonia and nitrite, *Can. J. Fish. Aquat. Sci.*, 38, 1105, 1981.

13. **Spotte, S. and Anderson, G.**, Plasma cortisol changes in seawater-adapted mummichogs (*Fundulus heteroclitus*) exposed to ammonia, *Can. J. Fish. Aquat. Sci.*, 46, 2065, 1989.
14. **Lang, T., Peters, G., Hoffmann, R., and Meyer, E.**, Experimental investigations on the toxicity of ammonia: effects on ventilation frequency, growth, epidermal mucous cells, and gill structure of rainbow trout *Salmo gairdneri*, *Dis. Aquat. Org.*, 3, 159, 1987.
15. **Smart, G. R.**, Investigations of the toxic mechanisms of ammonia to fish gas exchange in rainbow trout (*Salmo gairdneri*) exposed to acutely lethal concentrations, *J. Fish Biol.*, 12, 93, 1978.
16. **Arillo, A., Uva, B., and Vallarino, M.**, Renin activity in rainbow trout (*Salmo gairdneri*) and effects of environmental ammonia, *Comp. Biochem. Physiol. [A]*, 68, 307, 1981.
17. **Lloyd, R. and Orr, L. D.**, The diuretic response by rainbow trout to sub-lethal concentrations of ammonia, *Water Res.*, 3, 335, 1969.
18. **Smart, G.**, The effect of ammonia exposure on gill structure of the rainbow trout, *J. Fish Biol.*, 8, 471, 1976.
19. **Wood, C. M. and Patrick, M. L.**, Methods for assessing kidney and urinary bladder function in fish, in *Biochemistry and Molecular Biology of Fishes, Vol. 3, Analytical Techniques*, Hochachka, P. W. and Mommsen, T. P., Eds., Elsevier, New York, 1994, in press.
20. **Cooper, A. J. L. and Plum, F.**, Biochemistry and physiology of brain ammonia, *Physiol. Rev.*, 67, 440, 1987.
21. **Tremblay, G. C. and Bradley, T. M.**, L-Carnitine protects fish against acute ammonia toxicity, *Comp. Biochem. Physiol. [C]*, 101, 349, 1992.
22. **Meijer, A. J., Lamers, W. H., and Chamuleau, R. A. F. M.**, Nitrogen metabolism and ornithine cycle function, *Physiol. Rev.*, 70, 701, 1990.
23. **Wood, C. M.**, Ammonia and urea metabolism and excretion, in *The Physiology of Fishes*, Evans, D. H., Ed., CRC Press, Boca Raton, FL, 1993, 379.
24. **Wood, C. M., Perry, S. F., Wright, P. A., Bergman, H. L., and Randall, D. J.**, Ammonia and urea dynamics in the Lake Magadi tilapia, a ureotelic teleost fish adapted to an extremely alkaline environment, *Respir. Physiol.*, 77, 1, 1989.
25. **Wilkie, M. P. and Wood, C. M.**, Nitrogenous waste excretion, acid-base regulation, and ionoregulation in rainbow trout (*Oncorhynchus mykiss*) exposed to extremely alkaline water, *Physiol. Zool.*, 64, 1069, 1991.
26. **Arillo, A., Margiocco, C., Medlodia, F., Mensi, P., and Schenone, G.**, Ammonia toxicity mechanism in fish: studies on rainbow trout (*Salmo gairdneri* Rich.), *Ecotoxicol. Environ. Safety*, 5, 316, 1981.
27. **Iwata, K.**, Nitrogen metabolism in the mudskipper, *Periophthalmus cantonensis*: changes in free amino acids and related compounds in various tissues under conditions of ammonia loading, with special reference to its high ammonia tolerance, *Comp. Biochem. Physiol. [A]*, 91, 499, 1988.
28. **Saha, N. and Ratha, B. K.**, Induction of ornithine-urea cycle in a freshwater teleost, *Heteropneustes fossilis*, exposed to high concentrations of ammonium chloride, *Comp. Biochem. Physiol. [B]*, 108B, 315, 1994.
29. **Plisetskaya, E. M.**, Brain and heart glycogen content in some vertebrates and effect of insulin, *Endocrinol. Exp.*, 2, 251, 1968.
30. **Foster, G. D., Youson, J. H., and Moon, T. W.**, Carbohydrate metabolism in the brain of the adult lamprey, *J. Exp. Zool.*, 267, 27, 1993.
31. **Hayashi, S., Tang, M. Q., Hirakawa, T., and Yamada, S.**, Effect of high concentration of ammonia on cultured hepatocytes of eel (*Anguilla japonica*), *Comp. Biochem. Physiol. [C]*, 105C, 17, 1993.
32. **Walsh, P. J., Tucker, B. C., and Hopkins, T. E.**, Effects of confinement/crowding on ureogenesis in the Gulf toadfish *Opsanus beta*, *J. Exp. Biol.*, 191, 195, 1994.
33. **Iwata, K. and Kakuta, I.**, A comparison of catalytic properties of glutamate dehydrogenase from liver and muscle between amphibious *Periophthalmus cantonensis* and water-breathing gobid fishes *Tridentiger obscurus obscurus*, *Bull. Jap. Soc. Sci. Fish.*, 49, 1903, 1983.
34. **Campbell, J. W., Aster, P. L., and Vorhaben, J. E.**, Mitochondrial ammoniagenesis in liver of the channel catfish *Ictalurus punctatus*, *Am. J. Physiol.*, 244, R709, 1983.
35. **Morii, H., Nishikata, K., and Tamura, O.**, Ammonia and urea excretion from mudskipper fishes *Periophthalmus cantonensis* and *Boleophthalmus pectinirostris* transferred from land to water, *Comp. Biochem. Physiol. [A]*, 63, 23, 1979.
36. **Gordon, M. S., Wilson, W. N., and Alice, Y. Y.**, Aspects of the physiology of terrestrial life in amphibious fishes. III. The Chinese mudskipper, *Periophthalmus cantonensis*, *J. Exp. Biol.*, 72, 57, 1978.
37. **Danulat, E. and Kempe, S.**, Nitrogenous waste excretion and accumulation of urea and ammonia in *Chalcalburnus tarichi* (Cyprinidae) endemic to the extremely alkaline Lake Van (Eastern Turkey), *Fish Physiol. Biochem.*, 9, 377, 1992.
38. **Cohen, N. S., Kyan, F. S., Kyan, S. S., Cheung, C.-W., and Raijman, L.**, The apparent Km of ammonia for carbamoyl phosphate synthetase (ammonia) in situ, *Biochem. J.*, 229, 205, 1985.
39. **Evans, D. H., More, K. J., and Robbins, S. L.**, Modes of ammonia transport across the gill epithelium of the marine teleost fish *Opsanus beta*, *J. Exp. Biol.*, 144, 339, 1989.
40. **Wright, P. A., Randall, D. J., and Perry, S. F.**, Fish gill water boundary layer: a site of linkage between dioxide and ammonia excretion, *J. Comp. Physiol. [B]*, 158, 627, 1989.

41. **Heisler, N.,** Mechanisms of ammonia elimination in fishes, in *Animal Nutrition and Transport Processes 2. Transport, Respiration and Excretion: Comparative and Environmental Aspects*, Truchot, J. P. and Lahlou, B., Eds., Karger, Basel, 1990, 137.
42. **Brett, J. R. and Zala, C. A.,** Daily pattern of nitrogen excretion and oxygen consumption of sockeye salmon (*Oncorhynchus nerka*) under controlled conditions, *J. Fish. Res. Bd. Can.*, 32, 2479, 1975.
43. **Sayer, M. D. J.,** An investigation into the pattern of nitrogen excretion in *Blennius pholis*, *Comp. Biochem. Physiol. [A]*, 89, 359, 1988.
44. **Fisher-Scherl, T., Veeser, A., Hoffmann, R. W., Kühnhauser, C., Negele, R.-D., and Ewringmann, T.,** Morphological effects of acute and chronic atrazine exposure in rainbow trout (*Oncorhynchus mykiss*), *Arch. Environ. Contam. Toxicol.*, 20, 454, 1991.
45. **Prasad, T. A. V., Srinivas, T., and Reddy, D. C.,** Modulations in nitrogen metabolism in the hepatic and neuronal tissues of fish, *Tilapia mossambica* exposed to atrazine, *Biochem. Int.*, 23, 271, 1991.
46. **Claiborne, J. B. and Evans, D. H.,** Ammonia and acid-base balance during high ammonia exposure in a marine teleost (*Myoxocephalus octadecimspinosus*), *J. Exp. Biol.*, 140, 89, 1988.
47. **Palmijans, V., Krishnaswamy, P. R., Dumville, G., and Meister, A.,** Studies on the mechanism of glutamine synthesis: isolation and properties of the enzyme from sheep brain, *Biochemistry*, 1, 153, 1962.
48. **Deuel, T. F., Louie, M., and Lerner, A.,** Glutamine synthetase from rat liver, *J. Biol. Chem.*, 253, 6111, 1978.
49. **Mommsen, T. P. and Walsh, P. J.,** Urea synthesis in fishes: evolutionary and biochemical perspectives, in *Biochemistry and Molecular Biology of Fishes*, Vol. 1, Hochachka, P. W. and Mommsen, T. P., Eds., Elsevier, New York, 1991, 137.
50. **Wood, C. M., Bergman, H. L., Laurent, P., Maina, J. N., Narahara, A., and Walsh, P. J.,** Urea production, acid-base regulation and their interactions in the Lake Magadi tilapia, a unique teleost adapted to a highly alkaline environment, *J. Exp. Biol.*, 189, 13, 1994.
51. **Barber, M. L. and Walsh, P. J.,** Interactions of acid-base status and nitrogen excretion and metabolism in the ureogenic teleost *Opsanus beta*, *J. Exp. Biol.*, 185, 87, 1993.
52. **Singh, R. A. and Singh, S. N.,** Tissue distribution, effect of starvation and seasonal variation of arginase in the freshwater teleost, *Clarias batrachus* (L.), *Biochem. Arch.*, 4, 329, 1988.
53. **Hayashi, S., Fujiwara, S., and Noguchi, T.,** Degradation of uric acid in fish liver peroxisomes, *J. Biol. Chem.*, 264, 3211, 1989.
54. **Wright, P. A., Pärt, P., and Wood, C. M.,** Ammonia and urea excretion in the tidepool sculpin (*Oligocottus maculatus*): sites of excretion, effects of reduced salinity and mechanisms of urea transport, *Fish Physiol. Biochem.*, in press, 1995.
55. **Masoni, A. and Payan, P.,** Urea, inulin and para-amino-hippuric acid (PAH) excretion by the gills of the eel, *Anguilla anguilla* L. *Comp. Biochem. Physiol. [A]*, 47, 1241, 1974.
56. **Wood, C. M., Hopkins, T. E., Hogstrand, C., and Walsh, P. J.,** Pulsatile urea excretion in the ureogenic toadfish *Opsanus beta*; an analysis of rates and routes, *J. Exp. Biol.*, in press, 1995.
57. **Walsh, P. J., Wood, C. M., Perry, S. F., and Thomas, S.,** Urea transport by hepatocytes and red blood cells of selected elasmobranch and teleost fishes, *J. Exp. Biol.*, 193, 321, 1994.
58. **Danulat, E.** Biochemical-physiological adaptations of teleosts to highly alkaline, saline lakes, in *Biochemistry and Molecular Biology of Fishes, Vol. 5, Environmental and Ecological Biochemistry*, Hochachka, P. W. and Mommsen, T. P., Eds., Elsevier Science, Amsterdam, 1995, in press.
59. **Wood, C. M., Munger, R. S., and Toews, D. P.,** Ammonia, urea and H^+ distribution and the evolution of ureotelism in amphibians, *J. Exp. Biol.*, 144, 215, 1989.
60. **Masoni, A. and García-Romeu, F.,** Accumulation et excrétion de substances organiques par les cellules a chlorure de la branchie d'*Anguilla anguilla* L. adaptée a l'eau de mer, *Z. Zellforsch.*, 133, 389, 1972.
61. **Maina, J. N.,** A morphometric analysis of chloride cells in the gills of the teleosts *Oreochromis alcalicus* and *Oreochromis niloticus* and a description of presumptive urea-excreting cells in *O. alcalicus*, *J. Anat.*, 175, 131, 1991.
62. **Grubinko, V. V., Yakovenko, B. V., and Yavonenko, A. F.,** Effects of starvation on arginase activity and urea contents in the carp, *Cyprinus carpio*, *J. Ichthyol.*, 27(6), 107, 1988.
63. **Walsh, P. J., Danulat, E., and Mommsen, T. P.,** Variation in urea excretion in the Gulf toadfish, *Opsanus beta*, *Mar. Biol.*, 106, 323, 1990.
64. **Saha, N., Chakravorty, J., and Ratha, B. K.,** Diurnal variation in renal and extra-renal excretion of ammonia-N and urea-N in a freshwater air-breathing teleost, *Heteropneustes fossilis* (Bloch), *Proc. Indian Acad. Sci. (Anim. Sci.)*, 97, 529, 1988.
65. **Haywood, G. P., Isaia, J., and Maetz, J.,** Epinephrine effects on branchial water and urea flux in rainbow trout, *Am. J. Physiol.*, 232, R110, 1977.
66. **Griffith, R. W., Pang, P. K. T., and Benedetto, L. A.,** Urea tolerance in the killifish, *Fundulus heteroclitus*, *Comp. Biochem. Physiol. [A]*, 62, 327, 1979.
67. **Bieter, R. N.,** The action of some diuretics upon the aglomerular kidney, *J. Pharmacol. Exp. Ther.*, 43, 399, 1931.

68. **Schreckenbach, K., Spangenberg, R., and Krug, S.,** Die Ursache der Kiemennekrose, *Z. Binnenfisch.*, 22, 257, 1975.
69. **Halstead, B. G. and Tash, J. C.,** Unusual diel pHs in water as related to aquatic vegetation, *Hydrobiology*, 96, 217, 1982.
70. **Randall, D. J. and Wright, P. A.,** The interaction between carbon dioxide and ammonia excretion and water pH in fish, *Can. J. Zool.*, 67, 2936, 1989.
71. **Randall, D. J. and Lin, H.,** Effects of water pH on gas and ion transfer across fish gills, in *Fish Ecophysiology*, Rankin J. C. and Jensen F. B., Eds., Chapman & Hall, London, 1993, 265.
72. **Wright, P. A., Iwama, G. K., and Wood, C. M.,** Ammonia and urea excretion in Lahontan cutthroat trout (*Oncorhynchus clarki henshawi*) adapted to the highly alkaline Pyramid Lake (pH 9.4), *J. Exp. Biol.*, 175, 153, 1993.
73. **Randall, D. J., Wood, C. M., Perry, S. F., Bergman, H., Maloiy, G. M. O., Mommsen, T. P., and Wright, P. A.,** Urea excretion as a stategy for survival in a fish living in a very alkaline environment, *Nature*, 337, 165, 1989.
74. **Wilkie, M. P., Wright, P. A., Iwama, G. K., and Wood, C. M.,** The physiological responses of the Lahontan cutthroat trout (*Oncorhynchus clarki henshawi*), a resident of highly alkaline Pyramid Lake (pH 9.4), to challenge at pH 10, *J. Exp. Biol.*, 175, 173, 1993.
75. **Wilkie, M. P., Wright, P. A., Iwama, G. K., and Wood, C. M.,** The physiological adaptations of the Lahontan cutthroat trout, (*Oncorhynchus clarki henshawi*) following transfer from well water to the highly alkaline waters of Pyramid Lake, Nevada (pH 9.4), *Physiol. Zool.*, 67, 355, 1994.
76. **Chiu, Y. N., Austic, R. E., and Rumsey, G. L.,** Urea cycle activity and arginine formation in rainbow trout (*Salmo gairdneri*), *J. Nutr.*, 116, 1640, 1986.
77. **Milligan, C. L., McDonald, D. G., and Prior, E. T.,** Branchial acid and ammonia fluxes in response to alkalosis and acidosis in two marine teleosts: coho salmon (*Oncorhynchus kisutch*) and starry flounder(*Platichthys stellatus*), *Physiol. Zool.*, 64,169, 1991.
78. **Haywood, G. P.,** Ammonia toxicity in teleost fishes: a review, *Can. Tech. Rep. Fish. Aquat. Sci.*, 1177, 1, 1983.
79. **Iwata, K., Kakuta, I., Ikeda, M., Kimoto, S., and Wada, N.,** Nitrogen metabolism in the mudskipper, *Periophthalmus cantonensis*: a role of free amino acids in detoxication of ammonia produced during its terrestrial life, *Comp. Biochem. Physiol. [A]*, 68, 589, 1981.
80. **Kormanik, G. A.,** Ion and osmoregulation in prenatal elasmobranchs: evolutionary implications, *Am. Zool.*, 32, 294, 1992.
81. **Mommsen, T. P., Danulat, E., and Walsh, P. J.,** Metabolic actions of glucagon and dexamethasone in liver of the ureogenic teleost *Opsanus beta*, *Gen. Comp. Endocrinol.*, 85, 316, 1992.
82. **Matty, A.J.,** *Fish Endocrinology*, Croom Helm Ltd. Beckenham, Kent, 1985, 1.
83. **Cottelli, F., Andronico, F., Brivio, M., and Lora Lamia, C.,** Structure and composition of the fish egg chorion (*Carassius auratus*), *J. Ultrastr. Mol. Struct. Res.*, 99, 70, 1988.
84. **Davenport, J., Lonning, S., and Kjorsvik, E.,** Osmotic and structural changes during early development of eggs and larvae of the cod, *Gadus morhua* L, *J. Fish Biol.*, 19, 317, 1981.
85. **Lønning, S., Kjørsvik, E., and Davenport, J.,** The hardening process of the egg chorion of the cod, *Gadus morhua* and lumpsucker, *Cyclopterus lumpus*, *J. Fish Biol.*, 24, 505, 1984.
86. **Rombough, P. J.,** Respiratory gas exchange, aerobic metabolism and effects of hypoxia during early life, in *Fish Physiology*, Vol. 11A, Hoar, W. S. and Randall, D. J., Eds., Academic Press, San Diego, 1988, 59.
87. **Smith, S.,** Studies in the development of the rainbow trout (*Salmo irideus*) I. The heat production and nitrogenous excretion, *J. Exp. Biol.*, 23, 357, 1947.
88. **Riazi, A., Fremont, L., and Gozzelino, M. T.,** Characterization of egg yolk proteins from rainbow trout *Salmo gairdneri* (Rich.), *Comp. Biochem. Physiol. [B]*, 89, 399, 1988.
89. **Kamler, E.,** *Early Life History of Fish: an Energetics Approach*, Chapman & Hall, London, 1992.
90. **Holeton, G. F.,** Respiratory and circulatory responses of rainbow trout larvae to carbon monoxide and to hypoxia, *J. Exp. Biol.*, 55, 683, 1971.
91. **Peterson, R. H.,** Pectoral fin and opercular movements of Atlantic salmon (*Salmo salar*) alevins, *J. Fish. Res. Bd. Can.*, 32, 643, 1975.
92. **McDonald, D. G. and McMahon, B. R.,** Respiratory development in Arctic char *Salvelinus alpinus* under conditions of normoxia and chronic hypoxia, *Can. J. Zool.*, 55, 1461, 1977.
93. **Rice, S. D. and Stokes, R. M.,** Metabolism of nitrogenous wastes in the eggs and alevins of rainbow trout, *Salmo gairdneri* Richardson, in *The Early Life History of Fish*, Blaxter, J. H. S., Ed., Springer-Verlag, New York, 1974, 325.
94. **Wright, P. A., Felskie, A., and Anderson, P. M.,** Induction of ornithine-urea cycle enzymes and nitrogen metabolism and excretion in rainbow trout (*Oncorhynchus mykiss*) during early life stages, *J. Exp. Biol.*, 198, 127, 1995.
95. **Griffith, R. W.,** Guppies, toadfish, lungfish, coelacanths and frogs: a scenario for the evolution of urea retention in fishes, *Environ. Biol. Fish.*, 32, 199, 1991.
96. **Wright, P. A. and Wood, C. M.,** Muscle ammonia stores are not determined by pH gradients, *Fish Physiol. Biochem.*, 5, 159, 1988.
97. **Solbé, J. F. and Shurben, D. G.,** Toxicity of ammonia to early life stages of rainbow trout (*Salmo gairdneri*), *Water Res.*, 23, 127, 1989.

98. **Bergerhouse, D. L.** Lethal effects of elevated pH and ammonia on early life stages of walleye, *North Am. J. Fish. Mgmt.*, 12, 356, 1992.
99. **Robinette, H. R.**, Effect of selected sublethal levels of ammonia on the growth of channel catfish (*Ictalurus punctatus*), *Prog. Fish Cult.*, 38, 26, 1976.
100. **Broderius, S., Drummond, R., Fiandt, J., and Russom, C.**, Toxicity of ammonia to early life stages of the smallmouth bass at four pH values, *Environ. Toxicol. Chem.*, 4, 87, 1985.
101. **Bader, J. A. and Grizzle, J. M.**, Effects of ammonia on growth and survival of recently hatched channel catfish, *J. Aquat. Anim. Health*, 4, 17, 1992.
102. **Paynter, K. T., DiMichele, L., Hand, S. C., and Powers, D. A.**, Metabolic implication of Ldh-B genotype during early development in *Fundulus heteroclitus*, *J. Exp. Zool.*, 257, 24, 1991.
103. **Finn, R. N., Fyhn, H. J., and Evjen, M. S.**, Respiration and nitrogen metabolism of Atlantic halibut eggs (*Hippoglossus hippoglossus*), *Mar. Biol.*, 108, 11, 1991.
104. **Ronnestad, I., Fyhn, H. J., and Gravningen, K.**, The importance of free amino acids to the energy metabolism of eggs and larvae of turbot (*Scophthalmus maximus*), *Mar. Biol.*, 114, 517, 1992.
105. **Ronnestad, I., Groot, E. P., and Fyhn, H. J.**, Compartmental distribution of free amino acids and protein in developing yolk-sac larvae of Atlantic halibut (*Hippoglossus hippoglossus*), *Mar. Biol.*, 116, 349, 1993.
106. **Wright, P. A., Heming, T. A., and Randall, D. J.**, Downstream pH changes in water flowing over the gills of rainbow trout, *J. Exp. Biol.*, 126, 499, 1986.
107. **Lin, H. and Randall, D. J.**, H+-ATPase activity in crude homogenates of fish gill tissue: Inhibitor sensitivity and environmental and hormonal regulation, *J. Exp. Biol.*, 180, 163, 1993.
108. **Lin, H. and Randall, D. J.**, Evidence for the presence of an electrogenic proton pump on the gill epithelium, *J. Exp. Biol.*, 161, 119, 1991.
109. **Randall, D. J., Lin, H., and Wright, P. A.**, Gill water flow and the chemistry of the boundary layer, *Physiol. Zool.*, 64, 26, 1991.
110. **Dépêche, J., Gilles, R., Daufresne, S., and Chapello, H.**, Urea content and urea production via the ornithine-urea cycle pathway during development of two teleost fishes, *Comp. Biochem. Physiol. [A]*, 63, 51, 1979.
111. **Wourms, J. P. and Lombardi, J.**, Reflections on the evolution of piscine viviparity, *Am. Zool.*, 32, 276, 1992.
112. **Schindler, J. F. and Hamlett, W. C.**, Maternal-embryonic relations in viviparous teleosts, *J. Exp. Zool.*, 266, 378, 1993.
113. **Lombardi, J. and Wourms, J. P.**, The trophotaenial placenta of a viviparous goodeid fish. III. Protein uptake by trophotaeniae, the embryonic component, *J. Exp. Zool.*, 236, 165, 1985.
114. **Schindler, J. F. and Greven, H.**, Protein-gold transport in the endocytic complex of trophotaenial absorptive cells in the embryos of a goodeid teleost, *Anat. Rec.*, 233, 387, 1992.
115. **Veith, W. J.**, Reproduction in the live-bearing teleost *Clinus superciliosus*, *S. Afr. J. Zool.*, 14, 208, 1979.
116. **Veith, W. J.**, Viviparity and embryonic adaptations in the teleost *Clinus superciliosus*, *Can. J. Zool.*, 58, 1, 1980.
117. **deVlaming, V., Baltz, D., Anderson, S., Fitzgerald, R., Delahunty, G., and Barkley, M.**, Aspects of embryo nutrition and excretion among viviparous embiotocid teleosts: potential endocrine involvements, *Comp. Biochem. Physiol. [A]*, 76, 189, 1983.
118. **Wourms, J. P., Grove, B. D., and Lombardi, J.**, The maternal-embryonic relationships in viviparous fishes, in *Fish Physiology: The Physiology of Developing Fish*, Vol. 11B, Hoar, W. S. and Randall, D. J., Eds., Academic Press, New York, 1988, 1.
119. **Grove, B. D. and Wourms, J. P.**, The follicular placenta of the viviparous fish *Heterandia formosa* I. Ultrastructure and development of the embryonic absorptive surface, *J. Morphol.*, 209, 265, 1991.
120. **Boehlert, G. W. and Yoklavich, M. M.**, Reproduction, embryonic energetics, and the maternal-fetal relationship in the viviparous genus *Sebastes* (Pisces: Scorpaenidae), *Biol. Bull.*, 167, 354, 1984.
121. **Dygert, P. H. and Gunderson, D. R.**, Energy utilization by embryos during gestation in viviparous copper rockfish, *Sebastes caurinus*, *Environ. Biol. Fish.*, 30, 165, 1991.
122. **Wourms, J. P.**, Reproduction and development of *Sebastes* in the context of the evolution of piscine viviparity, *Environ. Biol. Fish.*, 30, 111, 1991.
123. **Shimizu, M., Kusakari, M., Yoklavich, M. M., Boehlert, G. W., and Yamada, J.**, Ultrastructure of the epidermis and digestive tract in *Sebastes* embryos, with special reference to the uptake of exogenous nutrients, *Environ. Biol. Fish.*, 30, 155, 1991.
124. **Kristoffersen, R., Broberg, S., and Pekkarinen, M.**, Histology and physiology of embryotrophe formation, embryonic nutrition and growth in the eel-pout *Zoarces viviparus*, *Ann. Zool. Fennici*, 10, 467, 1973.
125. **Korsgaard, B.**, Trophic adaptations during early intraovarian development of embryos of *Zoarces viviparus* (L.), *J. Exp. Mar. Biol. Ecol.*, 98, 141, 1986.
126. **Korsgaard, B.**, The chemical composition of follicular and ovarian fluids of the pregnant blenny *Zoarces viviparus* (L.), *Can. J. Zool.*, 61, 1101, 1983.
127. **Korsgaard, B. and Andersen, F. O.**, Embryonic nutrition, growth and energetics in *Zoarces viviparus* as indication of a maternal-fetal trophic relationship, *J. Comp. Physiol. [B]*, 155B, 437, 1985.

128. **Korsgaard, B.,** Amino acid uptake and metabolism by embryos of the blenny *Zoarces viviparus*, *J. Exp. Biol.*, 171, 315, 1992.
129. **Yoklavich, M. M. and Boehlert, G. W.,** Uptake and utilization of ^{14}C-glycine by embryos of *Sebastes melanops*, *Environ. Biol. Fish.*, 30, 147, 1991.
130. **Korsgaard, B.,** Nitrogen distribution and excretion during embryonic post-yolk sac development in *Zoarces viviparus*, *J. Comp. Physiol. [B]*, 164B, 42, 1994.
131. **Dépêche, J. and Schoffeniels, E.,** Changes in electrolytes, urea and free amino acids of *Poecilia reticulata* embryos following high salinity adaptation of the viviparous female, *Biochem. Syst. Ecol.*, 3, 111, 1975.
132. **Kormanik, G.,** Nitrogen budget in developing embryos of the spiny dogfish *Squalus acanthias*, *J. Exp. Biol.*, 144, 583, 1989.
133. **Kaushik, S. J., Dabrowski, K., and Luquet, P.,** Patterns of nitrogen excretion and oxygen consumption during ontogenesis of common carp (*Cyprinus carpio*), *Can. J. Fish. Aquat. Sci.*, 39, 1095, 1982.
134. **Schindelmeiser, J. and Greven, H.,** Nitrogen excretion in intra- and extrauterine larvae of the ovoviviparous salamander, *Salamandra salamandra* (L.) (Amphibia: Urodela), *Comp. Biochem. Physiol. [A]*, 70A, 563, 1981.
135. **Dépêche, J.,** Acquisition et limites et l'autonomie trophique embryonnaire au cours du développement du téléostéen vivipare (*Poecilia reticulata*), *Bull. Biol. Fr. Belg.*, 110, 45, 1976.
136. **Evans, D. H., Oikari, A., Kormanik, G. A., and Mansberger, L.,** Osmoregulation in prenatal dogfish, *Squalus acanthias*, *J. Exp. Biol.*, 101, 295, 1982.
137. **Cornish, D. A. and Veith, W. J.,** Embryonic adaptations and nutrition in the viviparous teleost *Clinus dorsalis*, *S. Afr. J. Zool.*, 21, 79, 1986.
138. **Read, L. J.,** Ornithine-urea cycle enzymes in early embryos of the dogfish *Squalus suckleyi* and the skate *Raja binoculata*, *Comp. Biochem. Physiol.*, 24, 669, 1968.
139. **Perlman, D. F. and Goldstein, L.,** Nitrogen Metabolism, in *Physiology of Elasmobranch Fishes*, Shuttleworth T. J. Ed., Springer-Verlag, Berlin, 1988, 253.
140. **Sheehan, R. J. and Lewis, W. M.,** Influence of pH and ammonia salts on ammonia toxicity and water balance in young channel catfish, *Trans. Am. Fish. Soc.*, 115, 891, 1986.
141. **Levi, G., Morisi, G., Coletti, A., and Catanzaro, R.,** Free amino acids in fish brain: normal levels and changes upon exposure to high ammonia concentrations *in vivo*, and upon incubation of brain slices, *Comp. Biochem. Physiol. [A]*, 49, 623, 1974.

Chapter 17

Uptake and Effects of Nitrite and Nitrate in Animals

Frank B. Jensen

CONTENTS

INTRODUCTION

Nitrite (NO_2^-) and nitrate (NO_3^-) are recognized as fundamental components of the nitrogen cycle in ecosystems but are rarely considered in relation to the metabolism and turnover of nitrogen in animals. These inorganic nitrogen compounds can, however, contribute significantly to the whole animal N-pool in certain environments or situations. Nitrite and nitrate are taken up by animals directly from the environments or via the food. Additionally, some nitrite and nitrate are endogenously produced.

Toxicological studies have extensively documented that nitrite is toxic to animals. An understanding of the mechanisms of nitrite toxicity demands insight into the sublethal and lethal physiological effects induced by nitrite. Pathophysiological focus has been on the ability of nitrite to oxidize hemoglobin to methemoglobin and on the possible role of nitrite in the formation of mutagenic and carcinogenic compounds. Recent research on aquatic animals has revealed several additional effects, which could be important in the toxicity of nitrite. Nitrate, in contrast to nitrite, is relatively nontoxic, but an elevated nitrate load may produce potential harmful effects via an endogeneous conversion of nitrate to nitrite.

The purpose of this chapter is to summarize current knowledge on uptake and physiological effects of nitrite and nitrate in animals and to point out some relevant questions for future studies. Prime focus is on aquatic vertebrates (fish) and invertebrates (crustaceans) but terrestrial mammals are also considered.

0-8493-8411-7/95/$0.00+$.50

NITRITE AND NITRATE IN AQUATIC ENVIRONMENTS

Nitrite concentrations are typically below 1 μ*M* in unpolluted waters. Elevated concentrations are developed in certain hypoxic habitats, in habitats receiving nitrogeneous effluents from sewage plants and industries, and in aquaculture systems, as a result of an imbalance in either bacterial denitrification or nitrification processes.[1] Anaerobic denitrifying bacteria reduce nitrate to N_2 with the intermediate production of nitrite. This production of nitrite raises nitrite concentrations in the oxygen minimum zone of the oceans and in other severely hypoxic or anoxic habitats.[2]

Aerobic nitrifying bacteria oxidize ammonia to nitrite (*Nitrosomonas* sp.) and nitrite to nitrate (*Nitrobacter* sp.). Upon discharge of ammonia to ecosystems, a transient build up of nitrite is likely to occur, since *Nitrosomonas* tends to grow faster than *Nitrobacter*, and since *Nitrobacter* is inhibited by an elevated level of free NH_3.[3] A similar argument applies to the establishment of nitrification in biological filters in recirculated aquaculture systems with fish. Ammonia excreted from the fish results in an initial peak in water ammonia concentration, which is succeeded by a decline in ammonia and a rise in nitrite concentrations upon growth of *Nitrosomonas*. The nitrite concentration may increase to values of 1 m*M* or above for a few days before growth of *Nitrobacter* decreases the nitrite concentration.[4,5] Nitrate accumulates to reach concentration in the millimolar range, if a denitrifying step is not included in the system.[4,5]

Water nitrate concentrations span the micromolar to the millimolar range. The primary source of elevated nitrate levels in aquatic ecosystems and in groundwater is agricultural use of fertilizers.[6] Other minor sources include combustion of fossil fuels, which emit nitrogen oxides (NO and NO_2) that are oxidized to nitric acid in the atmosphere and return to the surface as H^+ and NO_3^- in acid rain.[7] The WHO guideline for maximum nitrate concentration in drinking water is 10 mg NO_3^--N l^{-1} (0.7 m*M*). This concentration is exceeded in many agricultural areas.[6]

UPTAKE OF NITRITE IN AQUATIC ANIMALS

When freshwater fish and crayfish are exposed to nitrite they accumulate it in the extracellular fluid to concentrations that far exceed the ambient concentration (Figure 1).[8-14] The uptake of nitrite is

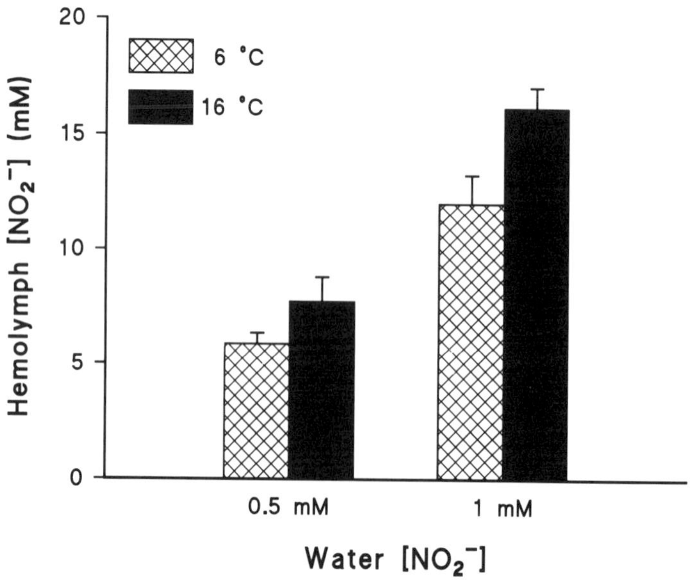

Figure 1 Hemolymph nitrite concentrations in crayfish *Astacus astacus* after 7 d of exposure to 0.5 and 1 m*M* ambient nitrite at two acclimation temperatures (6 and 16°C). Means ± SEM, *N* = 6 to 10 in each group. Water [Cl^-]: 1.3 m*M*. Water [Ca^{++}]: 3.5 m*M*. Water pH: approximately 8.3 (at 16°C). (From Jeberg, M. V. and Jensen, F. B., *Aquat. Toxicol.*, 29, 65, 1994.)

against a concentration and electrochemical gradient and appears to occur via the active chloride uptake mechanism in the gills. Thus, Cl^- and NO_2^- influx measurements reveal that nitrite is a competitive inhibitor of chloride uptake and vice versa.[13,15] The common uptake route for Cl^- and NO_2^- explains why an elevated ambient chloride level protects against nitrite accumulation and toxicity, and why fish species with high branchial chloride uptake rates (e.g., rainbow trout, perch, and pike) are more sensitive to nitrite than species with low chloride uptake rates (e.g., carp, tench, and eel).[8,14-16]

Additional evidence for nitrite entering freshwater animals via the branchial chloride uptake mechanism can be provided by studying nitrite uptake in situations where the branchial chloride uptake mechanism is modulated. One such situation is environmental hypercapnia. Acid–base compensation of a respiratory acidosis induced by environmental hypercapnia is partly or fully mediated by a reduced branchial Cl^- uptake, that increases the extracellular "strong ion difference," leading to an elevated extracellular $[HCO_3^-]$ and pH recovery.[17] Thus, if the nitrite and chloride influx is via the same route, nitrite uptake can be expected to decrease during hypercapnia. Preliminary data on crayfish *Astacus astacus* support this hypothesis. Following 24 h of nitrite exposure, the hemolymph nitrite concentration was significantly lower in hypercapnic than in normocapnic crayfish (Figure 2). Since the extracellular acidosis was only partially compensated after 24 h of hypercapnia, the difference in nitrite accumulation may be even larger during longer exposures.

The possibility exists that some nitrite enters aquatic animals via diffusion of HNO_2. The pK of nitrous acid is, however, about 3.3, whereby only a very small fraction of nitrite will be protonated at natural water pH.[1] In crayfish accumulating nitrite to 16 m*M* when exposed to 1 m*M* NO_2^- at 16°C in water with a pH of 8.3 (Figure 1), the hemolymph pH is about 7.9 to 8.[12] The lower pH and higher NO_2^- inside than outside the animals imply that the HNO_2 gradient in the animals is outward rather than inward. A contribution of HNO_2 diffusion to nitrite uptake has not been convincingly demonstrated, and therefore it must be assumed to be insignificant in most cases.[18]

By being hyperosmotic to the environment, freshwater animals inevitably have an osmotic influx of water that must be eliminated via the urine. Nitrite should be filtered in the kidney, as is other small molecules, which offers the possibility of an urinary elimination of nitrite. Freshwater teleost and crayfish are known to reabsorb NaCl in the kidney and to excrete a dilute urine. The high degree

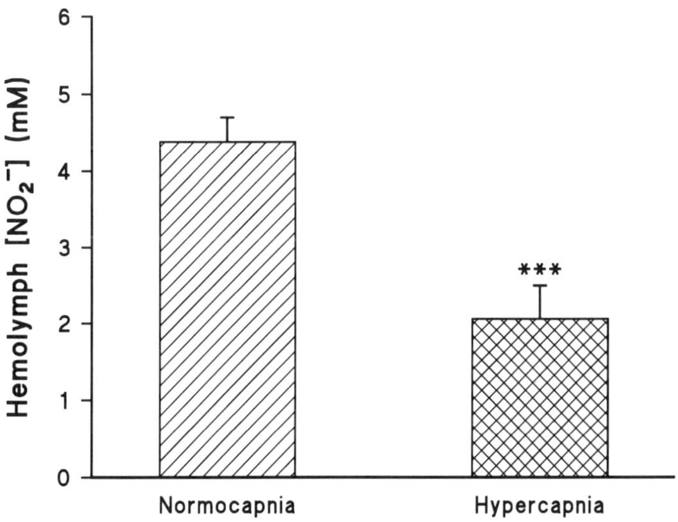

Figure 2 Hemolymph nitrite concentrations in crayfish *(A. astacus)* exposed to 1 m*M* ambient nitrite for 24 h under normocapnic (hemolymph P_{CO_2} = 3.5 ± 0.2 mmHg, *N* = 11) and hypercapnic (hemolymph P_{CO_2} = 10.2 ± 0.3 mmHg, *N* = 9) conditions at 15°C. Means ± SEM. Asterisks indicate a significant difference between values (P <0.001). (From Jensen, F. B. and Bach, A., unpublished data, 1994).

of nitrite accumulation that is observed in the extracellular fluid of fish and crayfish suggests that NO_2^- has an affinity for the kidney Cl^- uptake mechanism so that filtered nitrite is reabsorbed. This idea has not been experimentally investigated, however.

The degree of nitrite accumulation in a given species depends on the ambient nitrite level, exposure time, and various physicochemical conditions in the environment. The ambient chloride level (or the ratio between ambient chloride and nitrite concentrations) is particularly important. Nitrite can be accumulated to the millimolar range during exposure to water nitrite concentrations in the micromolar range, if water chloride levels are also in the micromolar range.[9] Conversely, exposure to nitrite concentrations in the millimolar range can be tolerated for long periods, if water chloride concentrations are adequately elevated.[8] The protective role of a high ambient chloride concentration means that nitrite uptake and toxicity in general is lower in marine than in freshwater animals.[19-23] Nitrite does enter seawater animals but reported extracellular concentrations are below or close to the ambient concentration.[19,20,23] One exception is marine shrimp which quickly accumulate nitrite in the hemolymph to higher concentrations than in the water.[24] In addition to chloride, ambient bicarbonate and calcium also provide some protection against nitrite toxicity.[18] Few studies have addressed the influence of temperature. In crayfish *A. astacus* an increase in acclimation temperature significantly increases nitrite accumulation (Figure 1).[25]

Species differences in nitrite uptake can result from species differences in branchial chloride uptake rates. Additionally, it is possible that the relative affinity of nitrite and chloride for the uptake sites varies between species. Nitrite uptake appears to be unusually low in largemouth bass and striped bass.[10,26] The combined action of a low branchial permeability, a low Cl^- uptake rate, and a low affinity of nitrite for the uptake mechanism would result in the apparent exclusion of nitrite seen in these species.

Under the same ambient conditions, freshwater crayfish accumulate nitrite faster and to much higher concentrations than freshwater fish. After four days of exposure to 1 m*M* nitrite at 16°C in water with 1.3 m*M* chloride, crayfish *A. astacus* have hemolymph nitrite concentrations of about 14 m*M*, whereas rainbow trout surviving in the same type of water for 4 d have plasma nitrite concentrations of about 2.5 m*M*.[23] The high uptake rate in crayfish could result from a comparatively high branchial Cl^- uptake rate, and may additionally relate to a relatively high Cl^-/Cl^- self-exchange component in crayfish gills, which changes to NO_2^-(in)/Cl^-(out) exchange in a nitrite-containing environment.[23,25,27] During continued sublethal nitrite exposure, steady extracellular nitrite concentrations are reached in crayfish, suggesting that a balance is established between uptake and loss.[12,25] Loss of nitrite from the extracellular compartment is composite of passive efflux to the environment across the gills, excretion (urinary loss), influx to intracellular compartments, and conversion of nitrite to other compounds.

UPTAKE OF NITRATE IN AQUATIC ANIMALS

Nitrate has received considerable interest in relation to problems with environmental eutrophication. However, uptake and effects of nitrate in aquatic animals has received surprisingly little attention. One reason may be that the toxicity of nitrate is very low in most aquatic species, with ambient nitrate concentrations of several tens of millimolars required to elevate mortality in short-term toxicity tests.[28,29] A noticeable exception is marine shrimp *Penaeus monodon* larvae, where a significant mortality occurs with nitrate levels as low as 16 μM.[30]

The extensive uptake of nitrite in freshwater teleosts and crustaceans raises the question whether nitrate, which typically is present in aquatic habitats at higher concentrations than nitrite, is similarly taken up and accumulated. This question was recently investigated in crayfish *A. astacus* and in rainbow trout. Following exposure to 1 m*M* ambient nitrate in hard freshwater, the extracellular nitrate concentration increased significantly, with higher levels being reached in rainbow trout than in crayfish (Figure 3). The extracellular nitrate concentrations, however,

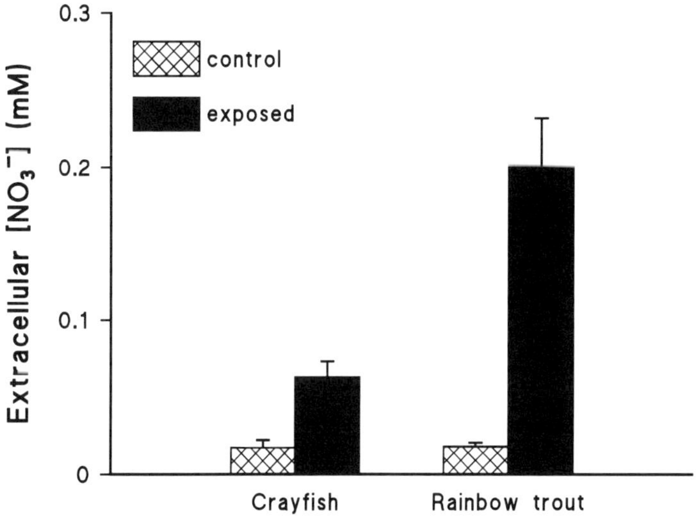

Figure 3 Extracellular nitrate concentrations in crayfish *(A. astacus)* and rainbow trout *(Oncorhynchus mykiss)* following 7 d of exposure to water with NO_3^- concentrations of 28 μM ("control") or 1 m*M* ("exposed"). Means ± SEM (*N* = 6 and 5 for crayfish and rainbow trout groups, respectively). Temperature: 15°C. Water [Cl^-]: 1.3 m*M*. Water [Ca^{++}]: 3.5 m*M*. (Crayfish: From Jensen, F. B., unpublished data, 1993; Rainbow trout: From Stormer, J., Rankin, J. C., and Jensen, F. B., unpublished data, 1993.).

stayed well below the ambient level in both species even after 7 d of exposure (Figure 3). This finding suggests that there is no extensive uptake of nitrate via the branchial Cl^- uptake mechanism, and that passive entry is also low. The minor uptake of nitrate could be fundamental to the low toxicity of nitrate. It also suggests that sublethal effects may be small. This question, however, requires further investigation.

Even though freshwater animals in general appear to have a low nitrate uptake, the uptake may depend on environmental pH. Thus, in contrast with the low nitrate uptake observed in crayfish *A. astacus* in hard freshwater at pH 8.3 (Figure 3), it has been observed that hemolymph nitrate concentrations increase well above ambient levels in crayfish *Procambarus clarki* in water acidified to pH 4 with nitric acid (0.1 m*M*).[31]

DIETARY INTAKE

Dietary intake of nitrate and nitrite is of prime concern in humans because of the potential health effects. A number of vegetables, including spinach, celery, radish, and carrot have high nitrate levels and also contain some nitrite.[32] Likewise, nitrite is used as a food additive to preserve meat, prevent botulism, and/or to improve meat color. This practice was introduced around 1700 by addition of saltpetre (nitrate) that is reduced to nitrite *in situ*.[33] Intake of nitrate via drinking water contributes significantly to the total nitrate intake. The use of fertilizers for agricultural purposes has led to nitrate contamination of groundwater, increasing nitrate concentrations in drinking water above recommended levels in many places.[6]

Health consequences in connection with a high nitrate intake are not due to the nitrate ion itself but to the reduction of nitrate to nitrite, which occurs via microorganisms in salivary fluid in the mouth as well as in the gastrointestinal tract.[32,34] Individuals with an achlorhydric stomach frequently have elevated gastic nitrite concentrations, since a reduced acid production and elevated stomach pH improves the survival of nitrate-reducing bacteria.[32,35]

The significance of dietary intake of nitrite or nitrate remains to be elucidated in aquatic animals. One issue is the consequences for predators of ingesting large amounts of nitrite when feeding on nitrite-contaminated prey. A further relevant topic is the uptake of nitrite across the intestine in marine fishes, who require high drinking rates in their osmoregulation.

ENDOGENOUS PRODUCTION

In addition to the direct and/or dietary entry of nitrite and nitrate into animals, some nitrite and nitrate can be produced endogenously. Activated mammalian macrophages synthesize nitrite and nitrate from L-arginine with the nitrogens of nitrite and nitrate originating from the guanido nitrogens of arginine.[36] The synthesis may involve an initial conversion of L-arginine to nitric oxide and L-citrulline via nitric oxide synthase and a subsequent oxidative metabolism of NO to nitrite and nitrate[37] (see also chapter 18). Following activation of macrophages by bacterial lipopolysaccharide both blood and urinary levels of nitrate increase.[38] The production of nitric oxide, nitrite, and nitrate is of interest in relation to the cytotoxic actions of activated macrophages.

UPTAKE INTO INTRACELLULAR COMPARTMENTS

Nitrite penetrates cell membranes and is bioaccumulated not only in the extracellular space but also in gill, liver, brain, and muscle tissues of fish.[9] The transport of nitrite across the red blood cell membrane attracts particular interest, since it is fundamental to the subsequent intracellular oxidation of hemoglobin to methemoglobin by nitrite.

In carp red cells, the membrane permeation of nitrite is strongly oxygenation and pH dependent. Nitrite rapidly enters deoxygenated red cells, whereas entry into oxygenated cells is limited at physiological pH.[39,40] When pH is lowered, nitrite also enters oxygenated red cells but at a lower rate than into deoxygenated cells.[40] The nitrite entry is not inhibited by DIDS, suggesting that entry via the band 3 anion exchanger is not important.[39] The extracellular nitrite concentration is much lower than the chloride and bicarbonate concentrations, and therefore competition for the external anion exchange sites does not favor nitrite entry through this route.[40] Nitrite appears to enter carp red cells primarily via conductive transport. This mode of transport accounts for the oxygenation and pH dependencies of the nitrite influx via the influences of oxygenation and pH on the red cell membrane potential. An influx via conductive transport is also consistent with the observed kinetics of nitrite entry and with a linear increase in influx rate with increasing extracellular nitrite concentrations.[23,39,40]

EFFECTS ON RESPIRATORY PIGMENTS

HEMOGLOBIN

Methemoglobin Formation

Nitrite-induced oxidation of functional hemoglobin (Hb, with heme irons in oxidation state II) to methemoglobin (metHb, with heme irons in oxidation state III) is well documented in both mammals and fish.[1,18,41] MetHb cannot bind oxygen whereby a severe methemoglobinemia strongly impedes blood O_2 transport.

At the molecular level, oxidation of oxygenated Hb by nitrite is characterized by an initial lag phase followed by an autocatalytic increase in reaction rate.[41-43] The oxidation mechanism is complex and has been the subject of much research with human Hb. The process proceeds via a series of intermediate reactions involving free radicals, with the radical NO_2 promoting the rapid oxidation of oxyHb during the autocatalytic phase.[42,44,45] The overall stoichiometry for the reaction is[42]:

$$4HbO_2 + 4NO_2^- + 4H^+ \longrightarrow 4Hb^+ + 4NO_3^- + O_2 + 2H_2O \quad (1)$$

The oxidation of mammalian Hb by nitrite is much slower in the absence than in the presence of oxygen, and the reaction mechanisms are different.[41] With rainbow trout Hb, a lowering of HbO_2 saturation (i.e., not complete absence of oxygen) suffices to strongly decrease the rate of metHb

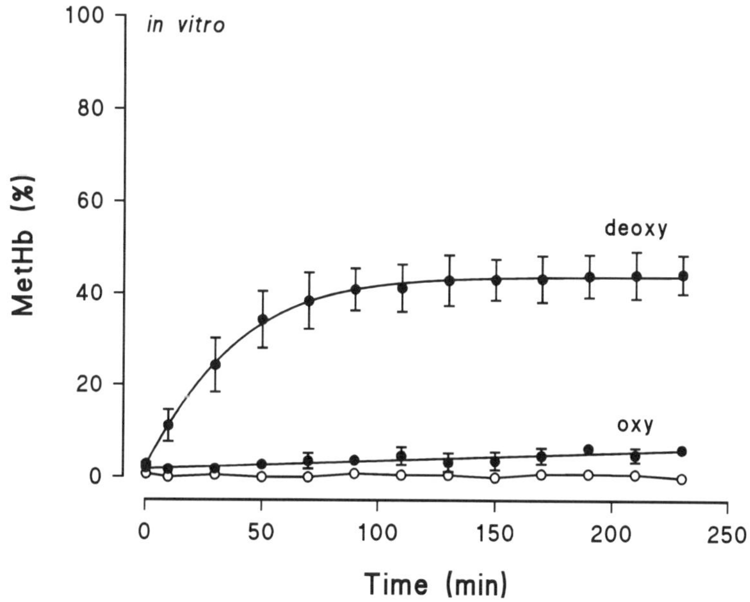

Figure 4 Methemoglobin (MetHb) formation in oxygenated and deoxygenated carp *(Cyprinus carpio)* blood (closed symbols) following addition of nitrite to a concentration of 3 m*M* at time zero. Open symbols refer to control experiments with blood in the absence of nitrite. Blood pH: 8.14. Temperature: 15°C. (From Jensen, F. B., *J. Exp. Biol.*, 171, 349, 1992.)

formation.[43] Nitrite-induced oxidation of trout Hb is also slowed down by increased concentrations of red cell organic phosphates, inorganic ions, NADH and catecholamines, thus showing that the oxidation process is under multifactorial control.[43]

At the red cell level, the rate of nitrite entry into the cells has a dominant influence on metHb formation. Thus, even though nitrite-induced metHb formation is faster with oxygenated than with deoxygenated Hb at the molecular level, metHb formation is faster and more pronounced with deoxygenated than with oxygenated carp red cells after nitrite addition (Figure 4).[39,40] This difference evidently originates in the more rapid entry of nitrite into deoxygenated than into oxygenated carp red cells (see above). The metHb level stabilizes at an elevated quasi steady level thereby suggesting that a balance is developed between nitrite-induced oxidation of hemoglobin and reduction of metHb to functional Hb via red cell metHb reductase systems (Figure 4).[39,40] Although several reductase systems are known, NADH metHb reductase is the most important in both mammals and fish.[41,46,47] The capacity for enzymatic reduction of metHb shows species differences which are likely to influence the degree of methemoglobinemia following nitrite administration.[41]

When fish are exposed to nitrite *in vivo*, the blood metHb content increases in parallel with the consecutive increase in plasma nitrite concentration (Figure 5). At any given time, a quasi balance may exist between nitrite-induced oxidation and enzymatic reduction of Hb but the continuing increase in nitrite load unavoidably forces metHb levels upwards — with metHb levels above 80% of the total Hb eventually being reached (Figure 5).

In mammals, metHb levels fluctuate if the nitrite load fluctuates. Rats receiving 0.5% $NaNO_2$ as drinking water have a daily variation in metHb content between 4% and about 80%, resulting from a nocturnal drinking activity and an effective reduction of oxidized Hb via metHb reductase between drinking events.[48]

MetHb formation in humans in association with a high dietary intake of nitrate primarily concerns infants younger than 6 months. Infants have a higher stomach pH that allows survival of bacteria reducing nitrate to nitrite. Additionally, their hemoglobin is more susceptible to oxidation by nitrite and they have an immature capacity for enzymatic metHb reduction.[6,41]

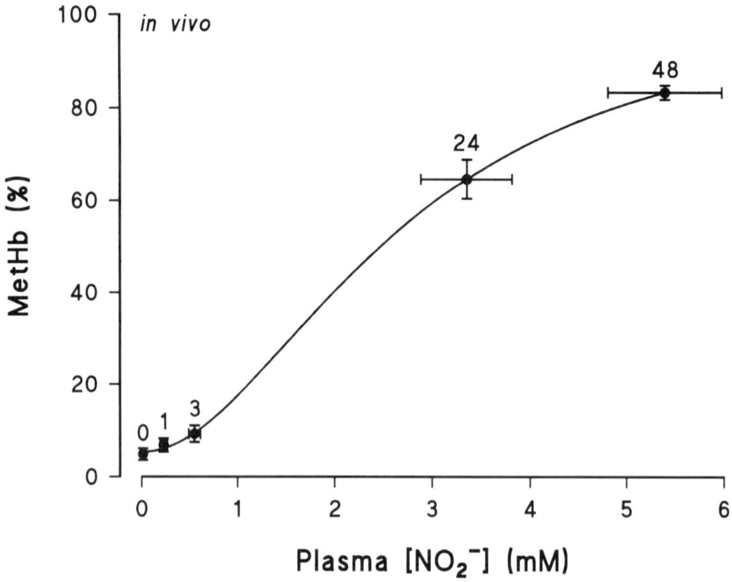

Figure 5 Blood methemoglobin content as a function of plasma nitrite concentration in carp *(Cyprinus carpio)* during *in vivo* exposure to 1 m*M* ambient nitrite. Numbers at points specify the duration (in h) of nitrite exposure. Means ± SEM (*N*=7). Water [Cl^-]: 0.3 m*M*. Water [Ca^{++}]: 1.7 m*M*. Temperature: 15°C. (From Jensen, F. B., Andersen, N. A., and Heisler, N., *J. Comp. Physiol. B.*, 157, 533, 1987.)

The severe methemoglobinemia that develops in nitrite-exposed fish strongly reduces the arterial O_2 content and eventually hampers O_2 delivery to the tissues.[11] Tissue O_2 shortage stimulates intracellular lactic acid production, which is reflected by an increased extracellular lactate concentration (Figure 6). Limitations in blood O_2 transport are particularly critical in situations with additional stress such as exercise or environmental hypoxia, where an optimized O_2 transport is required. Swimming performance in chinook salmon is impaired by high metHb levels.[49] At rest, however, fish can accommodate relatively high levels of unfunctional Hb without mortality. Mortality during nitrite exposure can be associated with both high and only moderately elevated metHb levels.[1,9] Thus, even though methemoglobinemia is an important component of nitrite toxicity, other effects are likely to contribute.

Oxygen Affinity and Total Hb Content

Methemoglobin tends to assume the quaternary R ("oxyHb") structure of hemoglobin, which has a higher O_2 affinity than the T ("deoxyHb") structure.[50,51] This suggests that the oxygen affinity of unoxidized heme groups should be elevated in metHb-containing blood as is indeed observed in mammalian blood.[41,52] In nitrite-exposed carp, however, the opposite response is seen (i.e., a substantial decrease in oxygen affinity).[11,39] This is a cellular effect originating mainly in nitrite-induced red cell shrinkage which decreases oxygen affinity by increasing the cellular concentrations and complex binding of hemoglobin and organic phosphates.[11,39] Further contributions to the decreased oxygen affinity may come from a lowered red cell pH and from binding of nitrite or nitrate to the Hb.[39] The reduced blood oxygen affinity significantly lowers the arterial oxygen content in nitrite-exposed carp.[11,53]

Beta-adrenergic stimulation of fish red cells elicits a rapid and adaptive increase in blood oxygen affinity in hypoxia and other stress situations via an activation of Na^+/H^+ exchange across the red cell membrane that elevates red cell pH and volume.[54] The catecholamine-stimulated H^+ efflux from trout red cells is strongly inhibited by nitrite-induced methemoglobinemia.[55]

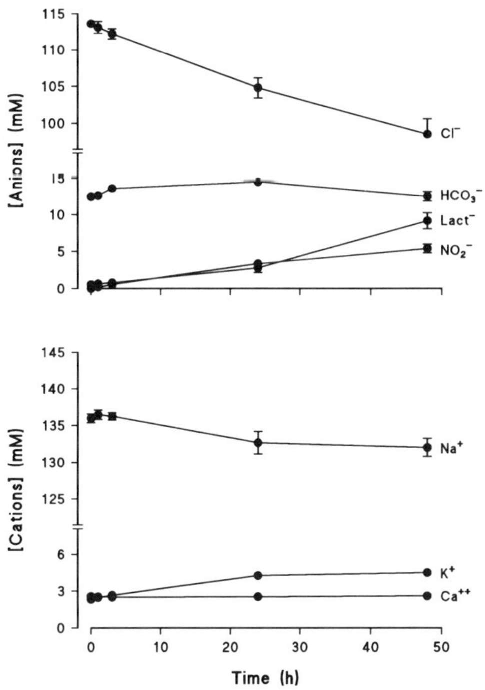

Figure 6 Time-dependent changes in plasma concentrations of anions (chloride, bicarbonate, lactate and nitrite) and cations (sodium, potassium and calcium) in carp exposed to 1 m*M* ambient nitrite at time zero. Means ± SEM (*N* = 7). Water [Cl^-]: 0.3 m*M*. Water [Ca^{++}]: 1.7 m*M*. Temperature: 15°C. (From Jensen, F. B., Andersen, N. A., and Heisler, N., *J. Comp. Physiol. B*, 157, 553, 1987.)

A diminished β-adrenergic response may further compromise blood oxygen transport in nitrite-exposed fish, especially if nitrite exposure occurs in combination with environmental hypoxia.

The total blood Hb content is reduced in nitrite-exposed fish and in mammals who are chronically treated with nitrite.[19,20,39,48,56] This effect may be related to the general deterioration of red cell functions, causing an increased removal of unfunctional red cells from the circulation.

OTHER HEME PROTEINS

In addition to hemoglobin, several other heme proteins are affected by nitrite. Nitrite impedes cellular O_2 transport and consumption by inactivating or inhibiting cellular myoglobin and mitochondrial cytochrome oxidase.[41,57,58] Nitrite also interacts with the heme of cytochrome P-450 and inhibits cytochrome P-450 mediated mixed-function oxidase activity in rabbit liver microsomes under anaerobic conditions.[59] This may influence drug metabolism in nitrite-intoxicated animals, where liver hypoxia is likely to occur. The synthesis of eicosanoids, which are "local hormones" governing many physiological processes, also appears to be influenced by nitrite.[40] This could be via an effect on prostaglandin synthetase, which requires a heme group for activity.

HEMOCYANIN

Crustacean hemocyanin (Hc) reversibly binds oxygen to binuclear copper sites with the copper atoms changing oxidation state upon binding and release of oxygen (Cu^ICu^I in deoxyHc and $Cu^{II}O_2^{2-}Cu^{II}$ in oxyHc). Nitrite can react with crayfish hemocyanin forming metHc. The reaction, however, occurs predominantly at low pH and in the presence of a large excess of nitrite and appears to be unimportant at physiological pH.[12,60,61] The hemolymph Hc concentration is decreased in nitrite-exposed crayfish.[12] Oxygen transport by hemocyanin is, however, generally much less affected by nitrite than is the case with oxygen transport by hemoglobin. This conclusion is supported by the finding of low lactate concentrations in nitrite-exposed crayfish.[12,13]

CARDIOVASCULAR EFFECTS

Nitrite relaxes vascular smooth muscle in mammals, leading to vasodilation.[62] The response resembles that evoked by nitrovasodilators such as nitroglycerin and organic nitrates, which are used therapeutically in humans. The cellular mechanism involves a stimulation of guanylate cyclase, probably via nitric oxide that is metabolically cleaved from the nitrovasodilators, with subsequent accumulation of cGMP that stimulates vascular smooth muscle relaxation in veins and arteries[63,64] (see also Chapter 18). Nitric oxide is recognized as an important endogenously produced physiological modulator that apart from acting as a local regulator of blood flow also inhibits platelet aggregation and influences microvascular permeability.[65,66] Since NO may be generated from nitrite, it is conceivable that effects induced by nitric oxide may be mimicked by nitrite.

A vasodilation-induced decrease in blood pressure can be counteracted by an increased cardiac pumping. Also, an increase in cardiac output would compensate for the substantial decrease in blood O_2 capacitance caused by nitrite-induced methemoglobinemia. It remains to be established whether cardiac activity is changed in nitrite-exposed fish. The topic deserves study not only with regard to the requirement for blood O_2 transport but also to the possibility that O_2 lack and electrolyte changes (see below) have a negative influence on myocardial function.

EFFECTS ON ELECTROLYTE BALANCE

Chloride homeostasis in freshwater fish and crayfish depends on an active Cl^- uptake across the gills that balances passive and urinary Cl^- loss. During nitrite exposure, the branchial Cl^- uptake is partly shifted to NO_2^- uptake, whereas the passive Cl^- loss persists. A net Cl^- efflux, resulting in Cl^- depletion, is therefore likely to occur. A major reduction in extracellular chloride levels has indeed been observed in nitrite-exposed carp and crayfish.[11-13]

The decrease in the extracellular chloride concentration is larger than the rise in nitrite concentration in both carp and crayfish.[11-13] In carp, the decrease in chloride concentration is stoichiometrically balanced by increases in nitrite and lactate concentrations and changes in the bicarbonate concentration (Figure 6), with the sum of anions staying constant.[11] In crayfish *A. astacus*, accumulation of nitrite to about 10 m*M* is associated with a decrease in the extracellular chloride concentration by about 40 m*M*.[12] In this case lactate levels do not increase. An analysis of the anion and cation changes on basis of the requirement for electroneutrality suggests that some unknown anion must increase in concentration during nitrite exposure. Our recent results show that hemolymph nitrate concentrations increase to very high levels in nitrite-exposed *A. astacus*, proposing that a significant fraction of the nitrite is converted to nitrate inside the animal.[67] Some nitrate is also formed from nitrite in mammals in the reaction between oxyHb and nitrite (see above) as well as by hepatocytes.[42,48,68]

Nitrite exposure increases the number of gill chloride cells in rainbow trout.[69,70] This response may be both beneficial by partly offsetting the chloride depletion via an increased chloride uptake capacity, and maladaptive by increasing nitrite uptake. The proliferation of chloride cells could be mediated by cortisol, which increases in concentration in nitrite-exposed fish.[71,72]

Nitrite-induced perturbations of cation regulation notably concerns potassium. In carp, a sustained elevation of the extracellular K^+ level is induced by nitrite (Figure 6).[11] Potassium is lost from the red cells and other intracellular compartments thereby elevating the extracellular level.[39] In crayfish, nitrite exposure induces a large decrease in muscle K^+ content (Figure 7). Therefore, potassium efflux from intracellular compartments may be a general response to nitrite.[12] Species and tissue specific differences can be anticipated, however. Potassium is lost from crayfish muscle but not from the midgut gland.[25]

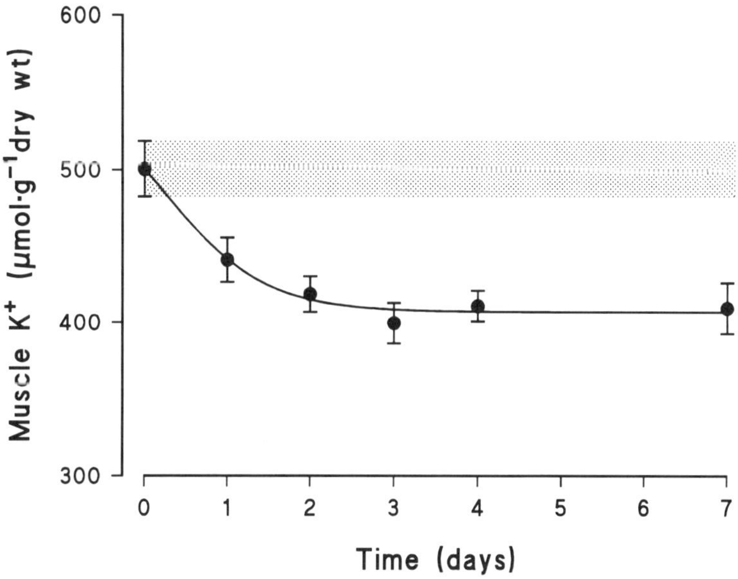

Figure 7 Potassium content in tail muscle of crayfish *(A. astacus)* during exposure to 0.8 m*M* ambient nitrite at 15°C. (From Jensen, F. B., *Aquat. Toxicol.*, 18, 51, 1990.)

The mechanism underlying cellular K^+ efflux has been investigated in carp red cells.[39,40] Nitrite-induced methemoglobinemia stimulates a KCl cotransport mechanism, leading to K^+ and Cl^- loss that drags osmotically obliged water and leads to red cell shrinkage.[40] The KCl cotransporter that is activated closely resembles that involved in cell volume regulation.[73] A net KCl loss may also apply to crayfish muscle. Thus, whereas the muscle K^+ content decreases, other muscle cations are unaffected, suggesting that the positive charge of potassium is balanced by a parallel loss of Cl^- rather than by a parallel uptake of cations.[12]

The extracellular sodium concentration decreases in nitrite-exposed carp and crayfish. The sodium change is small in carp (Figure 6), and somewhat larger in crayfish, but it is always smaller than the parallel occurring decrease in chloride concentration.[11-13] Extracellular calcium does not change during sublethal nitrite-exposure in either carp (Figure 6) or *A. astacus*, whereas a severe nitrite accumulation in crayfish *Pacifastacus leniusculus* raises the hemolymph calcium concentration, perhaps due to exoskeletal buffering of a parallel occurring acidosis.[11-13]

The disturbances in ionic regulation may contribute significantly to the toxicity of nitrite. Thus, chloride depletion only can be tolerated to a certain extent, and changes in extracellular and/or intracellular K^+ levels predictably affect membrane potentials, neurotransmission, skeletal and heart muscle functions, and intracellular metabolism.

ACID–BASE BALANCE

Extracellular acid–base changes during sublethal nitrite exposure in carp are characterized by a minor alkalosis, resulting from a transient increase in plasma $[HCO_3^-]$ and a small decrease in the arterial CO_2 tension.[11] The decrease in arterial P_{CO_2} is accompanied by a rise in P_{O_2}, suggesting hyperventilation during nitrite exposure. This hyperventilation is larger in combination with environmental hypoxia.[74] Even though extracellular pH increases, the red cell pH is unchanged, delineating a reduced distibution ratio of H^+ across the red cell membrane and a relative red cell acidosis.[39]

The finding of an extracellular alkalosis rather than an acidosis is interesting on basis of the gradual increase in plasma lactate levels (Figure 6). The production of lactic acid and the transfer of H^+ and lactate from intra- to extracellular compartments appears to occur at a rate that does not exceed the capacity of the branchial acid–base regulatory mechanisms to mediate the transfer of excess hydrogen ions to the water. A branchial transfer of acid–base equivalents via Cl^-/HCO_3^- exchange would explain the decrease in plasma Cl^- that is associated with the rise in plasma lactate.[11]

Sublethal nitrite exposure in *A. astacus* elicits a small respiratory alkalosis, whereas a more severe nitrite accumulation in *P. leniusculus* elevates hemolymph P_{CO_2} and decreases pH.[12,13] The hemolymph lactate concentration remains unchanged in both species, supporting the view that oxygen transport is only moderately affected by nitrite in crustaceans (see above).

N-NITROSO COMPOUNDS

Nitrite participates in chemical reactions with amines and amides to yield N-nitroso compounds, many of which are toxic, mutagenic, and/or carcinogenic in animals.[32,35] Nitrosation is favored by an acid environment, such as found in the stomach, where nitrite is converted to nitrous acid, forming dinitrogen trioxide that subsequently may react with a secondary amine to form a N-nitrosamine:

$$2NO_2^- + 2H^+ \longrightarrow 2HNO_2$$

$$2HNO_2 \longrightarrow N_2O_3 + H_2O$$

$$R_2NH + N_2O_3 \longrightarrow R_2N - N = O + HNO_2 \qquad (2)$$

Formation of N-nitroso compounds from nitrite and amines is well documented in gastric fluid of mammals.[32] A high intake of nitrate and nitrite may increase the risk of gastric and other cancers in humans but the issue is subject to controversy. Some studies have related a high nitrate intake to gastric cancer, whereas other investigations have observed a negative correlation between nitrate exposure and gastric cancer.[34] Vitamin C is a known inhibitor of endogenous formation of nitroso compounds, whereby a high nitrate intake via vitamin C-rich vegetables may be less risky than a high nitrate intake via drinking water.[32,35]

The mutagenic and carcinogenic activity of N-nitroso compounds is related to the metabolism of N-nitrosamines and nonenzymatic decomposition of N-nitrosamides into reactive carbonium ions, which can alkylate cellular DNA molecules and cause base substitutions during DNA replication.[32,35]

N-nitroso compounds can be found in various tissues, including muscle of nitrite-exposed rainbow trout.[75] In addition to the potential danger of tumor formation by nitroso compounds, nitrite, or some of its derivatives, induces structural damage to liver mitochondria in fish, and it also appears to repress the immune system.[76,77]

RECOVERY FROM NITRITE INTOXICATION

Elevation of ambient nitrite concentrations is often a transient phenomenon in aquatic habitats. It is therefore important to examine to what extent nitrite-induced effects can be reversed upon return to nitrite-free water. When nitrite-intoxicated fish or crayfish are transferred to clean water, the extracellular nitrite concentration declines. The rate of nitrite off loading is similar to or somewhat lower than the preceding accumulation rate.[11,13,19,74] A major part of the nitrite is probably lost across the gills. It is conceivable, however, that some nitrite is eliminated in the urine. The relative involvement of the gill and kidney routes to nitrite loss is unknown.

In fish, the off loading of nitrite is paralleled by a return of metHb levels toward control values.[19,21,74,78] Postexposure effects such as a decrease in the total Hb content can develop, however.[78]

Several of the nitrite-induced effects that are described in this chapter are of relatively recent discovery, and their reversal upon removal of the nitrite load has not been fully investigated. Future research is needed to improve our knowledge of nitrite-induced effects and their interaction and contribution to nitrite toxicity. Additionally, more effort should be directed toward studying the reversibility of nitrite-induced effects and possible nitrite-detoxification mechanisms.

REFERENCES

1. **Eddy, F. B. and Williams, E. M.,** Nitrite and freshwater fish, *Chem. Ecol.*, 3, 1, 1987.
2. **Anderson, J. J., Okubo, A., Robbins, A. S., and Richards, F. A.,** A model for nitrite and nitrate distributions in oceanic oxygen minimum zones, *Deep Sea Res.*, 29, 1113, 1982.
3. **Balmelle, B., Nguyen, K. M., Capdeville, B., Cornier, J. C., and Deguin, A.,** Study of factors controlling nitrite build-up in biological processes for water nitrification, *Water Sci. Tech.*, 26, 1017, 1992.
4. **Collins, M. T., Gratzek, J. B., Shotts, E. B., Jr., Dawe, D. L., Campbell, L. M., and Senn, D. R.,** Nitrification in an aquatic recirculating system, *J. Fish. Res. Bd. Can.*, 32, 2025, 1975.
5. **Avnimelech, Y., Weber, B., Hepher, B., Milstein, A., and Zorn, M.,** Studies in circulated fish ponds: organic matter recycling and nitrogen transformation, *Aquacult. Fish. Man.*, 17, 231, 1986.
6. **Bouchard, D. C., Williams, M. K., and Surampalli, R. Y.,** Nitrate contamination of groundwater: sources and potential health effects, *J. AWWA*, 84, 85, 1992.
7. **Mason, J.,** The causes and consequences of surface water acidification, in *Acid Toxicity and Aquatic Animals*, Morris, R., Taylor, E. W., Brown, D. J. A., and Brown, J. A., Eds., Cambridge University Press, Cambridge, 1989, 1.
8. **Bath, R. N. and Eddy, F. B.,** Transport of nitrite across fish gills, *J. Exp. Zool.*, 214, 119, 1980.
9. **Margiocco, C., Arillo, A., Mensi, P., and Schenone, G.,** Nitrite bioaccumulation in *Salmo gairdneri* Rich. and hematological consequences, *Aquat. Toxicol.*, 3, 261, 1983.
10. **Palachek, R. M. and Tomasso, J. R.,** Toxicity of nitrite to channel catfish (*Ictalurus punctatus*), tilapia (*Tilapia aurea*), and largemouth bass (*Micropterus salmoides*): evidence for a nitrite exclusion mechanism, *Can. J. Fish. Aquat. Sci.*, 41, 1739, 1984.
11. **Jensen, F. B., Andersen, N. A., and Heisler, N.,** Effects of nitrite exposure on blood respiratory properties, acid-base and electrolyte regulation in the carp (*Cyprinus carpio*), *J. Comp. Physiol. B*, 157, 533, 1987.
12. **Jensen, F. B.,** Sublethal physiological changes in freshwater crayfish, *Astacus astacus*, exposed to nitrite: haemolymph and muscle tissue electrolyte status, and haemolymph acid-base balance and gas transport, *Aquat. Toxicol.*, 18, 51, 1990.
13. **Harris, R. R. and Coley, S.,** The effects of nitrite on chloride regulation in the crayfish *Pacifastacus leniusculus* Dana (Crustacea: Decapoda), *J. Comp. Physiol. B*, 161, 199, 1991.
14. **Gutzmer, M. P. and Tomasso, J. R.,** Nitrite toxicity to the crayfish *Procambarus clarkii*, *Bull. Environ. Contam. Toxicol.*, 34, 369, 1985.
15. **Williams, E. M. and Eddy, F. B.,** Chloride uptake in freshwater teleosts and its relationship to nitrite uptake and toxicity, *J. Comp. Physiol. B*, 156, 867, 1986.
16. **Perrone, S. J. and Meade, T. L.,** Protective effect of chloride on nitrite toxicity to coho salmon (*Oncorhynchus kisutch*), *J. Fish. Res. Bd. Can.*, 34, 486, 1977.
17. **Truchot, J.-P.,** *Comparative Aspects of Extracellular Acid-Base Balance*, Springer-Verlag, Berlin, 1987.
18. **Lewis, W. M., Jr. and Morris, D. P.,** Toxicity of nitrite to fish: a review, *Trans. Am. Fish. Soc.*, 115, 183, 1986.
19. **Eddy, F. B., Kunzlik, P. A., and Bath, R. N.,** Uptake and loss of nitrite from the blood of rainbow trout, *Salmo gairdneri* Richardson, and Atlantic salmon, *Salmo salar* L. in fresh water and in dilute sea water, *J. Fish Biol.*, 23, 105, 1983.
20. **Scarano, G., Saroglia, M. G., Gray, R. H., and Tibaldi, E.,** Hematological responses of sea bass *Dicentrarchus labrax* to sublethal nitrite exposures, *Trans. Am. Fish. Soc.*, 113, 360, 1984.
21. **Almendras, J. M. E.,** Acute nitrite toxicity and methemoglobinemia in juvenile milkfish (*Chanos chanos* Forsskal), *Aquaculture*, 61, 33, 1987.
22. **Ary, R. D., Jr. and Poirrier, M. A.,** Acute toxicity of nitrite to the blue crab (*Callinectes sapidus*), *Prog. Fish-Cult.*, 51, 69, 1989.
23. **Jensen, F. B.,** Physiological effects of nitrite in teleosts and crustaceans, in *Toxicology of Aquatic Pollution. Physiological, Cellular and Molecular Approaches*, Taylor, E. W., Ed., Cambridge University Press, Cambridge, 1995, in press.
24. **Chen, J.-C. and Chen, S.-F.,** Accumulation of nitrite in the haemolymph of *Penaeus monodon* exposed to ambient nitrite, *Comp. Biochem. Physiol.*, 103C, 477, 1992.
25. **Jeberg, M. V. and Jensen, F. B.,** Extracellular and intracellular ionic changes in crayfish *Astacus astacus* exposed to nitrite at two acclimation temperatures, *Aquat. Toxicol.*, 29, 65, 1994.

26. **Mazik, P. M., Hinman, M. L., Winkelmann, D. A., Klaine, S. J., Simco, B. A., and Parker, N. C.,** Influence of nitrite and chloride concentrations on survival and hematological profiles of striped bass, *Trans. Am. Fish. Soc.*, 120, 247, 1991.
27. **Wheatly, M. G.,** Physiological responses of the crayfish *Pacifastacus leniusculus* to environmental hyperoxia. I. Extracellular acid-base and electrolyte status and transbranchial exchange, *J. Exp. Biol.*, 143, 33, 1989.
28. **Westin, D. T.,** Nitrate and nitrite toxicity to salmonoid fishes, *Prog. Fish-Cult.*, 36, 86, 1974.
29. **Tomasso, J. R. and Carmichael, G. J.,** Acute toxicity of ammonia, nitrite, and nitrate to the guadalupe bass, *Micropterus treculi*, *Bull. Environ. Contam. Toxicol.*, 36, 866, 1986.
30. **Muir, P. R., Sutton, D. C., and Owens, L.,** Nitrate toxicity to *Penaeus monodon* protozoea, *Mar. Biol.*, 108, 67, 1991.
31. **McMahon, B. R. and Stuart, S. A.,** The physiological problems of crayfish in acid waters, in *Acid Toxicity and Aquatic Animals*, Morris, R., Taylor, E. W., Brown, D. J. A., and Brown, J. A., Eds., Cambridge University Press, Cambridge, 1989, 171.
32. **Hotchkiss, J. H., Helser, M. A., Maragos, C. M., and Weng, Y.-M.,** Nitrate, nitrite, and N-nitroso compounds. Food safety and biological implications, in *Food Safety Assessment*, Finley, J. W., Robinson, S. F., and Armstrong, D. J., Eds., *ACS Symposium Series*, 484, 400, 1992.
33. **Lauer, K.,** The history of nitrite in human nutrition: a contribution from German cookery books, *J. Clin. Epidemiol.*, 44, 261, 1991.
34. **Forman, D., Al-Dabbagh, S., and Doll, R.,** Nitrates, nitrites and gastric cancer in Great Britain, *Nature*, 313, 620, 1985.
35. **Bartsch, H. and Montesano, R.,** Relevance of nitrosamines to human cancer, *Carcinogenesis*, 5, 1381, 1984.
36. **Iyengar, R., Stuehr, D. J., and Marletta, M. A.,** Macrophage synthesis of nitrite, nitrate, and N-nitrosamines: precursors and role of the respiratory burst, *Proc. Natl. Acad. Sci. U.S.A.*, 84, 6369, 1987.
37. **Vanin, A. F., Mordvintcev, P. I., Hauschildt, S., and Mülsch, A.,** The relationship between L-arginine-dependent nitric oxide synthesis, nitrite release and dinitrosyl-iron complex formation by activated macrophages, *Biochim. Biophys. Acta*, 1177, 37, 1993.
38. **Stuehr, D. J. and Marletta, M. A.,** Mammalian nitrate biosynthesis: mouse macrophages produce nitrite and nitrate in response to *Escherichia coli* lipopolysaccharide, *Proc. Natl. Acad. Sci. U.S.A.*, 82, 7738, 1985.
39. **Jensen, F. B.,** Nitrite and red cell function in carp: control factors for nitrite entry, membrane potassium ion permeation, oxygen affinity and methaemoglobin formation, *J. Exp. Biol.*, 152, 149, 1990.
40. **Jensen, F. B.,** Influence of haemoglobin conformation, nitrite and eicosanoids on K^+ transport across the carp red blood cell membrane, *J. Exp. Biol.*, 171, 349, 1992.
41. **Kiese, M.,** *Methemoglobinemia: A Comprehensive Treatise*, CRC Press, Cleveland, OH, 1974.
42. **Kosaka, H. and Tyuma, I.,** Mechanism of autocatalytic oxidation of oxyhemoglobin by nitrite, *Environ. Health Perspect.*, 73, 147, 1987.
43. **Jensen, F. B.,** Influence of nucleoside triphosphates, inorganic salts, NADH, catecholamines, and oxygen saturation on nitrite-induced oxidation of rainbow trout haemoglobin, *Fish Physiol. Biochem.*, 12, 111, 1993.
44. **Spagnuolo, C., Rinelli, P., Coletta, M., Chiancone, E., and Ascoli, F.,** Oxidation reaction of human oxyhemoglobin with nitrite: a reexamination, *Biochim. Biophys. Acta*, 911, 59, 1987.
45. **Arduini, A., Mancinelli, G., Radatti, G. L., Hochstein, P., and Cadenas, E.,** Possible mechanism of inhibition of nitrite-induced oxidation of oxyhemoglobin by ergothioneine and uric acid, *Arch. Biochem. Biophys.*, 294, 398, 1992.
46. **Jaffé, E. R.,** Methaemoglobinaemia, *Clin. Haematol.*, 10, 99, 1981.
47. **Freeman, L., Beitinger, T. L., and Huey, D. W.,** Methemoglobin reductase activity in phylogenetically diverse piscine species, *Comp. Biochem. Physiol.*, 75B, 27, 1983.
48. **Imaizumi, K., Tyuma, I., Imai, K., Kosaka, H., and Ueda, Y.,** In vivo studies on methemoglobin formation by sodium nitrite, *Int. Arch. Occup. Environ. Health*, 45, 97, 1980.
49. **Brauner, C. J., Val, A. L., and Randall, D. J.,** The effect of graded methaemoglobin levels on the swimming performance of chinook salmon (*Oncorhynchus tshawytscha*), *J. Exp. Biol.*, 185, 121, 1993.
50. **Perutz, M. F., Fersht, A. R., Simon, S. R., and Roberts, G. C. K.,** Influence of globin structure on the state of the heme. II. Allosteric transitions in methaemoglobin, *Biochemistry*, 13, 2174, 1974.
51. **Perutz, M. F.,** Hemoglobin structure and respiratory transport, *Sci. Am.*, 239, 68, 1978.
52. **Darling, R. C. and Roughton, F. J. W.,** The effect of methemoglobin on the equilibrium between oxygen and hemoglobin, *Am. J. Physiol.*, 137, 56, 1942.
53. **Jensen, F. B., Nikinmaa, M., and Weber, R. E.,** Environmental perturbations of oxygen transport in teleost fishes: causes, consequences and compensations, in *Fish Ecophysiology*, Rankin, J. C. and Jensen, F. B., Eds., Chapman and Hall, London, 1993, chap. 6.
54. **Nikinmaa, M.,** *Vertebrate Red Blood Cells. Adaptations of Function to Respiratory Requirements*, Springer-Verlag, Berlin, 1990.
55. **Nikinmaa, M. and Jensen, F. B.,** Inhibition of adrenergic proton extrusion in rainbow trout red cells by nitrite-induced methaemoglobinaemia, *J. Comp. Physiol. B*, 162, 424, 1992.

56. **Brown, D. A. and Mcleay, D. J.,** Effect of nitrite on methemoglobin and total hemoglobin of juvenile rainbow trout, *Prog. Fish-Cult.*, 37, 36, 1975.
57. **Doeller, J. E. and Wittenberg, B. A.,** Myoglobin function and energy metabolism of isolated cardiac myocytes: effect of sodium nitrite, *Am. J. Physiol.*, 261, H53, 1991.
58. **Paitian, N. A., Markossian, K. A., and Nalbandyan, R. M.,** The effect of nitrite on cytochrome oxidase, *Biochem. Biophys. Res. Commun.*, 133, 1104, 1985.
59. **Duthu, G. S. and Shertzer, H. G.,** Effect of nitrite on rabbit liver mixed-function oxidase activity, *Drug Metab. Dispos.*, 7, 263, 1979.
60. **Gondko, R., Serafin, E., and Mazur, J.,** The kinetics of the reaction of thiocyanate and nitrite ions with *Orconectes limosus* oxyhemocyanin, *Z. Naturforsch.*, 40c, 677, 1985.
61. **Tahon, J.-P., van Hoof, D., Vinckier, C., Witters, R., de Ley, M., and Lontie, R.,** The reaction of nitrite with the haemocyanin of *Astacus leptodactylus*, *Biochem. J.*, 249, 891, 1988.
62. **Laustiola, K. E., Vuorinen, P., Pörsti, I., Metsä-Ketelä, T., Manninen, V., and Vapaatalo, H.,** Exogenous GTP enhances the effects of sodium nitrite on cyclic GMP accumulation, vascular smooth muscle relaxation and platelet aggregation, *Pharmacol. Toxicol.*, 68, 60, 1991.
63. **Bassenge, E. and Zanzinger, J.,** Nitrates in different vascular beds, nitrate tolerance, and interactions with endothelial function, *Am. J. Cardiol.*, 70, 23B, 1992.
64. **Abrams, J.,** Nitrates and nitrate tolerance in congestive heart failure, *Coron. Art. Dis.*, 4, 27, 1993.
65. **Ignarro, L. J.,** Biological actions and properties of endothelium-derived nitric oxide formed and released from artery and vein, *Circ. Res.*, 65, 1, 1989.
66. **Kubes, P.,** Nitric oxide-induced microvascular permeability alterations: a regulatory role for cGMP, *Am. J. Physiol.*, 265, H1909, 1993.
67. **Jensen, F. B.,** unpublished data, 1993.
68. **Kim, Y.-M. and Lancaster, J. R., Jr.,** Tetrahydrobiopterin-dependent nitrite oxidation to nitrate in isolated rat hepatocytes, *FEBS Lett.*, 332, 255, 1993.
69. **Gaino, E., Arillo, A., and Mensi, P.,** Involvement of the gill chloride cells of trout under acute nitrite intoxication, *Comp. Biochem. Physiol.*, 77A, 611, 1984.
70. **Williams, E. M. and Eddy, F. B.,** Anion transport, chloride cell number and nitrite-induced methaemoglobinaemia in rainbow trout (*Salmo gairdneri*) and carp (*Cyprinus carpio*), *Aquat. Toxicol.*, 13, 29, 1988.
71. **Perry, S. F. and Laurent, P.,** Environmental effects on fish gill structure and function, in *Fish Ecophysiology*, Rankin, J. C. and Jensen, F. B., Eds., Chapman and Hall, London, 1993, chap. 9.
72. **Tomasso, J. R., Davis, K. B., and Simco, B. A.,** Plasma corticosteroid dynamics in channel catfish (*Ictalurus punctatus*) exposed to ammonia and nitrite, *Can. J. Fish. Aquat. Sci.*, 38, 1106, 1981.
73. **Jensen, F. B.,** Regulatory volume decrease in carp red blood cells: mechanisms and oxygenation dependency of volume-activated potassium and amino acid transport, *J. Exp. Biol.*, 198, 155, 1995.
74. **Williams, E. M., Glass, M. L., and Heisler, N.,** Blood oxygen tension and content in carp, *Cyprinus carpio* L., during hypoxia and methaemoglobinaemia, *Aquacult. Fish. Man.*, 23, 679, 1992.
75. **DeFlora, S. and Arillo, A.,** Mutagenic and DNA damaging activity in muscle of trout exposed in vivo to nitrite, *Cancer Lett.*, 20, 147, 1983.
76. **Arillo, A., Gaino, E., Margiocco, C., Mensi, P., and Schenone, G.,** Biochemical and ultrastructural effects of nitrite in rainbow trout: liver hypoxia as the root of the acute toxicity mechanism, *Environ. Res.*, 34, 135, 1984.
77. **Hanson, L. A. and Grizzle, J. M.,** Nitrite-induced predisposition of channel catfish to bacterial diseases, *Prog. Fish-Cult.*, 47, 98, 1985.
78. **Scarano, G. and Saroglia, M. G.,** Recovery of fish from functional and haemolytic anaemia after brief exposure to a lethal concentration of nitrite, *Aquaculture*, 43, 421, 1984.

Chapter 18

Nitric Oxide Synthesis

Jon M. Fukuto

CONTENTS

INTRODUCTION

In 1987, reports confirming an original proposal by Furchgott[1] and Ignarro et al.[2] indicated that endothelial cells were capable of endogenous nitric oxide (NO) generation.[3,4] In the following year or so, it was discovered that activated macrophages were also capable of NO synthesis.[5-7] These provocative findings caused a tremendous amount of excitement and led to a virtual explosion of work pertaining to the biosynthetic origin, physiological functions, and biological fate of NO. The discovery of the diverse and ubiquitous physiological roles of NO has since caused many to change their perception of the nature of biological effector and/or messenger molecules. For example, NO has been proposed to be representative of a new class of neuronal messengers.[8,9] Interest in the chemistry and associated biochemistry of NO did not, however, begin in 1987 with its discovery in mammalian cells. For example, the possible intermediacy of NO in the bacterial reduction of nitrite (NO_2^-) and nitrate (NO_3^-) has been the subject of numerous investigations. Also, being a major component of air pollution, the gas phase chemistry and biological effects of NO and its degradation products have been studied intensely as well. Moreover, NO releasing nitrovasodilators such as nitroglycerin have been utilized in the treatment of coronary artery disease for over 100 years. Thus, in spite of the recent excitement regarding the physiology, pharmacology, and biochemistry of NO, it is certainly not a new molecule to those involved in these other fields of endeavor. This review, however, will focus on the chemistry and biochemistry associated with NO synthesis in mammalian systems. In order to put NO in the proper perspective and to provide a basis for further discussion, it is worthwhile to first present a brief description of the biologically relevant chemistry as well as

0-8493-8411-7/95/$0.00+$.50

Figure 1 Lewis dot formalism for NO.

some of the physiological roles of NO. Thus, relevant aspects of NO chemistry and physiology will be reviewed prior to discussing its biosynthetic mechanism and regulation.

THE CHEMISTRY OF NO

Before discussing the biological utility and fate of NO, as well as its mechanism of biosynthesis, it is essential to have a reasonable understanding of its unique and fascinating chemistry. One of the most outstanding chemical features of NO is that it is a paramagnetic, radical species. Using the Lewis dot formalism to depict the bonding scheme, it is clear that NO has an unpaired electron (Figure 1). In spite of the fact that NO has an unpaired electron, it does not possess the type of reactivity normally associated with free radicals. By analogy with typical organic free radicals, it may be expected that NO should dimerize or possibly react with activated hydrogens; however, it does not. At room temperature and pressure, NO does not even have the tendency to dimerize and exists predominantly in the monomeric form. The reason for this lack of reactivity lies in the nature of the molecular orbital which the unpaired electron occupies. A simple molecular orbital diagram for NO shows that the unpaired electron resides in an antibonding orbital (Figure 2). Therefore, the dimerization of two NO molecules via the unpaired electron would not change the overall bond order (NO will have a bond order of 2.5 whether monomeric or dimeric).

In the series of redox-related nitrogen oxides, NO occupies a unique and central position. Figure 3 depicts the relationship of NO with several nitrogen oxide species. A two electron oxidation of ammonia (NH_3) gives hydroxylamine (NH_2OH) and a further two electron oxidation of hydroxylamine generates nitroxyl (HNO). NO can then be generated from HNO by a one electron oxidation, and further oxidation by a single electron results in nitrite (NO_2^-) formation. Thus, unlike the other

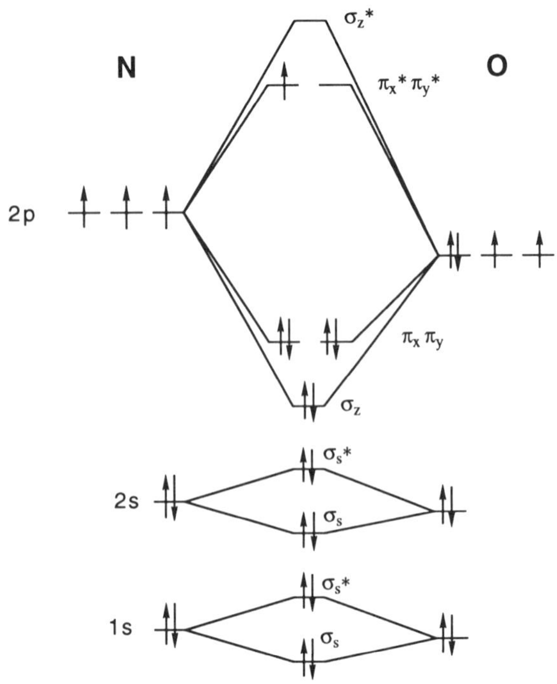

Figure 2 Molecular orbital diagram for NO. No mixing of s and p orbitals is assumed.

$$\underset{N(V)}{NO_3^-} \underset{-2e^-}{\overset{+2e^-}{\rightleftharpoons}} \underset{N(III)}{NO_2^-} \underset{-1e^-}{\overset{+1e^-}{\rightleftharpoons}} \underset{N(II)}{NO} \underset{-1e^-}{\overset{+1e^-}{\rightleftharpoons}} \underset{N(I)}{HNO} \underset{-2e^-}{\overset{+2e^-}{\rightleftharpoons}} \underset{N(-I)}{NH_2OH} \underset{-2e^-}{\overset{+2e^-}{\rightleftharpoons}} \underset{N(-III)}{NH_3}$$

Figure 3 Redox scheme for several nitrogen oxides.

nitrogen oxides shown in Figure 3, it can be seen that NO is an odd electron species whose oxidation state is intermediate between fully reduced nitrogen, NH_3 (N^{-III}) and fully oxidized nitrogen, NO_3^- (N^V). Under physiological conditions, the generation of NO or other nitrogen oxides may result in the eventual formation of a complex mixture of several redox-related nitrogen oxides. The possibility of interconversion of the nitrogen oxides under physiological conditions must always be considered when interpreting the biological effects associated with any nitrogen oxide species.[10,11]

Under anaerobic conditions, NO can be stable, and degassed aqueous solutions of NO can last almost indefinitely. In the presence of molecular oxygen (O_2), however, NO is unstable and, depending on conditions, decomposes to a variety of oxidized nitrogen oxide species. In the gas phase, the reaction of NO with O_2 results in the formation of nitrogen dioxide (NO_2) (Reaction 1).

$$2NO + O_2 \longrightarrow 2NO_2 \qquad (1)$$

The generation of NO_2 from NO and O_2 exhibits overall third order kinetics — first order in O_2 and second order in NO. The rate expression for this process has been shown to be $-d[NO]/dt = 2k_g[NO]^2[O_2]$. In biologically relevant aqueous solutions, the rate expression is very similar; $-d[NO]/dt = 4k_{aq}[NO]^2[O_2]$ with $k_{aq} = 2 \times 10^6\ M^{-2} \cdot s^{-1}$.[12] Therefore, since the rate of NO loss is second order in itself, at low concentrations, it can have an appreciable and physiologically relevant lifetime (>100 sec). At high concentrations, however, NO reacts rapidly with O_2. It should be mentioned that, like NO, NO_2 is also a free radical, paramagnetic species. NO_2 is considerably more reactive than NO (in a free radical sense), and exists predominantly as a diamagnetic dimer (N_2O_4) below its boiling point (21°C).

Unlike its gas phase chemistry, NO_2 is short lived in aqueous solutions. Two NO_2 molecules can react with water to give equal amounts of nitrite and nitrate (Reaction 2). Indeed, the addition of NO_2 to water results in the formation of 1 equivalent of NO_2^- and 1 equivalent of NO_3^-.

$$2NO_2 + H_2O \longrightarrow NO_2^- + NO_3^- + 2H^+ \qquad (2)$$

However, when NO_2 is generated in aqueous solution via the autooxidation of NO, the terminal product is almost entirely NO_2^- with little formation of NO_3^-.[13] This apparent paradox can be reconciled if the rapid reaction of NO with NO_2 is considered. The lack of NO_3^- generation under conditions where NO_2 is formed from NO oxidation is due to the rapid trapping of NO_2 by NO to give dinitrogen trioxide (N_2O_3). N_2O_3 then reacts with water to generate NO_2^- exclusively (Reactions 3 and 4).

$$NO + NO_2 \longrightarrow N_2O_3 \qquad (3)$$

$$N_2O_3 + H_2O \longrightarrow 2NO_2^- + 2H^+ \qquad (4)$$

NO also reacts rapidly with superoxide (O_2^-) to generate an unstable species known as peroxynitrite (^-OONO) (Reaction 5).[14] The rate of this bimolecular reaction is reported to be near diffusion controlled ($k = 6.7 \times 10^9\ mol^{-1}s^{-1}$).[15]

$$NO + O_2^- \longrightarrow {}^-OONO \qquad (5)$$

^-OONO has a pKa of 6.8 and is therefore substantially protonated at physiological pH to give peroxynitrous acid (HOONO). Isomerization of ^-OONO or HOONO leads to the eventual formation of NO_3^-. This isomerization was proposed to occur via the formation of hydroxyl radical ($\cdot OH$) and NO_2 as transient intermediates in NO_3^- generation (Reaction 6).[16] Thus, physiological generation of $\cdot OH$ and NO_2 via the decomposition of ^-OONO may be expected to be cytotoxic since both are potent oxidants.

$$^-OONO + H^+ \longrightarrow HOONO \longrightarrow \cdot OH + NO_2 \longrightarrow NO_3^- + H^+ \quad (6)$$

It is more likely, however, that HOONO itself is a cytotoxic entity, which does not form $\cdot OH$ or NO_2, as it appears to possess "hydroxyl radical-like" reactivity itself.[17] The toxicology and related physiological implications of the reaction of NO with $\cdot O_2^-$ will be addressed briefly below.

It has been reported that NO is capable of oxidizing sulfhydryl groups under basic conditions. The net products of this reaction are the corresponding disulfide and hyponitrous acid (Reactions 7, 8, and 9).[18]

$$RS^- + \cdot NO \longrightarrow RS - {}^{\cdot}NO^- \quad (7)$$

$$RS - {}^{\cdot}NO^- + H^+ \longrightarrow RS - {}^{\cdot}NOH \quad (8)$$

$$2RS - {}^{\cdot}NOH \longrightarrow RSN(OH) - N(OH)SR \longrightarrow RSSR + HON = NOH \quad (9)$$

Also of potential interest, sulfhydryl groups are also capable of reducing NO_2 to NO through the presumed intermediacy of a nitrosothiol (Reactions 10, 11, and 12).[18]

$$2RSH + NO_2 \longrightarrow RS\cdot + RSN(OH)_2 \quad (10)$$

$$RSH(OH)_2 \longrightarrow RS - NO + H_2O \quad (11)$$

$$RS - NO \longrightarrow RS\cdot + \cdot NO \quad (12)$$

Reactions 7, 8, 9, 10, 11, and 12 are somewhat speculative and certainly other possible mechanisms can be envisioned. However, these results indicate that NO and NO_2 may be capable of modifying protein thiols which may have significant toxicological implications.

Acidification of a primary NO decomposition product, NO_2^-, results in the generation of a potent nitrosating agent, nitrous acid (HONO). HONO is a weak acid ($pK_a = 3.34$), and therefore exists to some extent at lower pH. Reaction of HONO with sulfhydryl groups (or other appropriate nucleophiles) can give the corresponding nitrosothiol, which, by Reaction 12, can result in NO formation. Nitrosation of thiols and other nucleophiles can also occur via reaction with ferric-nitrosyl heme complexes.[19] Reaction of NO with a ferric hemeprotein results in an Fe(III)-NO complex. The subsequent chemistry of this complex can be more accurately predicted by envisioning this complex as a ferrous-nitrosonium complex [Fe(II)-NO^+] whereby NO acts as a reducing agent for the ferric iron. Attack of this complex by an appropriate nucleophile results in nitrosation of the nucleophile and overall reduction of the metal (see Reactions 13 and 14).

$$Fe(III) + NO \longrightarrow Fe(III) - NO \longleftrightarrow Fe(II) - NO^+ \quad (13)$$

$$Fe(II) - NO^+ + RSH \longrightarrow RS - NO + Fe(II) + H^+ \quad (14)$$

Again, considering the ubiquity of hemeproteins, this process may be physiologically relevant.

A particularly important aspect of NO chemistry which is likely responsible for a large portion of its physiological utility is that it is, in general, a good ligand for metalloproteins. For example,

NO forms strong complexes with a variety of reduced hemeproteins.[20] NO not only has a high affinity for ironhemes, but it also binds in a manner distinct from other ligands such as CO or O_2.[21] In fact, specific activation of the enzyme guanylate cyclase (the hemoprotein responsible for elevating intracellular cGMP levels), by NO may be due to its unique ability, upon coordination, to cause the release of a catalytically active ligand on iron.[22] This "trans-releasing" effect does not occur with other ligands such as CO, O_2, or isocyanides and, as expected, these ligands do not serve as activators of the enzyme. NO also forms complexes with nonheme iron and may play a definite role in the cytotoxicity of NO (discussed later).

This brief discussion of the chemistry of NO is by no means comprehensive and serves only as a preface to its other more biological and biochemical aspects. There are many other nitrogen oxide species which were not mentioned and many other more thorough and detailed reviews of the chemistry of nitrogen oxides are available.[23-28]

THE PHYSIOLOGY OF NO

The biological importance of NO in a variety of functions can be attested to by the amount of current interest and work being performed on NO-mediated physiological events. In fact, NO was chosen as "Molecule of the Year" in 1992 by the editors of *Science* on the basis of its newly discovered and defined roles in physiology.[29] Although the focus of this review is to discuss the biosynthesis of NO, it is worthwhile to briefly discuss the reasons for NO synthesis in mammalian systems. For a more comprehensive treatment of the topic, several other excellent reviews covering various aspects of NO physiology are available.[30-39]

ENDOTHELIUM-DERIVED RELAXING FACTOR (EDRF)

In 1980, a now classic paper was published describing the release of a vasorelaxing chemical substance from acetylcholine-stimulated endothelial cells.[40] This substance was coined endothelium-derived relaxing factor (EDRF) by Furchgott and Zawadzki at that time since its chemical identity was not known. Shortly after its discovery, a significant amount of work indicated that EDRF was a short-lived species under physiological conditions[41] that elicited smooth muscle relaxation by activating the enzyme guanylate cyclase, resulting in an elevation of intracellular cGMP levels.[42-44] EDRF was found to be inactivated by hemoglobin, methylene blue,[45] and superoxide.[46,47] Based partially on the similarities between authentic NO with EDRF, NO was proposed to be EDRF by several groups.[1-4] It was then determined that the amino acid L-arginine was the precursor to NO synthesis in endothelial cells.[48] It should be mentioned that several reports describe inconsistencies with the idea that EDRF is purely NO.[49-56] Attempts to reconcile some of the experimental inconsistencies regarding the identification of EDRF as NO have been made.[30] Others have proposed that an NO-containing protein, iron complex, or amino acid intermediate can act as circulating carriers which are capable of NO release and may, in fact, be responsible for a significant amount of EDRF activity.[57-62] Regardless, it is generally accepted that EDRF is NO or closely related species capable of NO release, and that NO is involved in the signal transduction system leading to smooth muscle relaxation.

NO IN THE VASCULAR SYSTEM

NO, like EDRF, appears to play a vital role in the maintenance of vascular tone (see References 63 and 64 for recent discussions). In endothelial cells, NO production is regulated by controlling intracellular Ca^{+2} levels. A variety of stimuli (e.g., acetylcholine, bradykinin, electrical current, shear force) cause Ca^{+2} influx into the cell, which results in an activation of NO biosynthesis. Being a noncharged, nonpolar species, NO is then capable of rapidly diffusing into neighboring smooth muscle cells. Activation of the enzyme guanylate cyclase within the smooth muscle cells by NO results in an accelerated conversion of GTP to cGMP. Thus, NO is a crucial participant in the cGMP signal transduction system. The role of cGMP in signal transduction for mediating, among other things, smooth muscle relaxation is beyond the scope of this review and has been recently discussed in detail elsewhere.[65,66]

Drugs which act by releasing NO have been utilized for many years as clinical vasorelaxants.[67] For example, nitroglycerin is representative of a class of vasorelaxants known as the nitrovasodilators. It has been used in the treatment of angina pectoris since 1879. With the discovery of endogenous NO synthesis, it became clear that the biological activity of these compounds was based on an existing NO-cGMP signal transduction system. NO release from nitrovasodilators often requires a bioactivation step and due to the diversity of structure and chemistry, the mechanisms of NO release may vary considerably.[68]

NO AND PLATELETS

Platelet aggregation *in vitro*, induced by a variety of agents, is inhibited by exogenous NO,[69] or that released from fresh or cultured endothelial cells.[70] NO also causes disaggregation of preformed platelet aggregates.[70] The contribution of NO released from leukocytes[71] to the regulation of platelet function in vivo remains to be established.[64] NO also inhibits adhesion of platelets[72] and monocytes[73] to endothelial cells. NO and prostacyclin (which increases cAMP levels) synergize with each other as inhibitors of platelet aggregation and inducers of disaggregation.[74] The ability of NO to inhibit these types of cellular adhesion may play a significant role in the circulatory system, especially in areas such as the capillary beds or cardiac valves where no smooth muscle cells are present.[37]

NO IN THE IMMUNE SYSTEM

In studies originally focused on a possible role of nitrogen oxides in cancer, urinary NO_3^- excretion in mammals was found to occur even when fed a low NO_3^- diet.[75] The source of NO_3^- was determined not to be intestinal bacteria since germ-free animals continued to excrete NO_3^-.[76] Significantly, infection markedly increased the level of urinary NO_3^- [77] and exposure of macrophages to infectious agents such as lipopolysaccharides caused an increase in NO_3^- and NO_2^- production.[78] Thus, it appears that macrophages can become activated as a result of infection and can enhance the production of NO_3^- and NO_2^-. It was later found that oxidation of a terminal guanidinium nitrogen on the amino acid arginine was the precursor to NO_3^- and NO_2^- formation in macrophages.[79] Finally, it was shown that NO was synthesized from arginine oxidation and that NO_3^- and NO_2^- formation was the result of NO decomposition.[5,6]

Thus, NO synthesis by activated macrophages represents an immune system response to infection[80,81] or tumor formation.[7,82] The cytotoxic and/or cytostatic activity of NO has been proposed to be multifaceted. For example, NO can interfere with cell function by binding and inhibiting iron-sulfur proteins.[83] Likewise, NO in conjunction with another product of macrophage activation, O_2^-, can combine to form a potent oxidant, peroxynitrite, ^-OONO. The cytotoxicity of peroxynitrite has been demonstrated and may contribute to the overall effect of macrophage-released NO.[84-86]

NO AND THE CENTRAL NERVOUS SYSTEM (CNS)

It has been known for some time that glutamate and other excitatory amino acid neurotransmitters such as L-aspartate, kainate, and N-methyl-D-aspartate (NMDA) cause a Ca^{+2}-dependent increase in cGMP in the CNS.[87,88] Further reports demonstrated that NO was capable of cerebellar guanylate cyclase activation,[89] that a low molecular weight species was present in rat forebrain and was capable of guanylate cyclase activation,[90] and that arginine was the precursor to the guanylate cyclase activating factor.[91] It was then shown that activation of NMDA receptors resulted in the release of a factor with striking similarity to EDRF or NO[92]; finally, the formation of NO was demonstrated in rat forebrain.[93] Thus, it appears that NO is critically involved in the glutamate-mediated signal transduction system.[94] That is, postsynaptic receptor activation results in an influx of Ca^{+2}, which in turn stimulates NO synthesis. NO can then diffuse into the presynaptic terminal or astrocyte process to activate guanylate cyclase elevating intracellular cGMP levels in those cells. The role of cGMP in the CNS is complex and has been recently reviewed.[95] Based on its ability to freely diffuse back to the presynaptic terminal after its transmitter elicited synthesis in the postsynaptic terminal, it has been postulated that NO may serve as the retrograde messenger required for long-term potentiation or memory formation.[92,96-101]

NO IN THE PERIPHERAL NERVOUS SYSTEM

Nonadrenergic, noncholinergic (NANC) neurons innervate a variety of smooth muscle systems including the gastrointestinal tract, corpus cavernosum, and esophagus. The L-arginine-NO pathway has been implicated in NANC-mediated events with NO possibly acting as the neurotransmitter. The actual identity of the NANC neurotransmitter remains a matter of significant speculation and controversy, however.[102-105] Regardless, there seems to be little doubt that even if NO is not the actual NANC neurotransmitter, it is a common mediator involved in NANC signal transduction.

OTHER ASPECTS OF NO PHYSIOLOGY/TOXICOLOGY

From the brief discussion above, it should be clear that NO participates in a remarkable array of seemingly unrelated physiological events and functions. A comprehensive listing of the proposed physiological roles of NO is beyond the scope of this review, but NO is involved in the function of a variety of other tissues and organs besides those mentioned above and other roles for NO are sure to be discovered.[30,31] Of significant note are the toxicological implications of NO synthesis. Since NO has multiple roles as an effector molecule (i.e., cytotoxic action) and messenger molecule (i.e., activator of guanylate cyclase), the possibility exists that one function may adversely impact the other. For example, hypoxic-ischemic brain injury appears to be mediated by glutamate since glutamate or NMDA-receptor antagonists are protective. Presumably, glutamate induces NO production which, under the conditions of ischemia, can be cytotoxic and cause neuronal cell death.[106,107] NO production by macrophages can be especially toxic to pancreatic islet cells and thus be a cause of Type 1 diabetes.[81] Also, endotoxin-induced hypotension may result from an induction of NO synthesis in smooth muscle tissue (and possibly macrophages).[30]

The diverse aspects of NO physiology and toxicology provide ample impetus for studies aimed at elucidating the mechanism and regulation of NO biosynthesis since pharmacological manipulation of NO biosynthesis can have obvious therapeutic implications. Thus, the remainder of this review will concentrate on the current state of understanding of the enzymology and chemical mechanism of NO biosynthesis with the goal of providing a basic introduction for those new to the field.

THE ENZYMOLOGY OF NO BIOSYNTHESIS

The source of NO in mammalian systems is arginine,[5,6,48,93] with NO (and citrulline in a 1:1 stoichiometry[108]) generation being the result of a 5-electron oxidation of arginine's terminal guanidinium nitrogen (Figure 4). This reaction is carried out by a class of enzymes known collectively as the nitric oxide synthases (NOS). NOS activity is physiologically ubiquitous, having been

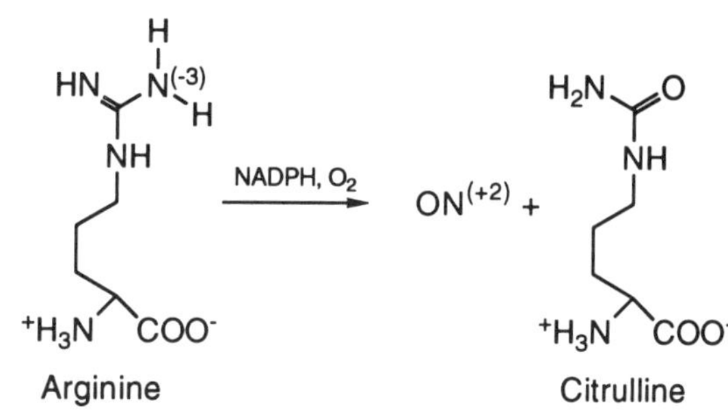

Figure 4 Conversion of arginine to citrulline and NO. Numbers in parentheses indicate the oxidation state of nitrogen.

demonstrated in liver, brain, heart, artery, vein, adrenals, lung, and spleen and in cells such as macrophages, endothelial cells, platelets, fibroblasts, neutrophils, astrocytes, hepatocytes, mast cells, adrenocarcinoma cells, neuroblastoma, mesangial, and epithelial cells.[109] It is likely that other cells and tissues will be found to possess NOS activity as well. Just as there are a great variety of physiological NO functions, there are also a variety of NOS isoforms. The NOS isoforms have a strict stereospecific requirement for the L-arginine isomer and have virtually identical cofactor and prosthetic group requirements.[109] They are all ironheme proteins[110-113] that utilize O_2[114] and NADPH[115,116] and require FAD, FMN,[117-120] and tetrahydrobiopterin (H_4B).[121] Although the absolute requirement for FMN has not been unequivocally established in all well-characterized isoforms, it is likely to be a common prosthetic group for the entire family of NOS (the roles of these cofactors and prosthetic groups will be discussed in detail later). These enzymes, however, differ significantly with regard to their tissue or cellular origin, biophysical properties, and in their mechanisms of regulation. In general terms, NOS can be divided into an inducible class (iNOS) and a constitutive class (cNOS). A variety of designations have been given to various isoforms depending on their origin and properties (such as nNOS or type Ia for the neuronal-based constitutive enzyme, type II for the soluble inducible isoform from macrophages, or type III for the particulate constitutive endothelial cell enzyme)[122]; however for the sake of simplicity only, the iNOS and cNOS designations will be generally used here.

NOS has been purified from a variety of sources such as rat cerebellum (soluble form),[123,124] rat cerebellum (particulate form),[125] porcine cerebellum,[126] bovine brain,[127] human cerebellum,[128] bovine aortic endothelial cells,[129] macrophages,[118,119,130] and rat liver.[131,132] Table 1 lists some of the more well-characterized NOS isoforms and some of their properties.

REGULATION OF NOS

One of the major differences between iNOS and cNOS is how they are regulated. The constitutive isoforms are, in general, activated by Ca^{+2} via the calcium binding protein calmodulin (CaM).[123] That is, intracellular Ca^{+2} influx results in CaM/Ca^{+2} binding to cNOS causing enzyme activation. The inducible isoforms are, however, not subject to regulation by Ca^{+2}. Interestingly, iNOS from macrophages contains tightly bound CaM which can be considered as more of an enzyme subunit.[133] The possible role of Ca^{+2} in the binding of CaM to macrophage iNOS is not currently known, but the addition of Ca^{+2} does not significantly alter macrophage iNOS enzyme activity. Another iNOS from rat liver is activated by CaM addition, but is relatively insensitive to Ca^{+2},[131] further supporting the general belief that iNOS isoforms are not regulated by Ca^{+2} (although they may require CaM).

Table 1 Representative NOS Isoforms

Type/tissue or cell	Properties	Reference
Inducible/mouse macrophages	Soluble, $K_m$arg = 2.8–16 m*M* 250–260 kDa (dimer), CaM containing	118, 119
Inducible/rat liver	Soluble, $K_m$arg = 11 m*M* 135 kDa (monomer), activated by CaM	131
Constitutive/bovine endothelial cells	Particulate, $K_m$arg = 2.9 m*M* 135 kDa (monomer), Ca^{+2}/CaM dependent	129
Constitutive/rat cerebellum	Soluble, $K_m$arg = 2.2 m*M* 279 kDa (dimer), Ca^{+2}/CaM dependent	124
Constitutive/porcine cerebellum	Soluble, $K_m$arg = not reported 200 kDa (monomer), Ca^{+2}/CaM dependent	126
Constitutive/human cerebellum	Soluble, $K_m$arg = not reported 160 kDa (monomer), Ca^{+2}/CaM dependent	128
Constitutive/rat cerebellum	Particulate, $K_m$arg = 15.5 m*M* 150 kDa (monomer), Ca^{+2}/CaM dependent	125

The mechanism by which CaM controls the activity of cNOS has been shown to involve the regulation of the transfer of electrons within the protein. That is, CaM binding to cNOS allows electrons (from NADPH) to flow from the bound flavins, FAD and FMN, to the ironheme.[134] In the case of iNOS, CaM binds either semi-irreversibly, or in a Ca^{+2} independent fashion, such that electron flow is continuous and not regulated by added Ca^{+2}.

Cloning of complementary DNA from rat cerebellum cNOS indicated that the enzyme contains sites for phosphorylation.[135] Others have demonstrated that purified rat brain cNOS can indeed be phosphorylated by Ca^{+2}/CaM-dependent protein kinase II and protein kinase C (PKC) and that enzyme activity changes as a function of phosphorylation.[136] Exclusive phosphorylation of rat and porcine brain cNOS by protein kinase A with no changes in enzyme activity has also been reported.[137] Phosphorylation of rat brain cNOS by cAMP-dependent kinase, PKC, and Ca^{+2}/CaM-dependent protein kinase occurs at different serine sites with a significant decrease in enzyme activity occurring after PKC activation in intact cells.[138] The physiological role of phosphorylation by the various kinases is not entirely understood and requires further study.

Regulation of NO levels derived from iNOS appears to primarily result from an induction of enzyme protein synthesis. Substrate or cofactor (especially H_4B) availability may play a role in iNOS regulation in macrophages and smooth muscle[109,139-141] as well as in cNOS,[142,143] although probably not to the fine degree exhibited by cNOS and Ca^{+2}. There are currently no other known mechanisms of regulation.

BIOPHYSICAL PROPERTIES OF NOS

NOS exists in both soluble and particulate isoforms (Table 1). For example, a particulate cNOS represents the major isoform found in endothelial cells, and association with membranes appears to be the result of N-terminal myristoylation.[144-146] Also, both particulate and soluble forms have been characterized in rat brain (although the cellular origins of these isoforms has not been fully elucidated).

Soluble cNOS purified from rat brain is a homodimer consisting of 155 kDa subunits.[124] iNOS from macrophages is also a homodimer made up of approximately 130 kDa subunits.[118,119] Studies with iNOS indicate that only the dimeric protein is active and, although this has not yet been demonstrated, it is likely that the dimeric protein is the active form for cNOS as well. Detailed studies with iNOS from macrophage regarding the relationship between the protein's quaternary structure, activity, and binding to cofactors or prosthetic groups indicate that the inactive monomeric protein retains the ability to bind CaM, FAD, and FMN. However, the monomers did not bind heme or H_4B. The active, dimeric protein contains all the components mentioned above and, significantly, dimerization of the subunits was promoted by heme, H_4B, and arginine.[147] The intimate kinetics of monomer association-dissociation are currently unknown and, if elucidated, may provide insight into possible mechanisms of regulation (and will certainly aid in experimental manipulation of the protein in *in vitro* situations). Also, nonionic detergents increase the activity of purified cNOS from rat brain possibly by stabilizing the active dimeric protein or inhibiting dissociation of a cofactor or prosthetic group.[148]

THE MECHANISM OF NO BIOSYNTHESIS

The conversion of arginine to citrulline and NO by NOS is a remarkable reaction from chemical and biochemical perspectives. The unprecedented conversion of a guanidine functionality (arginine) to a urea (citrulline) and an odd electron nitrogen oxide (NO) is formally a five-electron oxidation of nitrogen. This is presumably a multistep process carried out by a single protein. The mechanism by which NOS catalyzes this process is presently not entirely understood. As mentioned earlier, NOS utilizes O_2 and NADPH and requires FAD, FMN, and H_4B. Thiol participation in NOS catalysis has been proposed as well,[117] but the thiol requirement for NO production and/or biological activity in endothelial cells has been questioned.[149] Not surprisingly, NOS also requires a metal in the form of an iron heme prosthetic group.

THE RELATIONSHIP BETWEEN NOS AND CYTOCHROME P450

The fact that NOS is an ironhemeprotein which utilizes NADPH and O_2 and requires FAD and FMN makes it strikingly similar to the more well-studied and well-characterized cytochrome P450 monooxygenase system (450). In fact, NOS has been considered to be a type of self-sufficient P450.[110,111] The cytochromes P450 are a family of enzymes capable of oxidizing a variety of xenobiotic and endogenous substrates (see Reference 150 for a recent review). In mammalian cells, the P450 system consists of an NADPH binding and FMN- and FAD-containing P450 reductase protein and an ironheme-O_2 activating P450 protein. These two proteins associate in the endoplasmic reticulum membrane where the P450 reductase is responsible for transferring electrons from NADPH through FAD-FMN to the ironheme of the P450 protein. At first, it may seem paradoxical that an oxidizing enzyme requires reducing equivalents, until it is realized that reductive activation of iron-bound molecular oxygen results in the formation of water and a highly oxidizing oxygen species which oxidizes the enzyme substrate. The overall reaction is shown below (S – H denotes a substrate).

$$\mathrm{NADPH} + \mathrm{O_2} + \mathrm{S-H} + 2\mathrm{H^+} \longrightarrow \mathrm{NADP^+} + \mathrm{H_2O} + \mathrm{S-OH} \qquad (15)$$

Thus, the reducing equivalents from NADPH are used to reduce molecular oxygen to water and an "oxygen atom" equivalent species which is highly reactive. A generalized schematic of the proposed catalytic cycle for P450 is shown in Figure 5. Briefly, substrate binding to the resting state ferric heme enzyme (1) to give a ferric heme-SH complex (2), promotes reduction of P450 by NADPH (via P450 reductase) to generate a ferrous heme-SH (3), which can then bind O_2 to give a ferrous heme-O_2-SH complex (5). The ferrous heme intermediate can also bind carbon monoxide, instead of O_2 to give an inhibitory CO-complex (4). Further reduction of the ferrous heme-O_2-SH complex by P450 reductase generates an iron-peroxo species (6). Alternatively, superoxide (O_2^-) may be released from (5) resulting in nonproductive "uncoupling" of the system. The iron-peroxo species is short lived and decomposes in the presence of protons to give water and the ultimate oxidant, a high-oxidation state iron-oxo complex (7) (this species is depicted as an Fe^V complex in Scheme 2

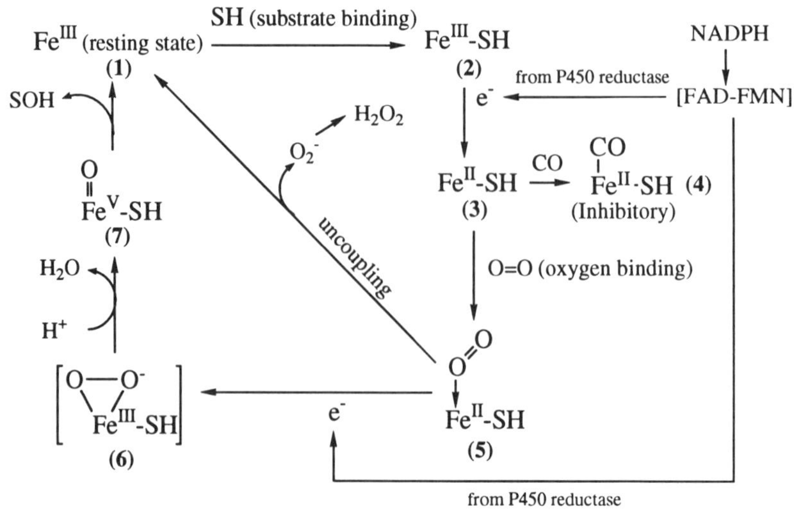

Figure 5 Catalytic cycle for P450. Fe represents an ironheme and S-H represents the substrate.

although it would probably be more accurately represented as an Fe^{IV}-porphyrin radical cation). Oxidation of substrate occurs and the oxidized product then dissociates from the active site thus regenerating the resting state Fe^{III} enzyme (1).

Whereas P450 is a two protein system, NOS is a single protein containing all relevant functions for P450-like catalysis, that is, both the reductase (electron transfer) and oxygen-activating domains. Significantly, there are bacterial P450 enzymes which are also "self-sufficient" in that both the reductase and heme components are contained in a single sequence.[151] A variety of similarities exist between the proposed catalytic chemistry of the P450 and NOS. For example, analogous to P450, NOS is an oxygenase which incorporates an oxygen atom from O_2 into citrulline[152] and NO.[114] Also, the ironheme of NOS has similar spectral characteristics to P450[110-112] and, when reduced, is inhibited by CO presumably by forming a ferrous heme-CO complex.[110,112,113] The FMN and FAD appear to serve as an electron conduit for the NADPH mediated reduction of the ironheme.[134] Also, like P450, NOS is capable of uncoupling to generate O_2^- and hydrogen peroxide.[153-155] Moreover, similar to documented reactivity of P450 reductase, NOS is able to reduce cytochrome c.[156] In fact, NOS has been found to have significant sequence homology with P450 reductase.[157]

POSSIBLE MECHANISMS OF NOS CATALYSIS

If indeed NOS converts L-arginine to L-citrulline and NO by a P450-like process, it would be expected that the initial step would be an N-hydroxylation to give N-hydroxy-L-arginine (one can envision SH in Figure 5 as the guanidinium function of arginine). This is the case as N-hydroxy-L-arginine is an intermediate in NO biosynthesis from L-arginine.[158-163] The stoichiometry of NADPH consumption for the conversion of L-arginine to N-hydroxy-L-arginine has been calculated to be 1 NADPH utilized per N-hydroxy-L-arginine produced and, interestingly, only 0.5 NADPH utilized for the conversion of N-hydroxy-L-arginine to NO[158] (Figure 6). The consumption of one equivalent of NADPH per N-hydroxy-L-arginine produced is consistent with a typical P450 process in that a two-electron oxidation of the substrate is carried out using a two-electron activation of molecular oxygen. P450 catalyzed N-hydroxylation is a well-known reaction,[164] and it has been recently shown that P450 is capable of N-hydroxylation of a guanidine containing compound.[165] Thus, the first step in NO biosynthesis appears to have some precedence in P450 chemistry. However, the subsequent step(s) in the process are not as straightforward or intuitively obvious.

The conversion of N-hydroxy-L-arginine to L-citrulline and NO is formally a three-electron oxidation of nitrogen. This odd electron requirement makes this step unprecedented in P450 chemistry since typical P450 catalyzed oxidations are even-electron processes. Although a peroxidatic P450 catalyzed oxidation of N-hydroxy-L-arginine has been reported to generate citrulline and nitrogen oxides,[166] unequivocal evidence for NO formation was not given. Chemical oxidation of model N-hydroxy guanidines reveals that HNO, and not NO, is the typical nitrogen oxide product.[167,168] This is not surprising since HNO would be the even, two-electron oxidation product. Interestingly, HNO has biological activity similar to NO in vascular systems,[169] and can be easily

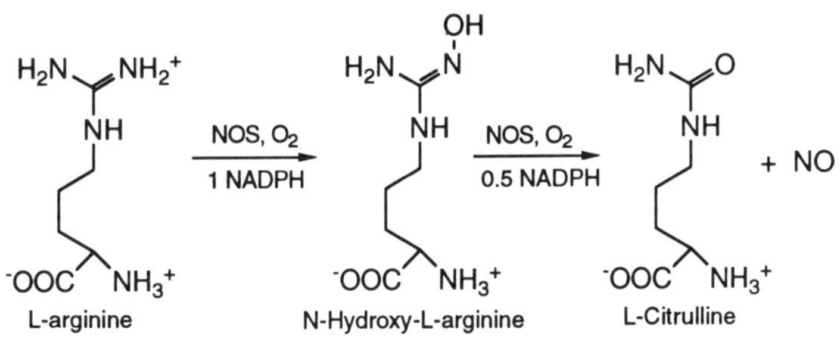

Figure 6 N-Hydroxy-L-arginine as an intermediate in the conversion of L-arginine to NO and L-citrulline.

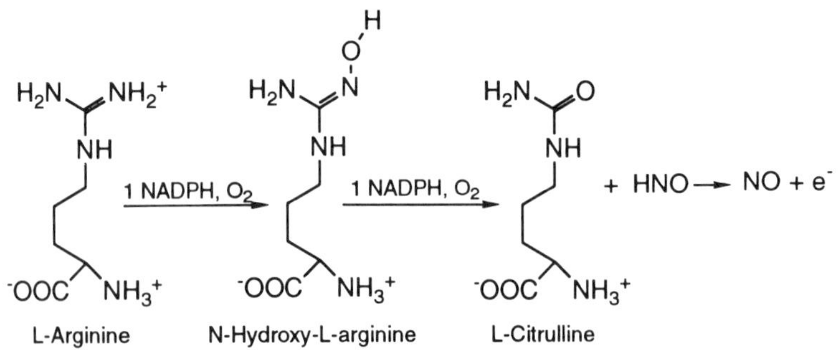

Figure 7 Possible HNO formation from NOS catalyzed oxidation of arginine.

converted to NO by a variety of relatively weak physiological oxidants.[170] Thus, HNO has been considered as a possible intermediate in NO biosynthesis.[109,169]

The odd electron requirement for NO synthesis necessitates a one-electron oxidation to occur at sometime during its formation since the biosynthetic precursors are all even electron species. Thus, proposed mechanistic schemes for NO generation from N-hydroxy-L-arginine all utilize a one-electron oxidation step at various times during the catalytic cycle. For example, as previously mentioned, if HNO were the immediate NOS nitrogen oxide product, a single electron oxidation (by a nonspecific or NOS mediated process) could give NO[109] (Figure 7). Donation of an electron from a biosynthetic intermediate back to the enzyme would account for the observed NADPH stoichiometry of 0.5 NADPH (one-electron) per NO formed from N-hydroxy-L-arginine. Other investigators have proposed that the biosynthetic intermediate, N-hydroxy-L-arginine, can act as a one-electron reducing agent (an electron being transferred to the enzyme) which then, as a radical cation, undergoes further NOS catalyzed oxidation.[171] Therefore, since the enzyme can receive an electron from the N-hydroxy-L-arginine intermediate, it only requires a single electron from NADPH for further P450-like catalysis. Marletta also proposed a similar mechanism whereby, again, N-hydroxy-L-arginine serves as a one-electron reducing agent for NOS and subsequent reactions can be viewed as similar to those performed by the cytochrome P450 aromatase.[172] (In aromatase, a nucleophilic dioxygen intermediate is the proposed oxidant.) Others have proposed a mechanism in which the N-hydroxy-L-arginine intermediate acts as a one-electron reducing agent for NOS prior to oxidation by a typical P450-like iron-oxo species.[173] Below is a generalized schematic of some of the proposed mechanisms for N-hydroxy-L-arginine oxidation (Figure 8). Pathway 1-2-3-4-5 (Figure 8) represents a mechanism by which N-hydroxy-L-arginine initially acts as a one-electron reducing agent following a one-electron donation from NADPH and binding of O_2. Subsequent attack of the radical cation intermediate by the ferrous-O_2^- species would then give an alkylperoxy-iron compound which may then decompose to give L-citrulline and arginine. Pathway 1-2-3-6-7-8 also relies on a reduction of the enzyme by N-hydroxy-L-arginine and NADPH. Addition of two protons (Step 6) and loss of water then generates the potent ferryl-oxo species (reminiscent of the ultimate oxidant generated in the proposed P450 catalytic cycle, Figure 5) which then oxidizes the radical cation intermediate to give a hydroxy-nitroso species which can then decompose to give NO and L-citrulline. Pathway 1-2-3-6-9-10 is similar except that epoxidation of the radical cation intermediate leads to oxazirine formation which then decomposes to L-citrulline and NO. All the mechanisms proposed in Figure 7 are highly speculative and are given primarily to indicate the necessity for a single-electron oxidation at some point during the catalytic cycle to account for the radical product, NO. In these mechanisms, N-hydroxy-L-arginine is shown as a reducing agent for the ferrous-O_2 complex instead of the ferric enzyme since it has been reported that addition of N-hydroxy-L-arginine to ferric NOS does not result in reduction to the ferrous form.[172]

So far, all mechanistic speculation has borrowed substantially from proposed P450 chemistry, considering the previously mentioned similarities between the two enzyme systems. It should be

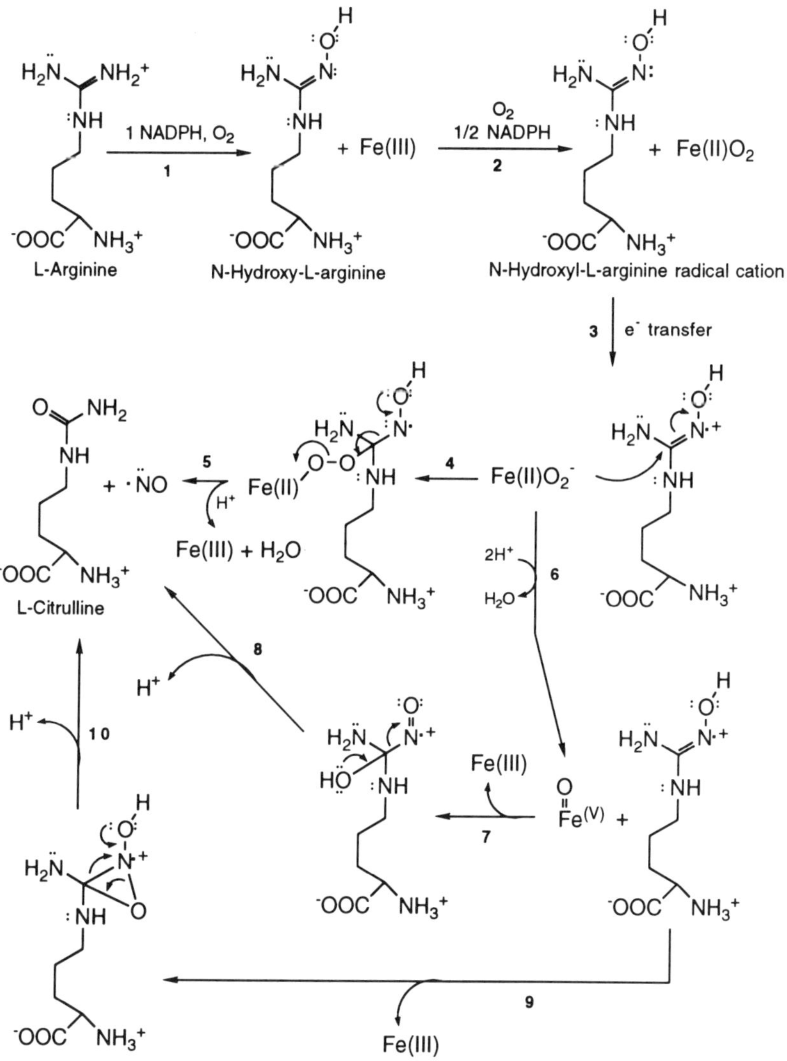

Figure 8 Possible mechanisms for the NOS catalyzed conversion of N-hydroxy-L-arginine to citrulline and NO.

realized, however, that the above discussions of mechanism have neglected a very important factor in NOS catalysis which definitely distinguishes it from P450: NOS has a requirement for H_4B while P450 does not. By analogy with P450, possible roles for ironheme, NADPH, FAD, and FMN in NOS catalysis can be envisioned and, in fact, mechanisms for NO biosynthesis (such as those shown in Figures 7 and 8) can be drawn independent of H_4B. Currently, the role of H_4B in NOS catalysis remains one of the foremost mysteries in the field.

THE POSSIBLE ROLE OF H_4B IN NOS CATALYSIS

There is no question that H_4B is required for maximum activity in all NOS isoforms.[120,121,174] H_4B is utilized by other oxidative enzyme systems as well and has been studied extensively. For example, H_4B is a required cofactor by the class of monooxygenases responsible for neurotransmitter biosynthesis represented by phenylalanine hydroxylase.[175] Its role in those systems has been proposed to be a direct activator of molecular oxygen via the formation of a 4a-hydroperoxybiopterin (Figure 9).[176] H_4B dependent monooxygenation of substrate by, for example, phenylalanine hy-

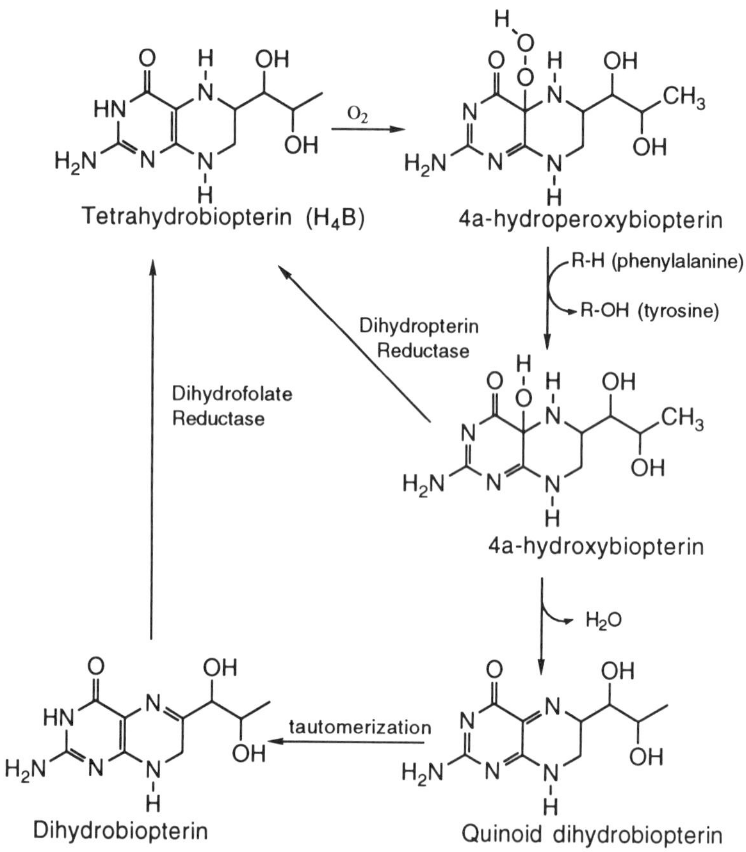

Figure 9 Redox scheme for H_4B in oxygenase enzymes.

droxylase generates an oxidized quinoid dihydropterin (H_2B-q), which has to be recycled by a reductase enzyme for further enzyme turnover. The recycling of H_2B-q to H_4B occurs when H_2B-q dissociates from the enzyme. Similar to cytochrome P450, the H_4B catalyzed process results in a net two-electron oxidation of substrate. These H_4B requiring enzyme systems also possess a nonheme iron located at the active site and substrate oxidation likely involves its participation as well.

Several studies have investigated the possible nature of the H_4B requirement by NOS. For example, it was found that citrulline formation in rat cerebellar NOS was not stoichiometric with H_4B utilization (>15 equivalents of citrulline produced per equivalent of H_4B consumed).[177] Moreover, exogenously added H_4B had little effect on the initial rate of the reaction. This study also indicated that H_4B was not recycled by NOS as measured by coupling H_4B synthesis with phenylalanine hydroxylase activity (this assumes, however, that H_4B must dissociate from NOS). Therefore, it was suggested that H_4B was not stoichiometrically redox active, but instead may be acting as an allosteric modulator or functioning to protect NOS by keeping a crucial catalytic component in the reduced state. It was, however, later shown that H_4B copurifies with both constitutive[120] and inducible[174] NOS and, since H_4B recycling can occur utilizing NADPH in the NOS system,[120] it is possible that redox cycling of H_4B occurs while still bound to the NOS enzyme. It is possible that enzyme-bound H_4B can redox cycle during enzyme turnover in a manner analogous to that shown in Figure 9, utilizing NADPH and significant levels of "free" H_4B or H_2B are not generated. Among other roles, H_4B may have an allosteric function since it promotes the formation of the active dimeric protein from its inactive monomers.[147] Using purified NOS from macrophages and H_4B analogs, other investigators have proposed a redox role for H_4B in NOS catalysis.[174] As alluded to earlier, this study proposes that H_2B-H_4B recycling may occur on the NOS protein without prior release of an oxidized biopterin intermediate.

To date, the function of H_4B in NOS catalysis remains an enigma. It seems unusual that NOS would have two reducing agents, H_4B and NADPH, "on board." Certainly, if the mechanistic analogy with P450 is correct, reductive oxygen activation by the ironheme is accomplished utilizing NADPH. Thus, the reducing equivalents of H_4B would likely have another role (of course the analogy with P450 may not be entirely correct). If, indeed, H_4B is redox active, it remains to be determined what the electron acceptor is, what it is used for, and why this acceptor will not utilize NADPH. It should be noted that the chemistry needed for NO production from arginine is likely to be multistep and significantly more complex than the single-step, two-electron oxidations typically mediated by P450. There is nothing prohibiting NOS from having chemistry distinct from P450 for one, or more, steps. H_4B may, indeed, be utilized by NOS in the direct generation of an active oxidant to carry out the requisite chemistry on arginine or a biosynthetic intermediate. Regardless, a significant amount of work is necessary to ascertain the role(s) of H_4B in NOS catalysis.

ALTERNATIVE SUBSTRATES AND INHIBITORS OF NOS

Studies on alternative substrates and inhibitors have helped elucidate the active site binding parameters and catalytic mechanisms of NOS. A number of NOS inhibitors have been discovered which act at a variety of sites on NOS. For example, diphenyleneiodonium compounds can be used on either iNOS or cNOS to irreversibly block the NADPH binding site.[178] Calmodulin inhibitors inhibit the constitutive NOS without affecting the inducible isoform.[179] As previously mentioned, CO represses iNOS and cNOS by binding the ironheme and blocking oxygen activation. Substituted imidazoles inhibit NOS presumably by binding the iron heme as well,[180] or by a combination of iron heme binding and interference with the role of calmodulin.[181] As with cytochrome P450,[182] NO has been found to inhibit cNOS presumably by blocking dioxygen binding to the iron heme.[183] Significantly, selective inhibition of iNOS has been observed for aminoguanidine,[184] and thus may represent a class of compounds of significant therapeutic value.

A variety of arginine analogs have been studied as NOS inhibitors and, due to their relative specificity, they constitute the majority of experimentally utilized NOS inhibitors. Interestingly, several arginine analogs show some isoform specificity and indicate that iNOS and cNOS may have slightly different substrate binding domains.[185,186] N-methyl-L-arginine was the first arginine-related NOS inhibitor discovered[187] and is one of the most widely utilized and studied analog. It is both a competitive and mechanism based, irreversible inhibitor of macrophage NOS.[188] N-methyl-L-arginine is processed like L-arginine, in that N-hydroxylation occurs, to give N^G-methyl-N^G-hydroxy-L-arginine, followed by L-citrulline formation.[171] Inhibition of macrophage NOS by N^G-methyl-N^G-hydroxy-L-arginine has both reversible and irreversible components.[189] The identification of the metabolic species responsible for NOS inhibition may provide significant insight into the chemistry performed by NOS since it would represent a non-NO product. N-Allyl and N-cyclopropyl-L-arginine analogs have also been studied as potential irreversible, suicide inhibitors of NOS.[190] The N-cyclopropyl analog is a reversible inhibitor whereas the N-allyl analog inhibited NOS by both reversible and irreversible pathways.

Several recent studies on the mechanism of inhibition by N-nitro-and N-methyl-L-arginine of endothelial cell[191] and brain NOS[192] indicate that both compounds bind NOS in a competitive and reversible manner. However, prolonged incubation times (>3 min) of the N-nitro analog cause reversible enzyme inactivation. Kinetic data indicate that both the association and dissociation of N-nitro-L-arginine with NOS is slow compared to that of the N-methyl analog. In addition, consistent with earlier work,[171,188] N-methyl-L-arginine was oxidized by NOS and a slow, turnover-dependent enzyme inactivation correlated with citrulline and NO formation. N-Nitro-L-arginine was not, however, metabolized by NOS suggesting a turnover-independent inactivation mechanism. Thus, based on the differential processing of the two analogs by NOS, the mechanism(s) of inactivation by N-methyl- and N-nitro-L-arginine are different.

Surprisingly, a compound with no structural similarity to arginine, phencyclidine, was reported to be metabolized by rat brain NOS resulting in suicide inhibition.[193] Phencyclidine and its analogs are also bioactivated by P450 which results in its mechanism-based irreversible inactivation.[194] These observations provide further evidence for the possible mechanistic link between P450 and NOS catalysis. This also raises the possibility that NOS may not have an overly specific substrate requirement and may metabolize arginine-unrelated xenobiotics. Thus, it is possible that NOS may be another drug metabolizing enzyme system and that drug-mediated inhibition of NOS may play a role in otherwise unwanted drug side effects.

CONCLUSION

Clearly, NO biosynthesis is an important topic due to its ubiquitous and diverse biological functions. Pharmacological manipulation of physiological NO levels may prove to be therapeutically useful in the treatment of a variety of diseases and disorders. Although a fair amount is understood regarding the reaction pathway of NO biosynthesis (L-arginine → N-hydroxy-L-arginine → citrulline and NO), and the NOS enzyme from a variety of sources has been relatively well characterized, the chemical mechanism of NO production remains less well understood. The discovery of this enzyme occurred only a few years ago (1989) and, considering the complexity of the chemistry and the wide array of cofactors and prosthetic groups which must be considered, credit must be given to those referenced herein that we know as much as we do in such a short time. Certainly, ongoing research in this area will provide a more substantial understanding of NO biosynthesis in the future.

REFERENCES

1. **Furchgott, R. F.,** Studies on relaxation of rabbit aorta by sodium nitrite: the basis for the proposal that the acid-activatable inhibitory factor from retractor penis in inorganic nitrite and the endothelium-derived relaxing factor is nitric oxide, in *Vasodilation: Vascular Smooth Muscle, Peptides, Autonomic Nerves and Endothelium,* Vanhoutte, P. M., Ed., Raven Press, New York, 1988, 401.
2. **Ignarro, L. J., Byrns, R. E., and Woods, K. S.,** Biochemical and pharmacological properties of endothelium-derived relaxing factor and its similarity to nitric oxide radical, in *Vasodilation: Vascular Smooth Muscle, Peptides, Autonomic Nerves and Endothelium,* Vanhoutte, P. M., Ed., Raven Press, New York, 1988, 427.
3. **Ignarro, L. J., Buga, G. M., Wood, K. S., Byrns, R. E., and Chaudhuri, G.,** Endothelium-derived relaxing factor produced and released from artery and vein is nitric oxide, *Proc. Natl. Acad. Sci. U.S.A.,* 84, 9265, 1987.
4. **Palmer, R. M. J., Ferrige, A. G., and Moncada, S.,** Nitric oxide release accounts for the biological activity of endothelium-derived relaxing factor, *Nature (London),* 327, 524, 1987.
5. **Marletta, M. A., Yoon, P. S., Iyengar, R., Leaf, C. D., and Wishnok, J. S.,** Macrophage oxidation of L-arginine to nitrite and nitrate: nitric oxide is an intermediate, *Biochemistry,* 21, 8706, 1988.
6. **Hibbs, J. B., Taintor, R. R., Vavrin, Z., and Rachlin, E. M.,** Nitric oxide: a cytotoxic activated macrophage effector molecule, *Biochem. Biophys. Res. Commun.,* 157(1), 87, 1988.
7. **Stuehr, D. J. and Nathan, C. F.,** Nitric oxide: a macrophage product responsible for cytostasis and respiratory inhibition in tumor target cells, *J. Exp. Med.,* 169, 1543, 1989.
8. **Snyder, S. H.,** Nitric oxide: first in a new class of neurotransmitters, *Science,* 257, 494, 1992.
9. **Verma, A., Hirsch, D. J., Glatt, C. E., Ronnett, G. V., and Snyder, S. H.,** Carbon monoxide: a putative neural messenger, *Science,* 259, 381, 1993.
10. **Stamler, J. S., Singel, D. J., and Loscalzo, J.,** Biochemistry of nitric oxide and its redox active forms, *Science,* 258, 1898, 1992.
11. **Lipton, S. A., Choi, Y.-B., Pan, Z.-H., Lei, S. Z., Chen, H.-S. V., Sucher, N. J., Loscalzo, J., Singell, D. J., and Stamler, J. S.,** A redox based mechanism for neuroprotective and neurodestructive effects of nitric oxide and related nitroso-compounds, *Nature (London),* 364, 626, 1993.
12. **Ford, P. C., Wink, D. A., and Stanbury, D. M.,** Autoxidation kinetics of aqueous nitric oxide, *FEBS Lett.,* 326(1-3), 1, 1993.
13. **Ignarro, L. J., Fukuto, J. M., Griscavage, J. M., Rogers, N. E., and Byrns, R. E.,** Oxidation of Nitric Oxide in Aqueous Solution to Nitrite but not Nitrate: Comparison of Enzymatically Formed Nitric Oxide from L-Arginine, *Proc. Natl. Acad. Sci. U.S.A.,* 90(17), 8103, 1993.

14. **Blough, N. V. and Zafiriou, O. C.,** Reaction of superoxide with nitric oxide to form peroxynitrite in alkaline aqueous solution, *Inorg. Chem.*, 24, 3502, 1985.
15. **Huie, R. E. and Padmaja, S.,** The reaction of NO with superoxide, *Free Rad. Res. Commun.*, 18(4), 195, 1993.
16. **Mahoney, L. R.,** Evidence for the formation of hydroxyl radicals in the isomerization of pernitrous acid to nitric acid in aqueous solution, *J. Am. Chem. Soc.*, 92(17), 5262, 1970.
17. **Koppenol, W. H., Moreno, J. J., Pryor, W. A., Ischiropoulos, H., and Beckman, J. S.,** Peroxynitrite, a cloaked oxidant formed by nitric oxide and superoxide, *Chem. Res. Toxicol.*, 5, 834, 1992.
18. **Pryor, W. A., Church, D. F., Govidan, C. K., and Crank, G.,** Oxidation of thiols by nitric oxide and nitrogen dioxide: synthetic utility and toxicological implications, *J. Org. Chem.*, 47, 156, 1982.
19. **Wade, R. S. and Castro, C. E.,** Redox reactivity of iron(III) porphyrins and heme proteins with nitric oxide. Nitrosyl transfer to carbon, oxygen, nitrogen and sulfur, *Chem. Res. Toxicol.*, 3, 289, 1990.
20. **Case, G. D., Dixon, J. S., and Schooley, J. C.,** Interactions of blood metalloproteins with nitrogen oxides and oxidant air pollutants, *Environ. Res.*, 20, 43, 1979.
21. **Traylor, T. G. and Sharma, V. S.,** Why NO?, *Biochemistry*, 31(11), 2847, 1992.
22. **Traylor, T. G., Duprat, A. F., and Sharma, V. S.,** Nitric oxide-triggered heme-mediated hydrolysis: a possible model for biological reactions of NO, *J. Am. Chem. Soc.*, 115, 810, 1993.
23. **Bonner, F. T. and Hughes, M. N.,** The aqueous solution chemistry of nitrogen in low positive oxidation states, *Comments Inorg. Chem.*, 7(4), 215, 1988.
24. **McCleverty, J. A.,** Reactions of nitric oxide coordinated to transition metals, *Chem. Rev.*, 79(1), 53, 1979.
25. **Schwartz, S. E. and White, W. H.,** Kinetics of reactive dissolution of nitrogen oxides into aqueous solution, in *Trace Atmospheric Constituents, Properties, Transformation and Fates*, Schwartz, S. E., Ed., Wiley and Sons, New York, 1983, 1.
26. **Vosper, A. J.,** Nitrogen oxides and oxyacids, in *Int. Rev. Sci., Inorg. Chem., Series 2,* Sowerty, D. B., Ed., University Park Press, Baltimore, 1975, 123.
27. **Gilbert, J. R. and Thomas, J. H.,** Reactions of the oxides of nitrogen, in *Comprehensive Chemical Kinetics*, Vol. 6, Bamford, C. H. and Tipper, C. F. H., Eds., Elsevier, New York, 1972, chapter 2.
28. **Cotton, F. A. and Wilkinson, G.,** *Advanced Inorganic Chemistry*, 5th ed., Part 2, Wiley and Sons, New York, 1988, chapter 10.
29. **Culotta, E. and Koshland, D. E. Jr.,** Molecule of the year, NO news is good news, *Science*, 258, 1862, 1992.
30. **Moncada, S., Palmer, R. M. J., and Higgs, E. A.,** Nitric oxide: physiology, pathophysiology and pharmacology, *Pharmacol. Rev.*, 43(2), 109, 1991.
31. **Nathan, C.,** Nitric oxide as a secretory product of mammalian cells, *FASEB J.*, 6, 3051, 1992.
32. **Feldman, P. L, Griffith, O. W., and Stuehr, D. J.,** The surprising life of nitric oxide, *Chem. Eng. News*, Dec. 20, 26, 1993.
33. **Gibaldi, M.,** What is nitric oxide and why are so many people studying it, *J. Clin. Pharmacol.*, 33, 488, 1993.
34. **Bredt, D. S. and Snyder, S. H.,** Nitric oxide, a novel neuronal messenger, *Neuron*, 8, 3, 1992.
35. **Lancaster, J. R.,** Nitric oxide in cells, *Am. Sci.*, 80, 248, 1992.
36. **Calver, A., Collier, J., and Vallance, P.,** Nitric oxide and cardiovascular control, *Exp. Physiol.*, 78, 303, 1993.
37. **Feng, Q. and Hedner, T.,** Endothelium-derived relaxing factor (EDRF) and nitric oxide (NO). I. Physiology, pharmacology and pathophysiological implications, *Clin. Physiol.*, 10, 407, 1990a.
38. **Feng, Q. and Hedner, T.,** Endothelium-derived relaxing factor (EDRF) and nitric oxide (NO). II. Physiology, pharmacology and pathophysiological implications, *Clin. Physiol.*, 10, 503, 1990b.
39. **Stefanovic-Racic, M., Stadler, J., and Evans, C. H.,** Nitric oxide and arthritis, *Arthritis Rheum.*, 36(8), 1036, 1993.
40. **Furchgott, R. F. and Zawadzki, J. V.,** The obligatory role of endothelial cells in the relaxation of arterial smooth muscle by acetylcholine, *Nature (London)*, 288, 373, 1980.
41. **Griffith, T. M., Edwards, D. H., Lewis, M. J., Newby, A. C. and Henderson, A. H.,** The nature of endothelium-derived relaxing factor, *Nature (London)*, 308, 645, 1984.
42. **Rapoport, R. M. and Murad, F.,** Agonist-induced endothelium-dependent relaxations in rat thoracic aorta may be mediated through cGMP, *Circ. Res.*, 52, 352, 1983.
43. **Furchgott, R. F., Cherry, P. D., and Zawadzki, J. V.,** Endothelial cells as mediators of vasodilation of arteries, *J. Cardiovasc. Pharmacol.*, 6 (Suppl. 2), 336, 1984.
44. **Diamond, J. and Chu, E. S.,** Possible role for cyclic GMP in endothelium-dependent relaxation of rabbit aorta by acetylcholine: comparison with nitroglycerin, *Res. Commun. Chem. Pathol. Pharmacol.*, 41, 369, 1983.
45. **Martin, W., Villani, G. M., Jothianandan, D., and Furchgott, R. F.,** Selective blockade of endothelium-dependent and glyceryl-trinitrite-induced relaxation by hemoglobin and methylene blue in the rabbit aorta, *J. Pharmacol. Exp. Ther.*, 232, 708, 1985.
46. **Rubanyi, G. M. and Vanhoutte, P. M.,** Superoxide anions and hyperoxia inactivate endothelium-derived relaxing factor, *Am. J. Physiol.*, 250, H822, 1986.
47. **Gryglewski, P. J., Palmer, R. M., and Moncada, S.,** Superoxide anion is involved in the breakdown of endothelium-derived relaxing factor, *Nature (London)*, 320, 454, 1986.

48. **Palmer, R. M. J., Ashton, D. S., and Moncada, S.,** Vascular endothelial cells synthesize nitric oxide from L-arginine, *Nature (London)*, 333, 664, 1988.
49. **Long, C. L., Shikano, K., and Berkowitz, B. A.,** Anion exchange resins discriminate between nitric oxide and EDRF, *Eur. J. Pharmacol.*, 142, 317, 1987.
50. **Greenberg, S. S., Wilcox, D. E., and Rubanyi, G. M.,** Endothelium-derived relaxing factor released from canine femoral artery by acetylcholine cannot be identified as free nitric oxide by electron paramagnetic spectroscopy, *Circ. Res.*, 67(6), 1447, 1990.
51. **Cocks, T. M., Angus, J. A., Cambell, J. H., and Cambell, G. R.,** Release and properties of endothelium-derived relaxing factor (EDRF) from endothelial cells in culture, *J. Cell. Physiol.*, 123, 310, 1985.
52. **Shikano, K., Ohlstein, E. H., and Berkowitz, B. A.,** Differential selectivity of endothelium-derived relaxing factor and nitric oxide in smooth muscle, *Br. J. Pharmacol.*, 92, 483, 1987.
53. **Myers, P. R., Guerra, R., and Harrison, D. G.,** Release of NO and EDRF from cultures bovine aortic endothelial cells, *Am. J. Physiol.*, 256, H1030, 1989.
54. **Marshall, J. J., Wei, E. P., and Kontos, H. A.,** Independent blockade of cerebral vasodilation from acetylcholine and nitric oxide, *Am. J. Physiol.*, 255, H847, 1988.
55. **Myers, P. R., Guerra, R., and Harrison, D. G.,** Release of multiple endothelium-derived relaxing factors from porcine coronary arteries, *J. Cardiovasc. Pharmacol.*, 20, 392, 1992.
56. **Rosenblum, W. I.,** Endothelium-derived relaxing factor in brain vessels is not nitric oxide, *Stroke*, 23(10), 527, 1992.
57. **Rubanyi, G. M., Johns, A., Wilcox, D., Bates, F. N., and Harrison, D.,** Evidence that a nitrosothiol, but not nitric oxide, may be identical with endothelium-derived relaxing factor, *J. Cardiovasc. Pharmacol.*, 17 (Suppl. 3), 41, 1991.
58. **Vedernikov, Y. P., Mordvintcev, P. I., Malenkova, I. V., and Vanin, A. F.,** Similarity between the vasorelaxing activity of dinitrosyliron cysteine complexes and endothelium-derived relaxing factor, *Eur. J. Pharmacol.*, 211, 313, 1992.
59. **Vanin, A. F.,** Endothelium-derived relaxing factor is a nitrosyl iron complex with thiol ligands, *FEBS Lett.*, 289(1), 1, 1991.
60. **Stamler, J., Simon, D. I., Osborne, J. A., Mullins, M. E., Jaraki, O., Michel, T., Singel, D. J., and Loscalzo, J.,** S-Nitrosylation of proteins with nitric oxide: synthesis and characterization of biologically active compounds, *Proc. Natl. Acad. Sci. U.S.A.*, 89, 444, 1992.
61. **Stamler, J. S., Simon, D. I., Jaraki, O., Osborne, J. A., Francis, S., Mullins, M., Singel, D., and Loscalzo, J.,** S-Nitrosylation of tissue-type plasminogen activator confers vasodilatory and antiplatelet properties on the enzyme, *Proc. Natl. Acad. Sci. U.S.A.*, 89, 8087, 1992.
62. **Stamler, J. S., Jaraki, O., Osborne, J., Simon, D. I., Keaney, J., Vita, J., Singel, D., Valeri, C. R., and Loscalzo, J.,** Nitric oxide circulates in mammalian plasma primarily as an S-nitroso adduct of serum albumin, *Proc. Natl. Acad. Sci. U.S.A.*, 89, 7674, 1992.
63. **Ignarro, L. J.,** Nitric oxide-mediated vasorelaxation, *Thromb. Haemost.*, 70(1), 148, 1993.
64. **Radomski, M. W. and Moncada, S.,** Regulation of vascular homeostasis by nitric oxide, *Thromb. Haemost.*, 70(1), 36, 1993.
65. **Waldman, S. A. and Murad, F.,** Cyclic GMP synthesis and function, *Pharmacol. Rev.*, 39(3), 163, 1987.
66. **Schmidt, H. H. H. W., Lohman, S. M., and Walter, U.,** The nitric oxide and cGMP signal transduction system: regulation and mechanism of action, *Biochim. Biophys. Acta*, 1178, 153, 1993.
67. **Ahlner, J., Andersson, R. G. G., Torfgard, K., and Axelsson, K. L.,** Organic nitrate esters: clinical use and mechanisms of action, *Pharmacol. Rev.*, 43(3), 351, 1991.
68. **Harrison, D. G. and Bates, J. N.,** The nitrovasodilators, new ideas about old drugs, *Circulation*, 87(5), 1461, 1993.
69. **Mellion, B. T., Ignarro, L. J., Ohlstein, E. M., Pontecorvo, E. G., Hyman, A. L. and Kadowitz, P. J.,** Evidence for the inhibitory role of guanosine 3′, 5′-monophophate in ADP-induced human platelet aggregation in the presence of nitric oxide and related vasodilators, *Blood*, 57, 946, 1987.
70. **Radomski, M. W. and Moncada, S.,** Role of nitric oxide in endothelial cell platelet interactions, in *Antithrombotics, Pathophysiological Rationale for Pharmacologic Interventions*, Herman, A. G., Ed., Kluwer Academic Publishers, Drodrecht, 1991, 27.
71. **McCall, T. B., Boughton-Smith, N. K., Palmer, R. M. J., Whittle, B. J. R., and Moncada, S.,** Synthesis of nitric oxide from L-arginine by neutrophils. Release and interaction with superoxide ion, *Biochem. J.*, 261, 293, 1989.
72. **Radomski, M. W., Palmer, R. M. J., and Moncada, S.,** The role of nitric oxide and cGMP in platelet adhesion to vascular endothelium, *Biochem. Biophys. Res. Commun.*, 148, 1482, 1987.
73. **Bath, P. M. W., Hassall, D. G., Gladwin, A. M., Palmer, R. M. J., and Martin, J. F.,** Nitric oxide and prostacyclin. Divergence of inhibitory effects on monocyte chemotaxis and adhesion to endothelium *in vitro*, *Atheroscler. Thromb.*, 11, 254, 1991.
74. **Radomski, M. W., Palmer, R. M. J., and Moncada, S.,** The anti-aggregating properties of vascular endothelium: Interaction between prostacyclin and nitric oxide, *Br. J. Pharmacol.*, 92, 639, 1987.
75. **Green, L. C., Wagner, D. A., De Luzuriage, R. K., Istfan, N., Young, V. R., and Tannenbaum, S. R.,** Nitrate biosynthesis in man, *Proc. Natl. Acad. Sci. U.S.A.*, 78, 7764, 1981.

76. **Green, L. C., Tannenbaum, S. R., and Goldman, R.,** Nitrate synthesis in the germfree and conventional rat, *Science*, 212, 56, 1981.
77. **Wagner, D. A., Young, V. R., and Tannenbaum, S. R.,** Incorporation of $^{15}NH_3$ into nitrate is enhanced by endotoxin treatment, *Proc. Natl. Acad. Sci. U.S.A.*, 80, 4518, 1983.
78. **Stuehr, D. J. and Marletta, M. A.,** Mammalian nitrate biosynthesis: Mouse macrophages produce nitrite and nitrate in response to Escherichia coli lipopolysaccharide, *Proc. Natl. Acad. Sci. U.S.A.*, 82, 7738, 1985.
79. **Iyengar, R., Stuehr, D. J., and Marletta, M. A.,** Macrophage synthesis of nitrite, nitrate and nitrosamines: precursors and role in respiratory burst, *Proc. Natl. Acad. Sci. U.S.A.*, 84, 6369, 1987.
80. **James, S. L. and Hibbs, J. B.,** The role of nitrogen oxides as effector molecules of parasite killing, *Parasitol. Today*, 6(9), 303, 1990.
81. **Kolb, H. and Kolb-Bachofen, V.,** Nitric oxide: A pathogenic factor in autoimmunity, *Immunol. Today*, 13(5), 157, 1992.
82. **Keller, R. and Keist, R.,** Abilities of activated macrophages to manifest tumoricidal activity and to generate reactive nitrogen intermediates: a comparative study *in vitro* and ex vivo, *Biochem. Biophys. Res. Commun.*, 164(3), 968, 1989.
83. **Lancaster, J. R. and Hibbs, J. B.,** EPR demonstration of iron-nitrosyl complex formation by cytotoxic activated macrophages, *Proc. Natl. Acad. Sci. U.S.A.*, 87, 1223, 1990.
84. **Beckman, J. S. and Crow, J. P.,** Pathological implications of nitric oxide, superoxide and peroxynitrite formation, *Biochem. Soc. Trans.*, 21, 330, 1993.
85. **Radi, R., Beckman, J. S., Bush, K., and Freeman, B. A.,** Peroxynitrite oxidation of sulfhydryls, *J. Biol. Chem.*, 266(7), 4244, 1991.
86. **Radi, R., Beckman, J. S., Bush, K. M., and Freeman, B. A.,** Peroxynitrite-induced membrane lipid peroxidation: the cytotoxic potential of superoxide and nitric oxide, *Arch. Biochem. Biophys.*, 288(2), 481, 1991.
87. **Ferrendelli, J. A., Chang, M. M., and Kinscherf, D. A.,** Elevation of cyclic GMP levels in the central nervous system by excitatory and inhibitory amino acids, *J. Neurochem.*, 22, 535, 1974.
88. **Garthwaite, J.,** Excitatory amino acid receptors and guanosine 3′, 5′-cyclic monophosphate in incubated slices of immature and adult rat cerebellum, *Neuroscience*, 7(10), 2491, 1982.
89. **Miki, N., Kawabe, Y., and Kuriyama, K.,** Activation of cerebral guanylate cyclase by nitric oxide, *Biochem. Biophys. Res. Commun.*, 75, 851, 1977.
90. **Deguchi, T.,** Endogenous activating factor for guanylate cyclase in synaptosomal -soluble fraction of rat brain, *J. Biol. Chem.*, 252, 7617, 1977.
91. **Deguchi, T. and Yoshioka, M.,** L-Arginine identified as an endogenous activator for soluble guanylate cyclase from neuroblastoma cells, *J. Biol. Chem.*, 257, 10147, 1982.
92. **Garthwaite, J., Charles, S. L., and Chess-Williams, R.,** Endothelium-derived relaxing factor release on activation of NMDA receptors suggests a role as intercellular messenger in the brain, *Nature (London)*, 336, 385, 1988.
93. **Knowles, R. G., Palacios, M., Palmer, R. M. J., and Moncada, S.,** Formation of nitric oxide from L-arginine in the central nervous system: a transduction mechanism for stimulation of the soluble guanylate cyclase, *Proc. Natl. Acad. Sci. U.S.A.*, 86, 5159, 1989.
94. **Bredt, D. S. and Snyder, S. H.,** Nitric oxide mediates glutamate-linked enhancement of cGMP levels in cerebellum, *Proc. Natl. Acad. Sci. U.S.A.*, 86, 9030, 1989.
95. **Garthwaite, J.,** Glutamate, nitric oxide and cell-cell signaling in the nervous system, *Trends Neurosci.*, 14(2), 60, 1991.
96. **O'Dell, T. J., Hawkins, R. D., Kandel, E. R., and Arancio, O.,** Tests of the roles of two diffusable substances in long-term potentiation: evidence for nitric oxide as a possible early retrograde messenger, *Proc. Natl. Acad. Sci. U.S.A.*, 88, 11285, 1991.
97. **Schuman, E. M. and Madison, D. V.,** A requirement for the intercellular messenger nitric oxide in long-term potentiation, *Science*, 254, 1503, 1991.
98. **Böhme, G. A., Bon, C., Stutzman, J.-M., Doble, A., and Blanchard, J.-C.,** Possible involvement of nitric oxide in long-term potentiation, *Eur. J. Pharmacol.*, 199, 379, 1991.
99. **Bon, C., Böhme, G. A., Doble, A., Stutzman, J.-M., and Blanchard, J.-C.,** A role for nitric oxide in long-term potentiation, *Eur. J. Neurosci.*, 4, 420, 1992.
100. **Haley, J. E., Wilcox, G. L., and Chapman, P. F.,** The role of nitric oxide in hippocampal long-term potentiation, *Neuron*, 8, 211, 1992.
101. **Dawson, T. M., Dawson, V. L., and Snyder, S. H.,** A novel neuronal messenger molecule in brain: The free radical, nitric oxide, *Ann. Neurol.*, 32(3), 297, 1992.
102. **Li, C. G. and Rand, M. J.,** Evidence for a role of nitric oxide in the neurotransmitter system mediating relaxation of the rat anococcygeus muscle, *Clin. Exp. Pharmacol. Physiol.*, 16, 933, 1989.
103. **Gillespie, J. S., Liu, X., and Martin, W.,** The effect of L-arginine and NG-monomethyl-L-arginine on the response of the rat anoccoccygeus to NANC nerve stimulation, *Br. J. Pharmacol.*, 98, 1080, 1989.
104. **Bult, H., Boeckxstaens, G. E., Pelckmans, P. A., Jordeans, F. H., VanMaercke, Y. M., and Herman, A. G.,** Nitric oxide as an inhibitory non-adrenergic, non-cholinergic neurotransmitter, *Nature (London)*, 345, 346, 1990.

105. **Hobbs, A. J., Tucker, J. F., and Gibson, A.,** Differentiation by hydroquinone of relaxations induced by exogenous and endogenous nitrites in non-vascular smooth muscle: role of superoxide anions, *Br. J. Pharmacol.*, 104, 645, 1991.
106. **Beckman, J. S.,** The double-edged role of nitric oxide in brain function and superoxide-mediated injury, *J. Dev. Physiol.*, 15, 53, 1991.
107. **Dawson, V. L., Dawson, T. M., London, E. D., Bredt, D. S., and Snyder, S. H.,** Nitric oxide mediates glutamate neurotoxicity in primary cortical cultures, *Proc. Natl. Acad. Sci. U.S.A.*, 88, 6368, 1991.
108. **Bush, P. A., Gonzalez, N. E., Griscavage, J. M., and Ignarro, L. J.,** Nitric oxide synthase from cerebellum catalyzes the formation of equimolar quantities of nitric oxide and citrulline from L-arginine, *Biochem. Biophys. Res. Commun.*, 185(3), 960, 1992.
109. **Stuehr, D. J. and Griffith, O. W.,** Mammalian nitric oxide synthases, in *Advances in Enzymology*, Meister, A., Ed., Wiley and Sons, New York, 1992, 287.
110. **White, K. A. and Marletta, M. A.,** Nitric oxide synthase is a cytochrome P-450 type hemoprotein, *Biochemistry*, 31(29), 6627, 1992.
111. **Stuehr, D. J. and Ikeda-Saito, M.,** Spectral characterization of brain and macrophage nitric oxide synthase, *J. Biol. Chem.*, 267(29), 20547, 1992.
112. **Klatt, P., Schmidt, K., and Mayer, B.,** Brain nitric oxide synthase is a haemoprotein, *Biochem. J.*, 288, 15, 1992.
113. **McMillan, K., Bredt, D. S., Hirsch, D. J., Snyder, S. H., Clark, J. E., and Masters, B. S. S.,** Cloned, expressed rat cerebellar nitric oxide synthase contains stoichiometric amounts of heme, which binds carbon monoxide, *Proc. Natl. Acad. Sci. U.S.A.*, 89, 11141, 1992.
114. **Leone, A. M., Palmer, R. M. J., Knowles, R. G., Francis, P. L., Ashton, D. S., and Moncada, S.,** Constitutive and inducible nitric oxide synthases incorporate molecular oxygen into both nitric oxide and citrulline, *J. Biol. Chem.*, 266(35), 23790, 1991.
115. **Palmer, R. M. J. and Moncada, S.,** A novel citrulline-forming enzyme implicated in the formation of nitric oxide by vascular endothelial cells, *Biochem. Biophys. Res. Commun.*, 158(1), 348, 1989.
116. **Stuehr, D. J., Kwon, N. S., Gross, S. S., Thiel, B. A., Levi, R., and Nathan, C. F.,** Synthesis of nitrogen oxides from L-arginine by macrophage cytosol: Requirement for inducible and constitutive components, *Biochem. Biophys. Res. Commun.*, 161(2), 420, 1989.
117. **Stuehr, D. J., Kwon, N. S., and Nathan, C. F.,** FAD and GSH participate in macrophage synthesis of nitric oxide, *Biochem. Biophys. Res. Commun.*, 168(2), 558, 1990.
118. **Hevel, J. M., White, K. A., and Marletta, M. A.,** Purification of the inducible murine macrophage nitric oxide synthase, *J. Biol. Chem.*, 266(34), 22789, 1991.
119. **Stuehr, D. J., Cho, H. J., Kwon, N. S., Weise, M. F., and Nathan, C. F.,** Purification and characterization of the cytokine-induced macrophage nitric oxide synthase: an FAD- and FMN-containing flavoprotein, *Proc. Natl. Acad. Sci. U.S.A.*, 88, 7773, 1991.
120. **Mayer, B., John, M., Heinzel, B., Werner, E. R., Wachter, H., Schultz, G., and Böhme, E.,** Brain nitric oxide synthase is a biopterin- and flavin-containing multi-functional oxido-reductase, *FEBS Lett.*, 288(1,2), 187, 1991.
121. **Kwon, N. S., Nathan, C. F., and Stuehr, D. J.,** Reduced biopterin as a cofactor in the generation of nitrogen oxides by murine macrophages, *J. Biol. Chem.*, 264(34), 20496, 1989.
122. **Forstermann, U., Schmidt, H. H. H. W., Pollock, J. S., Sheng, H., Mitchell, J. A., Warner, T. D., Nakane, M., and Murad, F.,** Isoforms of nitric oxide synthase, *Biochem. Pharmacol.*, 42(10), 1849, 1991.
123. **Bredt, D. S. and Snyder, S. H.,** Isolation of nitric oxide synthetase, a calmodulin-requiring enzyme, *Proc. Natl. Acad. Sci. U.S.A.*, 87, 682, 1990.
124. **Schmidt, H. H. H. W., Pollock, J. S., Nakane, M., Gorsky, L. D., Förstermann, U., and Murad, F.,** Purification of a soluble isoform of guanylyl cyclase-activating-factor synthase, *Proc. Natl. Acad. Sci. U.S.A.*, 88, 365, 1991.
125. **Hiki, K., Hattori, R., Kawai, C., and Yui, Y.,** Purification of insoluble nitric oxide synthase from rat cerebellum, *J. Biochem.*, 111, 556, 1992.
126. **Mayer, B., John, M., and Böhme, E.,** Purification of a Ca^{+2}/calmodulin-dependent nitric oxide synthase from porcine cerebellum, *FEBS Lett.*, 277(1-2), 215, 1990.
127. **Ohshima, H., Oguchi, S., Adachi, H., Iida, S., Suzuki, H., Sugimura, T., and Esumi, H.,** Purification of nitric oxide synthase from bovine brain: immunological characterization and tissue distribution, *Biochem. Biophys. Res. Commun.*, 183(1), 238, 1992.
128. **Schmidt, H. H. H. W. and Murad, F.,** Purification and characterization of a human NO synthase, *Biochem. Biophys. Res. Commun.*, 181(3), 1372, 1991.
129. **Pollock, J. S., Forstermann, U., Mitchell, J. A., Warner, T. D., Schmidt, H. H. H. W., Nakane, M., and Murad, F.,** Purification and characterization of particulate endothelium-derived relaxing factor synthase from cultured and native bovine aortic endothelial cells, *Proc. Natl. Acad. Sci. U.S.A.*, 88, 10480, 1991.
130. **Yui, Y., Hattori, R., Kosuga, K., Eizawa, H., Hiki, K., and Kawai, C.,** Purification of nitric oxide synthase from rat macrophages, *J. Biol. Chem.*, 266(19) 12544, 1991.

131. **Evans, T., Carpenter, A., and Cohen, J.,** Purification of a distinctive form of endotoxin-induced nitric oxide synthase from rat liver, *Proc. Natl. Acad. Sci. U.S.A.*, 89, 5361, 1992.
132. **Iida, S., Ohshima, H., Oguchi, S., Hata, T., Suzuki, H., Kawasaki, and Esumi, H.,** Identification of inducible calmodulin-dependent nitric oxide synthase in the liver of rats, *J. Biol. Chem.*, 267(35), 25385, 1992.
133. **Cho, H. J., Xie, Q.-W., Calatcay, J., Mumford, R. A., Swiderek, K. M., Lee, T. D., and Nathan, C.,** Calmodulin is a subunit of nitric oxide synthase from macrophages, *J. Exp. Med.*, 176, 599, 1992.
134. **Abu-Soud, H. M. and Stuehr, D. J.,** Nitric oxide synthases reveal a role for calmodulin in controlling electron transfer, *Proc. Natl. Acad. Sci. U.S.A.*, 90, 10769, 1993.
135. **Bredt, D. S., Hwang, P. M., Glatt, C. E., Lowenstein, C., Reed, R. R., and Snyder, S. H.,** Cloned and expressed nitric oxide synthase structurally resembles cytochrome P450 reductase, *Nature (London)*, 351, 714, 1991.
136. **Nakane, M., Mitchell, J., Forstermann, U., and Murad, F.,** Phosphorylation by calcium-calmodulin-dependent protein kinase II and protein kinase C modulates the activity of nitric oxide synthase, *Biochem. Biophys. Res. Commun.*, 180(3), 1396, 1991.
137. **Brune, B. and Lapetina, E. G.,** Phosphorylation of nitric oxide by protein kinase A, *Biochem. Biophys. Res. Commun.*, 181(2), 921, 1991.
138. **Bredt, D. S., Ferris, C. D., and Snyder, S. H.,** Nitric oxide synthase regulatory sites, *J. Biol. Chem.*, 267(16), 10976, 1992.
139. **Gross, S. S. and Levi, R.,** Tetrahydrobiopterin synthesis — an absolute requirement for cytokine-induced nitric oxide generation by vascular smooth muscle, *J. Biol. Chem.*, 267(36), 25722, 1992.
140. **Jorens, P. G., Vanoverveld, F. J., Bult, H., Vermeire, P. A., and Herman, A. G.,** Pterins inhibit nitric oxide synthase activity in rat alveolar macrophages, *Br. J. Pharmacol.*, 107(4), 1088, 1992.
141. **Kilbourne, R. G. and Griffith, O. W.,** Inhibition of inducible nitric oxide synthase with inhibitors of tetrahydrobiopterin biosynthesis, *J. Natl. Cancer Inst.*, 84(21), 1672, 1992.
142. **Wernerfelmayer, G., Werner, E. R., Fuchs, D., Hausen, A., and Reibnegger, G.,** Pteridine biosynthesis in human endothelial cells. Impact on nitric oxide-mediated formation of cGMP, *J. Biol. Chem.*, 268(3), 1842, 1993.
143. **Nunokawa, Y., Ishihara, T., Kanai, T., and Noguchi, T.,** (6R)-5,6,7,8-Tetrahydro-L-biopterin modulates nitric oxide activity in the rat cerebellum, *Arch. Biochem. Biophys.*, 298(2), 726, 1992.
144. **Pollock, J. S., Klinghofer, V., Forstermann, U., and Murad, F.,** Endothelial nitric oxide synthase is myristoylated, *FEBS Lett.*, 309(3), 402, 1992.
145. **Busconi, L. and Michel, T.,** Endothelial nitric oxide synthase. N-terminal myristoylation determines subcellular localization, *J. Biol. Chem.*, 268(12), 8410, 1993.
146. **Sessa, W. C., Barber, C. M., and Lynch, K. R.,** Mutation of N-myrisoylation site converts endothelial cell nitric oxide synthase from a membrane to a cytosolic protein, *Circ. Res.*, 72(4), 921, 1993.
147. **Baek, K. J., Thiel, B. A., Lucas, S., and Stuehr, D. J.,** Macrophage nitric oxide synthase subunits, *J. Biol. Chem.*, 268(28), 21120, 1993.
148. **Komori, Y., Chiang, K. T., and Fukuto, J. M.,** The effect of non-ionic detergents on the activity and/or stability of rat brain nitric oxide synthase, *Arch. Biochem. Biophys.*, 307(2), 311, 1993.
149. **Murphy, M. E., Piper, H. M., Watanabe, H., and Sies, H.,** Nitric oxide production by cultured aortic endothelial cells in response to thiol depletion and replenishment, *J. Biol. Chem.*, 266(29), 19378, 1991.
150. **Watanabe, Y. and Groves, J. T.,** Molecular mechanisms of oxygen activation by P-450, in *The Enzymes*, Vol. 20, 3rd ed., Sigman, D. S., Ed., Academic Press, San Diego, 1992, 405.
151. **Nahri, L. O. and Fulco, A. J.,** Characterization of a catalytically self-sufficient 119,000-dalton cytochrome P-450 monooxygenase induced by barbiturates in *bacillus megaterium*, *J. Biol. Chem.*, 261, 7160, 1986.
152. **Kwon, N. S., Nathan, C. F., Gilker, C., Griffith, O. W., Matthews, D. E., and Stuehr, D. J.,** L-Citrulline production from L-arginine by macrophage nitric oxide synthase, *J. Biol. Chem.*, 265(23), 13442, 1990.
153. **Pou, S., Pou, W. S., Bredt, D. S., Snyder, S. H., and Rosen, G. M.,** Generation of superoxide by purified brain nitric oxide synthase, *J. Biol. Chem.*, 267(34), 24173, 1992.
154. **Mayer, B., Heinzel, B., Klatt, P., John, M., and Böhme, E.,** Nitric oxide synthase-catalyzed activation of oxygen and reduction of cytochromes: reaction mechanisms and possible physiological implications, *J. Cardiovasc. Pharmacol.*, 20 (Suppl. 12), 54, 1992.
155. **Heinzel, B., John, M., Klatt, P., Böhme, E., and Mayer, B.,** Ca^{+2}/calmodulin-dependent formation of hydrogen peroxide by brain nitric oxide synthase, *Biochem. J.*, 281, 627, 1992.
156. **Klatt, P., Heinzel, B., John, M., Kastner, M., Böhme, E., and Mayer, B.,** Ca^{+2}/calmodulin-dependent cytochrome c reductase activity of brain nitric oxide synthase, *J. Biol. Chem.*, 267(16), 11374, 1992.
157. **Bredt, D. S., Hwang, P. M., Glatt, C. E., Lowenstein, C., Reed, R. R., and Snyder, S. H.,** Cloned and expressed nitric oxide synthase structurally resembles cytochrome P450 reductase, *Nature*, 351, 714, 1991.
158. **Stuehr, D. J., Kwon, N. S., Nathan, C. F., Griffith, O. W., Feldman, P. L., and Wiseman, J.,** N^{ω}-Hydroxy-L-arginine is an intermediate in the biosynthesis of nitric oxide from L-arginine, *J. Biol. Chem.*, 266(10), 6259, 1991.
159. **Wallace, G. C. and Fukuto, J. M.,** Synthesis and Bioactivity of N^{ω}-hydroxyarginine: a possible intermediate in the biosynthesis of nitric oxide from arginine, *J. Med. Chem.*, 34(5), 1746, 1991.

160. **Wallace, G. C., Gulati, P., and Fukuto, J. M.**, N^G-Hydroxy-L-arginine: a novel arginine analog capable of causing vasorelaxation in bovine intrapulmonary artery, *Biochem. Biophys. Res. Commun.*, 176(1), 528, 1991.

161. **Zembowicz, A., Hecker, M., MacArthur, H., Sessa, W. C., and Vane, J. R.**, Nitric oxide and another potent vasodilator are formed from N^G-hydroxy-L-arginine by cultured endothelial cells, *Proc. Natl. Acad. Sci. U.S.A.*, 88, 11172, 1991.

162. **Pufahl, R. A., Nanjappan, P. G., Woodard, R. W., and Marletta, M. A.**, Mechanistic probes of N-hydroxylation of L-arginine by inducible nitric oxide synthase from murine macrophages, *Biochemistry*, 31, 6822, 1992.

163. **Klatt, P., Schmidt, K., Uray, G., and Mayer, B.**, Multiple catalytic functions of brain nitric oxide synthase, *J. Biol. Chem.*, 268(20), 14781, 1993.

164. **Cho, A. K. and Lindeke, B., Vol. Eds.**, Biotransformation of organic nitrogen compounds, in *Progress in Basic and Clinical Pharmacology*, Lomax, P. and Vesell, Series Eds., Karger, Basel, 1988.

165. **Clement, B., Schultze-Mosgau, M.-H., and Wohlers, H.**, Cytochrome P450 dependent N-hydroxylation of a guanidine (debrisoquine), microsomal catalyzed reduction and further oxidation of the N-hydroxy-guanidine metabolite to the urea derivative, *Biochem. Pharmacol.*, 46(12), 2249, 1993.

166. **Boucher, J. L., Genet, A., Vadon, S., Delaforge, M., and Mansuy, D.**, Formation of nitrogen oxides and citrulline upon oxidation of N-hydroxy-L-arginine by hemeproteins, *Biochem. Biophys. Res. Commun.*, 184(3), 1158, 1992.

167. **Fukuto, J. M, Wallace, G. C., Hszieh, R., and Chaudhuri, G.**, Chemical oxidation of N-hydroxyguanidine compounds, *Biochem. Pharmacol.*, 43(3) 607, 1992.

168. **Fukuto, J. M., Stuehr, D. J., Feldman, P. L., Bova, M. P., and Wong, P.**, Peracid oxidation of an N-hydroxyguanidine compound: a chemical model for the oxidation of N-hydroxy-L-arginine by nitric oxide synthase, *J. Med. Chem.*, 36, 2666, 1993.

169. **Fukuto, J. M., Chiang, K., Hszieh, P., Wong, P., and Chaudhuri, G.**, The pharmacological activity of nitroxyl: a potent vasodilator with activity similar to nitric oxide and/or endothelium-derived relaxing factor, *J. Pharmacol. Exp. Ther.*, 263(2), 546, 1992.

170. **Fukuto, J. M., Hobbs, A. J., and Ignarro, L. J.**, Conversion of nitroxyl (HNO) to nitric oxide in biological systems: The role of physiological oxidants and relevance to the biological activity of HNO, *Biochem. Biophys. Res. Commun.*, 196(2), 707, 1993.

171. **Feldman, P. L., Griffith, O. W., Hong, H., and Stuehr, D. J.**, Irreversible inactivation of macrophage and brain nitric oxide synthase by L-N^G-methylarginine requires NADPH-dependent hydroxylation, *J. Med. Chem.*, 36, 491, 1993.

172. **Marletta, M. A.**, Nitric oxide synthase structure and mechanism, *J. Biol. Chem.*, 268(17) 12231, 1993.

173. **Klatt, P., Schmidt, K., Uray, G., and Mayer, B.**, Multiple catalytic functions of brain nitric oxide synthase, *J. Biol. Chem.*, 268(20), 14781, 1993.

174. **Hevel, J. M. and Marletta, M. A.**, Macrophage nitric oxide synthase: relationship between enzyme-bound tetrahydrobiopterin and synthase activity, *Biochemistry*, 31, 7160, 1992.

175. **Shiman, R.**, Phenylalanine hydroxylase and dihydropterin reductase in *Folates and Pterins*, Blakely, R. L. and Benkovic, S. J., Eds., Wiley and Sons, New York, 1984, 179.

176. **Dix, T. A. and Benkovic, S. J.**, Mechanism of oxygen activation by pteridine dependent monooxygenases, *Acc. Chem. Res.*, 21, 101, 1988.

177. **Giovanelli, J., Campos, K. L., and Kaufman, S.**, Tetrahydrobiopterin, a cofactor for rat cerebellar nitric oxide synthase, does not function as a reactant in the oxygenation of arginine, *Proc. Natl. Acad. Sci. U.S.A.*, 88, 7091, 1991.

178. **Stuehr, D. J., Fasehun, O. A., Kwon, N. S., Gross, S. S., Gonzalez, J. A., Levi, R., and Nathan, C. F.**, Inhibition of macrophage and endothelial cell nitric oxide synthase by diphenyleneiodonium and its analogs, *FASEB J.*, 5, 98, 1991.

179. **Schini, V. B. and Vanhoutte, P. M.**, Inhibitors of calmodulin impair the constitutive but not the inducible nitric oxide synthase activity in the rat aorta, *J. Pharmacol. Exp. Ther.*, 261(2), 553, 1992.

180. **Wolff, D. J., Datto, G. A., Samatovicz, R. A., and Tempsick, R. A.**, Calmodulin-dependent nitric-oxide synthase. Mechanism of inhibition by imidazole and phenylimidazoles, *J. Biol. Chem.*, 268(13), 9425, 1993.

181. **Wolff, D. J., Datto, G. A., and Samatovicz, R. A.**, The dual mode of inhibition of calmodulin-dependent nitric oxide synthase by antifungal imidazole agents, *J. Biol. Chem.*, 268(13), 9430, 1993.

182. **Wink, D. A., Osawa, Y., Darbyshire, J. F., Jones, C. R., Eshenaur, S. C., and Nims, R. W.**, Inhibition of cytochromes P450 by nitric oxide and a nitric oxide-releasing agent, *Arch. Biochem. Biophys.*, 300(1), 115, 1992.

183. **Rogers, N. E. and Ignarro, L. J.**, Constitutive nitric oxide synthase from rat cerebellum is reversibly inhibited by nitric oxide formed from L-arginine, *Biochem. Biophys. Res. Commun.*, 189(1), 242, 1992.

184. **Misko, T. P., Moore, W. M., Kasten, T. P., Nickols, G. A., Corbett, J. A., Tilton, R. G., McDaniel, M. L., Williamson, J. R., and Currie, M. G.**, Selective inhibition of the inducible nitric oxide synthase by aminoguanidine, *Eur. J. Pharmacol.*, 233, 119, 1993.

185. **Lambert, L. E., Whitten, J. P., Baron, B. M., Cheng, H. C., Doherty, N. S., and McDonald, I. A.**, Nitric oxide synthesis in the CNS, endothelium and macrophages differs in its sensitivity to inhibition by arginine analogs, *Life Sci.*, 48, 69, 1991.

186. **Gross, S. S., Stuehr, D. J., Aisaka, K., Jaffe, E. A., Levi, R., and Griffith, O. W.,** Macrophage and endothelial cell nitric oxide synthesis: Cell-type selective inhibition by N^G-aminoarginine, N^G-nitroarginine and N^G-methylarginine, *Biochem. Biophys. Res. Commun.*, 170(1), 96, 1990.
187. **Hibbs, J. B., Taintor, R. R., and Vavrin, Z.,** Macrophage cytotoxicity: Role for L-arginine deiminase and imino nitrogen oxidation to nitrite, *Science*, 235, 473, 1987.
188. **Olken, N. M., Rusche, K. M., Richards, M. K., and Marletta, M. A.,** Inactivation of macrophage nitric oxide synthase activity by N^G-methyl-L-arginine, *Biochem. Biophys. Res. Commun.*, 177(2), 828, 1991.
189. **Pufahl, R. A., Nanjappan, P. G., Woodard, R. W., and Marletta, M. A.,** Mechanistic probes of N-hydroxylation of L-arginine by inducible nitric oxide synthase from murine macrophages, *Biochemistry*, 31, 6822, 1992.
190. **Olken, N. M. and Marletta, M. A.,** N^G-Allyl- and N^G-cyclopropyl-L-arginine: Two novel inhibitors of macrophage nitric oxide synthase, *J. Med. Chem.*, 35, 1137, 1992.
191. **Mayer, B., Schmid, M., Klatt, P., and Schmidt, K.,** Reversible inactivation of endothelial cell nitric oxide synthase by N^G-nitro-L-arginine, *FEBS Lett.*, 333(1,2), 203, 1993.
192. **Klatt, P., Schmidt, K., Brunner, F., and Mayer, B.,** Inhibitors of brain nitric oxide synthase, *J. Biol. Chem.*, 269(3) 1674, 1994.
193. **Osawa, Y. and Davila, J. C.,** Phencyclidine, a psychomimetic agent and drug of abuse, is a suicide inhibitor of brain nitric oxide synthase, *Biochem. Biophys. Res. Commun.*, 194(3), 1435, 1993.
194. **Brady, J. F., Dokko, J., DiStefano, E. W., and Cho, A. K.,** Mechanism-based inhibition of cytochrome P450 (P450) by hetereocyclic analogs of phencyclidine (PCP), *Drug Metabol. Dispos.*, 15, 648, 1987.

INDEX

A

B

C

H

I

K

L

M

N

O

P

Q

R

S

T

U

V

W

X

Y